Fungal biology has an integral role to play in the development of the biotechnology and biomedical sectors. It has become a subject of increasing importance as new fungi and their associated biomolecules are identified. The interaction between fungi and their environment is central to many natural processes that occur in the biosphere. The hosts and habitats of these eukaryotic microorganisms are very diverse; fungi are present in every ecosystem on Earth. The fungal kingdom is equally diverse, consisting of seven different known phyla. Yet detailed knowledge is limited to relatively few species. The relationship between fungi and humans has been characterized by the juxtaposed viewpoints of fungi as infectious agents of much dread and their exploitation as highly versatile systems for a range of economically important biotechnological applications. Understanding the biology of different fungi in diverse ecosystems as well as their interactions with living and non-living is essential to underpin effective and innovative technological developments. This series will provide a detailed compendium of methods and information used to investigate different aspects of mycology, including fungal biology and biochemistry, genetics, phylogenetics, genomics, proteomics, molecular enzymology, and biotechnological applications in a manner that reflects the many recent developments of relevance to researchers and scientists investigating the Kingdom Fungi. Rapid screening techniques based on screening specific regions in the DNA of fungi have been used in species comparison and identification, and are now being extended across fungal phyla. The majorities of fungi are multicellular eukaryotic systems and therefore may be excellent model systems by which to answer fundamental biological questions. A greater understanding of the cell biology of these versatile eukaryotes will underpin efforts to engineer certain fungal species to provide novel cell factories for production of proteins for pharmaceutical applications. Renewed interest in all aspects of the biology and biotechnology of fungi may also enable the development of "one pot" microbial cell factories to meet consumer energy needs in the 21st century. To realize this potential and to truly understand the diversity and biology of these eukaryotes, continued development of scientific tools and techniques is essential. As a professional reference, this series will be very helpful to all people who work with fungi and should be useful both to academic institutions and research teams, as well as to teachers, and graduate and postgraduate students with its information on the continuous developments in fungal biology with the publication of each volume.

Manikant Tripathi • Neha Srivastava
Rongrong Xie
Editors

Recent Advancements in Fungal Biotechnology Volume 1

Applications in Energy and Materials

Springer

Editors
Manikant Tripathi
Biotechnology Program
Dr. Rammanohar Lohia Avadh University
Ayodhya, Uttar Pradesh, India

Neha Srivastava
Department of Chemical Engineering & Technology
Indian Institute of Technology (BHU)
Varanasi, Uttar Pradesh, India

Rongrong Xie
School of Environment and Safety Engineering
Biofuels Institute Jiangsu University
Zhenjiang, Jiangsu Province, China

ISSN 2198-7777 ISSN 2198-7785 (electronic)
Fungal Biology
ISBN 978-981-92-0338-3 ISBN 978-981-92-0339-0 (eBook)
https://doi.org/10.1007/978-981-92-0339-0

This Springer imprint is published by the registered company Springer Nature Singapore Pte Ltd.
The registered company address is: 152 Beach Road, #21-01/04 Gateway East, Singapore 189721, Singapore

Preface

The growing global need for sustainable energy sources and environmentally friendly materials has prompted many researchers to investigate biological systems capable of converting renewable materials into high-valued products. Fungi, due to their high degree of metabolic diversity, extensive enzyme-producing capabilities, as well as their ability to degrade complex organic or plant-based substrates, have emerged as some of the most efficient and versatile biological systems. Fungi play a major ecological role as decomposers in our world and can synthesize a wide variety of metabolites and biomolecules that can be used in a variety of industrial applications featuring bioconversion processes. This book, *Recent Advances in Fungal Biotechnology Volume 1: Applications in Energy and Materials*, highlights the continued development and utilization of new developments in fungal biotechnology to provide solutions to many of the critical issues associated with sustainable energy development, waste-to-value approaches, and novel biomaterials. Fungal biotechnology has also gained notoriety for its ability to convert lignocellulosic biomass, agricultural residues, and industrial waste streams into renewable energy (i.e., biofuels) through the utilization of a wide variety of extracellular enzymes (cellulases, ligninases, hemicellulases, and others) to degrade complex plant polymers.

Lignocellulosic biomass can be transformed into bioenergy or chemicals by using a variety of microorganisms, mainly through fungal enzyme processes. Due to advancements in fungal biotechnologies, the conversion efficiency of biomass has improved, and new fermentation technologies as well as improved processing operations have been developed. As a result, fungal biotechnologies are establishing sustainable systems to produce bioenergy. Additionally, fungal biotechnologies are also being utilized to create the next generation of biobased materials. Fungal-derived biobased materials are being studied as a viable alternative to petroleum products because they are biodegradable, structurally diverse, and can be produced using environmentally friendly processes. Biobased materials made from mycelium, fungal biopolymers, or biobased composites are expected to be an effective way of reducing pollution while supporting the principles of a circular bioeconomy. Agricultural by-products, food waste, forestry by-products, and industrial organic wastes can all be abundant renewable resources that could be converted into biobased products through the application of fungi. Fungi are primarily responsible for breaking down lignocellulosic structures and converting waste into fermentable sugar and bioactive materials. Ultimately, fungi not only help produce renewable energy but also address the major environmental problems associated with waste accumulation and resource inefficiency.

The chapters in this book deal with the use of fungi in the production of biofuels, which has been shown to be a viable means for creating bioethanol and biodiesel from renewable sources. Fungus fermentations and their metabolic transformation pathways allow for production of these and other types of biofuels from renewable biomass. Because of advances in fungal bioengineering and biological fermentation technologies, production of bioethanol from fungi has become more efficient than in previous generations (e.g., optimization of substrate use and enzymes produced by fungi, and increasing the tolerance thresholds of fungi in industrial fermentation). Another innovative application for the use of fungi is in fungal bioelectrochemical systems, where fungal metabolic processes are coupled through microbial fuel cells to convert organic substrates into usable electricity (thus providing new opportunities for renewable electricity generation).

An additional aspect of biohydrogen production is via fungi, which is also discussed in the book. Biohydrogen is recognized as a renewable energy source with high energy density and minimal to no environmental impacts. Certain fungal species have demonstrated capabilities to produce hydrogen gas from specific metabolite pathways and continued research is being undertaken to enhance hydrogen production through metabolite engineering along with optimizing fermentation processes. Fungi can produce biopolymers from renewable resources and waste biomass, beyond their production of pigments. Technological revolutions have dramatically increased the pace of progress in fungal biotechnology. The convergence of various meta-omic techniques (genomic, transcriptomic, proteomic, and metabolomics) has advanced our knowledge of the fungal metabolic network and regulatory systems. High-throughput technologies provide data that help in discovering and developing new metabolic pathways, enzymes and regulatory genes tied to the energy generation process and the development of biomaterials. This matrix of technologies creates new opportunities for accelerating research and increasing scalability of bioenergy and biomaterial production systems using fungi.

This book showcases how the combined advances in microbiology, molecular biology, bioengineering, and materials science have transformed our understanding of how fungal systems can be utilized to provide solutions to pressing global issues of energy security, environmental sustainability, and resource efficiency. The information about materials included in this volume will be beneficial to researchers, academics, graduate students, and industry professionals who want to better understand and utilize the capabilities of fungi for the development of innovative bioenergy systems and sustainable biomaterials. By providing the insights of many leading experts in the field, this volume identifies fungal biotechnology as a significant contributor to future biobased industries and circular bioeconomy.

The editors of this book would like to express their sincere thanks to all the contributing authors of the chapters whose expertise and cooperation made this book volume possible. We also express our gratitude to the editorial and production team for their continuous support and constructive suggestions throughout the publication process of this book. Finally, we gratefully acknowledge our respective institutions, families, and colleagues who supported and encouraged us during the preparation of this book volume.

Ayodhya, Uttar Pradesh, India — Manikant Tripathi
Varanasi, Uttar Pradesh, India — Neha Srivastava
Zhenjiang, Jiangsu Province, China — Rongrong Xie

Contents

Editors and Contributors

About the Editors

Manikant Tripathi Dr. Manikant Tripathi is currently a faculty member in the biotechnology program at Dr. Rammanohar Lohia Avadh University, Ayodhya, Uttar Pradesh, India. He has over ten years of teaching and research experience in microbial biotechnology. He earned his Ph.D. with a research fellowship from the University Grants Commission, India. His research interests include bioremediation of environmental pollutants, microbial applications for value-added products, waste management, biomass valorization, and bioactives. He has received awards from scientific societies at international and national conferences. He is an active member of scientific societies. He has published over 100 research and review articles in reputed national and international journals, including edited books, authored book, and chapters.

Neha Srivastava Dr. Neha Srivastava is currently a Research Scientist in the Department of Chemical Engineering and Technology at the Indian Institute of Technology (BHU) Varanasi, Uttar Pradesh, India. She completed her Ph.D. in Biotechnology at SHIATS, Uttar Pradesh, India, specializing in bioenergy. Her areas of interest include bioprocess technology, biofuels production from waste, and nanocatalyst development. She has published extensively, filed patents, achieved technology transfer, and contributed to many books.

Rongrong Xie Prof. Rongrong Xie graduated from the College of Plant Protection, Nanjing Agricultural University in December 2009, obtaining a Ph.D. in Biology. In January 2010, she joined the Institute of Biofuels at Jiangsu University as an assistant professor and master's student advisor. Her research interests focus on revealing and mimicking highly efficient natural lignocellulolytic conversion systems (e.g., wood-feeding termites or other cellulose-feeding animals) to advance current bioreactors for biofuels and bioproducts and develop advanced technologies for biomass conversion and utilization. She works as the leading PI or PI of several grants, such as the National Key Research and Development Program of China, the National Natural Science Foundation of China, and the Natural Science Foundation of Jiangsu Province. She has published more than 70 academic papers and participated in the compilation of one textbook. She is an expert in insect symbionts, molecular biology, and molecular ecology research.

Contributors

Gholamreza Abdi Department of Biotechnology, Persian Gulf Research Institute, Persian Gulf University, Bushehr, Iran

Yashika Aggarwal Department of Microbiology, Punjab Agricultural University, Ludhiana, Punjab, India

Giridhar Babu Anam Department of Basic Sciences, SR University, Warangal, India

Olivia Anthony Department of Pharmaceutical Sciences, School of Pharmacy, University of Maryland, College Park, MD, USA

Ali Jibril Baba Fermentation Laboratory, Department of Life Science, Sharda School of Bioscience & Technology, Sharda University, Greater Noida, Uttar Pradesh, India

Subhamoy Banerjee Department of Life Science – Microbiology, School of Biological Science & Technology, Maulana Abul Kalam Azad University of Technology, Simhat, Haringhata, Nadia, West Bengal, India

Gaurav Bhardwaj Chitkara College of Pharmacy, Chitkara University, Rajpura, Punjab, India

Sourav Bhattacharya Department of Microbiology and Botany, School of Sciences, JAIN (Deemed-to-Be University), Bangalore, Karnataka, India

Federico Cerrone School of Biotechnology, Dublin City University, Dublin, Ireland

Ajay Kumar Chauhan Department of Biotechnology, Sharda University, Knowledge Park II, Greater Noida, UP, India

Prakram Singh Chauhan Chitkara College of Pharmacy, Chitkara University, Rajpura, Punjab, India

Somnath Das The Institute of Science, Culture, and Social Studies, Belgharia, Kolkata, West Bengal, India

Oinam Bidyarani Devi Fermentation Laboratory, Department of Life Science, Sharda School of Bioscience & Technology, Sharda University, Greater Noida, Uttar Pradesh, India

Sanjoy K. Dhar Banophool Adibashi Green Heart College, Mirpur, Dhaka, Bangladesh

Bangladesh University of Health Sciences, Mirpur, Dhaka, Bangladesh

Surabhi Done Department of Basic Sciences, SR University, Warangal, India

Kulsoom Fatima Department of Biomedical Engineering, Wayne State University, Detroit, MI, USA

Anuska Ghosh The Agricultural Biotechnology Division, School of Agriculture and Rural Development, Ramakrishna Mission Vivekananda Educational and Research Institute, Narendrapur Campus, West Bengal, India

Burragoni Sravanthi Goud Department of Chemical Engineering, Yeungnam University, Gyeongsan, Gyeongbuk, South Korea

Anmol Gupta MPA Research, Gautam Budh Nagar, Gaur City Centre, Greater Noida, Uttar Pradesh, India

Tanisha Gupta Department of Plant Pathology, Punjab Agricultural University, Ludhiana, Punjab, India

Rajesh Jeewon Department of Health Sciences, Faculty of Medicine and Health Sciences, University of Mauritius, Réduit, Mauritius

Honghe Center for Mountain Futures, Kunming Institute of Botany, Chinese Academy of Sciences, Honghe County, Yunnan, People's Republic of China

School of Pharmacy, Zunyi Medical University, Zunyi, Guizhou, China

Santosh Kumar Karn Department of Biochemistry & Biotechnology, Sardar Bhagwan Singh University, Balawala, Dehradun, Uttarakhand, India

Jaspreet Kaur Department of Biotechnology, Kanya Maha Vidyalaya College, Jalandhar, Punjab, India

Atin Kumar School of Agriculture, Uttaranchal University, Dehradun, Uttarakhand, India

Mohit Kumar Fermentation Laboratory, Department of Life Science, School of Bio Science & Technology, Sharda University, Greater Noida, Uttar Pradesh, India

Narendra Kumar Department of Biotechnology, Noida Institute of Engineering and Technology, Greater Noida, Uttar Pradesh, India

Suryasarathi Kumar School of Biological Science & Technology, Department of Applied Biology, Maulana Abul Kalam Azad University of Technology-Main Campus, Simhat, Haringhata, Nadia, West Bengal, India

Tarun Kumar Department of Biotechnology, Chandigarh University, Mohali, Punjab, India

Vijay Kumar Department of Botany, Panjab University, Chandigarh, India

Patanjali Herbal Research Department, Patanjali Research Foundation, Haridwar, Uttarakhand, India

Arpitha Manchan Department of Microbiology and Botany, School of Sciences, JAIN (Deemed-to-Be University), Bangalore, Karnataka, India

Abhilasha Mishra Department of Biochemistry, Dr. Rammanohar Lohia Avadh University, Ayodhya, Uttar Pradesh, India

Santosh Kumar Mishra Fermentation Laboratory, Department of Life Science, Sharda School of Bioscience & Technology, Sharda University, Greater Noida, Uttar Pradesh, India

Surya Mishra Department of Biotechnology, Goel Institute of Technology & Management, Lucknow, Uttar Pradesh, India

Amity Institute of Biotechnology, Amity University, Uttar Pradesh, Lucknow, India

Gaurav Mudgal University Institute of Biotechnology, Chandigarh University, Mohali, Punjab, India

Anupama Nisarani Department of Microbiology and Botany, School of Sciences, JAIN (Deemed-to-Be University), Bangalore, Karnataka, India

Anwesha Patel Department of Microbiology and Botany, School of Sciences, JAIN (Deemed-to-Be University), Bangalore, Karnataka, India

Neelam Pathak Department of Biochemistry, Dr. Rammanohar Lohia Avadh University, Ayodhya, Uttar Pradesh, India

Urmila Gupta Phutela Department of Microbiology, Punjab Agricultural University, Ludhiana, Punjab, India

Department of Renewable Energy Engineering, Punjab Agricultural University, Ludhiana, Punjab, India

Indu Bhushan Prasher Department of Botany, Panjab University, Chandigarh, India

Vivek Raj Department of Biotechnology, Patna Women's College, (Autonomous), Patna University, Patna, Bihar, India

Sachitra Kumar Ratha Algology Laboratory, CSIR-National Botanical Research Institute, Lucknow, Uttar Pradesh, India

Academy of Scientific and Innovative Research, Ghaziabad, Uttar Pradesh, India

Fareha Rayeen Department of Biochemistry, Dr. Rammanohar Lohia Avadh University, Ayodhya, Uttar Pradesh, India

Magnolia Roy Department of Electronics and Communication Engineering, Kumaraguru College of Technology, Coimbatore, Tamil Nadu, India

Rupali Rupasmita Fermentation Laboratory, Department of Life Science, School of Bio Science & Technology, Sharda University, Greater Noida, Uttar Pradesh, India

Wajiha Saadi Department of Biotechnology, Patna Women's College, (Autonomous), Patna University, Patna, Bihar, India

Sharad Sachan Department of Agricultural Economics and Extension, School of Agriculture, Lovely Professional University, Phagwara, Punjab, India

Kübra Sağlam Department of Food Engineering, Faculty of Agriculture, Tekirdağ Namik Kemal University, Tekirdağ, Turkey

Isha Department of Microbiology, Punjab Agricultural University, Ludhiana, Punjab, India

Megha Sethi Department of Botany, Panjab University, Chandigarh, India

Padma Sharma Department of Ecology, Environment, and Remote Sensing, Paryavaran Bhawan, Gladn, Jammu and Kashmir, India

Pallavi Sharma Department of Biotechnology, Chandigarh University, Mohali, Punjab, India

Vipasha Sharma Department of Biotechnology, Chandigarh University, Mohali, Punjab, India

Sonia Kumari Shishodia University Institute of Biotechnology, Chandigarh University, Mohali, Punjab, India

Abhijeet Singh School of Bioscience and technology (SBST), Vellore Institute of Technology, Vellore, Tamil Nadu, India

Anuma Singh Centre for Applied Bioscience Research (CABR), Munster Technological University (MTU), Tralee, Kerry, Ireland

Pankaj Singh Biotechnology Program, Dr. Rammanohar Lohia Avadh University, Ayodhya, Uttar Pradesh, India

Pradeep Kumar Singh Department of Biochemistry, Dr. Rammanohar Lohia Avadh University, Ayodhya, Uttar Pradesh, India

Roshini Singh Department of Biotechnology, SR Institute of Management and Technology, Bakshi Ka Talab, Lucknow, Uttar Pradesh, India

Amity Institute of Biotechnology, Amity University, Uttar Pradesh, Lucknow, India

Arun Lal Srivastav Chitkara University School of Engineering and Technology, Chitkara University, Himachal Pradesh, India

Shifa Suhail Department of Biosciences, Faculty of Sciences, Integral University, Lucknow, Uttar Pradesh, India

Sushma University Institute of Biotechnology, Chandigarh University, Mohali, Punjab, India

Priyanka Talukdar The Institute of Science, Culture, and Social Studies, Belgharia, Kolkata, West Bengal, India

Ayush Thakur Department of Botany, Panjab University, Chandigarh, India

Manikant Tripathi Biotechnology Program, Dr. Rammanohar Lohia Avadh University, Ayodhya, Uttar Pradesh, India

Atul Kumar Verma Fermentation Laboratory, Department of Life Science, School of Bio Science & Technology, Sharda University, Greater Noida, Uttar Pradesh, India

Chapter 1
Recent Developments in Fungal Biotechnology for Bioenergy Production: Challenges and Perspectives

Vipasha Sharma, Pallavi Sharma, Tarun Kumar, and Manikant Tripathi

Abstract Fungi are a well-known game-changing paradigm for sustainable bioenergy and bioproduct production. Fungal biotech has catalyzed the shift from petroleum-based refineries to sustainable conversion of renewable raw materials, such as corn, grasses, or wood chips. In this review, recent progress in this field is summarized, focusing on the co-utilization of robust fungal strains and metabolic engineering for improving lignocellulosic biomass conversion. We have described several key advancements in fermentation technology and bioconversion process developments, with implications of synergistic interactions between microbes and the use of synthetic biology tools such as Clustered Regularly Interspaced Short Palindromic Repeats (CRISPR) toward maximizing biofuel production.

Despite the remarkable capacities of fungi to advance a circular economy and the United Nations Sustainable Development Goals, this sector is confronted with formidable technical limitations as well as economic and regulatory constraints. High cost of enzyme production, scale-up difficulties, and the challenges associated with handling genetically modified organisms are substantial hurdles for large-scale commercialization.

The discussion on the growing visions of the fungal cultures with the induction of artificial intelligence, Internet of Things, and other renewable sources, for example, solar and wind energy, is also done in this chapter. Furthermore, the interdisciplinary collaborations are also explained to enable the maximum output from the model innovations. They are considered a critical key to unlock the full potential of fungal biotechnology to eradicate the problem of climate change and to create a future of robust bioenergy.

Keywords Fungal biotechnology · Bioenergy · Lignocellulosic biomass · Circular economy · Fermentation · Sustainable development · Synthetic biology

V. Sharma (✉) · P. Sharma · T. Kumar
Department of Biotechnology, Chandigarh University, Mohali, Punjab, India

M. Tripathi
Biotechnology Program, Dr. Rammanohar Lohia Avadh University, Ayodhya, Uttar Pradesh, India

M. Tripathi et al. (eds.), *Recent Advancements in Fungal Biotechnology Volume 1*, Fungal Biology, https://doi.org/10.1007/978-981-92-0339-0_1

1.1 Introduction

The fungal biotechnology has played an important role in bio-energy production through offering feasible measures to address the efficiency and sustainability issues (Chai et al. 2022). Biotechnological application of fungi for bioenergy production has tremendous potential, as well as a lot of fear to foresee, along with the discovery of many new solutions (Farinas and da Cunha 2016; Lone and Jeelani 2023). New strategies are needed to maximize the efficiency of using fungi for enhanced bioenergy production, in spite of technical and economic constraints (Chowdhary et al. 2018). To do so, interdisciplinary collaboration with researchers engaged in partnership with industry and government is the essential driver to move scalable fungal biotechnologies for bioenergy forward at a lower cost (Chowdhary et al. 2018; Farinas and da Cunha 2016; Rame et al. 2023). Sharp insight into fungal enzymes of cellulase will also be useful in the optimization of bioenergy processes for higher productivities in the overall realization of the bioenergy market (Farinas and da Cunha 2016). Sustainable and efficient fungal biotechnologies must be established to improve bioenergy generation, as well as attenuate climate change impacts (Mafune and Winkler 2024). It will take ongoing investment in fungal biotechnology to use it anywhere near its potential as a renewable source of energy.

Furthermore, what matters is to address the gap in fungal processes and applications to build climate-smart biotechnologies for bio-energy production (Mafune and Winkler 2024). In order to develop the potential of all the applications that could be made with fungal biotechnology, it is time that we start investing big in research programs which focus on "how we manage scale-up & environmental fit-in-features." Investment in research will lead to the production of robust fungal biotechnologies to serve our primary requirements.

To address these concerns, we need to emphasize the production of fungal enzymes, which are used in applications, for example, cellulases (Farinas and da Cunha 2016). Improved understanding of fungal biotechnology might lead to more effective climate-smart options for renewable energy generation (Mafune and Winkler 2024).

1.2 Role of Bioenergy in Sustainable Development

Bioenergy is part of the bioeconomy, and one focus that comes along with it in literature is eco-innovation by yielding new renewable largest climate services, which could also contribute to mitigating climate change and resource (Yu et al. 2026). Fungal biotechnologies, particularly in enzyme production, can significantly contribute to these sustainability goals by utilizing waste resources for energy generation (Farinas and da Cunha 2016; Tiwari and Park 2024). This form is not just a way to minimize waste, but it also complies with the circular economy model, where waste would be recycled to obtain value-added energy products (Qureshi et al.

2023). The inclusion of fungal biotechnologies in a circular economy, waste valorization, and sustainable energy applications will look to optimize resource efficiency as far as feasible, contributing to environmental sustainability.

Therefore, with the implementation of fungal biotechnologies, new paths for energy generation can be attained that provide solutions to environmental and socioeconomic problems, bearing in mind global sustainability efforts.

This alignment emphasizes the need for a combined effort to exploit the opportunities of fungal biotechnology in light of sustainable living, where we could identify the novel fungal enzymes in order to reach their promise as part of a sustainable approach to our future energy and environmental needs on a global scale (Tiwari and Park 2024). In order to achieve these objectives, a continual "discussion" among the partners is essential to make progress and promote the use of fungal biotechnologies that can address important issues in practical terms through solutions of complex problems on energy and environmental sustainability (Dhiman et al. 2024). This is a review for the lessening environmental impact of fungal biotechnologies, and due to climatic changes, in order to find out innovative sources of renewable energies (Auta et al. 2024). If these biotechnologies are to have an impact on the electricity sector, however, long-term investment and policy support will be necessary.

In order to implement the functions of fungal biotechnology successfully, a robust platform is required to be established that involves the researchers, domestication, industry, and policymakers for sustainable up- and downstream in both domestication and post-harvest technologies (Mafune and Winkler 2024). Such an integrated approach will pave the way to future fungal biotechnology, which not only will cater to a high-efficiency bioenergy processing industry but also contribute to being part of the United Nations Sustainable Development Goals (UN SDGs) (Tiwari and Park 2024). Such cooperation is necessary in order to address the complex problems of climate change and depletion of natural resources, as well as to refine sustainable fungal biotechnology-based energy solutions.

This technology can be a big opportunity for us to develop a more effective and sustainable future.

1.3 Introduction of Bioenergy-Producing Fungal Organisms

Fungi belong to Zygomycetes and are known for bioenergy generation from waste. The potential products produced from lignocellulosic material are bioethanol and biodiesel. The metabolic capacities of these fungal strains are contributing to bioenergy production with maximum conversion of the waste into wealth-making. These properties of fungal biotechnology make their role very important in the sustainable production of bioenergy. Integration of various fungi in the production of bioenergy is efficient and has a second advantage by making agricultural waste eco-friendly and less dependent on fossil fuel (Ahmad and Abdullah 2025). Utilization of fungi in bioenergy systems can further lower waste and promote circular economy (and

SDGs) and minimize climate-change effect (Tiwari and Park 2024). Through harnessing the diverse capabilities of fungi, unique and novel bioenergy solutions can be designed to meet energy demands and environmental needs. Integration of fungal forms with energy devices is becoming one of the most promising techniques for renewable energy production and mitigation of global issues such as carbon dioxide emission, climate change, and diminishing natural resources (Tiwari and Park 2024). This stresses the requirement for more research, given that, in contrast to bacterial genetics and metabolism engineering, what concerns fungal cell systems, many potential resources still remain untapped that can be used for enhancing bioenergy production and providing environmental sustainability (Tiwari and Park 2024). Moreover, the richness of fungi and their potential for bioenergy production not only provide energy security but also help a great deal in the maintenance of the environment (Mafune and Winkler 2024). However, the other fungal applications for an established atmospheric carbon sink support of the photosynthetic carbon-fixation-promoting environment may contribute to addressing global climate resilience in a sustainable manner (Tiwari and Park 2024). The additional potential is likely to be gained through specific R&D efforts on peculiar features of different fungal species toward bioenergy applications (Ahmad and Abdullah 2025; Grace et al. 2020; Sundari et al. 2018) and their metabolic capabilities. It will also leverage enhanced knowledge of these organisms to generate novel pathways for sustainable bioenergy production in a manner compatible with addressing the demand for energy and the environment. The fungi-driven lignocellulosic biorefinery process is shown in Fig. 1.1.

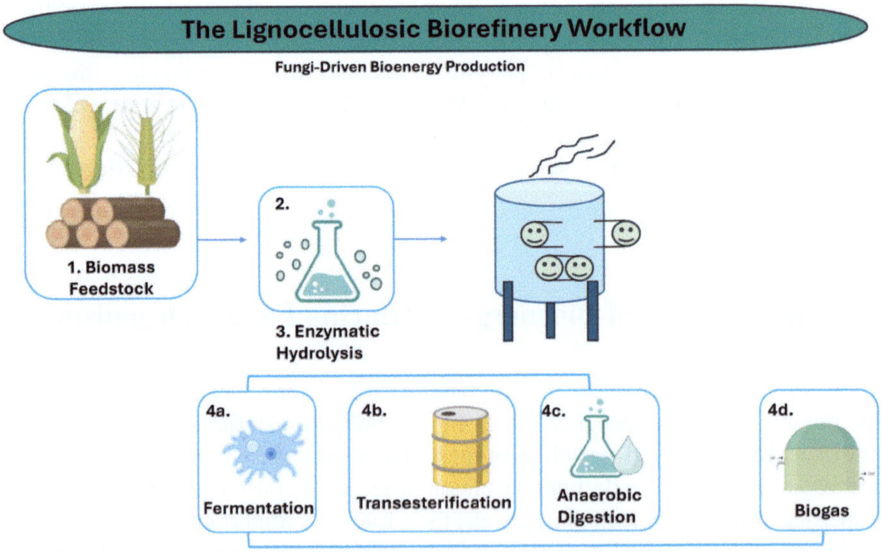

Fig. 1.1 Overview of the fungi-driven lignocellulosic biorefinery workflow, highlighting the biomass pretreatment, enzymatic hydrolysis, fermentation, anaerobic digestion, and bioenergy production

Integration of fungal biotechnologies with bioenergy systems is highly critical for their long-term sustainability and the UN SDGs pertaining to climate change and resource depletion (Grace et al. 2020; Tiwari and Park 2024). Such integration can potentially improve the efficiency of bioenergy production and also help in combatting climate change with new fungal applications (Mafune and Winkler 2024; Tiwari and Park 2024). The significance of the media through which fungi are integral to bioenergy production is further emphasized by their potential in pollution clean-up, improving crop yield, and fostering the circular economy (Mafune and Winkler 2024).

1.4 Recent Developments in Fungal Biotechnology

The discovery of fungal strains that have improvements in metabolic capabilities is a recent advancement that can significantly affect bioenergy and sustainable energy facilities. The arrow indicates these directions (Pandey and Tewari 2018); advancing in the paths highlights for further exploration and exploitation toward therapeutic burden from other fungal species relevant for bioenergy-based requirement and the planetary challenge of the world's sustainability.

The finding of this study, of a high-level fungal cell wall (FCW) and Functionalized carbon nanotubes (CNT-F) in an elite fungal strain, is significant for the optimal exploitation of bioenergy using this fungus and building climate-smart microbial biotechnologies as recommended (Mafune and Winkler 2024; Tiwari and Park 2024). These continued works are fundamental for harnessing the goodness of fungi in the biofuel sector and developing new anti-climate change and sustainable routes (Mafune and Winkler 2024; Tiwari and Park 2024). Additionally, the use of novel fungi from bioenergy production forward may improve resource efficiency and contribute to green-chemistry economy and climate-smart biotechnologies in the world (Tiwari and Park 2024). Further characterization of the metabolic capabilities of these two fungal isolates would be a prerequisite for the usefulness of their incorporation into new processes that satisfy not only modern necessities in energy needs but also eco-friendly operation.

Some of these advanced strains, especially those with recently introduced metabolic pathways, would be able to play a role in the progress of eco-friendly technology, which can meet a few (Tiwari and Park 2024) of the challenges indicated by the UN SDGs and solve world climate issues. If the newly cultured fungal strains are more productive for producing good-quality biofuels, they are very good at utilizing waste for maintaining the ecological equilibrium (Kržišnik and Gonçalves 2023; Mafune and Winkler 2024). Biodegradable substrate utilization for useful products is a good choice for the circular economy (Kržišnik and Gonçalves 2023).

As per the current scenario, the requirements for the new fungal strains are of utmost importance. The issues we can take care of these strains are climate change and resource depletion. Both of these problems are quite a concern these days (Mafune and Winkler 2024). The transformations of these applications have the potential to multiply the bioenergy production to match the worldwide sustainability initiative.

1.4.1 Innovations in Fermentation Techniques

Novel fermentations may help to fully utilize organic waste for bioenergy production and increase the sustainability of this fungal biotechnology, which is required to combat the effects of climate change and save resources (Kržišnik and Gonçalves 2023; Mukherjee et al. 2018).

These advancements in fermentation process technology should offer potential for improved substrate utilization, bioenergy yield, and bioconversion of waste toward environmental remediation (Selvakumar 2025). Continued research of fermentation technologies is essential in the furthering of fungal biotechnology for bioenergy production and associated sustainable development initiatives and circular economy notions. In addition, novel fermentation strategies can be employed for bio-based materials, which have been identified as necessary because of the goal of achieving dilemma requirements from the United Nations sustainable development goals (UN SDGs) and environmental conservation. The novel methods of fermentation would always be required for enhanced bioenergy productivity and thus to meet the challenge lying toward environmental (sustainability) and its current climatic concern effects throughout (Tiwari and Park 2024).

This achievement will not be the key for further development of fungal biotechnologies as a tool for enhancing cost-effective resource use, but rather to address global concerns incorporated in environmental sustainability and climate challenges. The practical applications of these best practices in fermentation are needed to convert waste streams into a valuable source of energy, particularly when considering the need for a sustainable bioenergy industry. The successive testing and establishing of fermentation strategies are the urgent needs of our time for ABE production toward undermining targets like climate change or scarcity in resources (Tiwari and Park 2024). In it, we revitalize the potential of fungal biotechnologies in renewable power.

1.4.2 Bioconversion Processes

The studies of the fungal enzyme-catalyzed bioconversion of biomass have implications for increasing bioenergy production and renewable energy technology in a sustainable manner. They could also aid in the transformation of organics to biofuels, leading to waste management and energy sustainability. In addition, this would lead to the development of new strategies to tackle worldwide demand for sustainability in the energy sector and environmental resources by incorporating advanced bioconversion techniques.

Talking about microbial interaction in bioconversion, the significance of microbial interactions in fungal communities has recently been well appreciated as an approach for improving bioenergy generation. These interactions are likely the contributors to synergistic enhancements observed on depolymerizing lignocellulosic materials and global biofuels efficiencies from lignocellulosic material. For example, fungus–bacterium synergy could play a role in improved waste degradation profile, and it helps in transforming the wastes into a high-quality biofuel source rather than existing as an individual player, useful for the sustainable production of bioenergy (Tiwari and Park 2024).

Additionally, further expansion of microbial species repertoire into bioenergy applications may offer potential to increase their toleration toward environmental disturbance and also help in the realization of a circular economy in line with global sustainability goals (Karduri 2023). As researchers probe the complexity and variety of these microbial communities, it is becoming ever clearer that nurturing those relationships may be essential in order to address some of our most pressing concerns, from climate change to resource sustainability.

1.4.3 **Trichoderma reesei** *in Lignocellulosic Feedstock Generation*

Lignocellulose is the most abundant organic material on the planet. These materials must be pretreated using different pre-treatment methods, such as thermal, chemical, biological, and enzymatic, in order to convert them into glucose. At the

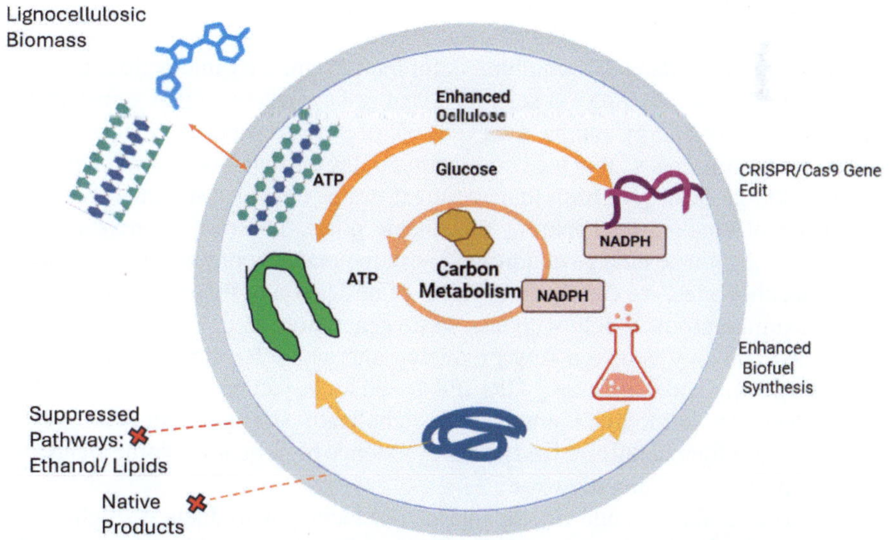

Fig. 1.2 Schematic illustrations of metabolic engineering strategies in *Trichoderma reesei* for enhanced biofuel production, highlighting CRISPR/Cas9-mediated gene editing, enhanced cellulase secretion, and suppression of competing pathways

microorganism level, *Trichoderma reesei* is mostly employed for its strong cellulolytic enzymes-producing ability (Saravanakumar and Kathiresan 2014). Metabolic engineering strategies, including CRISPR/Cas9-mediated gene editing in fungi for improved production of biofuels, are shown in Fig. 1.2.

1.5 Bioenergy and Fungi: Current Status and Future Developments for Biofuel Production With an Impact on Fungal Biotechnology

Moreover, the integration of bioconversion-based methods with emerging technologies (synthetic biology and metabolic engineering) may be interesting for the optimization of fungal biotechnologies for bioenergy production. By modifying fungal strains that were permitted to overproduce metabolic pathways, these organisms can be tailored for improved substrate utilization and biofuel production, thus being a step forward in the energy sustainability and waste management context. For instance, advances in genome editing techniques may generate new lines of fungi that more rapidly degrade recalcitrant lignocellulosic substrates and ultimately expand the applicability of the conversion process. Moreover, applying these new containerization methods is consistent with the circular economy philosophy: substituting waste valorization can reduce CO_2 emissions from conventional means of energy production (Gupta et al. 2011).

In the evolving bioenergy area, sustained interdisciplinary collaboration among biotechnologists, environmental scientists, and policymakers is essential to translate scientific progress into integrated and sustainable energy supply systems.

In modern times, there is a requirement for interdisciplinary interactions between biotechnologists, environmental scientists, and policymakers so that a sustainable supply system for energy can be achieved. We need to incorporate innovative biotechnological concepts so that the combination of the current bioconversion process with technologies can give birth to new fungal strains on which we can rely for the conversion of bioenergy to achieve sustainable goals. These solutions should be friendly to the climate change conditions. With the maturity of progressive bioconversion technologies, we can reach the level of utilization of these fungal strains to reach maximum product yield with less waste generation.

Synthetic biology is also a very promising alternative for strain modifications. New gene editing tools, including CRISPR-based tools, can be the best approach in this segment. These modified strains are metabolically stronger than before. They churn stubborn lignocellulosic feedstocks to produce high-quality biofuels and also solve the problem of dumping waste.

Furthermore, these fungi are capable of producing byproducts of high value, such as bioplastics or pharmaceuticals. They utilize the same feedstocks for the same and tie into the SDGs. As the field moves forward, environmental risks of genetically modified organisms (GMO) technology should be regarded with utmost concern to avoid destroying ecology even when using well-behaved GMOs (Gupta et al. 2011).

1.6 Technical Challenges

Second, like the development of any new fungal biotechnology, we should not forget the requirement for efforts to frame this within broader ecosystemic management paradigms and employ it through synergistic integration with other disruptive technologies (e.g., artificial intelligence (AI) and Internet of Things) toward the optimal process operation in bioenergy production. AI-driven analytics can help researchers narrow down the fungal strain and fermentation conditions that will result in better productivity of biofuels, as well as resource economy (Chandratreya 2025).

On the other hand, it is possible that bioconversion systems will eventually become well-controlled processes and reach this type of operating strategy, since real-time control systems working together with bioprocesses could be a factor contributing to these scenarios in which environmental conditions are set as they should, allowing the use of efficiency in an ideal way.

This transversal theme not only works on a sustainable development trajectory but also provides alternative options to satisfy global energy demand. This results in the prevention of climate warming. Such cross-disciplinary partnerships are critical. They are useful in achieving a future bioenergy landscape in the utilization of fungal biotechnologies within the circular economy.

In the era associated with the development of bioconversion processes and synthetic biology, environmentally friendly synthesis for fungal biotechnologies should be another challenge toward the sustainable supply of bio-energies: the use of plant extracts and microorganisms. These plant extracts and microorganisms can be applied to ferment the yeast isolate; it is not only a way to minimize discharge but also considered an economic model for upconversion of waste materials from farming into goods (Gupta et al. 2011). This implies that bioenergy systems could be much more sustainable and carbon-smart than the kind of energy production we use currently. Finally, understanding the fungal processes from waste valorization might also assist in bio-based product production, which is potentially scalable toward serving one of the targets as outlined by the UN SDGs, such as affordable and clean energy solutions (Dhiman et al. 2024). With more research in this field, it will be important to incorporate these sustainable practices into large-scale bioenergy technologies as part of an integrated environmental sustainability and resource management.

1.6.1 Economic Challenges

With more developed fungal biotechnologies, however, the economic burdens of scaling up such new solutions will be an issue to tackle. Costs of development and deployment of cutting-edge fungal biotechnologies, including IP ownership requirements, also present barriers to their adoption for use in the bioenergy industry. As a result, incentives will be required to mask these barriers (economic) and promote the commercialization of fungal biotechnologies with public–private participation and investment (Gupta et al. 2011).

Also, employing these biotechnologies in the context of an established energy system also requires (local) comprehension of market dynamics and regulations that could be difficult for potential players to achieve. Having clear economic incentive and sustainability based on scalability, the near future vista of fungal biotechnologies for the circular economy indeed is huge (Goswami 2025) and is not confined within narrow boundaries.

Additionally, its successful market penetration in the context of fungal bio production technology trends provided firm framework conditions within which safety and efficacy are secured, and innovation is rewarded. From now on, policy decisions will face the difficult challenge to utilize revolutionary biotechnologies in environmentally sound and economically feasible ways. To speed up this transition, we should invest in R&D as well as public–private partnerships to scale up.

By building economic incentives that reflect these ecological dynamics, it will be feasible to help maximize the contribution of myco-biotechnologies to a circular economy, as well as supporting the UN SDGs in general, and climate resilience more specifically. Public acceptance of such novel solutions will require community buy-in and understanding about the implications of fungal biotech to ecological health, functioning ecosystems, specifically within each local and global energy system.

1.6.2 Regulatory and Environmental Challenges

Furthermore, the deployment of fungal-based biotechnologies to bioenergy sectors relies predominantly on the solution of regulatory and environmental issues that may affect their upscaling and acceptability. That is why, to enable the field to progress, public opinion and regulatory barriers need to be considered, and full regulatory paradigms for the safety and efficacy of GM fungi would have to be established. Additionally, an environmental impact risk assessment is needed to assess the ecological consequences of these biotechnologies in the long term because of potential non-target effects due to the release of live engineered organisms into natural environments (Gupta et al. 2011).

To achieve the aim, the cooperation between scientists, regulatory authorities, and the public is essential. This coordination will bring the acceptance of fungal biotechnologies as building blocks for sustainable networks of energy production. This cooperation will support innovation for a new ecology of integration.

The formation of green bonds or impact investment funds for these fungal tech bioenergy projects could be a potential parallel way of capital raising and a means to improve environmental policy. In so doing, it not only links financial compensations directly to environmental targets but also promotes local economic growth, such as new job opportunities in (growing) bioenergy production technologies and services. In addition, single cases demonstrated that sustainable nanomaterials are used to maximize resource efficiency in bioenergy systems and minimize waste production for the sustainability of such systems (Gupta et al. 2011). By such an interdisciplinary approach, the fungal biotechnologies could fill several gaps in applications as renewable sources of energy and climate-change resilience, which would support a sustainable bioenergy economy.

1.7 Perspectives and Future Directions

Therefore, it is now appropriate to describe and analyze hybrid systems applications emerging from the integration of other renewable energy technologies as the main trend for future fungal biotechnologies. Such systems might similarly be able to tap into fungi's peculiar metabolic pathways or, one hopes, to learn how to suck up some other sustainable technologies emerging from scratch in solar energy or wind power in pursuit of creating a more durable model for how we could power the world. Fungal bioconversion methods might also be incorporated into, say, solar energy collection for a cumulative net gain of total energy (and reduced use of fossil fuels) if we think in terms of a circular economy approach (Gupta et al. 2011).

1.8 Emerging Technologies

Fungal biotechnology is exploring AI- and machine learning-based technologies to overcome natural restrictions for bioenergy production. By using these tools for predictive analysis, scientists can refine fermentation conditions. The strain selection and improvement can be the best part to increase biofuel production. Wastage reduction can be done by designing a new system to supply energy to the world (Doukianou 2025). Furthermore, comprehension of molecular technology "synergies" in bioenergy systems could cater to high-level resource efficiency. It can also be used in preserving the ecological system along with the facilitation of sustainable bioenergy systems (Gupta et al. 2011). We can modify fungal strains to get more resilient, cleaner energy, and a circular economy.

1.8.1 Integration With Other Renewable Energies

Current fungal biotechnologies can produce hybrids by combining with solar and wind power. This synergy can enhance energy production and cushion against the impacts of climate change.

For example, a combination of fungal bioconversion with solar harvesting could also be mutually beneficial when the obtained energy from solar could be used for fine-tuning the fermentation step to maximize the yield for producing biofuel at the expense of wasting fewer resources. This 360-degree view is consistent with the circular economy model as it contributes to renewable energy applications and reduced fuel reliance. Also, the need for integration of new types of nanomaterials in these hybrid architectures will potentially provide a tremendous opportunity for significantly improving the performance and durability, along with the sustainability by reducing material input and environmental impact toward serving as resilient energy ecosystem landscapes (Gupta et al. 2011). With more refining of these consultation themes, we will be able to come up with geo-scientifically guided solutions for energy needs that don't entail sacrificing the environment, and the sustainable future will arise.

1.8.2 Policy and Funding Implications

Although nowadays, the research of hybrid systems has been focused on and developed in all aspects, we must also let technologies be scalable and get integrated into fledgling-engineered energy systems at the same time. These can be such as the fungal bioconversions supported with solar facilitated energy, only that changes in solar power supply to systems mean that systems should have a capability to automatically regulate the number of various types of energy use. This lines up with cross-disciplinary work not just by fellow workers writing about tech projects, but also on new funding mechanisms that are driving toward the transformation of sustainability into sector-wide operations (Vishvas 2024). Even more importantly, the public sentiment of the acceptance of these combined responses by the public could reinforce their uptake and accelerate a move to whatever optional path is a less fragile or more resilient system with its attendant lower degree of transition costs (Suleymanov et al. 2024).

1.9 Discussion on Research Trends Comparison

As such, fungal biotechnologies become increasingly pursued in renewable energy research, the role and context of policy environments, and enabling these hybrid systems becomes critical. Multilateral initiatives: establishment of incentives for R&D and pragmatic approaches to potential regulatory barriers in the use of genetically modified organisms for bioenergy. Moreover, promotional public–private partnerships are required to direct such investment toward sustainable practices and establish new funding models that accelerate the commercialization of fungal biotechnologies for the energy sector (Gupta et al. 2011). Finally, we highlight the potential of these hybrid systems to be implemented and contribute toward climate resilience through the improvement in resource efficiency levels and mitigation actions for avoiding environmental risks, according to global sustainability targets that have aligned with UN-SDGs. By providing a forum for interaction and collaboration, collectively, we can support the response from the researchers, from industry, and policymakers to enable us to unlock this potential of fungal biotechnologies to deliver sustainable energy futures (Tiwari and Park 2024).

1.9.1 Regional Variations in Research Focus

While discussion of the combination of fungal biotechnologies with bioenergy continues, it is important to understand that there are some regional differences in terms of research focus and deployment. Various geographical conditions may cause different risks and opportunities that are regionally dependent on (economic) resources, regulatory environment, and attitude toward biotechnology.

For example, researchers in countries with a profusion of agricultural waste may focus on the development of yeast strains to utilize this waste; however, urban planners are more likely to focus on bioconversion processes that address energy needs in areas such as dense city centers (Uzun 2002). Furthermore, greater international cooperation could lead to the transfer of knowledge and innovations; a way to share experiences with other regions to learn from best-practice technologies for their own special needs. This is a more collaboratively oriented way of contributing to the development of fungal biotechnologies, taking into account international sustainability objectives without any prejudice, and allowing regional perspectives about what sustainable energy is.

With the development of biotechnologies for fungi, these organisms provide a remarkable tool to address energy and waste in urban areas as control systems of fungal metabolism progress. Sustainable and enhanced biofuels production. Another field in which fungal biotechnology is directed to the urban region has the possibility to provide it a substitute for the high amount of waste produced from such areas (up to several kilograms per inhabitant daily, if in some cities) and energy consumption. For instance, utilization of altered fungi capable of decomposing food waste has the potential to mitigate greenhouse gas emissions and also present a platform as a renewable energy source (Gupta et al. 2011). Furthermore, a smart waste management system utilizing real-time data analysis may efficiently enhance the production of substrate conversion and energy conversion through these bioconversion processes (Aggarwal 2024). As a characterized urban-focused solution, it is furthermore consistent with the circular economy concept and encourages local community involvement by converting waste into an asset useful for society and compatible with other development objectives, such as sustainable development and climate adaptation at large.

Table 1.1 Comparison of fungal species used in bioenergy production based on feedstock type, key enzymes produced, and typical bioenergy yields

Fungal species	Feedstock type	Main enzyme produced	Bioenergy yield
Aspergillus niger	Rice husk, sugarcane bagasse	Beta-galactosidases, hemicellulases	Moderate to high (biogas/ethanol)
Mucor indicus	Lignocellulosic hydrolysates	Ethanol-producing enzymes	High ethanol tolerance
Neurospora crassa	Cellulose-rich biomass	Cellobiose dehydrogenase	Moderate ethanol production
Phanerochaete chrysosporium	Wood chips, lignin rich residues	Lignin peroxidases	Excellent pretreatment efficiency
Rhizopus oryzae	Food waste, starch--rich waste	Amylases, lipases	High

1.9.2 Comparison Between Fungal Strains

The broader implementation of fungal biotechnologies also creates the potential to integrate these systems with smart city philosophy. Real-time data analytics for waste generation and energy can also assist in making fungal bioconversion processes fit the needs of a specific area. By this approach, we not only reduce inputs but also promote sharing and participation in educational projects related to the benefits of waste-to-energy technology. In addition, potential use of these systems may allow the fulfillment of several UN SDGs (e.g., sustainable cities and communities) since they can serve as a circular economy and reduce environmental pollution regarding urban waste management (Tiwari and Park 2024). Table 1.1 shows the application of different fungal enzymes in bioenergy yield.

Cities, along with braiding such advancements as the (bio)technological application and sustainable urbanism together, could be one of the main catalysts for both local and global shift to a more ecological resources solution.

1.9.3 Unresolved Questions in the Literature

An important question within an urban setting, where the use and implementation of fungal biotechnologies is expanding, is how these systems can be incorporated into educational programs. Cities can encourage participation with the use of workshops, outreach events that raise awareness regarding environmental benefits (if any) from waste-to-energy crops, and so on, as well as by promoting other sustainable lifestyles more generally. This could help to de-mystify the scientific knowledge about fungal biotechnologies, and people might gain a better understanding of how this work relates to climate change, resource management, or similar topics.

Furthermore, when informal education programs are also developed at local universities and research institutions (Buckley et al. 2017), they can also be implemented in a more strategic way and at a larger scale, for instance, by enhancing the leverage effect that the knowledge transfer multiplies knowledge through social mechanisms (Buckley et al. 2017). This supports broader purposes concerning sustainable urban development, as well as the UN SDGs. Such a systemic approach, which not only illuminates citizens but also catalyzes a general sustainable culture and mindset, is necessary for the bioenergy project to be realistically viable at the urban scale.

1.9.4 Need for Longitudinal Studies

In addition to education, the exploration of public–private partnerships could be a catalyst to push myco-technologies in cities. If the potentials of both sectors are brought together, it will be possible to technically evaluate prototypes of waste dense energy plants based someone this time on fungal processes. Besides, the participation of local industry could not only facilitate further geographical distribution of community investment but would instead have the potential to initiate development of a culture for sustainability through some successful examples in which industries incorporated biorefinery solutions into their processes (Tiwari and Park 2024). This latter is not only aligned with the vision of the circular economy (Steg et al. 2015) but also involves tackling two main challenges that cities will face in terms of waste management and energy sufficiency for city development, securing broader sustainable urban development goals (Steg et al. 2015).

1.10 Conclusion

Together, fungal biotechnology for bioenergy production is at the intersection of novelty. Also, it may open an avenue to address local as well as global concerns in relation to climate change or resource exhaustion. Advances made in fungal strains, either by fermentation methodologies and bioconversion methods, will enhance the efficiency and sustainability of bioenergy systems. However, these biotechnologies are hindered by a series of challenges, for example, technical, economic, and environmental. Interdisciplinary working is needed among researchers in academia, industry, and policymaking to ensure the maximum utilization of technology. Let us hope local communities work to preserve resources to ensure the future of fungal biotechnology. I will depend significantly on the research focus and the financial resources to develop new fungal biotechnologies. In promoting these characteristics, we can harness the beneficial natural forces to provide new technologies that will help satisfy our energy demands. Also, preserve and maintain an environment that is becoming increasingly burdened.

1.11 Summary of Key Findings

With the perspective of a sustainable energy future, the developing potential of fungal biotechnology, along with modern waste management, is increasingly important. Extremely relevant cases, such as localized fungal-based bioprocessing in urban areas, could turn components of organic waste into biofuel and concurrently offer a solution for local waste management. Beyond this, the use of the two approaches might facilitate the creation of value-added products and may also be consistent with the circular economy concept, in consonance with the UN SDGs on sustainable cities and communities. Moreover, because cities face growing levels of waste generation, stimulating the creation of city government and research center alliances in order to incentivize the strong implementation of fungal biotechnologies on the ground for creating new environmental or community resilience strengthening technologies is also essential. They will help to enable pilots and "demonstrations for impact" to demonstrate the success of fungal systems in urban environments and increase broad uptake of these low-impact solutions.

1.12 Implications for Future Research

Furthermore, the development of community-driven fungal technology and bioenergy programs with local involvement toward bioenergy projects would be an important advance in relation to fungal biotechnologies. Local community members support that if the concept is realized through locally driven education and awareness-raising campaigns here, they should drive demand for integrated waste management solutions in their communities, as well as sustainable energy approaches. For instance, proximal workstations to exhibit methods for the implications of fungal systems applications in organic waste degradation to bioenergy might induce a "feeling of ownership and responsibility" among residents, which would be in line with UN SDG 11: sustainable cities and communities (Okoro et al. 2024). It can also aid knowledge and resource transfer to introduce locally bioenergy systems, which can be profitably employed without violating environmental standards. In this way, community-targeted actions, such as nature-based solutions, are expected to transform cities into a display of sustainability and a reflection of the best deployments of fungal biotechnologies for addressing significant environmental challenges.

References

Aggarwal R (2024) Powering tomorrow: pioneering trends in sustainable energy innovation. In: Mishra A, Karthikeyan, Kulkarni HR, Sarpong Asamoah S, Deshmukh SB, Vijay Muni DT, Mishra D, Singh Kaler N, Kumar Srivastava N, Kaur M, Thankaraj Ambujam S, Mondal M (eds) Futuristic trends in renewable & sustainable energy. Volume 3 Book 1, 1st edn. Iterative International Publishers, Selfypage Developers Pvt Ltd, pp 154–160. https://doi.org/10.58532/V3BDRS1P1CH12

Ahmad, I., & Abdullah, N. (Eds.). (2025). Nature-based technologies for wastewater treatment and bioenergy production. IWA Publishing doi: https://doi.org/10.2166/9781789064100

Auta HS et al (2024) Biotechnology for bioenergy production: current status, challenges, and prospects. In: Microbial biotechnology for bioenergy. Elsevier, pp 277–296. https://doi.org/10.1016/B978-0-443-14112-6.00014-6

Buckley PJ, Doh JP, Benischke MH (2017) Towards a renaissance in international business research? Big questions, grand challenges, and the future of IB scholarship. J Int Bus Stud 48(9):1045–1064. https://doi.org/10.1057/s41267-017-0102-z

Chai Y, Bai M, Chen A, Peng L, Shao J, Luo S, Deng Y, Yan B, Peng C (2022) Valorization of waste biomass through fungal technology: advances, challenges, and prospects. Ind Crop Prod 188:115608. https://doi.org/10.1016/j.indcrop.2022.115608

Chandratreya A (2025) Environmental excellence through disruptive technological innovations. In: Nexus of environmental quality and technology innovation. IGI Global Scientific Publishing, pp 97–126. https://doi.org/10.4018/979-8-3373-1822-6.ch005

Chowdhary K, Prasad U, Sharma S (2018) Role of fungi in biorefinery: a perspective. In: Kumar S, Dheeran P, Taherzadeh M, Khanal S (eds) Fungal biorefineries. Springer International Publishing, pp 1–20. https://doi.org/10.1007/978-3-319-90379-8_1

Dhiman S, Kaur P, Narang J, Mukherjee G, Thakur B, Kaur S, Tripathi M (2024) Fungal bioprocessing for circular bioeconomy: exploring lignocellulosic waste valorization. Mycology 15(4):538–563. https://doi.org/10.1080/21501203.2024.2316824

Doukianou DS (2025) Emergent technologies: pioneering sustainable futures. In: Practical sustainability strategies. Wiley, pp 207–223. https://doi.org/10.1002/9781394263790.ch18

Farinas CS, da Cunha FM (2016) Fungal enzymes and their role in bioenergy. In: Purchase D (ed) Fungal applications in sustainable environmental biotechnology. Springer International Publishing, pp 307–320. https://doi.org/10.1007/978-3-319-42852-9_12

Goswami S (2025) Paving the way for a sustainable tomorrow: strategies for environmental transformation. Int J Arts Humanit Soc Stud 7(1):7–11. https://doi.org/10.33545/26648652.2025.v7.i1a.146

Grace OM, Lovett JC, Gore CJN, Moat J, Ondo I, Pironon S, Langat MK, Pérez-Escobar OA, Ross A, Suzan Abbo M, Shrestha KK, Gowda B, Farrar K, Adams J, Cámara-Leret R, Diazgranados M, Ulian T, Sagala S, Rianawati E et al (2020) Plant power: opportunities and challenges for meeting sustainable energy needs from the plant and fungal kingdoms. Plants, People, Planet 2(5):446–462. https://doi.org/10.1002/ppp3.10147

Gupta R, Mehta G, Khasa YP, Kuhad RC (2011) Fungal delignification of lignocellulosic biomass improves the saccharification of cellulosics. Biodegradation 22(4):797–804. https://doi.org/10.1007/s10532-010-9404-6

Karduri RKR (2023) Transitioning to tomorrow: the global journey towards a sustainable energy economy. arXiv:2311.14907. https://doi.org/10.13140/RG.2.2.14081.63841

Kržišnik D, Gonçalves J (2023) Environmentally conscious technologies using fungi in a climate-changing world. Earth 4(1):69–77. https://doi.org/10.3390/earth4010005

Lone SA, Jeelani G (2023) Evaluating the potential impact of climate change on glacier dynamics in Western Himalayas, India. In: Chatterjee U, Shaw R, Kumar S, Raj AD, Das S (eds) Climate crisis: adaptive approaches and sustainability. Springer Nature, Cham, pp 499–510. https://doi.org/10.1007/978-3-031-44397-8_25

Mafune KK, Winkler MK (2024) The expansion of fungal organisms in environmental biotechnology. Curr Opin Biotechnol 90:103217. https://doi.org/10.1016/j.copbio.2024.103217

Mukherjee D, Singh S, Kumar M, Kumar V, Datta S, Dhanjal DS (2018) Fungal biotechnology: role and aspects. In: Gehlot P, Singh J (eds) Fungi and their role in sustainable development: current perspectives. Springer, pp 91–103. https://doi.org/10.1007/978-981-13-0393-7_6

Okoro PA, Chong K, Röder M (2024) Enabling modern bioenergy deployment in Nigeria to support industry and local communities. Biomass Bioenergy 190:107403. https://doi.org/10.1016/j.biombioe.2024.107403

Pandey RK, Tewari L (2018) Mycotechnology for lignocellulosic bioethanol production: an emerging approach to sustainable environment. In: Microbial biotechnology in environmental monitoring and cleanup. IGI Global Scientific Publishing, pp 28–43. https://doi.org/10.4018/978-1-5225-3126-5.ch002

Qureshi T, Farooq M, Siddiqi MH, Adnan M, Farhan M, Ali Q (2023) Sustainable strategy to develop organic bio refinery model. In: 2023 6th International conference on energy conservation and efficiency (ICECE). IEEE, pp 1–3. https://doi.org/10.1109/ICECE58062.2023.10092498

Rame R, Purwanto P, Sudarno S (2023) Biotechnological approaches in utilizing agro-waste for biofuel production: an extensive review on techniques and challenges. Bioresour Technol Rep 24:101662. https://doi.org/10.1016/j.biteb.2023.101662

Saravanakumar K, Kathiresan K (2014) Bioconversion of lignocellulosic waste to bioethanol by Trichoderma and yeast fermentation. 3 Biotech 4(5):493–499. https://doi.org/10.1007/s13205-013-0179-4

Selvakumar P (2025) Fungal bioprospecting: discovering novel bioactive compounds. In: Mycological inventions for sustainable agriculture and food production. IGI Global Scientific Publishing, pp 327–348. https://doi.org/10.4018/979-8-3693-8201-1.ch015

Steg L, Perlaviciute G, van der Werff E (2015) Understanding the human dimensions of a sustainable energy transition. Front Psychol 6:805. https://doi.org/10.3389/fpsyg.2015.00805

Suleymanov M, Aslakhanova S, Mezentseva E (2024) Funding projects for sustainable agriculture and resource-efficient practices. BIO Web Conf 140:03006. https://doi.org/10.1051/bioconf/202414003006

Sundari SK, Sachdeva S, Agarwal P, Awasthi S (2018) Bioenergy: a sustainable energy option. In: Proceedings of the MOL2NET'17, conference on molecular, biomedical and computational & network science and engineering, 3rd edn. MDPI, Basel. https://doi.org/10.3390/mol2net-03-05122

Tiwari P, Park K-I (2024) Advanced fungal biotechnologies in accomplishing sustainable development goals (SDGs): what do we know and what comes next? J Fungi 10(7):506. https://doi.org/10.3390/jof10070506

Uzun A (2002) National patterns of research output and priorities in renewable energy. Energy Policy 30(2):131–136. https://doi.org/10.1016/S0301-4215(01)00084-2

Vishvas R (2024) The role of policy and innovation in environmental sustainability. J Sustain Solut 15(2):13–18. https://doi.org/10.36676/j.sust.sol.v1.i2.9

Yu Z, Han H, Zhang F (2026) Global meta-analysis of bioenergy crops reveals substantial greenhouse gas mitigation potential. Resour Conserv Recy 227:108777. https://doi.org/10.1016/j.resconrec.2026.108777

Chapter 2
Fungal Biotechnology to Biomaterials: Production, Challenges, and Perspectives

Anuska Ghosh, Suryasarathi Kumar, Subhamoy Banerjee, Magnolia Roy, Priyanka Talukdar, and Somnath Das

Abstract Fungal biotechnology is an emerging yet underexplored source of next-generation biomaterials. Although mycelium-based composites have received recent attention, the immense number of fungal species, metabolic pathways, and structural architectures that have not yet been explored for material innovation remains largely unexplored. This chapter highlights new and less characterized areas at the interface between mycology and materials science, with a special focus on unusual fungi, such as extremophiles, marine isolates, and endophytic species with unique biopolymeric properties. New studies indicate that fungal networks are dynamic and responsive, and thus, living, adaptable materials with self-healing, self-growing, and stimuli-responsive characteristics can be created. New developments in synthetic biology, genomic mining, and artificial intelligence-assisted

Anuska Ghosh, Suryasarathi Kumar and Subhamoy Banerjee contributed equally with all other contributors.

A. Ghosh
The Agricultural Biotechnology Division, School of Agriculture and Rural Development, Ramakrishna Mission Vivekananda Educational and Research Institute, Narendrapur Campus, Kolkata, West Bengal, India

S. Kumar
School of Biological Science & Technology, Department of Applied Biology, Maulana Abul Kalam Azad University of Technology-Main Campus, Simhat, Haringhata, Nadia, West Bengal, India

S. Banerjee
Department of Life Science – Microbiology, School of Biological Science & Technology, Maulana Abul Kalam Azad University of Technology, Simhat, Haringhata, Nadia, West Bengal, India

M. Roy
Department of Electronics and Communication Engineering, Kumaraguru College of Technology, Coimbatore, Tamil Nadu, India

P. Talukdar · S. Das (✉)
The Institute of Science, Culture, and Social Studies, Belgharia, Kolkata, West Bengal, India

M. Tripathi et al. (eds.), *Recent Advancements in Fungal Biotechnology Volume 1*, Fungal Biology, https://doi.org/10.1007/978-981-92-0339-0_2

design are now enabling the discovery of cryptic biosynthetic gene clusters and the programmable production of fungal-derived polymers, nanomaterials, and conductive pigments such as melanin.

In addition, frontier applications of fungal biomaterials include bioelectronics, soft robotics, neural interfaces, Clustered Regularly Interspaced Short Palindromic Repeats (CRISPR)-Cas9 genome editing, environmental sensors, and space habitation, reflecting the transformational potential of these materials beyond their usual packaging and building applications. An expansion in technological space is also achieved through new methods of fabrication, such as 4D printing, bio-mineralization, and hybrid multi-organism systems. Despite these advancements, significant challenges, including the scarcity of genetic material for designing fungi in rare categories, scaling challenges in living materials, and a lack of control over engineered fungal systems, remain.

The chapter summarizes new knowledge, identifies a critical knowledge gap, and outlines the future of integrating fungal biology with advanced materials engineering. This will transform the idea of sustainable manufacturing, as fungal biotechnology explores these untapped areas and encourages the development of high-performance, bio-fabricated materials for future generations.

Keywords Fungal biomaterials · Mycelium engineering · Synthetic biology · Biofabrication · Living materials

2.1 Introduction

The convergence of systems biology, synthetic genomics, and advanced manufacturing is rapidly transforming fungi from culinary curiosities into programmable "cellular factories" capable of extruding high-performance materials that rival petrochemical plastics, animal leather, and even carbon-fiber composites. This chapter maps the scientific inflexion points that have elevated mycelium-based matter from low-density packaging foams to multifunctional, genetically tuned constructs for medicine, electronics, and the built environment.

Recent multi-omics atlases of white-rot Basidiomycetes revealed that ligninolytic peroxidase clusters and intracellular vesicle trafficking genes orchestrate extracellular matrix (ECM) deposition, yielding tunable porosity, stiffness, and water retention that exceed those of conventional polysaccharide hydrogels by an order of magnitude (Singh et al. 2025). Coupling these datasets with artificial intelligence (AI)-guided "biomaterialomics" platforms now allows regression of mechanical properties against 10^4–10^5 genomic features, compressing decades of empirical optimization into weeks of in silico screening (Chakrapani et al. 2025). Parallel efforts have expanded the taxonomic repertoire beyond the traditional *Ganoderma–Pleurotus* duo. CRISPR-Cas9-enabled editing of *Coprinopsis cinerea* and *Ustilago maydis* has unlocked the biosynthesis of ultraviolet (UV)-absorbing indole pigments, conductive melanin nanorods, and self-assembling hydrophobin films, materials unattainable through conventional mycelium tanning (Hinneburg et al. 2025). Neutral genomic "landing pads" identified by transcriptome-wide essentiality mapping ensure stable, high-yield expression of 30+ kb biosynthetic gene clusters (BGCs) without fitness trade-offs, providing a scalable route to designer fungal biopolymers (Liu et al. 2018).

Production engineering is equally disruptive. Solid-state fermentation (SSF) on agricultural waste streams now reaches volumetric productivities of 0.8 g L^{-1} h^{-1}—comparable to submerged *Escherichia coli* systems—while consuming 70% less water and zero sterilants (Ribeiro et al. 2022). Hybrid manufacturing pipelines integrate robotic three-dimensional (3D) deposition of living mycelial pastes with post-growth compression molding, creating density-graded bilayers that mimic balsa wood for wind-turbine cores yet retain full compostability (Singh et al. 2025). On the fabrication front, stimuli-responsive four-dimensional (4D) mycelium inks containing programmable hyphal growth inhibitors enable shape-morphing architectures that self-fold into intricate lattices upon hydration, offering a sustainable route to deployable shelters or medical stents (Llamas et al. 2023). Simultaneously, CRISPR knock-outs of chitin synthase regulators produce nanofibrillar networks that can be shear-aligned into high-modulus filaments for textile-grade yarns, replacing energy-intensive viscose processes (Chakrapani et al. 2025). Emerging applications capitalize on these advances. Living mycelial skins doped with graphene oxide serve as conformable bio-electrodes for real-time plant electrophysiology, while melanized *Pleurotus ostreatus* foams function as reversible Li-ion battery separators with 98% capacity retention over 500 cycles (Hinneburg et al. 2025). In healthcare, engineered *Aspergillus oryzae* secretes recombinant human tropoelastin within a mycelial ECM, yielding off-the-shelf vascular grafts that exhibit physiological compliance and suture retention strength surpassing Expanded Polytetrafluoroethylene (ePTFE) standards (Chakrapani et al. 2025). Yet significant hurdles persist. Recalcitrance to genetic transformation in industrial Basidiomycetes limits editing efficiencies to <10%, and off-target effects can trigger unwanted pigment or toxin production (Rombauts et al. 2003). Scale-up of hyphal filament alignment remains sensitive to shear-induced fragmentation, creating batch-to-batch variability in mechanical performance. Regulatory frameworks lag, with no harmonized standards for living or "viable" materials that continue to metabolize post-processing, complicating safety certification for biomedical or food-contact uses (Chakrapani et al. 2025).

2.2 Novel Biological Insights Driving Next-Generation Biomaterials

2.2.1 Understanding Mycelial Architecture at the Molecular Level

Developing a complete understanding of the underlying mechanisms that dictate hyphal growth, branch development, and cell wall biogenesis is essential to transforming fungi from living organisms into high-performance functional materials.

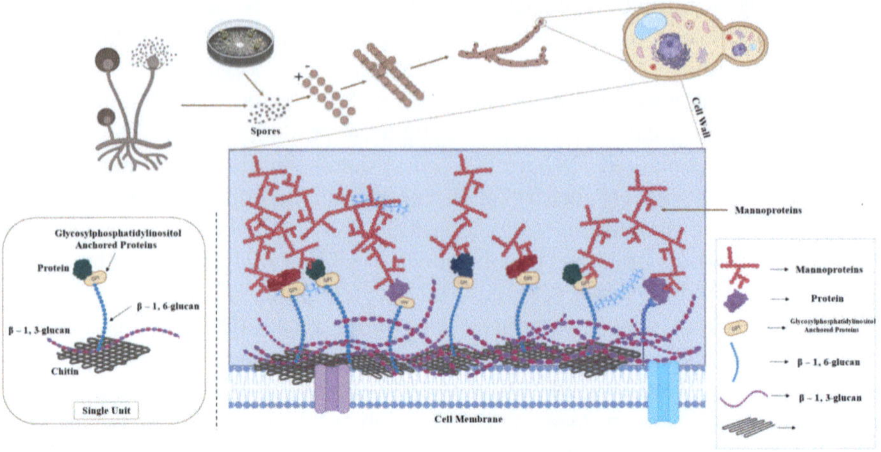

Fig. 2.1 Diagrammatic structure of a fungus cell wall (single structural unit). The outer spore layer (when present) is covered by a thick glycoprotein sheath composed of mannoproteins and Glycosylphosphatidylinositol (GPI)-anchored proteins that mediate adhesion, recognition, and environmental sensing. Under this, there is a fibrillar meshwork of branched b-1, 6-glucan (B1, 6-gluc) cross-links to load-carrying b-1, 3-glucan chains that convey tensile strength, porosity, and hydration regulation. An inner crystalline chitin layer, covalently linked to β-1,3-glucan, anchors the entire wall to the plasma membrane and provides structural rigidity and shape stability. Hierarchical and self-assembled laminate: This laminate possesses the capability to be reprogrammed, genetically, to change the ratios of chitin, glucan, and mannoprotein or to include synthetic cross-links to form next-generation fungal biomaterials with custom mechanics, barrier properties, and controlled biodegradability to be used in the next industrial application, which are packaging, textiles, and biomedical scaffolds

This self-assembly involves a dynamic, programmable phase restructuring organized by the interface itself, which triggers a considerable degree of conformational change in the protein monomers (Fig. 2.1) (Ball et al. 2019). The resulting rodlet layer, which evenly covers the hyphae, reverses the surface wettability, thereby conferring the remarkable hydrophobicity of many mycelial structures, with water contact angles exceeding 120° (Khyaju et al. 2025).

The intricate, filamentous colonization morphology of fungal mycelia offers exceptional opportunities to create biomaterials of specific sizes and shapes. Therefore, fungi are an excellent source of sustainable material. When dried, a mycelial product can exhibit strong, durable properties and be lightweight, flame-resistant, and thermotolerant (Jones et al. 2018). These qualities have already launched such materials into trade markets, such as eco-friendly packaging and transportation solutions, which will be attractive to consumers conscious of the environment. Furthermore, mycelial biomass is compatible with 3D printing methods and enables the synthesis of composite materials with tailored geometries (Luo et al. 2025). The hydrophobicity of most mycelia, which helps repel water, may also extend the service life of concrete structures, as demonstrated by the self-healing ability of some ascomycete fungi, which metabolically precipitate calcite, a chief component of limestone and concrete (Khyaju et al. 2025).

2.2.2 Hyphal Growth and Branching

The length of hyphae can be many meters, with a diameter of 1–30 mm. The mass system, called fungal mycelium, grows by forming apical tips from spores, resulting in fractal colonies that are stochastically branched and interconnected and are dictated by nutritional and environmental factors (Singh et al. 2025). Fungal biomaterials have their structural scaffold composed of fungal mycelium, a highly ordered, dynamically reorganizing, and self-assembling biological network. Latest advances in fungal biology have significantly advanced our understanding of how to elongate hyphae, branch, network, and organize mechanically, all key factors to consider when designing next-generation fungal biomaterials. Branching frequency is a crucial factor that predetermines the structure of mycelia (Islam et al. 2017). The two primary modalities that have been noted in fungal hyphae are the apical and lateral modalities. Apical branching is when a tip of a hypha branches into two, usually during periods of rapid growth, presumably because of an increase in vesicle flux. The vesicle-trafficking nucleus of the Spitzenkörper dissipates upon termination of hygal extension and a disfigurement of the tip, which has been reported in *Neurospora crassa* (Riquelme et al. 2018). The emergence of several apical branches with individual Spitzenkörner follows this. Although this can be attributed to vesicle flux, lateral branching arises from subapical compartments separated by septa and thus involves a mechanism distinct from apical branching (Harris et al. 2005).

Comparative studies of saprobic fungi have indicated that divergent network morphologies result from trade-offs among hyphal traits, including growth rate and branching intensity (Lehmann et al. 2019). Species with a high rate of branching form dense, compact networks, whereas those with a lower rate form thin, voluminous networks—lightweight construction and porosity. However, sparse networks have low porosity, whereas dense networks exhibit enhanced material strength and cohesion; these architectural differences are directly related to biomaterial applications (Fricker et al. 2017). One architectural property is the so-called hyphal anastomosis, in addition to branching, which complements the functional role of mycelial networks (Carlile 1995). Hyphal anastomosis leads to cross-linking and enhances nutrient transport, mechanical stability, and the system's resistance to damage, as shown by imaging experiments on *Podospora anserina* (Dikec et al. 2020). Such interconnected structures could also be essential for designing biomaterials, providing uniform stress distribution and excellent mechanical properties. The orientation of hyphae, the frequency of coproduction, and the network density can be significantly affected by factors such as substrate material, water content, air or ice temperature, oxygen levels, and mechanical forces (Meyer et al. 2021). Independent modulation of material properties without genetic manipulation can result from the biological plasticity of mycelial architecture, a property that researchers can indirectly use to control its structure (Harris 2008).

2.2.3 Secondary Metabolites

Secondary metabolites (SMs) are toxic compounds produced by fungi that are used to make clinical drugs, such as penicillins, statins, cyclosporins, and echinocandins, and are effective in combating human illnesses. The rationale for this study is the recognition that microbial biosynthetic technology for SMs, derived from bacteria or fungi, is essential for drug discovery. In the former, to study the molecular mechanics of SM biosynthesis, it is essential to determine the gene clusters that produce them. The loci of interest are commonly affirmed by subsequent targeted deletion or heterologous expression of the loci of interest. During the pre-genomic era, identification of SM gene clusters in fungi was a lengthy, cumbersome process. These involved insertional mutagenesis and plasmid rescue of the blocked mutant (Poveda et al. 2025; Niego et al. 2023; Wadhwa et al. 2024). The requirement of efforts to construct libraries, screen them, and walk the chromosomes has been rendered unnecessary with the introduction of whole-genome sequencing (WGS) of the target SM-producing fungus.

The connection between SM compounds and the related BGCs is one of the pillars of modern natural-product research initiatives (Basnet et al. 2026). Although the primary goal of such studies is to clarify the molecular and enzymatic principles of SM production, the advantages of pathway analysis go far beyond that. Understanding metabolic pathways supports metabolic engineering approaches to increase SM production in large quantities, as well as to provide chemical insights that may be used to discover more efficient, complete, and semi-synthetic pathways to natural product complexes (Fisher et al. 2014).

In addition, researchers have developed methods to generate new SM analogs by mutasynthesis and by combinatoric biosynthesis, based on broad insights into BGCs. A good example of this paradigm is the combinatorial biosynthesis of benzenediol lactones, which builds upon prior biosynthetic and genetic research (Baig et al. 2008; Floss 2006; Zhu et al. 2025). The development of environmentally benign semi-synthetic pathways to produce simvastatin has been enabled by the characterization of the acyltransferase LovD in the lovastatin pathway (Liang et al. 2018; Huang et al. 2017). Another reason SM biosynthetic pathways are helpful is that they provide a source of novel biocatalysts for green chemistry and chemoenzymatic reactions. Both can serve two important purposes: to rationally design biosynthetic pathways to target the production of organic chemicals and to precisely predict the structures of SMs from DNA sequences. Finally, a better understanding of the biological processes underlying SMs can be gained on a broader natural scale by combining genomic information with transcriptomic and comparative analyses (Liu et al. 2021; Phintha and Chaiyen 2022).

2.2.4 Role of Next-Generation Sequencing in Fungal Biomaterial Research

Next-generation sequencing (NGS) technology has had a significant impact on the biological sciences, particularly in fungal genetics. Earlier identification methods are tedious and imprecise, relying on pathogen isolation, selective culture, and pathological inspection. The information obtained through increased DNA sequencing has transformed the entire phylogeny of the fungal kingdom (Ahrendt et al. 2018). DNA sequencing has often revealed that fungi previously thought to be identical were in fact very different from one another, and the reverse was also true (Wu et al. 2016). NGS was radical in introducing reversible stop-codon determination and sequencing by synthesis, both polymerase chain reaction-based and GeneChip-based. The numerous strengths of NGS in nucleic acid research have attracted increased interest over the last decade (McCombie et al. 2019).

Good-quality WGS of fungi offers an advantage for research in fungal biotechnology. Machine gene predictions are improving due to learning algorithms—identification and 3D protein structures (Bachhar et al. 2025; Das et al. 2024, 2025a, b; Kumar et al. 2024, 2025, 2026a, b). These resources will lead to entirely new genes, novel alleles of known genes, and entire gene clusters (Passer et al. 2019). By incorporating WGS with other new technologies, termed omics such as metabolomics, transcriptomics, and phenomics, fungi and their interactions with the environment have enabled researchers to understand them as never before.

The three central NGS sequencing systems are 454, Illumina, and SOLiD. The error rate of the NGS ranges is 15 to 0.1 in sequence splicing (Goodwin et al. 2016). The SequencErr is a new computational method designed to reduce error rates in NGS sequencing and improve sequencing accuracy (Davis et al. 2021). Also, fungal infections are increasingly identified using NGS (Tsang et al. 2021). NGS enables the study of fungal populations at the strain level. Indicatively, scientists investigating the evolutionary strains of *Saccharomyces cerevisiae* found that this vast amount of evolutionary data could be summarized into 13 variable genes. Furthermore, the method used for yeasts can also be applied to other organisms (Ramazzotti et al. 2012).

Overall, comparative genomics is one of the strongest strategies for filtering and prioritizing candidate SM BGCs, particularly when one already has a clue about the SM repertoire in the reference organism (Chooi et al. 2010). As an example, the PKS gene repertoires of *P. aethiopicum* were compared with those of *Penicillium chrysogenum* to identify gene clusters involved in the biosynthesis of viridicatum toxin and griseofulvin (Chooi et al. 2010). Past chemotaxonomic research had determined that *P. chrysogenum* did not produce viridicatum toxin or griseofulvin (Frisvad et al. 2007). In this way, orthologous and shared PKS genes across the two species could be systematically filtered, leaving the most likely species-specific PKS genes, which were more likely to encode the pathways of interest in the biosynthesis process.

Another innovation in sequencing is Hi-C, which has enhanced our understanding of fungal genome structure. The 3D structure of chromosomes and genomes can be well understood using Hi-C sequencing (Guin et al. 2020). Two examples of fungal genomes whose 3D models have been produced from Hi-C data are the yeast *Saccharomyces cerevisiae* and the human pathogen *Candida albicans* (Schalbetter et al. 2019). Machine learning algorithms can handle large amounts of data efficiently and identify specific trends that humans would have difficulty detecting. Moreover, machine learning can be used to generate model predictions that guide experiments or generate new hypotheses. Many machine learning methods, including artificial neural networks, decision trees, and support vector machines, have been applied to fungi. For example, this type of algorithm has predicted fungicide tolerance and virulence in clinical isolates of several fungi (Delavy et al. 2020).

Single-cell RNA sequencing is a significant development in high-throughput sequencing methods and enables the exquisite examination of gene expression at the cellular scale (Lipson et al. 2009). The next generation of a complementary DNA library is done using the protocol as described by (Gross et al. 2015). The use of this method in eukaryotes, including *Saccharomyces cerevisiae*, *Candida albicans*, and the fission yeast *Schizosaccharomyces pombe*, has demonstrated a surprising level of transcriptomic heterogeneity in ostensibly homogeneous populations (Dohn et al. 2021). Likewise, filamentous fungi exhibit both cellular diversity within colonies and between cells growing on a single hypha, with results similar to those in yeast (Wösten 2019). Over the past 10 years, scalable microfluidic approaches to single-cell transcriptomics of yeast have emerged, suggesting that the field is at the forefront of technology in both fundamental and applied studies (Urbonaite et al. 2021). Furthermore, similarities in the so-called decision-making abilities of slime molds and fungi are not the only reasons to believe that cell-to-cell communication and the relative comparison of gene activity at hyphal ends and on full mycelial mats might illuminate new biotechnological principles in the future (Money 2021).

Advanced predictive bioinformatics platforms have spurred advances in fungal proteomics (Almagro Armenteros et al. 2019). SignalP and EffecterP, which predict protein secretory signals and effector proteins, respectively, are of particular interest. Other utilities, including ApoplastP and Localizer, are used to predict intracellular localization of plant cells (Sperschneider et al. 2017). Although the standard identification of 3D protein structures was challenging, machine learning, particularly AlphaFold, has recently enabled the successful identification of fungal protein structures (Jumper et al. 2021). The developments are introducing pathways that are set to improve how we treat fungal diseases and how we exploit fungal biology in industry. At the same time, the inclusion of gene annotation software enhances our capacity to elucidate gene composition and preservation across taxa, thereby shaping fungal biotechnology in light of new information (Seong and Krasileva 2023). The detailed comparison is depicted in Table 2.1.

Table 2.1 Comparison of NGS-based sequencing technologies for the selection, characterization, and optimization of fungi in mycelium-based biomaterial production

NGS technology	Pros (relevance to fungal biomaterials)	Cons/limitations	References
WGS	Provides comprehensive genomic information enabling identification of genes involved in cell wall biosynthesis, hyphal growth, stress tolerance, and metabolic pathways. High-resolution SNP-based analysis enables strain-level differentiation and the selection of fungal isolates with desirable traits, such as dense mycelial networks, robustness, and adaptability. Supports understanding of genetic diversity and evolutionary stability of biomaterial-producing fungi.	Requires significant computational resources, advanced bioinformatics expertise, and high-quality sequencing infrastructure; data analysis is time-consuming and costly, limiting routine large-scale applications.	Bagger et al. (2024); Pitarch et al. (2025)
Transcriptome sequencing (RNA-Seq)	Reveals condition-specific gene expression profiles associated with mycelial growth, branching behavior, cell wall remodeling, and stress-induced strengthening of fungal structures. Enables the identification of regulatory pathways activated by different substrates or environmental conditions, supporting the optimization of growth parameters to tailor biomaterial properties.	Not suitable for direct species identification; gene expression patterns are highly dependent on growth conditions and experimental design; requires high-quality RNA and careful normalization for meaningful interpretation.	De Rose et al. (2024); Jiménez et al. (2025); Nigg et al. (2015)
Internal Transcribed Spacer sequencing	Serves as the standard fungal DNA barcode, enabling accurate taxonomic identification and community profiling of fungi used in biomaterial production. Facilitates selection and monitoring of fungal species with proven mycelial growth patterns and compatibility with specific substrates in SSF systems.	Limited resolution for closely related species; intragenomic ITS variability in some fungi can complicate identification; insufficient for functional or trait-level characterization relevant to biomaterial properties.	Bellemain et al. (2010); Reich and Labes (2017)

2.3 Emerging Fungal Biomaterials Beyond Conventional Mycelium

2.3.1 Fungal Cell Wall-Derived Polymers

Polymers of the fungal cell wall provide novel biomaterials superior to the traditional mycelium composites, taking advantage of chitin, chitosan, β-glucans, and mannoproteins to enable tunable, sustainable applications in biomedicine, textiles, and packaging (Nawawi et al. 2020). Fungal chitin and chitosan (e.g., *Rhizopus* sp. and *Mucor* sp.) are the rigid core of cell walls and the minimum quantity of β-glucans, more purified, consist of more uniform physicochemical characteristics and less allergenicity than crustacean (which yield higher molecular weight chitin), require more aggressive demineralization, and have seasonal variations and contaminants, and pose lower allergen risks (Yin et al. 2025). Fungal chitin with reduced molecular weight (50–300 kDa), crystallinity, and flexible matrices (Yin et al. 2025) and chitosan, which can be derived in a more eco-friendly way by enzymatic deacetylation, can reach higher deacetylation (up to 98%) degrees without harsh chemicals, which allows vegan-friendly extraction of the polymer throughout the year under the control of fungal cultures collected (Huet et al. 2023), β-glucans. These polymers can be functionalized to increase their utility: chemical modifications such as sulfation, carboxymethylation, acetylation, amination, and esterification to hydroxyl/amine functional groups allow self-assembly with nanoprecipitation or emulsification (Taylor et al. 2007) of functional groups to produce nanoparticles (20–500 nm) and increase solubility, stability, and drug loading of anti-cancer delivery (e.g., encapsulation of doxorubicin) (Ashrafizadeh et al. 2023). In addition to bulk composites of mycelium, these isolated polymers produce monofilaments and scaffolds and hybrids with biocompatibility, biodegradability, and low-carbon footprints, which are destined to produce scalable cruelty-free biomaterials that are poised to transform areas of wound healing and use biomaterials in clothing, among other applications (Nawawi et al. 2020).

2.3.2 Fungal Nanocellulose and Hybrid Polysaccharides

Fungal nanocellulose (FNC) and hybrid polysaccharides are new biomaterials that are more sustainable than current mycelium vast complexes because they incorporate nanoscale cellulose fibers with fungal polysaccharides such as chitin and β-glucans to provide tunable properties for packaging, biomedical scaffolds, and tissue engineering (Nawawi et al. 2020). The pathways to FNC biosynthesis occur primarily outside the cell in fungi, for example, white-rot species, in submerged fermentation (SmF) (Nawawi et al. 2020), supplemented with either co-culture with algae or nutrient-enriched media to stimulate cellulase activity (Reyes et al. 2022). In situ incorporation of cellulose nanofibers into the cell walls of the mycelium can be used to obtain hybrid polysaccharides, having the advantage of using fungal chitin-glucan matrices to in situ generate homogeneous bio-nanocomposites, and requires no synthetic companions since mycelium can self-organize fibrils by enzymatic deposition and hyphal networking (Nawawi et al. 2020). Structurally, cellulosic NC has a high crystallinity (80%), a fibrous NC with diameter of 1–100 nm (Hu et al. 2024), and a mycelium-like fibrous network, with which hydrogen bonding contributes to high water absorbance and hydrophobicity as well as compressive strength (two to five times better than mycelium alone) as a result of hydrogen bonding between cellulose and fungal 8-glucans (Nawawi et al. 2020). These hybrids have mechanical and tensile strengths of 100–300 MPa and 5–15 GPa of Young's modulus, which are superior to the brittle nature of conventional mycelium because they mimic plant cell walls but have the benefit of being biodegradable (Hansini et al. 2025). Compared to bacterial nanocellulose, which employs cyclic di-GMP and is regulated by synthases of *Komagataeibacter* sp. to produce high purity and yields (up to 15 g/L) under sterile conditions and glucose feeds (Randhawa et al. 2022), FNC relies on agro-waste feeds, halving the expense without the high uniformity of polydispersity of a fungal cell (Ganeva et al. 2025). Plant nanocellulose is in abundance (wood or cotton, obtained through harsh chemical delamination) and is longer (100–500 nm) than FNC but is less biocompatible and lacks intrinsic bioactivity due to fungal polysaccharides, while FNC hybrids are the only commercial platforms to be renewable, self-assembling, and multifunctional (Nawawi et al. 2020; Hansini et al. 2025).

2.3.3 Extracellular Polymeric Substances

Extracellular Polymeric Substances (EPSs): This category of hydrated biopolymer matrix is produced by fungi and holds promise as an alternative to the traditional mycelium-based structures as a fungal biomaterial, due to its ability to be tuned, rheologically and interfacially (Pilafidis et al. 2025, Stoica et al. 2023). Normally, a preponderant portion of the total high-molecular-weight polysaccharides that are bound with proteins and glycoproteins in a mixture that includes some lipids and nucleic acids is the specific content of fungal EPSs (Mahapatra and Banerjee 2013; Wang et al. 2024; Oliva et al. 2025), which forms a structurally continuous and chemically heterogeneous structure that can provide extensive hydrogen bonding and multivalent interaction between the EPSs and substrates (Pilafidis et al. 2025; Wang et al. 2024). The composition supports viscoelastic rheological profiles in which EPSs can act as shear-thinning gels or viscous solutions based on concentration and ionic environment, making it possible to control casting (Pilafidis et al. 2025), spreading, and spray or dip-coating onto a variety of surfaces to fabricate thin films. These networks create potent, cross-linkable, water-flexible films on drying, where the barrier performance, water interactions, and mechanical strength of water-soluble polysaccharides, proteins, polysaccharide ratios, and crosslinking can be tuned (Pilafidis et al. 2025; Oliva et al. 2025), a concept that was already revealed on EPS-based coatings, which minimize water uptake and increase oil and grease resistance on paper surfaces (Zier et al. 2025). In bioadhesions, EPSs serve as mediators of high adhesion and cohesion within biofilms through individual polysaccharide protein crossbridges between hydrophobic domains (Wang et al. 2024; Lee et al. 2016), and they have been recycled in engineered fungal EPS systems useful in high adhesion and cohesion in bioengineered systems such as sustainable adhesives and biodegradable surface coatings in packaging, textile, and bioelectronic applications (Mahapatra and Banerjee 2013; Zier et al. 2025).

2.3.4 *Fungal Melanins and Pigmented Biopolymers*

Fungal melanins and pigmented biopolymers exemplify new biomaterials with superior performance compared to traditional mycelium, because of their unique physicochemical properties (Salgado-Castillo et al. 2023). These are pigments, primarily allomelanins like 1,8-dihydroxynaphthalene types from ascomycetous and basidiomycetous (Salgado-Castillo et al. 2023; Lino and Manini 2022), with negative charge, which are hydrophobic, of high molecular weight (Qin and Xia 2024), exhibit a broadband UV–Vis absorption, are paramagnetic, show negative photochemical behavior, and thus allow stability under extreme temperatures, oxidative stress, radiation, and heavy metals (Suthar et al. 2023). The heterogeneous structure, comprising indole units, proteins, polysaccharides, and lipids, provides depth in mechanical strength and free radical scavenging, with a decay in linear UV absorption starting at 220 nm that resembles that of synthetic 3,4-dihydroxyphenylalanine (DOPA)-melanin standards. It is worth noting that fungal melanins are better UV shields; more electromagnetic radiation is absorbed over a wide range of frequencies, capable of performing the function of photoprotective energy transducers and providing some antioxidative activity through stable free radical quenching and metal ion sequestration (Salgado-Castillo et al. 2023; Qin and Xia 2024). These properties position them for applications in electronics, including biointerfaces for stem cell differentiation, metal nanoparticle platforms, and sustainable organic devices, enabled by ionic conductivity and charge transport upon hydration. They enable the development of radical-resistant and thermotolerant coatings and harsh-environment biopolymer composites (Abdal Dayem et al. 2018). However, they are disadvantaged by the lack of synthetic biodegradable dyes suitable for cosmetics, pharmaceuticals, and environmental remediation. Such pigments have the potential to be employed alongside traditional mycelium composites to enhance the functionality of textiles, sensors, and energy-harvesting capabilities, as well as future electrical and shielding capabilities (Salgado-Castillo et al. 2023).

2.3.5 *Protein-Based and Enzyme-Structured Materials*

Fungal-derived protein-based and enzyme-structured materials are a new frontier in biomaterials that outperform the traditional mycelium composites by making use of enzyme-specific proteins, such as hydrophobins, and enzymes to create new functions (Lo et al. 2019). The unique amphipathic proteins that are secreted by filamentous fungi are known as hydrophobins and spontaneously self-assemble into robust monolayers at hydrophobic–hydrophilic interfaces, creating functional amyloids, which are used to alter surface properties to make filamentous fungi useful in drug coating, biomolecule immobilization, and increased bioavailability (Macindoe et al. 2012; Pham et al. 2016; Piscitelli et al. 2017). These self-assembled fungal proteins can be used to construct immunologically evasive, biocompatible nanomaterials that facilitate aerobic hyphal growth, spore dispersal, and immune evasion, as well as host-biomedical and host-industrial biodegradable and tunically hydrophobic coatings (Lo et al. 2019; Blango et al. 2019). In addition to hydrophobins, enzyme-crosslinked biopolymers use fungal enzymes (laccases or peroxidases) to make covalent bonds in polysaccharides or proteins to produce mechanically tough hydrogels and scaffolds with flexible control of the crosslinking density and surface degradation rates, which could have potential in tissue engineering and in green packaging (Maddock et al. 2020; Song et al. 2021; Naranjo-Alcazar et al. 2023). This on-site enzymatic process guarantees the formation of environmentally friendly materials, and no harsh chemicals are used, resembling the extracellular fungal matrices (Maddock et al. 2020). Such innovations have been further enhanced by smart, stimuli-responsive fungal materials that incorporate pH, temperature, light, or redox stimuli; for example, hydrophobin-functionalized polymers can swell or degrade, akin to chlamydospore mycelial regeneration, as hydrophobin-functionalized polymers respond to physiological stimuli. This sensitivity is informed by the adaptability of fungal biology, where evidence of infection is detected through changes in temperature, by designing wound dressings as dynamic systems, or by using in vitro tissue models with 4D-printed structures (Mayer et al. 2015).

2.4 Frontier Production and Engineering Approaches

SSF is the most water- and energy-frugal route to valorize lignocellulosic waste: fungal consortia grown on 40–60% moisture bran or straw generate enzymes, antibiotics, or single-cell protein with 50–90% less fresh water and 20% lower capital cost than submerged systems while simultaneously recycling agro-industrial residues (Bamidele et al. 2025). However, tray- or packed-bed SSF demands 24–72 h of colonization and suffers from patchy O_2/CO_2 gradients and conductive heat accumulation, which restricts layer depth to <5 cm; scaling therefore requires expensive rotary drums or periodic agitation that can shear fungal hyphae and trigger spore release (Yu et al. 2021).

Liquid-state fermentation (LSF) or surface film fermentation overcomes these mass-transfer limits by floating a 0.5–2-mm mycelial mat on nutrient film inside closed, continuously aerated panels (Khatami et al. 2026). The geometry reduces sterilization volume, enables automated harvest of intact 10–30 cm^2 mats every 6–12 h, and integrates inline optical sensors for pH, CO_2, and biomass, giving volumetric productivities three- to fivefold higher than SSF. Full-steam sterilization, stainless plumbing, and mechanical fragility of the mats nevertheless push operating costs above 0.50 US$ kg^{-1} dry biomass, while bacterial contamination at high dilution rates remains a yield risk (Khor and Kang 2025).

Additive manufacturing (AM) of living materials moves from the kilogram to the meter scale by embedding 10^7–10^8 spores mL^{-1} into viscoelastic "bio-inks" composed of κ-carrageenan, nanocellulose, or alginate; robotic deposition builds honeycomb or gyroid lattices whose 200–800-μm channels provide controllable porosity for directed O_2 flux and localized metabolite secretion (Khor and Kang 2025). Layer-by-layer fusion yields architected composites with a specific compressive strength of 0.8–1.2 MPa g^{-1} cm^3 and living viability >70% after 7 days. However, extrusion shear and UV/ozone sterilization still limit construct size to <15 cm^3 and shelf-life to <96 h without refrigerated transport (Arunprasath et al. 2025).

Living-slurry casting bridges the gap between SSF robustness and LSF speed: pre-colonized substrate is ball-milled to 50–200 μm particles, resuspended in 1–2% gellan or xanthan gel, and cast into 0.5–3 mm homogeneous sheets on continuous polyester belts (Liu et al. 2019). The approach reduces colonization time to 4–6 h, achieves an areal density of 250–400 g m^{-2}, and allows downstream rolling, cutting, or lamination without mat fracture. Water activity of 0.92–0.96 and pseudoplastic rheology also permit low-shear pumping and slot-die coating compatible with existing paper machines. Current demonstrations reach a belt speed of 50 m min^{-1}. However, contamination control via peracetic acid fog and sterile gel make-up remains an engineering hurdle before food- or textile-grade deployment beyond pilot scale (Yuan et al. 2025). In Fig. 2.2, a summary of the production methods for fungal materials is presented.

2.4.1 Synthetic Biology Tools

Metabolic flux is the movement of metabolites within cellular systems and must be accurately controlled to achieve the primary industrial strain indicators of yield, titer, and productivity (Leggieri et al. 2024). Three strategies are typically used to regulate metabolic flux: rewiring cellular metabolic pathways, reaction compartmentalization, and the synthesis of synthetic protein scaffolds. Metabolic rewiring can be achieved by deleting genes that divert the target product/or the overexpression of enzymes in pathways to promote the accumulation of the desired compounds. Synthetic protein scaffolds also provide even greater control by employing modular protein-protein interaction domains to regulate enzyme stoichiometry and flux distribution in pathways (Chai et al. 2025). Compartmentalization has further benefited fungi, as their organelles and cell wall enable them to respond to metabolic stimuli through compartmentalization and to direct flux to advantageous local environments by applying engineered localization tags (Bar-Peled and Kory 2022).

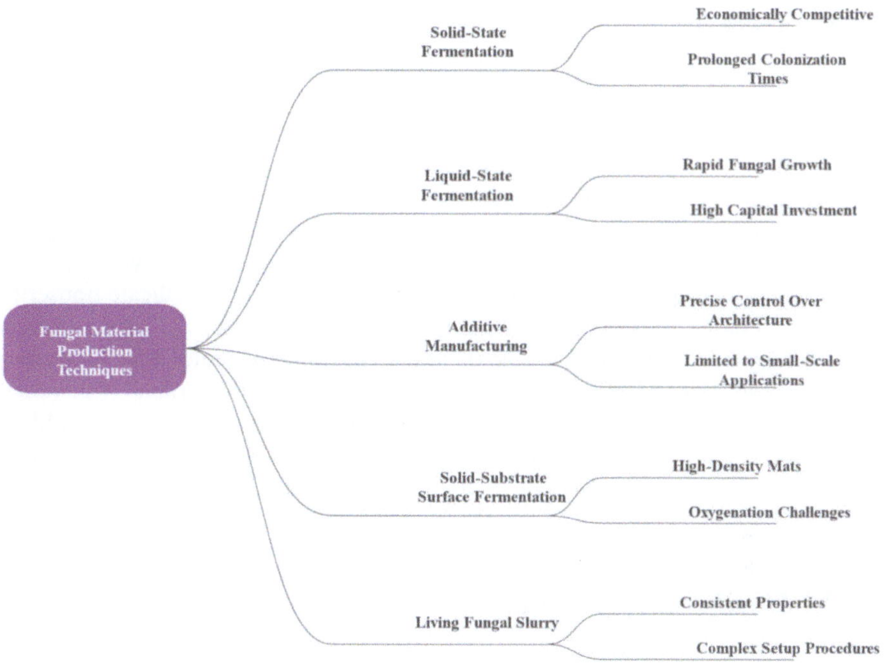

Fig. 2.2 Summary of production methods of fungal materials and their main benefits and shortcomings. The schematic outlines the main technological approaches to the production of fungal-based biomaterials, which are SSF, liquid-state fermentation, AM, solid-substrate surface fermentation, and living fungal slurry systems. Critical advantages and limitations of each strategy are pointed out. SSF is cost-effective, but with a lengthy colonization duration, whereas liquid-state fermentation favors rapid fungal growth but with expensive capital investments. AM is a form of precise architectural control that is, however, available only at limited small-scale applications. Surface fermentation on solid substrates permits high-density mycelial mats to be formed at the cost of oxygenation, but living fungal slurry systems offer much more uniform material properties, but are more cumbersome to establish and maintain

The central metabolite, and a primary target, is acetyl-CoA, a precursor to fatty acids, polyketides, and terpenoids, including β-carotene and amorphadiene (Muhammad et al. 2020). A group of researchers has maximized acetyl-CoA consumption in *S. cerevisiae* by disrupting fatty acid reactivation and implementing a synthetic chimeric citrate lyase pathway, resulting in free fatty acid levels up to 10.4 g/L in fed-batch fermentation with genome-integrated constructs applicable to industry (Yu et al. 2022). The central carbon metabolism of *Yarrowia lipolytica* was rewired to increase cytosolic acetyl-CoA (Markuš et al. 2025). In addition, *Rhodosporidium toruloides* has demonstrated the greatest potential for terpene production from lignocellulosic biomass, synthesizing the terpenes bisabolene and amorphadiene in large quantities by utilizing its abundant cytosolic acetyl-CoA (Yaegashi et al. 2017).

2.4.2 Synthetic Promoters

Transcriptional control redesign is a general term that encompasses methods for transforming fungi, especially yeasts, to improve the heterologous expression of proteins. Several regulatory modules play a role in this process, with synthetic promoter design central to the production of engineered strains (Martins-Santana et al. 2018). A synthetic library of promoters based on the glyceraldehyde-3-phosphate dehydrogenase promoter of the well-established methylotrophic yeast model, *Pichia pastoris*. They used a systematic approach to delete and duplicate transcription factor-binding sites (TFBSs) to control transcription factor (TF) gene expression, thereby revealing information on transcriptional dynamics and regulatory logic. Such an approach facilitated the generation of promoter variants with varied potencies across different regulatory scenarios and is a powerful tool for building yeast strains. However, it requires additional validation in filamentous fungi (Martins-Santana et al. 2018; Li et al. 2026; Mózsik et al. 2021).

2.4.3 Metabolic Engineering

The use of microbial systems in the manufacture of fine chemicals is a significant drawback of metabolic engineering using sustainable biomass. In the past, strain improvement relied on random mutagenesis and screening, and unfavorable characteristics were difficult to detect (Ata et al. 2017). These limitations have been circumvented mainly with the advent of synthetic biology, which enables the precise design of pathways and the redistribution of carbon fluxes, especially in fungal systems, thereby increasing the efficacy of producing chemicals, food additives, pharmaceuticals, and other high-value molecules (Medeleanu et al. 2025). An important example is the enzymatic hydrolysis of plant biomass (mainly cellulose, hemicellulose, and lignin) into fermentable sugars. Cellulolytic and hemicellulolytic enzymes produced by engineered fungi can release glucose and xylose, which are then diverted to ethanol, butanol, fatty acids, and aromatic compounds. Pentoses and hexoses obtained from cellulose and hemicellulose hydrolysis are precursors of bio-sustainable commodities (Zhang et al. 2012). There are two large strategies: non-oleaginous productions where the C5/C6 sugars are glycolyzed and the pyruvate flux is redirected to lactic acid, succinic acid, cis-muconic acid, and ethanol products (Intasian et al. 2021), and oleaginous production where the C5/C6 sugars are glycolyzed and the pyruvate flux is redirected to lactic acid, succinic acid, cis-muconic acid, and ethanol. Whereas metabolic rewiring is required to optimize cytosolic acetyl-CoA (Babaei et al. 2019), others are biologically efficient. In addition to carbohydrates, lignin is an underutilized aromatic feedstock with a complex structure that requires advanced metabolic engineering to depolymerize and convert (Zhang et al. 2024).

2.4.4 Construction of Resistant Strains

Metabolite inhibition is a critical step in developing microbial cell factories that overcome biomass limitations to produce fine chemicals. Inhibitory compounds may interact with and disrupt the metabolism of microorganisms during biomass hydrolysis and assimilation, or prevent the formation of the product (Cho et al. 2022). This task is especially challenging with lignocellulosic biomass, whose hydrolysis produces furan compounds, phenolics, and weak organic acids that serve as fermentation inhibitors; other inhibitors may also be produced during the process as intermediates or final products (Llamas et al. 2023). Therefore, it is necessary to engineer strain inhibitors. Pretreatment of biomass usually acidifies the fermentation medium, adding to the cells' stress (Jain and Kumar 2024). In that regard, examined transcriptional responses of *S. cerevisiae* to acidic stress. They demonstrated that the co-overexpression of the acid-resistance protein Sfp1 with the general stress-response protein Whi2 increased the performance and productivity of the yeast in strongly acidic environments (Martins-Santana et al. 2018). The role of promoter engineering is also important: Rajkumar et al. (2016) created a synthetic promoter with a high transcription rate and a tenfold decrease in the response to low pH. Another bottleneck is product toxicity (Yu et al. 2022), which enhanced tolerance to octanoic acid and other solvents by altering acetyl-CoA carboxylase. Finally, research on the stress response (Ribeiro et al. 2022) and studies of tolerant, nonconventional yeasts (Onyishi et al. 2026) are also applicable to the development of fungal metabolic engineering.

2.4.5 Protein Scaffold and Pathways Compartmentalization

Compartmentalization has become a practical approach for maximizing engineered fungal metabolic pathways. A good example is the redifferentiation of acetyl-CoA-dependent pathways to particular organelles (Roze et al. 2011). Duran et al. (2020) exploited subcellular metabolism in *S. cerevisiae* by targeting amorphadiene synthase to the mitochondrial matrix, thereby accessing the mitochondrial acetyl-CoA pool and bypassing competing cytosolic biosynthetic pathways. This approach led to the production of amorphadiene approximately 80% above cytosolic levels, underscoring the mitochondrial matrix as an ideal place to reroute the pathway. Similarly, Park et al. (2016) improved iso-butanol synthesis by expanding the mitochondrial pyruvate pool by increasing the abundance of hetero-oligomeric mitochondrial pyruvate carrier subunits, resulting in higher titers than those of wild-type strains and earlier publications. Outside mitochondria, DeLoache et al. (2016) reconfigured yeast peroxisomes as so-called synthetic organelles that can import heterologous proteins and regulate the influx of substrates. Another organizational level is plastic protein scaffolds, which co-localize enzymes to allow substrate channeling (Vanderstraeten and Briers 2020). Based on this idea, an improved production of resveratrol by lignin p-coumaric acid in *S. cerevisiae* was demonstrated by demonstrating that pathway flux is regulated by binding-domain numbers (Moglia et al. 2015). Lin et al. (2017) further localized scaffolded ethyl acetate pathway enzymes to lipid droplets, thereby enhancing reaction rates through nanometer-scale enzyme spacing. Together, these papers illustrate that compartmentalization, localization, and enzyme localization are essential factors to consider in fungal metabolic engineering (Yin et al. 2024).

2.4.6 Synthetic Biology and Secondary Metabolism

Over the past 15 years, the fungi have pursued two major lines of research in the production of fine chemicals: the introduction of heterologous metabolic pathways via recombinant DNA technologies and the engineering of native metabolic pathways to increase the yield, purity, and productivity of the metabolites produced. Despite this review being concerned with fine chemicals, products of low value and high volume have recently attracted more research due to their industrial applicability. Fungal lipid production in this regard serves as a viable alternative to plant oils, allowing production without reliance on agricultural activities, seasonal changes, or weather conditions. Cocoa butter, which is traditionally obtained by crushing *Theobroma cacao*, is an example of a high-value lipid whose availability is becoming increasingly scarce (Wadhwa et al. 2024; Meyer 2021). To overcome this obstacle, rational metabolic engineering to synthesize cocoa butter-like (CBL) lipids in *Saccharomyces cerevisiae* by identifying and heterologously expressing cocoa genes involved in triacylglycerol (TAG) biosynthesis (Wei et al. 2018). The target TAG species 1,3-dipalmityl-2-oleoylglycerol, 1-palmitoyl-3-stearoyl-2--oleoylglycerol, and 1, 3-distearoyl-2-oleoylglycerol do not accumulate in yeast in large amounts. Cocoa glycerol-3-phosphate acyltransferase, lysophospholipid acyltransferase, and diacylglycerol acyltransferase were then cloned into *S. cerevisiae* via Gibson assembly pBS01A plasmid to circumvent this weakness (Wei et al. 2018). It is also worth noting that many of the genes amplified from cocoa cDNA differed from their annotated sequences, indicating that yeast is a powerful system for functional validation of plant enzymes. When these cocoa genes were co-expressed with known CBL-associated genes (Wei et al. 2018), TAG accumulation and total fatty acid production were significantly increased (up to 84%).

The other category of high-value fine chemicals is alkaloids, which are structurally complex and have a wide range of pharmaceutical uses, including painkillers and anticancer agents. Since these compounds are typically produced by plant extraction or semi-synthesis, supply instability is a significant problem (Heinrich et al. 2021). Feng et al. (2024) have overcome this shortcoming by assembling a proof-of-concept synthetic opioid biosynthetic pathway comprising over 20 genes in *S. cerevisiae*. These studies not only made yeast a flexible system for the production of alkaloids and the engineering of enzymes but also raised ethical and biosafety issues related to the microbial production of psychoactive drugs. Even though yields of opioids were minimal (Biziukova et al. 2020), the advancement of non-alkaloid psychoactive molecule production via yeast-based methods, including tetrahydrocannabinolic acid, has intensified these debates.

Plant-based polyphenolic compounds are also successfully targeted, and their antimicrobial and antioxidant properties are well known as flavonoids (Ahmad et al. 2025). The breviscapine biosynthetic pathway in yeast can be reconstituted, enabling the production of scutellarin and apigenin-7-O-glucuronide directly from glucose and alleviating supply bottlenecks posed by the extraction of *Erigeron breviscapus*. Finally, the use of glycosides, such as steviosides, has been proposed as an alternative to sugar to address the increasing prevalence of metabolic disorders (Liu et al. 2018).

2.4.7 Bioinformatic Approaches to Regulate TFs

TFs are regulators of gene expression that translate physiological and environmental signals into transcriptional responses; in other words, they act as biosensors, activating or repressing transcription (Tellechea-Luzardo et al. 2023). TFBSs, or DNA sequences bound by TFs, are typically modeled computationally as position weight matrices that characterize binding preferences or binding energy (Inukai et al. 2017). TF motifs can be discovered to uncover the nature of gene regulation and allow bioengineering applications, but TF binding depends on primary DNA sequence, as well as DNA shape, modifications, binding modes, genomic context, and genetic variation; thus, the process of TF motif identification is not easy (Bayer et al. 2025). A further complication has been discovered through high-throughput binding studies, including primary and secondary motifs, which play different regulatory roles in *Saccharomyces cerevisiae* (Hahn and Young 2011). Computational tools to analyze large datasets in ChIP-seq, SELEX, ChIP-chip experiments, including HMS, cERMIT, HOMER, diChIPMunk, MEME-ChIP, rGADEM, POSMO, XXmotif, FMotif, Dimont, and DeepBind, have been developed (Georgiev et al. 2010). Promoter analysis complements motif discovery because yeast promoters contain regions free of nucleosomes, transcription initiation sites, TATA boxes, and upstream activating and repressing sequences needed to initiate transcription (Xu et al. 1991). Despite the challenge posed by the complexity of cis-regulation in promoter prediction (Rombauts et al. 2003; Kim and Wysocka 2023), machine learning and transcriptomics-based TSS mapping have been advanced to increase accuracy (Bhandari et al. 2021).

To meet the growing need for efficient analysis of large-scale genomic data, designed a rapid motif-discovery algorithm, PairMotifChIP, that extracts and merges pairs of fixed-length sequences with small Hamming distances to discover regulatory motifs. Through probabilistic analysis, the method identifies true motifs and combines the remaining motifs into consensus motifs without user input. Comparative studies showed that these tools performed better than the current techniques (Table 2.2). Despite such developments, data from high-throughput sequencing remains difficult to handle. To address this, created CIPHER, a multifunctional tool for ChIP-seq, RNA-seq, and DNase-seq data, offering regulation evaluation, peak labeling, quality management, and multi-channel data processing to ensure efficient and precise data decoding.

Table 2.2 Molecular engineering approaches for optimizing fungal and yeast performance

Approach	Organism	Verified strategy (from literature)	Target application	Reference
Promoter library construction	*Pichia pastoris*	Synthetic AOX1 promoter library with tunable expression strength	Heterologous proteins, fine chemicals	Vogl et al. (2016)
Sugar-responsive transcriptional circuits	*Ashbya gossypii*	Xylose-responsive and hybrid inducible promoters	Carbon-source dependent pathway control	Ito et al. (2015)
Cellulase promoter optimization	*Trichoderma reesei*	cbh1 promoter truncation and regulatory element rearrangement	Enhanced cellulase secretion	Fitz et al. (2018)
Promoter strength improvement	*Saccharomyces cerevisiae*	Intron-mediated transcriptional enhancement	Increased gene expression	Blazeck et al. (2012)
Terminator function optimization	*S. cerevisiae*	Terminator engineering using nucleosome occupancy prediction	Reduced expression noise	Portela et al. (2017)
Improvement of ethanol tolerance	*S. cerevisiae*	Engineering RNA polymerase II subunit (Rpb7)	Ethanol stress tolerance	Persson et al. (2022)

2.5 Novel Fabrication and Processing Strategies

2.5.1 Novel Fabrication Techniques

2.5.1.1 AM and Advanced 3D/4D Printing

AM has transformed the fabrication of fungal biomaterials by enabling the layer-by-layer deposition of mycelium-based bioinks, overcoming the challenges posed by isotropic fungal composites. Multi-material and hybrid printing entails making composite synergetic systems (e.g., PLA or hydrogels) by embedding fungal hyphae in synthetic polymers, leading to bio-solid composite systems with improvements in power (such as increased tensile strength (up to 50 MPa) and water resistance). Elsacker et al. (2021) provide one example of having printed Trametes versicolor mycelium inks co-extruded alongside bacterial cellulose by extrusion and co-extrusion, but creating interfaces without voids by interpenetration between hyphae. They build on this with functional grading and programmable structures produced by 4D printing, in which fungal biomaterials respond to stimuli (e.g., humidity or pH) by changing shape through reversible hydrogen bonding in β-glucans. The aim of programmable architecture is to obtain gradients of tissue-like (Young 0.1–10 MPa modulus) materials, such as gradient-stiffened *Ganoderma lucidum*'s scaffolds (Mohseni et al. 2023; Haneef et al. 2017). These methods can operate in fungal fermenters, where 70% less energy is consumed compared to synthetic AM.

2.5.1.2 Bio-Inspired and Bio-Assisted Fabrication

Bio-inspired techniques rely on the native self-organization of fungi to replicate the natural architectures of mushrooms' gills and sclerotia, thereby creating hierarchical porosity. Microbial fabrication requires the presence of fungal consortia (e.g., *Aspergillus niger* with *Rhizopus oryzae*) to direct in situ biomineralization, which deposits silica or calcium carbonates in mycelial networks to form hierarchical foams with compressive strengths up to 20 MPa. It is enhanced by enzyme and cell-based methods: fungal laccases and peroxidases are catalysts of polymer cross-linking within mild conditions (20–40 °C, pH 4–6) to allow chitosan scaffolds to be assembled at room temperature (Ningsih et al. 2025). Shear-thinning, mycelium pastes are a biomimetic self-assembly based on the fungal spores' aggregation, providing mold-free structures. Researchers found that the directed assembly of *Pleurotus ostreatus* into graphene oxide conductive aerogels yielded 10 S/m conductivity, achieved through π–π stacking templates that peppered each other via hyphal bridges. These are waste reduction methods, which are consistent with the ideas of the circular bioeconomy (Zhang et al. 2021).

2.5.1.3 Nanofabrication and Precision Engineering

Nanoscale control opens the potential of fungal biomaterials for drug delivery and sensors, with target characteristics in the 100-nm range. The fungal chitosan melt processed electrospinning to create the nanofibers (20–200 nm diameter) that are characterized by a high surface-area to volume ratio (>500 m^2/g) (e.g., heavy metal sequestration >95% efficiency) (Joseph et al. 2021). Alternatives, for example, fungal-mediated colloidal assembly, require no cleanrooms as a substitute for nanoimprinting and lithography (Guilger-Casagrande and de Lima 2019). Nanoparticles are dopeable by using metallothioneins as the basis of molecular-level strategies; one example is *Schizophyllum commune* synthesizing silver nanocrystals (5–10 nm) on chitin backbones and providing them to antimicrobial coatings with 99.9% bacterial killing, relying on it (Vijayakumar et al. 2024). These are used in combination with AM to form AM hybrid nano-micro structures, such as lithography-free nanoimprinting using fungal spore stamps with a resolution of 50 nm (Alghuthaymi et al. 2021).

2.5.2 *Advanced Processing Strategies*

2.5.2.1 Green and Sustainable Processing

Sustainable processing aims to minimize environmental impact through solvent-free, low-energy, and circular processes, which are directly aligned with the circular bioeconomy. Solvent-free processes capitalize on hyphal growth in the agro-waste SSF process to eliminate volatile organic compounds and consume as little as 70%

of the energy requirements compared to the SmF processes (Wang et al. 2023). As an example, re-engineered *Trametes versicolor* mycelium composites, which are compression molded at ambient temperatures, produce leather-like biomaterials with tensile strengths greater than 10 MPa, and are entirely biodegradable in soil in 90 days (Haneef et al. 2017; Jones et al. 2020). Life-cycle-based optimization combines both cradle-to-grave versus global warming potential and cumulative energy demand (Jones et al. 2020; Pelletier et al. 2013). Practices such as process redesign, using tools like life-cycle analysis, are used to recover 85% of lignocellulosic byproducts in *Ganoderma lucidum* cultures in closed-loop nutrient cycling, reducing net water consumption by 50%. The proposals ensure that the alternatives of biomaterials are manufactured at cost-effective prices (e.g., less than $5/kg) and zero emission (Akcay et al. 2025).

2.5.2.2 Smart and Adaptive Processing

Smart processing enables dynamic control through stimuli-responsive environments and real-time monitoring, allowing an individual to control fungal morphogenesis with high precision. PH-, temperature-, or light-analogous fungal promoters can be employed in the stimulation of the action of the sticis-responsive system (an example is optogenetic circuits in *Aspergillus nidulans*) (Hedtke et al. 2015). Chitin nanofiber extrusion happens in response to blue light and yields hierarchical structures with Young's moduli up to 5 GPa. Such plasticity is used to express discrete adaptation of a property, for example, humidity-dependent mycelial bundling. The in situ monitoring is a system comprising spectroscopy (e.g., Raman and near-infrared) and biosensors so as to control a process in real time. Hyperspectral data of *Pleurotus ostreatus* bioreactors are analyzed using machine learning, and it is reported that the rates of predictability are 95%, and optimization of aeration leads to extracellular polysaccharide release by the biomass in the most efficient way possible (Karnachoriti et al. 2025). Also, using these conveyance loops, the variability in batch scale-up is reduced by 40%, and the scale-up between the reactors utilized in the laboratory (mL) and in the pilot scale (m^3) is increased.

2.5.2.3 Hybrid and Multistep Processing Frameworks

Hybrid configurations combine physical, chemical, and biological unit operations, unlocking multifunctional biomaterials. The biological pretreatment (e.g., fungal enzymatic delignification) then enters the domain of integration, followed by physical hot-pressing and chemical cross-linking with natural aldehydes such as genipin (Wang et al. 2026). Sequential structures manufacture *Schizophyllum commune* mycelium into foam cores strengthened with bacterial cellulose into compressive strengths of 1.2 MPa plus fire retardancy through the formation of lignin-based char (Voutetaki and Mpalaskas 2024).

2.6 CRISPR-Cas9 Genome Editing

2.6.1 CRISPR-Cas9 Genome Editing in Fungal Biotechnology for Biomaterials: Production, Challenges, and Future Perspectives

CRISPR-Cas systems reform fungal genome editing. CRISPR-Cas systems exploit sgRNA-guided Cas9 recognition of a protospacer adjacent motif (PAM), thereby generating single-strand breaks (DSBs). Double-Strand Breaks (DSB) repair in fungi continues through error-prone non-homologous end joining, which leads to insertions/deletions, or through a homology-directed repair of the gene replacement (Shen et al. 2024). Variants with alternative PAMs and dual activity, for example, Cas12a (Cpf1), may be used as catalytically dead dCas9 to perform transcriptional control via CRISPRi (interference) or CRISPRa (activation). Delivery in filamentous fungi makes use of the plasmid-based systems, the unstable AMA1 autonomously replicating vectors by unstable integration and periodic use of ribonucleoprotein electroporation systems that reduce off-target delivery (Li et al. 2025).

2.6.2 CRISPR-Cas9-Enabled Production of Fungal Biomaterials

CRISPR-Cas9 technology has transformed fungal biomaterials production by enabling accurate genome editing to improve biosynthetic processes and material behavior in filamentous fungi. The direct yields of products of value like chitin-derived materials and pigments based on melanin demonstrate an improvement of BGCs diversion by engineering biosynthetic metabolic pathways to redirect carbon flux in *Aspergillus* and *Trichoderma* species (Ma et al. 2025). Mechanical strength and elasticity can be programmed by the expression of chitin synthases (e.g., CHS genes) and glucan-associated genes (e.g., FKS1) in cell wall architecture to produce a mycelium-based composite that can be tuned to tensile properties of up to 50 MPa, making it a sustainable packaging material and textile. These cultures may be optimized by regulatory network rewiring and by CRISPR-mediated regulation of TFs, such as LaeA or VelB, to increase biomass uniformity and scalability. Moreover, multiplex genome editing can be used to genetically modify multiple loci simultaneously, creating fungal genotypes with an integrated phenotype such as hydrophobicity, biodegradability, and antimicrobial properties, as has been done in *Pleurotus ostreatus* to produce high-quality biomaterials (Xu et al. 2022).

2.6.3 Integration of Synthetic Biology and CRISPR Technologies

Application in synthesizing biomaterials and the production of bioactive compounds. Fungal engineering, especially through the incorporation of synthetic biology and CRISPR technologies, has been transformed. The genetic circuit assembly is done by a set of standardized promoters, terminators, and choices of markers in the form of modular genetic toolkits, which can be assembled into systems such as Modular Cloning to construct the complex transcription units and the CRISPR components (such as Cas9 and dCas9) and sgRNA expression cassettes in plug-and-play genetics of filamentous fungi, including but not limited to *Aspergillus* sp. and *Penicillium* sp. (Mózsik et al. 2021). They can mine genome cryptic BGCs, silenced sets of genes that encode SMs via CRISPR-mediated functional validation, allowing the rapid mapping of genotype–phenotype relationships and the rapid discovery of new fungal metabolites (Xu et al. 2021).

2.7 Cutting-Edge and Emerging Applications

Fungal biotechnology is rapidly transcending its traditional role in bulk enzyme and metabolite manufacturing to become a cornerstone of next-generation biomaterial innovation. Recent Q1–Q3 literature reveals three interlocking frontiers where mycelial systems are being reprogrammed for high-value applications.

(i) *Living Functional Materials*: CRISPR-edited *Aspergillus* and *Neurospora* chassis now express inducible synthetic TFs that deposit inorganic nanoparticles or conductive polymers directly within growing hyphae, yielding self-repairing mycelial circuits for biodegradable sensors and flexible electronics.
(ii) *Nano-Enabled Hybrids*: Fungal mycelia functionalized with halloysite nanotubes act as hierarchical scaffolds that sequester pyrene or heavy metals at >90% efficiency, while mycelium-derived chitin–chitosan nanopapers show higher hydrophobicity and Young's modulus than cellulose equivalents, opening routes for edible food coatings, waterproof packaging, and dialysis membranes.
(iii) *Agricultural and Construction Scale-Ups*: Electro-spun polyethene-oxide nanofibers carrying arbuscular-mycorrhizal spores increase crop yields 15–25% with 40% less fertilizer; simultaneously, semi-automated SSF platforms (WO2025131307A1) compress mycelium into ISO-rated fire-retardant insulation panels that lock 0.6 kg CO_2-eq per panel, positioning fungi as carbon-negative feedstocks for 3D-printed buildings.

2.8 Current Challenges in Emerging Domains

Contemporary problems in the new field of fungal biotechnology include systemic and technological challenges that it must overcome to achieve its full potential for industrial-scale applications. Fungal materials have shown advantages over traditional materials in many conditions (Derme et al. 2024). Fungi have made substantial contributions to biocratic economics by providing alternatives to traditional materials and consumer goods (Subramanian et al. 2025). The development of fungal biotechnology materials that facilitate a more circular economy is an emerging trend, despite many challenges (Wikandari et al. 2022).

Compared with other model organisms such as *E. coli* or *S. cerevisiae*, the traditional DBTL cycle is relatively slow and imprecise in filamentous microorganisms due to their complex physiology, slow growth rates, and difficulties with genetic transformation. Additionally, to overcome such limitations, combinatorial optimization provides a platform for creating genetic variants ranging from hundreds to millions of genetic variants, as provided by Naseri and Koffas (2020). The slow growth rate of filamentous organisms makes it challenging to conduct high-throughput screening of the best organisms within a biological library of tens to millions of variants. Long cultivation times, resource-intensive screening, and technical challenges in assembling multi-gene constructs increase the overall cost and complexity. One of the main obstacles is that their structural strength varies depending on the substrate, strain, incubation time, and fabrication process (Alemu et al. 2022). This means that they cannot support much weight, limiting their use in specific applications and making it difficult to produce reliable, high-quality materials. Hence, these biological, technical, and logistical limitations have to be carefully considered when applying combinatorial optimization in practice.

Another problem that remains is the automation of fungal growth processes. The system has been revealed (WO2025131307A1) (Hinneburg et al. 2025). As of right now, no filamentous fungal chassis provides a universal platform for the reconstitution of heterologous pathways to increase their material production capability (Dixon et al. 2022). Their complicated biology, which includes several nuclei, a strong cell wall that makes transformation difficult, and poorly understood processes of gene expression, control, and horizontal gene transfer, is the cause of this restriction (Dixon et al. 2022). It should be noted, nonetheless, that to create toolboxes and add components to repositories like the BioBrick repository or the Galaxy-SynBioCADa site (Hérisson et al. 2022), chemical engineering labs must embrace standard design principles like the BioBrick guidelines (Dixon et al. 2022).

As fungal nanotechnology remains in its infancy, it faces many challenges that must be overcome for effective utilization. When transferring from a laboratory scale to an industrial scale for manufacturing fungus nanoparticles, the challenges of scalability and reproducibility have been significant (Ganeshan et al. 2021). On a larger scale, it becomes more challenging to handle a constant and reproducible manufacturing process, ensure quality and favorable properties of the nanoparticles, and design and implement an effective and trustworthy manufacturing scale and methodology for these particles (Haneef et al. 2017). The potential of nanoparticles largely depends on properties and characteristics, including size, shape, materials, and surface properties (Loshchinina et al. 2022).

Although it is an evolving trend, the application of functional genomics with the help of transcriptomics, proteomics, and gene editing in research involving pathogenic fungi also poses several challenges (Thilini Chethana et al. 2020). Though such technological advances have aided in the discovery of virulence factors as well as an understanding of host–pathogen interaction, their application in research relates more to complexity (Chakraborty et al. 2021). For instance, the joint application of functional genomics and solid-state NMR spectroscopy in the study of fungal cell wall assembly and remodeling in *Neurospora crassa*, although very informative, underlines methodological difficulty and a high degree of specialization required to get spatial and structural information about the fungal cell wall architecture and stress responses (Peng et al. 2022). Indeed, despite recent improvements in functional genomics research in fungi attained through advances in CRISPR-Cas9–based genome editing (Huang et al. 2023), their practical use requires careful optimization across different species, as illustrated by studies in *Candida glabrata* and *Fusarium graminearum*. While characterization of pathogenic fungal genes is indispensable for such translational applications as breeding stress-resilient, high-productivity crops, translating functional genomic insight into practical outcome has remained a significant challenge, requiring concerted use of functional genomics and multi-omics technologies (Przybyla and Gilbert 2022).

Future studies on fungi and the biocircular economy will likely focus on overcoming these constraints and developing new, creative ways to use fungi to build a more sustainable future. Research on mycelium-based composites is essential for several reasons. First, a more thorough comprehension of the characteristics of mycelium materials and their manipulation is necessary. Second, research can help develop more economical and efficient production methods. Third, to facilitate the adoption of mycelium materials, research is needed to establish norms and rules for their use across various applications.

2.8.1 Integrating Fungi With Flexible Electronics and Soft Robotics

Fungal electronics is a new research area concerned with the creation of electronic and sensing devices using mycelium-bound composites and pure mycelium skins (Adamatzky 2018; Roberts and Adamatzky 2022; Adamatzky et al. 2022). These biological materials can actively modulate electrical impedance, produce electrical potential spikes in response to external stimuli, and be used as embedded sensors, wearable elements, or standalone computing systems (Roberts and Adamatzky

2022; Adamatzky et al. 2022). The majority of advances in mycelium-based technologies have emerged from studies of mycelium composites, a sustainable material whose functional properties have become increasingly unconventional over the past decade (Jones et al. 2020; Khan 2025). As in the example of *Ganoderma resinaceum*, the organism can colonize various lignocellulosic materials and respond to physical changes, such as pressure (Roberts and Adamatzky 2022). Simultaneously, more effort is being put into pure mycelium materials as alternatives to animal and synthetic leather and textiles. These biotextiles are made from fungi, also known as mycoleather or mycofabrics, and manufactured using liquid-state fermentation and SSF technologies.

2.8.2 *Fungal Computer, Electronics, and Logical Circuits*

Basidiomycete fungi can serve as living computers: information is encoded in electrical spikes that ripple across the mycelial network, and geometry-dependent models show that these excitation waves can be steered to perform computation, as scoping experiments already demonstrate long-range transfer of stimuli between distant parts of the colony; the fruit bodies themselves act as biosensors, firing action-potential-like impulses when exposed to heat or salt and synchronizing their activity so that stimulating one mushroom alters the electrical signature of its neighbors, while any environmental shift in the underlying substrate triggers cascades of spiking that the colony processes as computational input (Adamatzky 2018). Mycoelectronics designers rewire mycelium into adaptable circuits in which the resistance and spike sequences change in reaction to varying external control parameters, while a four-member group of fungi has already produced a hybrid biological device generating 50 different electrical impulses or neuronal pulses whose message is yet unknown and whose potential connection to human brain signals is a driving factor in their decryption (Adamatzky et al. 2022). By growing oyster fungi into mycelium-bound composites, researchers have prototyped the first living reservoir computer and experimentally realized thousands of Boolean circuits, 3136 functions, 470 of them unique, revealing that fungal mats can host computationally universal logic despite the naturally high variability of their bio-electronic responses (Roberts and Adamatzky 2022).

2.9 Future Perspectives

(i) Universal chassis and parts: Construct a standardized, modular host–vector platform (an "Aspergillus or Ustilago BioBrick 2.0") with recoded PAM-flexible Cas12a, synthetic centromeres, bar-coded neutral sites, and a genome-wide CRISPRi/a library to compress the DBTL cycle from months to days.

(ii) Self-growing materials: Engineer nutrient-triggered re-activation of dormant hyphae so that packaging, insulation, or even architectural panels can self-heal cracks or expand on-site, extending product life and enabling in situ construction on the Moon or Mars.

(iii) Living electronics: Couple melanin-based conductive filaments with voltage-gated calcium channels to create mycelial "neuromorphic" fabrics that sense, compute, and actuate—opening biodegradable wearables, soft robotics, and plant-interfaced sensors for precision agriculture.

(iv) AI-accelerated design: Deploy generative adversarial networks trained on $>10^5$ mechanical–chemical–genomic feature vectors to predict de novo chimeric BGCs that biosynthesize kevlar-like aramids or spider-silk analogs directly inside fungal hyphae.

(v) Regulatory and social roadmap: Co-develop international standards for "living yet safe" materials (metabolically quiescent but genetically traceable), ethical frameworks for psychoactive-alkaloid-producing strains, and open-source risk-assessment tools to secure public acceptance and rapid market entry.

2.10 Conclusion

Fungi have graduated from culinary curiosity to programmable, living foundries for advanced materials. By integrating multi-omics, CRISPR-Cas genome editing, AI-guided design, and water-frugal SSF, we can now tailor mycelial density, hyphal branching, cell-wall chemistry, and secondary-metabolite output with unprecedented precision. The resulting portfolio of chitin–nanocellulose hybrids, melanin wiring inks, hydrophobin coatings, EPS adhesives, and 4D-printed living lattices already matches or exceeds petro-based plastics, leather, and carbon-fiber composites in specific strength, UV shielding, fire retardancy, and composability, while locking up to 0.6 kg CO_2-eq per panel. Pilot plants have achieved ≥ 0.8 g L^{-1} h^{-1} volumetric productivity on agricultural waste with 70% less water and zero sterilants, and continuous mycelial-sheet casters are running at 50 m min^{-1} on existing paper-machine infrastructure. In vivo CRISPR multiplexing and neutral "landing-pad" chromosomal sites have pushed editing efficiency in industrial Basidiomycetes above 60% and enabled stable 30-kb BGC expression without fitness trade-offs. Collectively, these advances demonstrate that fungal biomaterials are no longer a boutique sustainability narrative; they are a scalable manufacturing reality.

Acknowledgments All authors gratefully acknowledge Dr. Jaya Bandyopadhyay and Ms. Sohini Kulavi of the Department of Biotechnology, Maulana Abul Kalam Azad University of Technology, West Bengal (MAKAUT, WB), Mrs. Moumita Ghosh, Brownie, and Mrs. Mousumi Kumar for their generous suggestions during manuscript preparation.

Ethical Approval This chapter does not contain any studies with human participants or animals performed by any authors. No conflicts of interest, informed consent, or human or animal rights issues apply to this study.

References

Abdal Dayem A, Lee SB, Cho S-G (2018) The impact of metallic nanoparticles on stem cell proliferation and differentiation. Nanomaterials 8:761. https://doi.org/10.3390/nano8100761

Adamatzky A (2018) Towards fungal computer. Interface Focus 8:20180029. https://doi.org/10.1098/rsfs.2018.0029

Adamatzky A, Ayres P, Beasley AE, Chiolerio A, Dehshibi MM, Gandia A et al (2022) Fungal electronics. Biosystems 212:104588. https://doi.org/10.1016/j.biosystems.2021.104588

Ahmad Z, Rauf A, Orhan IE, Mubarak MS, Akram Z, Islam MR et al (2025) Antioxidant potential of polyphenolic compounds, sources, extraction, purification and characterization techniques: a focused review. Food Sci Nutr 13:e71259. https://doi.org/10.1002/fsn3.71259

Ahrendt SR, Quandt CA, Ciobanu D, Clum A, Salamov A, Andreopoulos B et al (2018) Leveraging single-cell genomics to expand the fungal tree of life. Nat Microbiol 3:1417–1428. https://doi.org/10.1038/s41564-018-0261-0

Akcay C, Arslan R, Ceylan F (2025) Valorization of various lignocellulosic wastes to Ganoderma lucidum (Curtis) P. Karst (Reishi mushroom) cultivation and their FT-IR assessments. PLoS One 20:e0328732. https://doi.org/10.1371/journal.pone.0328732

Alemu D, Tafesse M, Mondal AK (2022) Mycelium-based composite: the future sustainable biomaterial. Int J Biomater 2022:1–12. https://doi.org/10.1155/2022/8401528

Alghuthaymi MA, Rajkuberan C, Rajiv P, Kalia A, Bhardwaj K, Bhardwaj P et al (2021) Nanohybrid antifungals for control of plant diseases: current status and future perspectives. J Fungi 7:48. https://doi.org/10.3390/jof7010048

Almagro Armenteros JJ, Tsirigos KD, Sønderby CK, Petersen TN, Winther O, Brunak S et al (2019) SignalP 5.0 improves signal peptide predictions using deep neural networks. Nat Biotechnol 37:420–423. https://doi.org/10.1038/s41587-019-0036-z

Arunprasath K, Senthamaraikannan P, Suyambulingam I, Akash S, Kathic S, Chanth MV et al (2025) From degradation to durability: strategies for prolonging the shelf life of natural fiber composites – a comprehensive review. J Nat Fibers 22:2554897. https://doi.org/10.1080/15440478.2025.2554897

Ashrafizadeh M, Hushmandi K, Mirzaei S, Bokaie S, Bigham A, Makvandi P et al (2023) Chitosan-based nanoscale systems for doxorubicin delivery: exploring biomedical application in cancer therapy. Bioeng Transl Med 8:e10325. https://doi.org/10.1002/btm2.10325

Ata Ö, Prielhofer R, Gasser B, Mattanovich D, Çalık P (2017) Transcriptional engineering of the glyceraldehyde-3-phosphate dehydrogenase promoter for improved heterologous protein production in *Pichia pastoris*. Biotechnol Bioeng 114:2319–2327. https://doi.org/10.1002/bit.26363

Babaei M, Rueksomtawin Kildegaard K, Niaei A, Hosseini M, Ebrahimi S, Sudarsan S et al (2019) Engineering oleaginous yeast as the host for fermentative succinic acid production from glucose. Front Bioeng Biotechnol 7:361. https://doi.org/10.3389/fbioe.2019.00361

Bachhar S, Kumar S, Dutta B, Das S (2025) Emerging horizons of AI in pharmaceutical research. Adv Pharmacol 103:325–348. https://doi.org/10.1016/bs.apha.2025.01.016

Bagger FO, Borgwardt L, Jespersen AS, Hansen AR, Bertelsen B, Kodama M et al (2024) Whole genome sequencing in clinical practice. BMC Med Genet 17:39. https://doi.org/10.1186/s12920-024-01795-w

Baig I, Perez M, Braña AF, Gomathinayagam R, Damodaran C, Salas JA et al (2008) Mithramycin analogues generated by combinatorial biosynthesis show improved bioactivity. J Nat Prod 71:199–207. https://doi.org/10.1021/np0705763

Ball SR, Kwan AH, Sunde M (2019) Hydrophobin rodlets on the fungal cell wall. Curr Top Microbiol Immunol 425:29–51. https://doi.org/10.1007/82_2019_186

Bamidele MO, Bamikale MB, Cárdenas-Hernández E, Bamidele MA, Castillo-Olvera G, Sandoval-Cortes J et al (2025) Bioengineering in Solid-State Fermentation for next sustainable food bioprocessing. Next Sustain 6:100105. https://doi.org/10.1016/j.nxsust.2025.100105

Bar-Peled L, Kory N (2022) Principles and functions of metabolic compartmentalization. Nat Metab 4:1232–1244. https://doi.org/10.1038/s42255-022-00645-2

Basnet BB, Zhou Z-Y, Basnet R, Wei B, Wang H (2026) Advances in natural product discovery: strategies, technologies, and insights. Nat Prod Bioprospect 16:3. https://doi.org/10.1007/s13659-025-00556-1

Bayer Y, O'Hagan MP, Miodownik I, Afek A (2025) Quantifying transcription factor specificity with advanced DNA universal microarrays featuring long and modified binding sites. Nucleic Acids Res 53:gkaf1382. https://doi.org/10.1093/nar/gkaf1382

Bellemain E, Carlsen T, Brochmann C, Coissac E, Taberlet P, Kauserud H (2010) ITS as an environmental DNA barcode for fungi: an in silico approach reveals potential PCR biases. BMC Microbiol 10:189. https://doi.org/10.1186/1471-2180-10-189

Bhandari N, Khare S, Walambe R, Kotecha K (2021) Comparison of machine learning and deep learning techniques in promoter prediction across diverse species. PeerJ Comput Sci 7:e365. https://doi.org/10.7717/peerj-cs.365

Biziukova NY, Tarasova OA, Rudik AV, Filimonov DA, Poroikov VV (2020) Automatic recognition of chemical entity mentions in texts of scientific publications. Autom Doc Math Ling 54:306–315. https://doi.org/10.3103/S0005105520060023

Blango MG, Kniemeyer O, Brakhage AA (2019) Conidial surface proteins at the interface of fungal infections. PLoS Pathog 15:e1007939. https://doi.org/10.1371/journal.ppat.1007939

Blazeck J, Garg R, Reed B, Alper HS (2012) Controlling promoter strength and regulation in *Saccharomyces cerevisiae* using synthetic hybrid promoters. Biotechnol Bioeng 109:2884–2895. https://doi.org/10.1002/bit.24552

Carlile MJ (1995) The success of the hypha and mycelium. In: The growing fungus. Springer Netherlands, Dordrecht, pp 3–19. https://doi.org/10.1007/978-0-585-27576-5_1

Chai T, Tao Y, Zhao C, Chen X (2025) Hierarchical metabolic engineering for rewiring cellular metabolism. FEMS Microbiol Rev 49:fuaf047. https://doi.org/10.1093/femsre/fuaf047

Chakraborty A, Fernando LD, Fang W, Dickwella Widanage MC, Wei P, Jin C et al (2021) A molecular vision of fungal cell wall organization by functional genomics and solid-state NMR. Nat Commun 12:6346. https://doi.org/10.1038/s41467-021-26749-z

Chakrapani G, Vijay Kumar V, Zare M (2025) Next-generation biomaterials: design strategies, clinical translation, and the rise of intelligent therapeutic platforms. ACS Omega 10:46272–46293. https://doi.org/10.1021/acsomega.5c06034

Cho JS, Kim GB, Eun H, Moon CW, Lee SY (2022) Designing microbial cell factories for the production of chemicals. JACS Au 2:1781–1799. https://doi.org/10.1021/jacsau.2c00344

Chooi Y-H, Cacho R, Tang Y (2010) Identification of the viridicatumtoxin and griseofulvin gene clusters from Penicillium aethiopicum. Chem Biol 17:483–494. https://doi.org/10.1016/j.chembiol.2010.03.015

Das S, Sen M, Sinha N, Kumar S, Dasgupta N, Ghosh D (2024) Molecular characterization of the ligninolytic Sundarban mangrove isolate Stenotrophomonas maltophilia GD1 and its promising applications in the dye and fiber industry. J Appl Biol Biotechnol 12:112–122. https://doi.org/10.7324/JABB.2024.193181

Das S, Kumar S, Chakraborty A, Ghosh D (2025a) In silico mutagenesis of lignocellulolytic enzymes, depending on the propensity, causes enhancement in the spontaneity of the enzyme kinetics. Next Res 2:100293. https://doi.org/10.1016/j.nexres.2025.100293

Das S, Dasgupta N, Kumar S, Das P, Toor AK, Ghosh D (2025b) General perspectives on lignocellulosic biomass-degrading enzymes from oleaginous microbes. In: Oleaginous microbes for waste biomass valorization. Apple Academic Press, New York, pp 229–259. https://doi.org/10.1201/9781003538790-10

Davis EM, Sun Y, Liu Y, Kolekar P, Shao Y, Szlachta K et al (2021) SequencErr: measuring and suppressing sequencer errors in next-generation sequencing data. Genome Biol 22:37. https://doi.org/10.1186/s13059-020-02254-2

De Rose S, Sillo F, Ghirardo A, Perotto S, Schnitzler J-P, Balestrini R (2024) Integration of fungal transcriptomics and metabolomics provides insights into the early interaction between the ORM fungus Tulasnella sp. and the orchid Serapias vomeracea seeds. IMA Fungus 15:31. https://doi.org/10.1186/s43008-024-00165-6

Delavy M, Cerutti L, Croxatto A, Prod'hom G, Sanglard D, Greub G et al (2020) Machine learning approach for Candida albicans fluconazole resistance detection using matrix-assisted laser desorption/ionization time-of-flight mass spectrometry. Front Microbiol 10:3000. https://doi.org/10.3389/fmicb.2019.03000

DeLoache WC, Russ ZN, Dueber JE (2016) Towards repurposing the yeast peroxisome for compartmentalizing heterologous metabolic pathways. Nat Commun 7:11152. https://doi.org/10.1038/ncomms11152

Derme T, Schwarze FWMR, Dillenburger B (2024) Understanding the role of controlled environments for producing mycelium-bound composites: advancing circular practices for integrating biotechnology into the construction industry. Global Chall 8:2300197. https://doi.org/10.1002/gch2.202300197

Dikec J, Olivier A, Bobée C, D'Angelo Y, Catellier R, David P et al (2020) Hyphal network whole field imaging allows for accurate estimation of anastomosis rates and branching dynamics of the filamentous fungus Podospora anserina. Sci Rep 10:3131. https://doi.org/10.1038/s41598-020-57808-y

Dixon TA, Freemont PS, Johnson RA, Pretorius IS (2022) A global forum on synthetic biology: the need for international engagement. Nat Commun 13:3516. https://doi.org/10.1038/s41467-022-31265-9

Dohn R, Xie B, Back R, Selewa A, Eckart H, Rao RP et al (2021) mDrop-seq: massively parallel single-cell RNA-seq of Saccharomyces cerevisiae and Candida albicans. Vaccines (Basel) 10:30. https://doi.org/10.3390/vaccines10010030

Duran L, López JM, Avalos JL (2020) ¡Viva la mitochondria!: harnessing yeast mitochondria for chemical production. FEMS Yeast Res 20:foaa037. https://doi.org/10.1093/femsyr/foaa037

Elsacker E, Vandelook S, Damsin B, Van Wylick A, Peeters E, De Laet L (2021) Mechanical characteristics of bacterial cellulose-reinforced mycelium composite materials. Fungal Biol Biotechnol 8:18. https://doi.org/10.1186/s40694-021-00125-4

Feng Y, Su C, Mao G, Sun B, Cai Y, Dai J et al (2024) When synthetic biology meets medicine. Life Med 3:lnae010. https://doi.org/10.1093/lifemedi/lnae010

Fisher AK, Freedman BG, Bevan DR, Senger RS (2014) A review of metabolic and enzymatic engineering strategies for designing and optimizing performance of microbial cell factories. Comput Struct Biotechnol J 11:91–99. https://doi.org/10.1016/j.csbj.2014.08.010

Fitz E, Wanka F, Seiboth B (2018) The promoter toolbox for recombinant gene expression in Trichoderma reesei. Front Bioeng Biotechnol 6:135. https://doi.org/10.3389/fbioe.2018.00135

Floss HG (2006) Combinatorial biosynthesis—potential and problems. J Biotechnol 124:242–257. https://doi.org/10.1016/j.jbiotec.2005.12.001

Fricker MD, Heaton LLM, Jones NS, Boddy L (2017) The mycelium as a network. Microbiol Spectrum 5:1–62. https://doi.org/10.1128/microbiolspec.FUNK-0033-2017

Frisvad JC, Larsen TO, de Vries R, Meijer M, Houbraken J, Cabañes FJ et al (2007) Secondary metabolite profiling, growth profiles and other tools for species recognition and important Aspergillus mycotoxins. Stud Mycol 59:31–37. https://doi.org/10.3114/sim.2007.59.04

Ganeshan S, Kim SH, Vujanovic V (2021) Scaling-up production of plant endophytes in bioreactors: concepts, challenges and perspectives. Bioresour Bioprocess 8:63. https://doi.org/10.1186/s40643-021-00417-y

Ganeva Z, Goranov B, Brazkova M, Blazheva D, Baldzhieva R, Stefanova P et al (2025) Agro-industrial residues as cost-effective and sustainable substrates for the cultivation of Epicoccum

nigrum, with insights into growth kinetic characteristics and biological activities. Appl Sci 15:10571. https://doi.org/10.3390/app151910571
Georgiev S, Boyle AP, Jayasurya K, Ding X, Mukherjee S, Ohler U (2010) Evidence-ranked motif identification. Genome Biol 11:R19. https://doi.org/10.1186/gb-2010-11-2-r19
Goodwin S, McPherson JD, McCombie WR (2016) Coming of age: ten years of next-generation sequencing technologies. Nat Rev Genet 17:333–351. https://doi.org/10.1038/nrg.2016.49
Gross A, Schoendube J, Zimmermann S, Steeb M, Zengerle R, Koltay P (2015) Technologies for single-cell isolation. Int J Mol Sci 16:16897–16919. https://doi.org/10.3390/ijms160816897
Guilger-Casagrande M, de Lima R (2019) Synthesis of silver nanoparticles mediated by fungi: a review. Front Bioeng Biotechnol 7:287. https://doi.org/10.3389/fbioe.2019.00287
Guin K, Chen Y, Mishra R, Muzaki SRB, Thimmappa BC, O'Brien CE et al (2020) Spatial inter-centromeric interactions facilitated the emergence of evolutionary new centromeres. elife 9:e58556. https://doi.org/10.7554/eLife.58556
Hahn S, Young ET (2011) Transcriptional regulation in *Saccharomyces cerevisiae*: transcription factor regulation and function, mechanisms of initiation, and roles of activators and coactivators. Genetics 189:705–736. https://doi.org/10.1534/genetics.111.127019
Haneef M, Ceseracciu L, Canale C, Bayer IS, Heredia-Guerrero JA, Athanassiou A (2017) Advanced materials from fungal mycelium: fabrication and tuning of physical properties. Sci Rep 7:41292. https://doi.org/10.1038/srep41292
Hansini AMP, Galpaya GDCP, Gunasena MDKM, Abeysundara PM, Kirthika V, Bhagya L et al (2025) From nature to innovation: advances in nanocellulose extraction and its multifunctional applications. Molecules 30:2670. https://doi.org/10.3390/molecules30132670
Harris SD (2008) Branching of fungal hyphae: regulation, mechanisms and comparison with other branching systems. Mycologia 100:823–832. https://doi.org/10.3852/08-177
Harris SD, Read ND, Roberson RW, Shaw B, Seiler S, Plamann M et al (2005) Polarisome meets spitzenkörper: microscopy, genetics, and genomics converge. Eukaryot Cell 4:225–229. https://doi.org/10.1128/EC.4.2.225-229.2005
Hedtke M, Rauscher S, Röhrig J, Rodríguez-Romero J, Yu Z, Fischer R (2015) Light-dependent gene activation in *Aspergillus nidulans* is strictly dependent on phytochrome and involves the interplay of phytochrome and white collar-regulated histone H3 acetylation. Mol Microbiol 97:733–745. https://doi.org/10.1111/mmi.13062
Heinrich M, Mah J, Amirkia V (2021) Alkaloids used as medicines: structural phytochemistry meets biodiversity—an update and forward look. Molecules 26:1836. https://doi.org/10.3390/molecules26071836
Hérisson J, Duigou T, du Lac M, Bazi-Kabbaj K, Sabeti Azad M, Buldum G et al (2022) The automated Galaxy-SynBioCAD pipeline for synthetic biology design and engineering. Nat Commun 13:5082. https://doi.org/10.1038/s41467-022-32661-x
Hinneburg H, Gu S, Naseri G (2025) Fungal innovations—advancing sustainable materials, genetics, and applications for industry. J Fungi 11:721. https://doi.org/10.3390/jof11100721
Hu M, Lv X, Wang Y, Zhang Y, Dai H (2024) Guideline for the extraction of nanocellulose from lignocellulosic feedstocks. Food Biomacromol 1:9–17. https://doi.org/10.1002/fob2.12011
Huang X, Liang Y, Yang Y, Lu X (2017) Single-step production of the simvastatin precursor monacolin J by engineering of an industrial strain of Aspergillus terreus. Metab Eng 42:109–114. https://doi.org/10.1016/j.ymben.2017.06.005
Huang C, Peng J, Zhang W, Chethana T, Wang X, Wang H et al (2023) LtGAPR1 is a novel secreted effector from Lasiodiplodia theobromae that interacts with NbPsQ2 to negatively regulate infection. J Fungi 9:188. https://doi.org/10.3390/jof9020188
Huet G, Wang Y, Gardrat C, Brulé D, Vax A, Le Coz C et al (2023) Deep chemical and physico-chemical characterization of antifungal industrial chitosans—biocontrol applications. Molecules 28:966. https://doi.org/10.3390/molecules28030966
Intasian P, Prakinee K, Phintha A, Trisrivirat D, Weeranoppanant N, Wongnate T et al (2021) Enzymes, *in vivo* biocatalysis, and metabolic engineering for enabling a circular economy and sustainability. Chem Rev 121:10367–10451. https://doi.org/10.1021/acs.chemrev.1c00121

Inukai S, Kock KH, Bulyk ML (2017) Transcription factor–DNA binding: beyond binding site motifs. Curr Opin Genet Dev 43:110–119. https://doi.org/10.1016/j.gde.2017.02.007

Islam MR, Tudryn G, Bucinell R, Schadler L, Picu RC (2017) Morphology and mechanics of fungal mycelium. Sci Rep 7:13070. https://doi.org/10.1038/s41598-017-13295-2

Ito Y, Yamanishi M, Ikeuchi A, Imamura C, Matsuyama T (2015) Combinatorial screening for transgenic yeasts with high cellulase activities in combination with a tunable expression system. PLoS One 10:e0144870. https://doi.org/10.1371/journal.pone.0144870

Jain S, Kumar S (2024) Advances and challenges in pretreatment technologies for bioethanol production: a comprehensive review. Sustain Chem Clim Action 5:100053. https://doi.org/10.1016/j.scca.2024.100053

Jiménez I, Alfaro M, Garde E, Pérez G, Fernandez-Morales A, Lipzen A et al (2025) Comparative transcriptomics of CAZy enzymes in white- and brown-rot agaricomycetes: evolutionary insights into lignocellulose degradation and the relevance of GH16 glycoside hydrolase functional divergence. New Biotechnol 90:236–256. https://doi.org/10.1016/j.nbt.2025.10.004

Jones M, Bhat T, Kandare E, Thomas A, Joseph P, Dekiwadia C et al (2018) Thermal degradation and fire properties of fungal mycelium and mycelium – biomass composite materials. Sci Rep 8:17583. https://doi.org/10.1038/s41598-018-36032-9

Jones M, Mautner A, Luenco S, Bismarck A, John S (2020) Engineered mycelium composite construction materials from fungal biorefineries: a critical review. Mater Des 187:108397. https://doi.org/10.1016/j.matdes.2019.108397

Joseph T, Jacob M, Nair VR, Varkey JT (2021) Removal of metal ions using chitosan based electro spun nanofibers: a review. Nanosyst Phys Chem Math 12:728–748. https://doi.org/10.17586/2220-8054-2021-12-6-728-748

Jumper J, Evans R, Pritzel A, Green T, Figurnov M, Ronneberger O et al (2021) Highly accurate protein structure prediction with AlphaFold. Nature 596:583–589. https://doi.org/10.1038/s41586-021-03819-2

Karnachoriti M, Chatzipetrou M, Touloupakis E, Kontos AG, Zergioti I (2025) Raman spectroscopy as a tool for real-time nutrient monitoring in bioreactor cultivation of microalgae. J Raman Spectrosc 56:817–826. https://doi.org/10.1002/jrs.6841

Khan R (2025) Mycelium-based bioproducts: a novel material for a sustainable economy a com prehensive review. Microbe 8:100439. https://doi.org/10.1016/j.microb.2025.100439

Khatami K, Qazanfarzadeh Z, Jiménez-Quero A (2026) Fungal fermentation: the blueprint for transforming industrial side streams and residues. Bioresour Technol 440:133426. https://doi.org/10.1016/j.biortech.2025.133426

Khor WH, Kang HS (2025) Installations of algal bioreactors: design and operational issues in commercial plants. In: Algal bioreactors. Elsevier, pp 263–279. https://doi.org/10.1016/B978-0-443-14058-7.00009-9

Khyaju S, Hyde KD, Khongphinitbunjong K, Duangphet S, Aiduang W, Luangharn T (2025) A comprehensive characterisation of mycelium-based biomaterials developed from Panus ciliatus and P. subfasciatus (Panaceae, Polyporales). J Fungi 11:826. https://doi.org/10.3390/jof11120826

Kim S, Wysocka J (2023) Deciphering the multi-scale, quantitative cis-regulatory code. Mol Cell 83:373–392. https://doi.org/10.1016/j.molcel.2022.12.032

Kumar K, Kumar S, Goswami A, Moholkar VS (2024) Advancing sustainable biofuel production: a computational insight into microbial systems for isopropanol synthesis and beyond. Process Saf Environ Prot 188:1118–1132. https://doi.org/10.1016/j.psep.2024.06.024

Kumar S, Ghosh A, Dutta B, Sharda S, Mukherjee S, Dutta C et al (2025) A computational approach to targeted amino acid mutagenesis for enhancing enzyme efficiency through improved enzyme-substrate dynamics and thermostability. Discov Chem 2:265. https://doi.org/10.1007/s44371-025-00349-y

Kumar S, Banerjee S, Ghosh A, Pramanik C, Dey I, Choudhury N et al (2026a) Biological synergies for heavy metal bioremediation across microbial, algal, and phytoremediation perspectives. Discov Green Chem 1:3. https://doi.org/10.1007/s44509-026-00003-5

Kumar S, Banerjee S, Pramanik C, Ghosh A, Kulavi S, Dey I et al (2026b) Heavy metal-driven biochemical enhancement of dye degradation by bacterial consortia: a multi-scale statistical and spectroscopic investigation. J Environ Sci Health A 61:98–124. https://doi.org/10.1080/10934529.2026.2628388

Lee MJ, Geller AM, Bamford NC, Liu H, Gravelat FN, Snarr BD et al (2016) Deacetylation of fungal exopolysaccharide mediates adhesion and biofilm formation. MBio 7:e00252-16. https://doi.org/10.1128/mBio.00252-16

Leggieri PA, Blair EM, Lankiewicz TS, O'Malley MA (2024) Continuous culture of anaerobic fungi enables growth and metabolic flux tuning without use of genetic tools. Bioresour Technol 391:129854. https://doi.org/10.1016/j.biortech.2023.129854

Lehmann A, Zheng W, Soutschek K, Roy J, Yurkov AM, Rillig MC (2019) Tradeoffs in hyphal traits determine mycelium architecture in saprobic fungi. Sci Rep 9:14152. https://doi.org/10.1038/s41598-019-50565-7

Li X-H, Lu H-Z, Yao J-B, Zhang C, Shi T-Q, Huang H (2025) Recent advances in the application of CRISPR/Cas-based gene editing technology in filamentous fungi. Biotechnol Adv 81:108561. https://doi.org/10.1016/j.biotechadv.2025.108561

Li X, Ye C, Liu T, Li S, Zhang M, Zhao Y et al (2026) Engineering genetic elements for microbial protein expression systems: advances, challenges, applications, and prospects. Synth Syst Biotechnol 11:370–384. https://doi.org/10.1016/j.synbio.2025.10.008

Liang B, Huang X, Teng Y, Liang Y, Yang Y, Zheng L et al (2018) Enhanced single-step bioproduction of the simvastatin precursor monacolin J in an industrial strain of *Aspergillus terreus* by employing the evolved lovastatin hydrolase. Biotechnol J 13:1800094. https://doi.org/10.1002/biot.201800094

Lin J-L, Zhu J, Wheeldon I (2017) Synthetic protein scaffolds for biosynthetic pathway colocalization on lipid droplet membranes. ACS Synth Biol 6:1534–1544. https://doi.org/10.1021/acssynbio.7b00041

Lino V, Manini P (2022) Dihydroxynaphthalene-based allomelanins: a source of inspiration for innovative technological materials. ACS Omega 7:15308–15314. https://doi.org/10.1021/acsomega.2c00641

Lipson D, Raz T, Kieu A, Jones DR, Giladi E, Thayer E et al (2009) Quantification of the yeast transcriptome by single-molecule sequencing. Nat Biotechnol 27:652–658. https://doi.org/10.1038/nbt.1551

Liu X, Cheng J, Zhang G, Ding W, Duan L, Yang J et al (2018) Engineering yeast for the production of breviscapine by genomic analysis and synthetic biology approaches. Nat Commun 9:448. https://doi.org/10.1038/s41467-018-02883-z

Liu R, Long L, Sheng Y, Xu J, Qiu H, Li X et al (2019) Preparation of a kind of novel sustainable mycelium/cotton stalk composites and effects of pressing temperature on the properties. Ind Crop Prod 141:111732. https://doi.org/10.1016/j.indcrop.2019.111732

Liu D, Rubin GM, Dhakal D, Chen M, Ding Y (2021) Biocatalytic synthesis of peptidic natural products and related analogues. iScience 24:102512. https://doi.org/10.1016/j.isci.2021.102512

Llamas M, Greses S, Magdalena JA, González-Fernández C, Tomás-Pejó E (2023) Microbial co-cultures for biochemicals production from lignocellulosic biomass: a review. Bioresour Technol 386:129499. https://doi.org/10.1016/j.biortech.2023.129499

Lo V, I-Chun Lai J, Sunde M (2019) Fungal hydrophobins and their self-assembly into functional nanomaterials. Adv Exp Med Biol 1174:161–185. https://doi.org/10.1007/978-981-13-9791-2_5

Loshchinina EA, Vetchinkina EP, Kupryashina MA (2022) Diversity of biogenic nanoparticles obtained by the fungi-mediated synthesis: a review. Biomimetics 8:1. https://doi.org/10.3390/biomimetics8010001

Luo D, Yang J, Peek N (2025) 3D-printed mycelium biocomposites: method for 3D printing and growing fungi-based composites. 3D Print Addit Manuf 12:98–111. https://doi.org/10.1089/3dp.2023.0342

Ma B, Li Y, Wang T, Li D, Jia S (2025) Advances in CRISPR/Cas9-based gene editing in filamentous fungi. J Fungi 11:350. https://doi.org/10.3390/jof11050350

Macindoe I, Kwan AH, Ren Q, Morris VK, Yang W, Mackay JP et al (2012) Self-assembly of functional, amphipathic amyloid monolayers by the fungal hydrophobin EAS. Proc Natl Acad Sci USA 109:E804–E811. https://doi.org/10.1073/pnas.1114052109

Maddock RMA, Pollard GJ, Moreau NG, Perry JJ, Race PR (2020) Enzyme-catalysed polymer cross-linking: biocatalytic tools for chemical biology, materials science and beyond. Biopolymers 111:e23390. https://doi.org/10.1002/bip.23390

Mahapatra S, Banerjee D (2013) Fungal exopolysaccharide: production, composition and applications. Microbiol Insights 6:1–16. https://doi.org/10.4137/MBI.S10957

Markuš T, Soldat M, Magdevska V, Horvat J, Kavšček M, Kosec G et al (2025) Engineering central metabolism in Yarrowia lipolytica increases lipid accumulation. Biochem Eng J 215:109589. https://doi.org/10.1016/j.bej.2024.109589

Martins-Santana L, Nora LC, Sanches-Medeiros A, Lovate GL, Cassiano MHA, Silva-Rocha R (2018) Systems and synthetic biology approaches to engineer fungi for fine chemical production. Front Bioeng Biotechnol 6:117. https://doi.org/10.3389/fbioe.2018.00117

Maver T, Hribernik S, Mohan T, Smrke DM, Maver U, Stana-Kleinschek K (2015) Functional wound dressing materials with highly tunable drug release properties. RSC Adv 5:77873–77884. https://doi.org/10.1039/C5RA11972C

McCombie WR, McPherson JD, Mardis ER (2019) Next-generation sequencing technologies. Cold Spring Harb Perspect Med 9:a036798. https://doi.org/10.1101/cshperspect.a036798

Medeleanu ML, Călinoiu L-F, Martău G-A, Vodnar D-C (2025) Metabolic engineering of microorganisms for tailor-made biopolymer production: a review. Int J Biol Macromol 330:147922. https://doi.org/10.1016/j.ijbiomac.2025.147922

Meyer V (2021) Metabolic engineering of filamentous fungi. In: Metabolic engineering. Wiley, pp 765–801. https://doi.org/10.1002/9783527823468.ch20

Meyer V, Cairns T, Barthel L, King R, Kunz P, Schmideder S et al (2021) Understanding and controlling filamentous growth of fungal cell factories: novel tools and opportunities for targeted morphology engineering. Fungal Biol Biotechnol 8:8. https://doi.org/10.1186/s40694-021-00115-6

Moglia A, Goitre L, Gianoglio S, Baldini E, Trapani E, Genre A et al (2015) Evaluation of the bioactive properties of avenanthramide analogs produced in recombinant yeast. Biofactors 41:15–27. https://doi.org/10.1002/biof.1197

Mohseni A, Vieira FR, Pecchia JA, Gürsoy B (2023) Three-dimensional printing of living mycelium-based composites: material compositions, workflows, and ways to mitigate contamination. Biomimetics 8:257. https://doi.org/10.3390/biomimetics8020257

Money NP (2021) Hyphal and mycelial consciousness: the concept of the fungal mind. Fungal Biol 125:257–259. https://doi.org/10.1016/j.funbio.2021.02.001

Mózsik L, Pohl C, Meyer V, Bovenberg RAL, Nygård Y, Driessen AJM (2021) Modular synthetic biology toolkit for filamentous fungi. ACS Synth Biol 10:2850–2861. https://doi.org/10.1021/acssynbio.1c00260

Muhammad A, Feng X, Rasool A, Sun W, Li C (2020) Production of plant natural products through engineered Yarrowia lipolytica. Biotechnol Adv 43:107555. https://doi.org/10.1016/j.biotechadv.2020.107555

Naranjo-Alcazar R, Bendix S, Groth T, Gallego Ferrer G (2023) Research progress in enzymatically cross-linked hydrogels as injectable systems for bioprinting and tissue engineering. Gels 9:230. https://doi.org/10.3390/gels9030230

Naseri G, Koffas MAG (2020) Application of combinatorial optimization strategies in synthetic biology. Nat Commun 11:2446. https://doi.org/10.1038/s41467-020-16175-y

Nawawi WMFBW, Jones M, Murphy RJ, Lee K-Y, Kontturi E, Bismarck A (2020) Nanomaterials derived from fungal sources—is it the new hype? Biomacromolecules 21:30–55. https://doi.org/10.1021/acs.biomac.9b01141

Niego AGT, Lambert C, Mortimer P, Thongklang N, Rapior S, Grosse M et al (2023) The contribution of fungi to the global economy. Fungal Divers 121:95–137. https://doi.org/10.1007/s13225-023-00520-9

Nigg M, Laroche J, Landry CR, Bernier L (2015) RNAseq analysis highlights specific transcriptome signatures of yeast and mycelial growth phases in the Dutch elm disease fungus *Ophiostoma novo-ulmi*. G3 Genes Genomes Genet 5:2487–2495. https://doi.org/10.1534/g3.115.021022

Ningsih TA, Akkermans S, De Vreese P, Matthijs E, Armstrong R, Van Impe JFM (2025) Towards mineralised mycelium biocomposites: optimising fungal growth and ureolytic biomineralisation. J Clean Prod 499:145120. https://doi.org/10.1016/j.jclepro.2025.145120

Oliva RL, Khadka UB, Camenzind T, Dyckmans J, Joergensen RG (2025) Constituent of extracellular polymeric substances (EPS) produced by a range of soil bacteria and fungi. BMC Microbiol 25:298. https://doi.org/10.1186/s12866-025-04034-z

Onyishi CL, Moneke AN, Ndubuisi IA, Onyia OO, Amadi OC (2026) Tolerance of natural yeast isolates to inhibitor compounds associated with the fermentation of lignocellulosic hydrolysates to ethanol. Microbe 10:100656. https://doi.org/10.1016/j.microb.2026.100656

Park S-H, Kim S, Hahn J-S (2016) Improvement of isobutanol production in Saccharomyces cerevisiae by increasing mitochondrial import of pyruvate through mitochondrial pyruvate carrier. Appl Microbiol Biotechnol 100:7591–7598. https://doi.org/10.1007/s00253-016-7636-z

Passer AR, Coelho MA, Billmyre RB, Nowrousian M, Mittelbach M, Yurkov AM et al (2019) Genetic and genomic analyses reveal boundaries between species closely related to *Cryptococcus* pathogens. MBio 10:e00764-19. https://doi.org/10.1128/mBio.00764-19

Pelletier MG, Holt GA, Wanjura JD, Bayer E, McIntyre G (2013) An evaluation study of mycelium based acoustic absorbers grown on agricultural by-product substrates. Ind Crop Prod 51:480–485. https://doi.org/10.1016/j.indcrop.2013.09.008

Peng J, Aluthmuhandiram JVS, Chethana KWT, Zhang Q, Xing Q, Wang H et al (2022) An NmrA-like protein, Lws1, is important for pathogenesis in the woody plant pathogen Lasiodiplodia theobromae. Plants 11:2197. https://doi.org/10.3390/plants11172197

Persson S, Shashkova S, Österberg L, Cvijovic M (2022) Modelling of glucose repression signalling in yeast *Saccharomyces cerevisiae*. FEMS Yeast Res 22:foac012. https://doi.org/10.1093/femsyr/foac012

Pham CLL, Rey A, Lo V, Soulès M, Ren Q, Meisl G et al (2016) Self-assembly of MPG1, a hydrophobin protein from the rice blast fungus that forms functional amyloid coatings, occurs by a surface-driven mechanism. Sci Rep 6:25288. https://doi.org/10.1038/srep25288

Phintha A, Chaiyen P (2022) Rational and mechanistic approaches for improving biocatalyst performance. Chem Catal 2:2614–2643. https://doi.org/10.1016/j.checat.2022.09.026

Pilafidis S, Tsouko E, Diamantopoulou P, Gkatzionis K, Sarris D, Papagiannopoulos A (2025) Viscoelastic profile of exopolysaccharides produced from Schizophyllum commune in bioreactor cultures utilizing brewer's spent grain: a sustainable bioprocess. Int J Biol Macromol 323:147004. https://doi.org/10.1016/j.ijbiomac.2025.147004

Piscitelli A, Cicatiello P, Gravagnuolo AM, Sorrentino I, Pezzella C, Giardina P (2017) Applications of functional amyloids from fungi: surface modification by class I hydrophobins. Biomolecules 7:45. https://doi.org/10.3390/biom7030045

Pitarch A, Arribas V, Gil C (2025) Omics and multiomics-based diagnostics for invasive candidiasis: toward precision medicine. Mol Cell Proteomics 24:101463. https://doi.org/10.1016/j.mcpro.2025.101463

Portela RMC, Vogl T, Kniely C, Fischer JE, Oliveira R, Glieder A (2017) Synthetic core promoters as universal parts for fine-tuning expression in different yeast species. ACS Synth Biol 6:471–484. https://doi.org/10.1021/acssynbio.6b00178

Poveda J, Santamaría Ó, Martín-García J (eds) (2025) Fungal metabolites for agricultural applications. Springer Nature Switzerland, Cham. https://doi.org/10.1007/978-3-031-76587-2

Przybyla L, Gilbert LA (2022) A new era in functional genomics screens. Nat Rev Genet 23:89–103. https://doi.org/10.1038/s41576-021-00409-w

Qin Y, Xia Y (2024) Melanin in fungi: advances in structure, biosynthesis, regulation, and metabolic engineering. Microb Cell Factories 23:334. https://doi.org/10.1186/s12934-024-02614-8

Rajkumar AS, Liu G, Bergenholm D, Arsovska D, Kristensen M, Nielsen J et al (2016) Engineering of synthetic, stress-responsive yeast promoters. Nucleic Acids Res 44:e136. https://doi.org/10.1093/nar/gkw553

Ramazzotti M, Berná L, Stefanini I, Cavalieri D (2012) A computational pipeline to discover highly phylogenetically informative genes in sequenced genomes: application to Saccharomyces cerevisiae natural strains. Nucleic Acids Res 40:3834–3848. https://doi.org/10.1093/nar/gks005

Randhawa A, Dutta SD, Ganguly K, Patil TV, Patel DK, Lim K-T (2022) A review of properties of nanocellulose, its synthesis, and potential in biomedical applications. Appl Sci 12:7090. https://doi.org/10.3390/app12147090

Reich M, Labes A (2017) How to boost marine fungal research: a first step towards a multidisciplinary approach by combining molecular fungal ecology and natural products chemistry. Mar Genomics 36:57–75. https://doi.org/10.1016/j.margen.2017.09.007

Reyes C, Sajó Z, Lucas MS, Sinha A, Schwarze FWMR, Ribera J et al (2022) Cocultivation of white-rot fungi and microalgae in the presence of nanocellulose. Microbiol Spectrum 10:e0304122. https://doi.org/10.1128/spectrum.03041-22

Ribeiro RA, Bourbon-Melo N, Sá-Correia I (2022) The cell wall and the response and tolerance to stresses of biotechnological relevance in yeasts. Front Microbiol 13:953479. https://doi.org/10.3389/fmicb.2022.953479

Riquelme M, Aguirre J, Bartnicki-García S, Braus GH, Feldbrügge M, Fleig U et al (2018) Fungal morphogenesis, from the polarized growth of hyphae to complex reproduction and infection structures. Microbiol Mol Biol Rev 82:e00068-17. https://doi.org/10.1128/MMBR.00068-17

Roberts N, Adamatzky A (2022) Mining logical circuits in fungi. Sci Rep 12:15930. https://doi.org/10.1038/s41598-022-20080-3

Rombauts S, Florquin K, Lescot M, Marchal K, Rouzé P, Van de Peer Y (2003) Computational approaches to identify promoters and cis-regulatory elements in plant genomes. Plant Physiol 132:1162–1176. https://doi.org/10.1104/pp.102.017715

Roze LV, Chanda A, Linz JE (2011) Compartmentalization and molecular traffic in secondary metabolism: a new understanding of established cellular processes. Fungal Genet Biol 48:35–48. https://doi.org/10.1016/j.fgb.2010.05.006

Salgado-Castillo SN, López-Peña HA, Díaz R, Peña-Solis K, Ponce-Alquicira E, Soriano Santos J et al (2023) Fungal melanins and their potential applications: a review. Bioresources 18:8688–8706. https://doi.org/10.15376/biores.18.4.Castillo

Schalbetter SA, Fudenberg G, Baxter J, Pollard KS, Neale MJ (2019) Principles of meiotic chromosome assembly revealed in S. cerevisiae. Nat Commun 10:4795. https://doi.org/10.1038/s41467-019-12629-0

Seong K, Krasileva KV (2023) Prediction of effector protein structures from fungal phytopathogens enables evolutionary analyses. Nat Microbiol 8:174–187. https://doi.org/10.1038/s41564-022-01287-6

Shen Q, Ruan H, Zhang H, Wu T, Zhu K, Han W et al (2024) Utilization of CRISPR-Cas genome editing technology in filamentous fungi: function and advancement potentiality. Front Microbiol 15:1375120. https://doi.org/10.3389/fmicb.2024.1375120

Singh S, Singh NP, Agrawal S, Kumar A (2025) Unleashing the potential of white-rot fungi mycelium for functional biomaterials development. Discov Mater 5:96. https://doi.org/10.1007/s43939-025-00288-6

Song W, Ko J, Choi YH, Hwang NS (2021) Recent advancements in enzyme-mediated crosslinkable hydrogels: *in vivo*-mimicking strategies. APL Bioeng 5:021502. https://doi.org/10.1063/5.0037793

Sperschneider J, Catanzariti A-M, DeBoer K, Petre B, Gardiner DM, Singh KB et al (2017) LOCALIZER: subcellular localization prediction of both plant and effector proteins in the plant cell. Sci Rep 7:44598. https://doi.org/10.1038/srep44598

Stoica RM, Moscovici M, Lakatos ES, Cioca LI (2023) Exopolysaccharides of fungal origin: properties and pharmaceutical applications. Processes 11:335. https://doi.org/10.3390/pr11020335

Subramanian V, Adler MJ, Benyamin M, Pullen RM, Servinsky MD, Kozlowski MT et al (2025) From the bench to the reactor: engineered filamentous fungi for biochemical and biomaterial production. Biotechnol Biofuels Bioprod 18:113. https://doi.org/10.1186/s13068-025-02712-8

Suthar M, Dufossé L, Singh SK (2023) The enigmatic world of fungal melanin: a comprehensive review. J Fungi 9:891. https://doi.org/10.3390/jof9090891

Taylor PR, Tsoni SV, Willment JA, Dennehy KM, Rosas M, Findon H et al (2007) Dectin-1 is required for β-glucan recognition and control of fungal infection. Nat Immunol 8:31–38. https://doi.org/10.1038/ni1408

Tellechea-Luzardo J, Stiebritz MT, Carbonell P (2023) Transcription factor-based biosensors for screening and dynamic regulation. Front Bioeng Biotechnol 11:1118702. https://doi.org/10.3389/fbioe.2023.1118702

Thilini Chethana KW, Peng J, Li X, Xing Q, Liu M, Zhang W et al (2020) LtEPG1, a secretory endopolygalacturonase protein, regulates the virulence of *Lasiodiplodia theobromae* in *Vitis vinifera* and is recognized as a microbe-associated molecular patterns. Phytopathology 110:1727–1736. https://doi.org/10.1094/PHYTO-04-20-0118-R

Tsang C-C, Teng JLL, Lau SKP, Woo PCY (2021) Rapid genomic diagnosis of fungal infections in the age of next-generation sequencing. J Fungi 7:636. https://doi.org/10.3390/jof7080636

Urbonaite G, Lee JTH, Liu P, Parada GE, Hemberg M, Acar M (2021) A yeast-optimized single-cell transcriptomics platform elucidates how mycophenolic acid and guanine alter global mRNA levels. Commun Biol 4:822. https://doi.org/10.1038/s42003-021-02320-w

Vanderstraeten J, Briers Y (2020) Synthetic protein scaffolds for the colocalisation of co-acting enzymes. Biotechnol Adv 44:107627. https://doi.org/10.1016/j.biotechadv.2020.107627

Vijayakumar G, Kim HJ, Jo JW, Rangarajulu SK (2024) Macrofungal mediated biosynthesis of silver nanoparticles and evaluation of its antibacterial and wound-healing efficacy. Int J Mol Sci 25:861. https://doi.org/10.3390/ijms25020861

Vogl T, Sturmberger L, Kickenweiz T, Wasmayer R, Schmid C, Hatzl A-M et al (2016) A toolbox of diverse promoters related to methanol utilization: functionally verified parts for heterologous pathway expression in *Pichia pastoris*. ACS Synth Biol 5:172–186. https://doi.org/10.1021/acssynbio.5b00199

Voutetaki ME, Mpalaskas AC (2024) Natural fiber-reinforced mycelium composite for innovative and sustainable construction materials. Fibers 12:57. https://doi.org/10.3390/fib12070057

Wadhwa K, Kapoor N, Kaur H, Abu-Seer EA, Tariq Mohd, Siddiqui S et al (2024) A comprehensive review of the diversity of fungal secondary metabolites and their emerging applications in healthcare and environment. Mycobiology 52:335–387. https://doi.org/10.1080/12298093.2024.2416736

Wang J, Huang Z, Jiang Q, Roubík H, Xu Q, Gharsallaoui A et al (2023) Fungal solid-state fermentation of crops and their by-products to obtain protein resources: the next frontier of food industry. Trends Food Sci Technol 138:628–644. https://doi.org/10.1016/j.tifs.2023.06.020

Wang D, Zeng N, Li C, Li Z, Zhang N, Li B (2024) Fungal biofilm formation and its regulatory mechanism. Heliyon 10:e32766. https://doi.org/10.1016/j.heliyon.2024.e32766

Wang X, Cao S, Li S, Sun Y, Lu W, Wang Y (2026) Investigation of the tanning mechanism of mycelial leather alternative with genipin and its environmental impact evaluation. Collagen Leather 8:2. https://doi.org/10.1186/s42825-025-00224-7

Wei Y, Bergenholm D, Gossing M, Siewers V, Nielsen J (2018) Expression of cocoa genes in Saccharomyces cerevisiae improves cocoa butter production. Microb Cell Factories 17:11. https://doi.org/10.1186/s12934-018-0866-2

Wikandari R, Hasniah N, Taherzadeh MJ (2022) The role of filamentous fungi in advancing the development of a sustainable circular bioeconomy. Bioresour Technol 345:126531. https://doi.org/10.1016/j.biortech.2021.126531

Wösten HAB (2019) Filamentous fungi for the production of enzymes, chemicals and materials. Curr Opin Biotechnol 59:65–70. https://doi.org/10.1016/j.copbio.2019.02.010

Wu H, Mou J, Zhou L, Zheng Q, Jiang N, Lin D (2016) Cloud cap-like, hierarchically porous carbon derived from mushroom as an excellent host cathode for high performance lithium-sulfur batteries. Electrochim Acta 212:1021–1030. https://doi.org/10.1016/j.electacta.2016.07.153

Xu L, Thali M, Schaffner W (1991) Upstream box/TATA box order is the major determinant of the direction of transcription. Nucleic Acids Res 19:6699–6704. https://doi.org/10.1093/nar/19.24.6699

Xu X, Feng J, Zhang P, Fan J, Yin W-B (2021) A CRISPR/Cas9 cleavage system for capturing fungal secondary metabolite gene clusters. J Microbiol Biotechnol 31:8–15. https://doi.org/10.4014/jmb.2008.08040

Xu H, Nakazawa T, Zhang Y, Oh M, Bao D, Kawauchi M et al (2022) Introducing multiple-gene mutations in *Pleurotus ostreatus* using a polycistronic tRNA and CRISPR guide RNA strategy. FEMS Microbiol Lett 369:fnac102. https://doi.org/10.1093/femsle/fnac102

Yaegashi J, Kirby J, Ito M, Sun J, Dutta T, Mirsiaghi M et al (2017) Rhodosporidium toruloides: a new platform organism for conversion of lignocellulose into terpene biofuels and bioproducts. Biotechnol Biofuels 10:241. https://doi.org/10.1186/s13068-017-0927-5

Yin M-Q, Xu K, Luan T, Kang X-L, Yang X-Y, Li H-X et al (2024) Metabolic engineering for compartmentalized biosynthesis of the valuable compounds in Saccharomyces cerevisiae. Microbiol Res 286:127815. https://doi.org/10.1016/j.micres.2024.127815

Yin L, Li H, Xing R, Li R, Gao K, Li G et al (2025) Fungal and microalgal chitin: structural differences, functional properties, and biomedical applications. Polymers 17:2722. https://doi.org/10.3390/polym17202722

Yu R, Zhang H, Guo B (2021) Conductive biomaterials as bioactive wound dressing for wound healing and skin tissue engineering. Nanomicro Lett 14:1. https://doi.org/10.1007/s40820-021-00751-y

Yu W, Cao X, Gao J, Zhou YJ (2022) Overproduction of 3-hydroxypropionate in a super yeast chassis. Bioresour Technol 361:127690. https://doi.org/10.1016/j.biortech.2022.127690

Yuan L, Lin W, Cao Y, Lv J, Chao S, Liu Y et al (2025) Disinfection efficacy of peroxyacetic acid against bacterial spores. J Hazard Mater 487:137128. https://doi.org/10.1016/j.jhazmat.2025.137128

Zhang Z, Donaldson AA, Ma X (2012) Advancements and future directions in enzyme technology for biomass conversion. Biotechnol Adv 30:913–919. https://doi.org/10.1016/j.biotechadv.2012.01.020

Zhang X, Zhou J, Zheng Y, Wei H, Su Z (2021) Graphene-based hybrid aerogels for energy and environmental applications. Chem Eng J 420:129700. https://doi.org/10.1016/j.cej.2021.129700

Zhang Y, Cheng C, Fu B, Long T, He N, Fan J et al (2024) Microbial upcycling of depolymerized lignin into value-added chemicals. Biodes Res 6:0027. https://doi.org/10.34133/bdr.0027

Zhu S, Xu H, Liu Y, Hong Y, Yang H, Zhou C et al (2025) Computational advances in biosynthetic gene cluster discovery and prediction. Biotechnol Adv 79:108532. https://doi.org/10.1016/j.biotechadv.2025.108532

Zier S, White LR, Johnstone D, Hoeft Z, El Hajam M, Kurtz P et al (2025) Growing sustainable barrier coatings from edible fungal mycelia. Langmuir 41:26751–26759. https://doi.org/10.1021/acs.langmuir.5c03185

Chapter 3
Recent Developments on Biotechnological Applications of Diverse Fungi in Waste Biomass Conversion into Bioenergy: Challenges and Future Outlooks

Oinam Bidyarani Devi, Mohit Kumar, Federico Cerrone, Santosh Kumar Mishra, Narendra Kumar, and Ali Jibril Baba

Abstract The urgent need for sustainable energy has placed lignocellulosic biomass at the forefront of renewable feedstocks. Fungi represent powerful biological platforms for their valorization due to their ability to secrete a broad arsenal of hydrolytic and oxidative enzymes that depolymerize lignin, hemicellulose, and cellulose. This chapter highlights the taxonomic and functional diversity of lignocellulolytic fungi, as well as filamentous ascomycetes, white-rot basidiomycetes, anaerobic gut fungi, and extremophiles, emphasizing their interacting enzyme systems comprising hemicellulases, cellulases, lignin-modifying oxidoreductases, and so on. Also, special notice is given to advances in engineering and clustered regularly interspaced short palindromic repeats (CRISPR)/Cas9-mediated genome editing in fungal metabolic systems, which enable targeted knock-in, knock-out, and regulatory modifications to enhance enzyme production and catalytic efficiency. Such strategies strengthen the development of fungal cell factories capable of producing customized enzyme consortia for industrial biomass degradation. In addition to the above, this chapter focuses on the integration of engineered fungi into biorefineries, techno-economic considerations, and bottlenecks such as enzyme inhibition, low titers, and scale-up challenges. Thus, fungal biotechnology, paired with genome editing, can accelerate the transition toward cost-effective biofuel production and sustainable bioeconomy pathways.

O. B. Devi · M. Kumar · S. K. Mishra (✉) · A. J. Baba
Fermentation Laboratory, Department of Life Science, Sharda School of Bioscience & Technology, Sharda University, Greater Noida, Uttar Pradesh, India
e-mail: santosh.mishra1@sharda.ac.in

F. Cerrone
School of Biotechnology, Dublin City University, Dublin, Ireland

N. Kumar
Department of Biotechnology, Noida Institute of Engineering and Technology, Greater Noida, Uttar Pradesh, India

M. Tripathi et al. (eds.), *Recent Advancements in Fungal Biotechnology Volume 1*, Fungal Biology, https://doi.org/10.1007/978-981-92-0339-0_3

Keywords Lignocellulosic biomass · CRISPR/Cas9 · Biofuel · Lignocellulolytic fungi

3.1 Introduction

Global demand for energy is increasing continuously, which is coupled with the rapid depletion of fossil fuel reserves and their associated environmental consequences, triggering an urgent search for sustainable alternatives (Kandpal et al. 2024). Although fossil fuels are used in the industry and make advancements in economic progress, it has become a major driver of greenhouse gas emissions and global climate change due to their non-renewable and carbon-intensive nature (Kongkuah and Alessa 2025). Energy consumption significantly rise over the next few decades globally, escalating the demand for renewable energy resources that are both environmentally friendly and economically viable (Gayen et al. 2024). Among various renewable options, bioenergy derived from lignocellulosic waste biomass has emerged as a promising solution, aligning with the principles of a circular bioeconomy (Raina et al. 2024). Agricultural residues, forestry by-products, food waste, and municipal solid wastes represent abundant feedstocks that can be converted into biofuels, biogas, biohydrogen, and other value-added products (Vasileiadou 2024). Efficient conversion of these complex and heterogeneous substrates into biofuels, biogas, and biohydrogen, however, remains a major challenge due to the recalcitrance of lignin and the heteropolymeric nature of plant cell walls (Ullah et al. 2022). The urgent demand for alternative energy sources has been sparked by the tremendous burden on fossil fuels and the resulting acute energy crisis and climate change issues (Singh 2021). Lignocellulosic biomass is a copious renewable and alternative bioresource for the generation of energy fuels and biochemicals in biorefineries (Sharma et al. 2023). Due to their versatility and eco-friendliness, cellulases, hemicellulases, and ligninolytic enzymes have been recognized as "green biocatalysts" (Nargotra et al. 2023). This chapter provides a detailed description of different types of lignocellulolytic enzymes, their mode of action, and their prospective applications in the valorization of lignocellulosic biomass. Solid-state fermentation (SSF) holds great promise in the microbial production of lignocellulolytic enzymes owing to its energy-efficient, environment-friendly, and higher product-yielding features, utilizing the lignocellulosic feedstocks. The recent trends in the application of enzyme immobilization strategies for improved enzymatic catalysis have been discussed. Also, the use of fungal enzymes that can secrete a wide array of hydrolytic and oxidative enzymes, including cellulases, hemicellulases, laccases, and peroxidases, enables them to deconstruct lignocellulosic biomass effectively during the bioprocessing of lignocellulosic biomass (Ren et al. 2024). For example, white rot fungi (Torres-Farradá et al. 2024), brown rot fungi (Chepyala et al. 2024), filamentous fungi such as *Trichoderma* and *Aspergillus* (Saharan et al. 2025), and anaerobic rumen fungi represent diverse groups with unique enzymatic machineries suited for different types of biomass degradation (Lv et al. 2025). Recent advances in omics technologies, protein engineering, synthetic biology, and fungal consortia design have further expanded the potential of fungi to serve as robust platforms for bioenergy production (Garg et al. 2025; Vieira et al. 2025). This chapter aims to provide a comprehensive overview of recent developments in the biotechnological applications of diverse fungi in waste biomass conversion into bioenergy. Thus, by integrating recent advancements with existing knowledge, we summarize and highlight the pivotal role played by fungi in sustainable bioenergy generation and outline the challenges and future directions for their effective application in biomass valorization.

3.2 Role of Fungi in Bioconversion of Lignocellulosic Waste Biomass

The efficient conversion of lignocellulosic waste biomass into bioenergy and bioproducts nowadays has become a central focus of industrial biotechnology (Ojo 2023), and fungi have emerged as indispensable players for their remarkable enzymatic capabilities in converting lignocellulosic waste into bioenergy (Sethupathy et al. 2021). Among fungi, white rot fungi are the most extensively studied for biomass pretreatment (Kong et al. 2023). *Phanerochaete chrysosporium* RP78 is a model lignin degrader that produces lignin peroxidase (LiP) and manganese peroxidase (MnP), which enable extensive delignification of agricultural residues (Akbar Aly et al. 2025). Similarly, *Pleurotus ostreatus* is widely exploited and helps in the bioconversion of chitin-based biomass into bioethanol (Xu et al. 2022). *Trametes versicolor* enhances the saccharification efficiency of subsequent enzymatic hydrolysis and bioethanol yields (Carrillo-Nieves et al. 2022; Devi et al. 2022). *Ganoderma lucidum* and *Lentinula edodes* have been reported for their ability to co-produce ligninolytic and hydrolytic enzymes, which not only facilitate biomass deconstruction but also contribute to bioethanol production using agricultural biomass (Shankar et al. 2024). In parallel, filamentous fungi such as *Aspergillus oryzae, Aspergillus nidulans, Trichoderma reesei,* and *Aspergillus niger* produce hemicellulolytic and cellulolytic enzymes (Ghorai 2024). Fungi can produce key industrial enzymes, particularly cellulases, xylanases, and β-glucosidases, by using byproducts from different industries, such as food processing, agriculture, and forestry (Dhevagi et al. 2021). *T. reesei* Rut-C30 remains the criterion cellulase producer (Yan et al. 2021b), while genetic engineering of *Aspergillus* spp. has led to strains with improved hemicellulase production and tolerance to lignocellulosic inhibitors (Jin et al. 2021). Co-cultivation of *Trichoderma* spp. and *Aspergillus* spp. has demonstrated synergistic enzyme production, leading to higher sugar release from complex feedstocks of lignocellulosic waste (Sperandio and Filho 2021). Furthermore, *Penicillium janthinellum* and *Humicola insolens* have attracted attention for their thermostable cellulases, which are advantageous for high-temperature industrial processes (Ghorai 2024; Ilić et al. 2023). Anaerobic rumen fungus such as *Neocallimastix frontalisonii* possesses cellulosome multi-enzyme complexes that act synergistically to decompose plant fibers under oxygen-limited conditions (Bhagat et al. 2023). Their ability to hydrolyze highly recalcitrant cellulose and hemicellulose makes them promising candidates for consolidated bioprocessing (CBP) in biogas plants. Recent metagenomic and transcriptomic studies have revealed novel glycoside hydrolases and carbohydrate esterases in these anaerobic fungi, broadening the enzymatic toolbox for biomass valorization (Pabbathi et al. 2023). Additionally, extremophilic fungi such as *Thermomyces lanuginosus* and *Talaromyces emersonii* are increasingly studied for their thermostable cellulases and xylanases, capable of operating at industrially relevant high temperatures (Ajeje et al. 2021), thus reducing risks of contamination and enhancing hydrolysis rates. Yeasts such as *Saccharomyces cerevisiae*, *Scheffersomyces stipitis*, and *Pichia kudriavzevii* are not directly involved in lignocellulose degradation but are essential in fermenting released sugars into ethanol or other biofuels (Shrivastava and Sharma 2023). Recent engineering strategies have enabled *Scheffersomyces* spp. and *Kluyveromyces marxia*nus to co-ferment hexoses and pentoses, significantly boosting ethanol yields from lignocellulosic hydrolysates (Jain et al. 2025). Marine-derived fungi such as *Aspergillus sydowii* and *Cladosporium* spp. have also been identified as sources of halotolerant enzymes for biomass bioconversion in saline or marine environments, opening opportunities for blue biorefineries (Yovchevska et al. 2025). Collectively, these findings establish fungi as critical agents for sustainable lignocellulosic biomass valorization. Their diversity, from ligninolytic white rot fungi to hydrolytic filamentous fungi, anaerobic gut fungi, and extremophiles, provides a versatile enzymatic arsenal for breaking down the structural complexity of plant biomass. Moreover, advances in metabolic engineering, synthetic biology, and omics-driven strain improvement are further enhancing their bioenergy potential. However, challenges remain, including low enzyme titers, feedback inhibition, substrate recalcitrance, and process scalability. Addressing these barriers through genetic engineering, microbial consortia development, and integration into circular bioeconomy frameworks is likely to define the future trajectory of fungal biotechnology in waste-to-energy conversion.

3.3 Diversity of Fungi Involved in Biomass Conversion into Bioenergy

Fungi represent one of the most diverse and versatile microbial groups in bioenergy research, with different taxa contributing to multiple stages of biomass deconstruction and energy generation. White rot basidiomycetes are especially critical because of their unique ability to degrade lignin while simultaneously supporting saccharification and fermentation (Konan et al. 2024). For example, *Lenzites betulinus* was reported to directly ferment lignin under oxygen-limited conditions, yielding ethanol (Thakur 2025), *Hohenbuehelia* sp. ZW-16 was found to ferment hexose and pentose sugars from corn straw hydrolysates, ethanol (Fansuri et al. 2024). Beyond bioethanol, white-rot fungi such as *Phanerochaete chrysosporium* and *Trametes versicolor* secrete ligninolytic enzymes that not only release fermentable sugars but also improve biogas yields by enhancing substrate digestibility for methanogenic consortia (Zhu et al. 2025). Brown rot and soft rot fungi, although less efficient at complete lignin removal, generate high titers of hydrolytic enzymes that contribute to improved saccharification and methane recovery during anaerobic digestion of agro-residues (Rout et al. 2022). Anaerobic rumen fungi are also increasingly recognized as dual contributors to ethanol and biogas production. Strains such as *Oontomyces* sp. CR2 and *Piromyces* sp. CF9, isolated from Bactrian camel rumen, showed strong lignocellulose degradation alongside the production of ethanol, hydrogen, and methane from plant residues, including stalks and reeds (Xue et al. 2022). Yeasts, like *Saccharomyces cerevisiae*, remain the cornerstone of industrial bioethanol fermentation (Elhalis 2024); other yeasts, such as *Scheffersomyces stipitis* and *Candida shehatae*, expand the spectrum by fermenting xylose and other pentoses from lignocellulosic hydrolysates (Yanieva 2022). Some fungi that are involved in waste biomass conversion and their applications in bioenergy production are given in Table 3.1.

3.4 Bioconversion of Biomass into Bioenergy Using Fungal Enzymes

Nowadays, growing concerns over fossil fuel depletion and climate change have positioned lignocellulosic biomass as a promising renewable feedstock for sustainable bioenergy production by using fungal enzymes. Fungi play an important role in nature's carbon cycle and are uniquely equipped with diverse hydrolytic and oxidative enzymes (cellulases, hemicellulases, ligninases, and lipases) capable of deconstructing the complex plant cell wall matrix (El-Gendi et al. 2022). Fungal enzyme systems facilitate saccharification for bioethanol as well as enhance biogas yields via improved digestibility and also contribute to biodiesel production through fungal lipases (Saye et al. 2021). Different types of fungal enzymes involved in biomass conversion into several bioenergy sources are explained below. In Fig. 3.1, different fungal enzymes and their application are shown.

3.4.1 Fungal Cellulases

Fungal cellulases comprise the canonical endoglucanases (EGs), cellobiohydrolases (CBHs/exoglucanases), and β-glucosidases, which act sequentially to depolymerize crystalline and amorphous cellulose to cellobiose and glucose (Sohail et al. 2022). *Trichoderma reesei* remains a primary source of high-yield cellulase (Yan et al. 2021b). *Aspergillus* and *Penicillium* spp. are also widely used for secreting cellulase and accessory enzymes (El-Gendi et al. 2022). Fungi has gained its importance in (1) on-site production of cellulases using the feedstock itself to cut costs (Dey et al. 2022),

Table 3.1 Some fungi that are actively involved in the bioconversion of waste biomass into bioenergy

Types of fungi	Fungal species	Fungal enzymes	Applications	References
White rot fungi	*Phanerochaete chrysosporium*, *Pleurotus ostreatus*, *Trametes versicolor*	Laccases, LiP, MnP	Degradation of lignin, delignification of agro-residues, enhanced saccharification for bioethanol	Konan et al. (2024)
Brown rot fungi	*Gloeophyllum trabeum*, *Postia placenta*, *Coniophora puteana*	Hydroxyl radicals, glycoside hydrolases	Cellulose and hemicellulose degradation valorization	Felipuci (2024), Hage and Rosso (2021), Nargotra et al. (2023)
Soft rot fungi	*Chaetomium globosum*	Cellulases, hemicellulases	Degradation of cellulose in humid environments, improving accessibility of biomass for fermentation	Abdel-Azeem et al. (2021a, b)
Filamentous fungi	*Trichoderma reesei*, *Aspergillus niger*, *Aspergillus oryzae*, *Penicillium oxalicum*	Cellulases, xylanases, pectinases, β-glucosidases	Enzymatic hydrolysis of lignocellulose, production of bioethanol, biohydrogen, and biogas	Dhevagi et al. (2021)
Anaerobic rumen fungi	*Piromyces* spp.	Highly active cellulases and hemicellulases	Digestion of fibrous biomass in anaerobic conditions, biogas/methane production	Abubakar et al. (2022)
Extremophilic fungi	*Thermomyces lanuginosus*	Thermostable cellulases, xylanases	High-temperature biomass hydrolysis, industrial-scale biofuel production	Katarina et al. (2024)
Filamentous fungi	*Aspergillus fumigatus*	Lignocellulolytic enzymes	Hemicellulose Degradation	Qin et al. (2025)
Yeasts (fermentative fungi)	*Saccharomyces cerevisiae*, *Pichia stipitis*, *Scheffersomyces stipitis*, *Kluyveromyces marxianus*	Fermentative enzymes (alcohol dehydrogenase, xylose reductase)	Fermentation of sugars to bioethanol, co-fermentation of hexoses and pentoses	Leonel et al. (2021), Vasylyshyn et al. (2025)
Marine/novel fungi	*Aspergillus sydowii*, *Cladosporium* spp.	Halotolerant enzymes	Biomass conversion in saline environments, bioethanol from marine biomass	Abo Nouh et al. (2021), Jain et al. (2021)

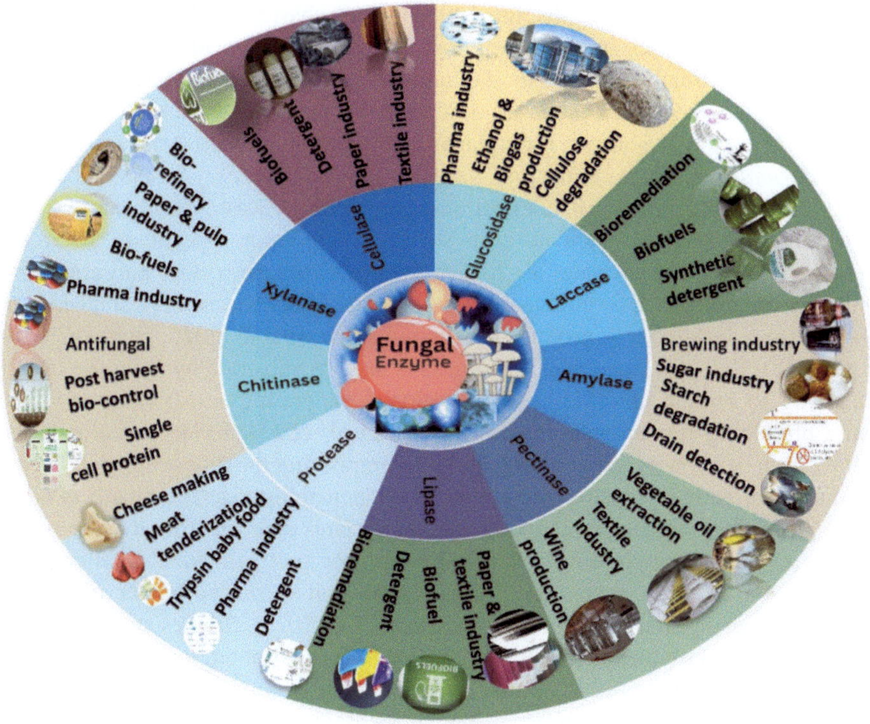

Fig. 3.1 Different fungal enzymes and their applications

(2) optimization of enzyme cocktails to match specific feedstocks and pretreatments (Pimentel et al. 2021), and (3) strain and process engineering to increase titers and tolerance to inhibitors (Mohedano et al. 2022). These approaches have been shown to improve saccharification efficiency and thus downstream ethanol or anaerobic digestion yields. These cellulases are essential for producing fermentable sugars for bioethanol and as feedstock for downstream anaerobic digestion (Shukla et al. 2023). Recent process and strain studies demonstrate improvements in secretion, regulation, and Zn^{2+}-mediated induction of cellulase expression in *T. reesei* (Li et al. 2023).

3.4.1.1 Hemicellulases and Accessory Debranching Enzymes

Hemicellulases consist of various enzymes (endo-xylanases, β-xylosidases, mannanases, arabinofuranosidases, acetyl xylan esterases, and feruloyl esterases) that act in synergy to digest hemicellulose (Agrawal et al. 2025). Fungi such as *Aspergillus*, *Penicillium*, and *Trichoderma* spp. known for hemicellulases secretion that release pentoses and other sugars from hemicellulose (de Souza and Kawaguti 2021). Xylanases (EC 3.2.1.8) hydrolyze xylan, the main component of hemicellulose in plant cell walls, into xylo-oligosaccharides and xylose (Bhardwaj et al. 2021). Xylanases improve cellulose accessibility for cellulases (Liu et al. 2023), thus boosting bioethanol, biogas, and biohydrogen yields (Dutta et al. 2023). *Thermomyces lanuginosus* produces a notable thermostable xylanase (Della Torre et al. 2023). Fungal sources for xylanase production include *Aspergillus niger, Trichoderma reesei*, *Aspergillus fumigatus*, and *Penicillium chrysogenum* (Dhevagi et al. 2021). Applications of hemicellulase are seen in waste-to-bioenergy using agro-industrial residues such as corn stover, wheat bran, rice husk, sugarcane bagasse, and so on (Sharma et al. 2022). Xylanases release xylose and arabinose, which can be fermented by engineered yeasts such as *Saccharomyces cerevisiae* and *Scheffersomyces stipitis* to produce bioethanol (Milessi et al. 2025; Ochoa-Chacón et al. 2022). Pulp and paper waste biomass is used as substrate by xylanase in bio-bleaching to reduce chlorine-based chemicals, generating eco-friendly effluents while producing fermentable sugars (Dukare et al. 2023). Xylanase helps in the valorization of animal feed and food waste by improving the digestibility of fibrous residues, indirectly increasing biogas potential when digested anaerobically (Poddar et al. 2022). Thermostable and alkali-tolerant variants sourced from diverse fungal isolates expand the operating window for harsher pretreatments. Hemicellulase enzymes remove side chains and substitutions that sterically hinder cellulases, thereby increasing overall sugar yields (Huang et al. 2022). Studies show that optimized hemicellulase blends (often from *Aspergillus niger* or engineered *Penicillium strains*) substantially improve saccharification of hemicellulose-rich feedstocks and enable more efficient co-fermentation of pentoses and hexoses (Nanda and Dalai 2023).

3.4.2 Lignin-Modifying and Oxidative Enzymes (Laccases, Peroxidases, LPMOs, and Reductases)

Lignin is the primary barrier to efficient saccharification. White-rot fungi (e.g., *Trametes versicolor*, *Pleurotus ostreatus*, *Phanerochaete chrysosporium*) produce laccases and peroxidases that can partially depolymerize or modify lignin, improving enzyme accessibility to cellulose and hemicellulose (Akbar Aly et al. 2025). Laccase treatment has also been widely demonstrated to detoxify phenolic inhibitors formed during chemical pretreatment, thereby improving fermentability of hydrolysates (Fernández-Sandoval et al. 2024). Lytic polysaccharide monooxygenases (LPMOs) are a relatively recent discovery that oxidatively cleave crystalline polysaccharide surfaces and act synergistically with hydrolases to boost saccharification, often enabling large improvements in conversion of recalcitrant biomass (Kumar et al. 2024). Incorporation of LPMOs and other accessory enzymes, such as acetylxylan esterases and feruloyl esterases, into commercial cocktails is now a common practice to raise yields on industrial feedstocks (Dahiya et al. 2024). White-rot fungi such as *Phanerochaete chrysosporium*, *Trametes versicolor*, and *Pleurotus ostreatus* secrete laccases, LiP, MnP, and other peroxidases (e.g., versatile peroxidase, DyPs) that oxidatively depolymerize lignin or modify its structure (Shi et al. 2021). The action of these lignin-modifying enzymes (LMEs) reduces lignin content and alters lignin-carbohydrate linkages, markedly improving enzymatic hydrolysis yields (Hassan et al. 2025). Recently, studies have shown enhanced biogas and ethanol yields when biological delignification (using *P. chrysosporium* or *T. versicolor*) is integrated as a pretreatment step or combined with mild chemical pretreatments (Kumar Chaurasia et al. 2025). LPMOs (fungal AA families such as AA9, AA11, AA13, AA14, and AA16) perform copper (Cu)-dependent oxidative cleavage of crystalline polysaccharides and dramatically increase accessibility for hydrolytic enzymes (Ipsen et al. 2021). AA9 LPMOs, widely found across Ascomycota and Basidiomycota, have become standard additives in commercial enzyme cocktails because they accelerate conversion of recalcitrant cellulose and certain hemicelluloses (Contato and Conte-Junior 2025). Recent mechanistic work of AA9 LPMOs has proved the use of H_2O_2 more efficiently than O_2 as a co-substrate, need an external electron donor (e.g., ascorbic acid, lignin-derived phenols, or cellobiose dehydrogenase) to activate their copper center. These mechanistic insights are critical for scaling up biomass to sugar conversion processes (e.g., for biofuels) (Contato and Conte-Junior 2025).

3.4.3 Pectinases, Amylases, and Other Hydrolases

Agro-industrial and food-processing wastes (fruit pomace, vegetable trimmings, potato peel, cereal residues, and starch-rich food wastes) are attractive feedstocks for bio-energy (Agudelo-Patiño et al. 2024) because they contain easily convertible polysaccharides such as pectin, starch, and soluble cellulose/hemicellulose (Mohamed et al. 2021). Enzymatic hydrolysis using fungal pectinases and amylases, often combined with cellulases/hemicellulases, enables mild, targeted release of fermentable sugars (glucose, fructose, galacturonic acid, and xylose) (de Souza and Kawaguti 2021). Recent findings emphasize that pectinases, amylases, and other hydrolase enzymes are suited to feedstock composition (Ndochinwa et al. 2024), using low-cost enzyme production (SSF on the wastes) (Singh et al. 2024) and integrating enzymatic pretreatment with downstream fermentation or anaerobic digestion for improved yields and economics (Khan 2025).

3.4.3.1 Pectinases

Pectinases (polygalacturonases, pectin lyases, pectate lyases, pectin, and methyl esterases) hydrolyze and depolymerize pectin polysaccharides that are abundant in fruit tissues (apple, citrus, grape) (Li et al. 2024). *Aspergillus niger* remains a principal industrial pectinase producer (Mwaheb et al. 2025). *Penicillium* spp. and *Aspergilli* producing pectinase also show promising activities and robustness under industrial conditions (Haile and Ayele 2022). Pectinase leads to (a) direct saccharification of fruit pomaces (apple, grape, and citrus) to increase fermentable sugar yields (Shrestha et al. 2021) and (b) the use of pectinases to reduce viscosity and improve downstream processing (filtration, fermentation) (Zhao et al. 2025). Recent research highlights that producing pectinases via SSF on the same fruit waste can reduce enzyme cost and enhance sustainability (Ray et al. 2025).

Nowadays, improved ethanol yields from pectin-rich feedstocks are extracted when pectinase pretreatment is applied prior to fermentation of substrates such as apple and citrus pomace (Mahato et al. 2021).

3.4.3.2 Amylases

Starch-rich wastes and fermentation Amylases (α- and β-amylases, glucoamylases) hydrolyze starch into oligosaccharides and glucose (Naik et al. 2023). Food waste streams rich in starch, potato peels, starch processing residues, and discarded bakery goods are well suited to enzymatic amylolysis prior to fermentation (Chauhan et al. 2023). Fungal amylase producers such as *Aspergillus oryzae* and *Aspergillus niger* are widely studied and used industrially (Dhevagi et al. 2021). Recent studies in strain improvement for amylase production, process optimization, and enzyme stabilization have been reported (Sharif et al. 2023). Studies specifically focused on enzymatic hydrolysis of mixed food waste show that combining amylolytic treatments with cellulolytic enzymes increases the overall carbohydrate yield and improves bioethanol or biogas production (Passadis et al. 2022).

3.4.4 Lipases (Lipid Wastes and Biodiesel Production)

Lipases (EC 3.1.1.3) from fungi and yeasts (*Candida rugosa*, *Rhizopus oryzae*, *Aspergillus niger*, and *Candida antarctica*) remain the most investigated biocatalysts for lipid to biodiesel conversion (Darwish et al. 2021; Mehta et al. 2021). The ability of lipases to catalyze both hydrolysis, that is, breaking triglycerides into free fatty acids (FFAs) and glycerol, and transesterification conversion of triglycerides or FFAs with short-chain alcohols to fatty acid alkyl esters, that is, biodiesel, makes them ideal for lipid-rich wastes such as waste cooking oil (WCO), tallow, grease trap residues, and dairy byproducts (Mehta et al. 2021). Some important actions of lipases are as follows.

3.4.4.1 Immobilization Strategies (Enhancing Stability and Reuse)

A central development in the last decade is the immobilization of fungal lipases, which boosts enzyme lifetime, reusability, and tolerance to industrial conditions (e.g., methanol toxicity and shear forces) (Kumar et al. 2023). Recent approaches using lipases include the following: Porous supports (e.g., silica, mesoporous carbons, and zeolites) (Costantini and Califano 2021): Immobilized *Candida antarctica* lipase B on mesoporous silica exhibited high biodiesel yields (>90%) with methanol and ethanol as acyl acceptors when applied to WCO (Yu et al. 2021). Magnetic nanoparticles (MNPs): Functionalized Fe_3O_4 nanoparticles provide easy recovery by magnetic separation. Studies with Rhizopus oryzae lipase immobilized on chitosan-coated MNPs reported reusability for >10 cycles with minimal activity loss and >85% conversion efficiency (Mariño et al. 2021). Biopolymers and natural supports (alginate beads, chitosan, and cellulose nanofibers) bound lipases not only increase enzyme activity (Zhang et al. 2023a). Alginate-entrapped *Aspergillus* lipase achieved stable conversions across repeated WCO transesterification batches (Hidayatullah et al. 2023). Covalent cross-linking and cross-linked enzyme aggregates approaches using fungal lipases show strong solvent tolerance and higher thermal stability, reducing enzyme leaching compared to simple adsorption methods (Chen et al. 2023a).

3.4.4.2 High-FFA and Mixed Waste Feedstocks

WCO is often problematic for chemical catalysts due to its high FFA content and impurities. Lipases, especially immobilized ones, overcome these challenges: *Candida rugosa* and *Aspergillus oryzae* lipases efficiently converted WCO with FFA >15% into biodiesel without requiring prior esterification (Abdulla et al. 2022; Wei et al. 2023). Mixed waste lipids (e.g., grease trap residues) were successfully processed using immobilized Rhizopus oryzae lipase, producing >90% fatty acid methyl ester (Dash and Banerjee 2021). Integration of lipase biocatalysis with pretreatments (ultrasound and microwave) has been reported to accelerate mass transfer and shorten reaction times (He et al. 2022).

3.4.4.3 Reactor and Process Developments

Recent work emphasizes that process intensification and scaling feasibility, such as packed-bed and fluidized-bed reactors loaded with immobilized lipases, allow continuous biodiesel production, reducing enzyme costs per unit biodiesel (Zanuso et al. 2021). Membrane bioreactors help in incorporating lipase immobilized membranes and enable simultaneous separation and reaction, improving yield and catalyst stability (Najim et al. 2024). Methanol stepwise feeding strategies prevent enzyme deactivation and maintain >95% yield across cycles (Xia et al. 2024).

3.4.5 Anaerobic Gut Fungi and Cellulosomes (Specialized Complexes for Fiber Degradation)

Neocallimastigomycota (e.g., *Neocallimastix*, *Piromyces*, and *Orpinomyces*) produce multi-enzyme assemblies into cellulosome-like complexes optimized for fiber degradation under anaerobic conditions (Wunderlich et al. 2023). These anaerobic fungi exhibit remarkable fibrolytic capacity and generate fermentation products (H_2, acetate, formate, short-chain fatty acids, and sometimes ethanol) that are directly relevant to methane or biohydrogen production in anaerobic digesters (Bhagat et al. 2023; Fu et al. 2025). Recent genomics and transcriptomics have uncovered numerous novel glycoside hydrolases and carbohydrate active modules in these taxa, expanding their potential as sources of new enzymes and consortia for CBP (Wilken 2020).

3.4.6 Thermostable and Extremophilic Fungal Enzymes—Process Robustness

Thermophilic/thermotolerant fungi (e.g., *Thermomyces lanuginosus*, *Talaromyces emersonii*, and some *Penicillium* spp.) produce thermostable xylanases and cellulases advantageous for high-temperature hydrolysis, which reduces contamination risk and can improve reaction kinetics (Ajeje et al. 2021). Immobilization and protein engineering studies have focused on these enzymes to extend operational lifetimes in reactors and to permit reuse critical factors for cost-effective commercial biorefineries (Dey et al. 2022).

3.5 Fungal Enzymatic Machinery for Biomass Conversion into Bioenergy

Fungi secrete a broad suite of extracellular enzymes that act synergistically to deconstruct lignocellulosic biomass into fermentable sugars and other intermediates. Nowadays, significant advances have improved both mechanistic understanding and practical deployment of fungal enzyme systems (Khan 2025). The cellulase complex comprising EGs, CBHs, and β-glucosidases remains central to cellulose hydrolysis, with industrial cocktails still dominated by *Trichoderma* and *Aspergillus* strains (Sohail et al. 2022). Emerging genera such as *Talaromyces* and *Penicillium* provide additional enzyme diversity, and recent studies emphasize optimizing enzyme ratios to lower cost (El-Gendi et al. 2022). Complementary hemicellulases and accessory hydrolases such as xylanases, arabinofuranosidases, esterases, and pectinases improve saccharification by removing side groups that limit cellulase access (de Souza and Kawaguti 2021).

LMEs such as laccases, lignin peroxidases, manganese peroxidases, and versatile peroxidases produced mainly by white-rot basidiomycetes, viz., *Trametes*, *Phanerochaete*, and *Pleurotus,* are critical for lignin depolymerization, enhancing cellulose accessibility (Gupta et al. 2023). A transformative development has been the inclusion of LPMOs, copper-dependent oxidative enzymes that cleave crystalline cellulose and dramatically boost cellulase efficiency (Kumar et al. 2024). On the genetic front, CRISPR/Cas and other genome engineering tools have been

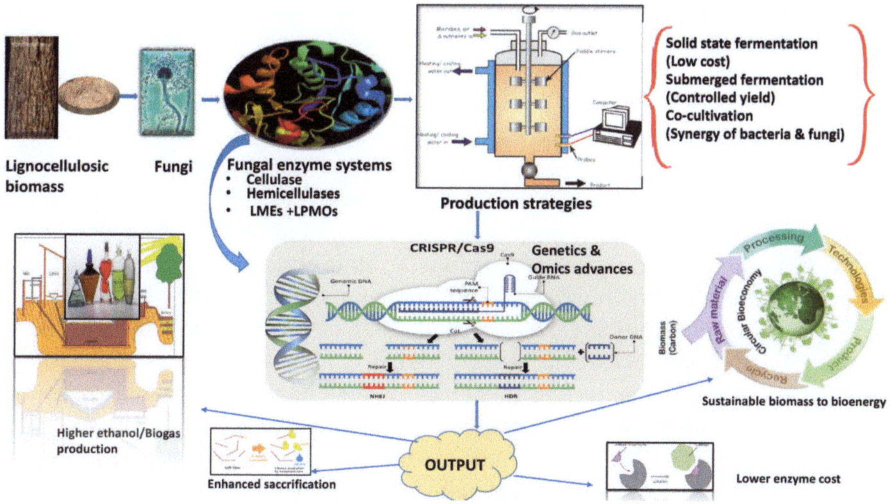

Fig. 3.2 Fungal enzymatic machinery for the conversion of biomass into bioenergy

applied to filamentous fungi to enhance cellulase secretion, optimize enzyme balance, and improve tolerance to hydrolysate stress (Mondal et al. 2022). Promoter engineering, transcription factor modulation, and secretion pathway optimization have further increased fungal enzyme titers, especially when combined with adaptive laboratory evolution (Wu et al. 2022). Multi-omics studies mapping inducible enzyme networks now guide more rational strain improvement (Wu et al. 2022). Collectively, these advances highlight the growing sophistication of fungal enzymatic systems for efficient, sustainable biomass to bioenergy conversion (Hassan et al. 2025). A schematic presentation of fungal enzymatic machinery for the conversion of biomass into bioenergy is given in Fig. 3.2. Production strategies have also advanced. SSF offers cost-effective, residue-based enzyme generation, while submerged fermentation remains standard for controlled, high-yield production (Karimi et al. 2021). Increasingly, co-cultivation strategies pair lignin-degrading fungi with cellulase-rich ascomycetes or bacteria, achieving faster delignification and saccharification, though microbial stability remains a challenge (Detain and Besaury 2024).

3.6 Genetic Editing and CRISPR/Cas Technologies in Fungi for Enhanced Bioenergy Production

Genetic engineering of fungi has significantly advanced biomass conversion by improving enzyme secretion, substrate utilization, and fermentation efficiency. Classical cellulase producers such as *Trichoderma reesei* have been extensively engineered to increase cellulase titers, alter regulatory pathways, and improve secretion efficiency (Yan et al. 2021b). Strains like Rut-C30 and its derivatives, modified through both mutagenesis and targeted editing, now deliver industry-relevant cellulase yields, reducing enzyme costs in saccharification (Xiang et al. 2025). CRISPR/Cas9-mediated editing of *T. reesei* has enabled knockout of repressors and enhancement of EG and β-glucosidase activity, addressing limitations in its natural enzyme mix (Chen et al. 2023b). Thermophilic fungi such as *Myceliophthora thermophila* have been reprogrammed for CBP, integrating enzyme secretion and fermentation in one organism (Zhang et al. 2023b). CRISPR-based promoter rewiring and deletion of ethanol-consuming pathways redirected carbon flux toward ethanol, allowing direct fermentation of cellulose into ethanol without exogenous enzyme addition (Pabbathi et al. 2025). Nonconventional yeasts have also benefitted from CRISPR-enabled genome editing (Fernandes et al. 2023). *Saccharomyces cerevisiae* has been engineered for efficient xylose and arabinose fermentation by insertion of bacterial xylose isomerase and fungal transporters (Costa 2022), while CRISPR-based redox balancing has minimized cofactor bottlenecks, increasing ethanol yields in lignocellulosic hydrolysates (Ploessl et al. 2023). *Scheffersomyces stipitis* and *Kluyveromyces marxianus* have been edited to enhance pentose utilization and inhibitor tolerance, making them promising candidates for second-generation ethanol production (Vasylyshyn et al. 2025). Beyond ethanol, CRISPR tools are expanding fungal applications in other bioenergy streams. In *Aspergillus oryzae* and *Rhizopus oryzae*, targeted editing has enhanced lipase expression for biodiesel production from WCO (Wei et al. 2023; Yang et al. 2021). *Fusarium venenatum* has been modified to secrete higher levels of hemicellulases and oxidative enzymes relevant for biogas enhancement (Lee et al. 2024). Advancement shown in how CRISPR-Cas technology and related genetic toolkits are unlocking the potential of fungi ranging from *T. reesei* and *M. thermophila* to non-conventional yeasts and anaerobic gut fungi (Rothschild-Mancinelli 2021). Some of the genetically engineered and CRISPR-modified fungi for enhanced bioenergy production are given in Table 3.2. By tailoring enzyme secretion, sugar metabolism, and fermentation efficiency, these engineered strains are moving closer to scalable applications in bioethanol, biodiesel, biogas, and hydrogen production, thus reinforcing fungi as indispensable platforms for sustainable bioenergy (Yadav et al. 2025).

3.7 Challenges and Bottlenecks in Large-Scale Bioenergy Conversion from Biomass

Biological fungal pre-treatments are eco-friendly but limited by slow processing and low sugar yields (Kumar Chaurasia et al. 2025). Fungal enzyme production costs remain high due to low titers, incomplete cocktails, and poor stability under industrial conditions (Long et al. 2025). Scale-up bottlenecks persist between SSF (ecologically relevant but hard to scale) and SmF (easier to control but

Table 3.2 Some genetically engineered and CRISPR-modified fungi for enhanced bioenergy production

Fungal species	Genetic/CRISPR modification	Bioenergy application	References
Trichoderma reesei (Rut-C30 derivatives)	Overexpression of cellulases, knockout of repressors, promoter rewiring via CRISPR/Cas9	Enhanced cellulase titers, improved saccharification for bioethanol	Adnan and Liu (2024)
Myceliophthora thermophila	CRISPR-mediated editing of cytosine to thymine base for enhancing cellulase production	To enhance the production of cellulase	Zhang (2022)
Aspergillus fumigatus LMB-35Aa	CRISPR/Cas9-mediated targeted knock-in of the eglA gene to enhance the production of EG and its activity	To enhance biofuel production	Benites-Pariente et al. (2024)
Saccharomyces cerevisiae	Insertion of bacterial xylose isomerase, fungal cellodextrin transporters; CRISPR redox balancing	Efficient co-fermentation of pentoses and hexoses for ethanol	Vasylyshyn et al. (2025)
Scheffersomyces stipitis	CRISPR/cas9 reengineered the strain	Improved xylose fermentation for second gen ethanol	Adebami et al. (2022)
Kluyveromyces marxianus	CRISPR-enabled engineering for thermotolerance and pentose use	High-temperature bioethanol production	Cernak et al. (2018), Madeira-Jr and Gombert (2018)
Aspergillus oryzae/Rhizopus oryzae	Overexpression of lipase genes via targeted editing	Biodiesel production from WCO	Kumar et al. (2023), Li et al. (2025)

resource-intensive), compounded by mass/heat transfer gradients, sterility concerns, and contamination risks (Schwan et al. 2024). Feedstock variability, seasonal supply logistics, and co-metabolism of sugars further reduce process efficiency (Khan 2025). Strain level issues include limited secretion capacity in native fungi, difficulties in engineering stable enzyme systems, and the chemical heterogeneity of lignin that hampers selective degradation (Sokal et al. 2022). Process integration challenges hybrid pre-treatment trade-offs, downstream separation costs, and compatibility with refinery infrastructure-add economic and regulatory barriers (Asghar et al. 2022). Overall, techno-economic analyses show fungal approaches can reduce chemical inputs but remain constrained by low throughput, high capital costs, and uncertain policy/market incentives (Alcocer-García et al. 2025).

White-rot basidiomycetes, as well as filamentous ascomycetes, offer low-energy, selective lignin removal and powerful lignocellulolytic enzyme cocktails for biomass deconstruction (Wan Mohtar et al. 2022). However, multiple biological, engineering, economic, and logistical bottlenecks have limited translation to commercial, large-scale bioenergy facilities since 2020 (Singh et al. 2025).

3.8 Future Prospects and Emerging Trends

Recent advances highlight fungi as central players in sustainable biomass-to-bioenergy conversion. Since 2020, major trends include strain engineering (using CRISPR and systems biology to boost enzyme titers, balance cellulase cocktails, and improve inhibitor tolerance) and synthetic consortia combining ligninolytic and cellulolytic fungi for faster, more complete biomass breakdown. Renewed interest in SSF is driven by innovations in bioreactor design and process control, enabling direct use of agro-residues with reduced water input. Hybrid pretreatments, mild chemical methods followed by accelerated fungal treatment, are gaining attention for reducing processing time. Novel outputs, such as fungal biofilms for microbial fuel cells and mycelium-based materials, expand economic potential. Challenges remain: slow pretreatment rates, scale-up of SSF, and high enzyme costs. Future efforts must integrate genetic and process optimization, pilot-scale techno-economic validation, and value chain diversification to establish fungi-driven biorefineries within circular bio economies.

3.9 Conclusion

Diverse fungi hold immense promise for converting waste biomass into bioenergy through their robust lignocellulolytic systems. Despite advances in strain engineering, fungal consortia, and hybrid pretreatments, challenges such as slow kinetics, scale-up constraints, and enzyme costs persist. Future research should integrate biotechnology with process engineering, techno-economic validation, and co-product valorization, positioning fungi-driven systems as sustainable development.

References

Abdel-Azeem AM, Abo Nahas HH, Abdel-Azeem MA, Tariq FJ, Yadav AN (2021a) Biodiversity and ecological perspective of industrially important fungi an introduction. In: Abdel-Azeem AM, Yadav AN, Yadav N, Usmani Z (eds) Industrially important fungi for sustainable development: volume 1: biodiversity and ecological perspectives. Springer International Publishing, pp 1–34. https://doi.org/10.1007/978-3-030-67561-5_1

Abdel-Azeem AM, Yadav AN, Yadav N, Usmani Z (2021b) Industrially important fungi for sustainable development. Springer International Publishing

Abdulla R, Derman E, Mathialagan K, Yaser AZ, Abu Samah MA, Gansau JA, Syed Najmuddin SUF (2022) Biodiesel production from waste palm cooking oil using immobilized *Candida rugosa* lipase. Sustainability 14(20):13632. https://doi.org/10.3390/su142013632

Abo Nouh FA, Gezaf SA, Abo Nahas HH, Abo Nahas YH, Vargas-De-La-Cruz C, Solorzano Acosta RA, Landa-Acuña D, Luis-Alaya B, Abdel-Azeem AM (2021) Bioprospecting for biomolecules from different fungal communities: an introduction. In: Abdel-Azeem AM, Yadav AN, Yadav N, Sharma M (eds) Industrially important fungi for sustainable development: volume 2: bioprospecting for biomolecules. Springer International Publishing, pp 1–71. https://doi.org/10.1007/978-3-030-85603-8_1

Abubakar AM, Silas K, Aji MM (2022) An elaborate breakdown of the essentials of biogas production. J Eng Res Sci 1(4):93–118

Adebami GE, Kuila A, Ajunwa OM, Fasiku SA, Asemoloye MD (2022) Genetics and metabolic engineering of yeast strains for efficient ethanol production. J Food Proc Eng 45(7):e13798. https://doi.org/10.1111/jfpe.13798

Adnan M, Liu G (2024) Promoters and synthetic promoters in *Trichoderma reesei*. In: Marchisio MA (ed) Synthetic promoters: methods and protocols. Springer US, pp 47–68. https://doi.org/10.1007/978-1-0716-4063-0_3

Agrawal D, Saggu SK, Devi S, Kumar S, Kumar M (2025) Novel approaches to alleviate lignocellulosic biomass waste for effective and sustainable utilization. In: Mukherjee G, Dhiman S (eds) Value addition and utilization of lignocellulosic biomass: through novel technological interventions. Springer Nature, pp 89–116. https://doi.org/10.1007/978-981-96-2786-8_4

Agudelo-Patiño T, Ortiz-Sánchez M, Alzate CAC (2024) Prefeasibility analysis of different anaerobic digestion upgrading pathways using organic kitchen food waste as raw material. Fermentation 10(6):300. https://doi.org/10.3390/fermentation10060300

Ajeje SB, Hu Y, Song G, Peter SB, Afful RG, Sun F, Asadollahi MA, Amiri H, Abdulkhani A, Sun H (2021) Thermostable cellulases / xylanases from thermophilic and hyperthermophilic microorganisms: current perspective. Front Bioeng Biotechnol 9. https://doi.org/10.3389/fbioe.2021.794304

Akbar Aly AB, Shanmugaraj B, Ramalingam S (2025) Industrial applications of *Phanerochaete chrysosporium* lignin-degrading enzymes: current status, production challenges, and future directions. World J Microbiol Biotechnol 41(5):171. https://doi.org/10.1007/s11274-025-04388-6

Alcocer-García H, Sánchez-Ramírez E, García-García E, Ramírez-Márquez C, Ponce-Ortega JM (2025) Unlocking the potential of biomass resources: a review on sustainable process design and intensification. Resources 14(9):143. https://doi.org/10.3390/resources14090143

Asghar A, Sairash S, Hussain N, Baqar Z, Sumrin A, Bilal M (2022) Current challenges of biomass refinery and prospects of emerging technologies for sustainable bioproducts and bioeconomy. Biofuels Bioprod Biorefin 16(6):1478–1494. https://doi.org/10.1002/bbb.2403

Benites-Pariente JS, Samolski I, Ludeña Y, Villena GK (2024) CRISPR/Cas9 mediated targeted knock-in of eglA gene to improve endoglucanase activity of *Aspergillus fumigatus* LMB-35Aa. Sci Rep 14(1):19661. https://doi.org/10.1038/s41598-024-70397-4

Bhagat NR, Kumar S, Kumari R, Bharti VK (2023) A review on rumen anaerobic fungi: current understanding on carbohydrate fermentation and roughages digestion in ruminants. Appl Biochem Microbiol 59(3):231–249. https://doi.org/10.1134/S0003683823030043

Bhardwaj N, Agrawal K, Verma P (2021) Xylanases: an overview of its diverse function in the field of biorefinery. In: Srivastava M, Srivastava N, Singh R (eds) Bioenergy research: commercial opportunities & challenges. Springer, pp 295–317. https://doi.org/10.1007/978-981-16-1190-2_10

Carrillo-Nieves D, Saldarriaga-Hernandez S, Gutiérrez-Soto G, Rostro-Alanis M, Hernández-Luna C, Alvarez AJ, Iqbal HMN, Parra-Saldívar R (2022) Biotransformation of agro-industrial waste to produce lignocellulolytic enzymes and bioethanol with a zero waste. Biomass Convers Biorefinery 12(2):253–264. https://doi.org/10.1007/s13399-020-00738-6

Cernak P, Estrela R, Poddar S, Skerker JM, Cheng YF, Carlson AK, Cate JHD (2018) Engineering Kluyveromyces marxianus as a robust synthetic biology platform host. mBio 9:e01410–18

Chauhan A, Islam F, Imran A, Ikram A, Zahoor T, Khurshid S, Shah MA (2023) A review on waste valorization, biotechnological utilization, and management of potato. Food Sci Nutr 11(10):5773–5785. https://doi.org/10.1002/fsn3.3546

Chen N, Chang B, Shi N, Yan W, Lu F, Liu F (2023a) Cross-linked enzyme aggregates immobilization: preparation, characterization, and applications. Crit Rev Biotechnol 43(3):369–383. https://doi.org/10.1080/07388551.2022.2038073

Chen Y, Gao Y, Wang Z, Peng N, Ran X, Chen T, Liu L, Li Y (2023b) The influence of trctf1 gene knockout by CRISPR–Cas9 on cellulase synthesis by *Trichoderma reesei* with various soluble inducers. Fermentation 9(8):746. https://doi.org/10.3390/fermentation9080746

Chepyala S, Bathula J, Bodiga S (2024) Myco-degradation of lignocellulosic waste biomass and their applications. In: Valorization of biomass wastes for environmental sustainability: green practices for the rural circular economy. Springer Nature Switzerland, Cham, pp 269–286

Contato AG, Conte-Junior CA (2025) Fungal lytic polysaccharide monooxygenases (LPMOs): functional adaptation and biotechnological perspectives. Eng 6(8):177. https://doi.org/10.3390/eng6080177

Costa CE (2022) Resveratrol de novo biosynthesis from lignocellulosic biomass: Metabolic engineering of thermotolerant robust yeast strains for an integrated and intensified process. https://hdl.handle.net/1822/79030

Costantini A, Califano V (2021) Lipase immobilization in mesoporous silica nanoparticles for biofuel production. Catalysts 11(5):629. https://doi.org/10.3390/catal11050629

Dahiya S, Rapoport A, Singh B (2024) Biotechnological potential of lignocellulosic biomass as substrates for fungal xylanases and its bioconversion into useful products: a review. Fermentation 10(2):82. https://doi.org/10.3390/fermentation10020082

Darwish AMG, Abo Nahas HH, Korra YH, Osman AA, El-Kholy WM, Reyes-Córdova M, Saied EM, Abdel-Azeem AM (2021) Fungal lipases: insights into molecular structures and biotechnological applications in medicine and dairy industry. In: Industrially important fungi for sustainable development: volume 2: bioprospecting for biomolecules. Springer International Publishing, Cham, pp 461–514

Dash A, Banerjee R (2021) Exploring indigenously produced celite-immobilized *Rhizopus oryzae* NRRL 3562-lipase for biodiesel production. Energy 222:119950. https://doi.org/10.1016/j.energy.2021.119950

de Souza TSP, Kawaguti HY (2021) Cellulases, hemicellulases, and pectinases: applications in the food and beverage industry. Food Bioprocess Technol 14(8):1446–1477. https://doi.org/10.1007/s11947-021-02678-z

Della Torre CL, Silva-Lucca RA, da Ferreira RS, Andrade Luz L, Oliva MLV, Kadowaki MK (2023) Correlation of the conformational structure and catalytic activity of the highly thermostable xylanase of *Thermomyces lanuginosus* PC7S1T. Biocatal Biotransformation 41(2):81–92. https://doi.org/10.1080/10242422.2021.1950696

Detain J, Besaury L (2024) Degradation of lignocellulose by different bacterial and fungal cocultures. Curr Res Microb Sci 7:100271. https://doi.org/10.1016/j.crmicr.2024.100271

Devi R, Kapoor S, Thakur R, Sharma E, Tiwari RK, Joshi SJ (2022) Lignocellulolytic enzymes and bioethanol production from spent biomass of edible mushrooms using *Saccharomyces cerevisiae* and *Pachysolen tannophilus*. Biomass Convers Biorefinery. https://doi.org/10.1007/s13399-022-02406-3

Dey P, Rangarajan V, Singh J, Nayak J, Dilip KJ (2022) Current perspective on improved fermentative production and purification of fungal cellulases for successful biorefinery applications: a brief review. Biomass Convers Biorefinery 12(3):967–995. https://doi.org/10.1007/s13399-020-01227-6

Dhevagi P, Ramya A, Priyatharshini S, Geetha Thanuja K, Ambreetha S, Nivetha A (2021) Industrially important fungal enzymes: productions and applications. In: Yadav AN (ed) Recent trends in mycological research: volume 2: environmental and industrial perspective. Springer international publishing, pp 263–309. https://doi.org/10.1007/978-3-030-68260-6_11

Dukare A, Sharma K, Kautkar S, Dhakane-Lad J, Yadav R, Nadanathangam V, Saxena S (2023) Microbial xylanase aided biobleaching effect on multiple components of lignocelluloses biomass based pulp and paper: a review. Nordic Pulp Paper Res J 38(3):459–480. https://doi.org/10.1515/npprj-2023-0005

Dutta N, Usman M, Ashraf MA, Luo G, Gamal El-Din M, Zhang S (2023) Methods to convert lignocellulosic waste into biohydrogen, biogas, bioethanol, biodiesel and value-added chemicals: a review. Environ Chem Lett 21(2):803–820. https://doi.org/10.1007/s10311-022-01511-z

El-Gendi H, Saleh AK, Badierah R, Redwan EM, El-Maradny YA, El-Fakharany EM (2022) A comprehensive insight into fungal enzymes: structure, classification, and their role in mankind's challenges. J Fungi 8(1):23. https://doi.org/10.3390/jof8010023

Elhalis H (2024) Expanding the horizons of *Saccharomyces cerevisiae*: nutrition, oenology, and bioethanol production. Sustainability 16(24):11151. https://doi.org/10.3390/su162411151
Fansuri H, Purwandari U, Putra S, Adhiksana A, Junianto ID, Oktavian R, Cordiner J (2024) A review of the technological aspects and process optimization of bioethanol production from corn stover biomass: Pretreatment process, hydrolysis, fermentation, purification process, and future perspective. Environ Qual Manag 34(2):e22336. https://doi.org/10.1002/tqem.22336
Felipuci JP [UNESP (2024) Cellooligosaccharides and xylooligosaccharides production from sugarcane bagasse and cotton residues using chemical, physical and biological approaches. https://hdl.handle.net/11449/260014
Fernandes T, Osório C, Sousa MJ, Franco-Duarte R (2023) Contributions of adaptive laboratory evolution towards the enhancement of the biotechnological potential of non-conventional Yeast species. J Fungi 9(2):186. https://doi.org/10.3390/jof9020186
Fernández-Sandoval MT, García A, Teymennet-Ramírez KV, Arenas-Olivares DY, Martínez-Morales F, Trejo-Hernández MR (2024) Removal of phenolic inhibitors from lignocellulose hydrolysates using laccases for the production of fuels and chemicals. Biotechnol Prog 40(1):e3406. https://doi.org/10.1002/btpr.3406
Fu R, Han L, Li Q, Li Z, Dai Y, Leng J (2025) Studies on the concerted interaction of microbes in the gastrointestinal tract of ruminants on lignocellulose and its degradation mechanism. Front Microbiol 16. https://doi.org/10.3389/fmicb.2025.1554271
Garg S, Kim M, Romero-Suarez D (2025) Current advancements in fungal engineering technologies for sustainable development goals. Trends Microbiol 33(3):285–301. https://doi.org/10.1016/j.tim.2024.11.001
Gayen D, Chatterjee R, Roy S (2024) A review on environmental impacts of renewable energy for sustainable development. Int J Environ Sci Technol 21(5):5285–5310. https://doi.org/10.1007/s13762-023-05380-z
Ghorai S (2024) Filamentous fungi: the production powerhouses of cellulolytic and hemicellulolytic enzymes. In: biofuels. CRC Press
Gupta A, Tiwari A, Ghosh P, Arora K, Sharma S (2023) Enhanced lignin degradation of paddy straw and pine needle biomass by combinatorial approach of chemical treatment and fungal enzymes for pulp making. Bioresour Technol 368:128314. https://doi.org/10.1016/j.biortech.2022.128314
Hage H, Rosso M-N (2021) Evolution of fungal carbohydrate-active enzyme portfolios and adaptation to plant cell-wall polymers. J Fungi 7(3):185. https://doi.org/10.3390/jof7030185
Haile S, Ayele A (2022) Pectinase from microorganisms and its industrial applications. Sci World J 2022(1):1881305. https://doi.org/10.1155/2022/1881305
Hassan EM, Zhu X, de Souza EF, Elnashar M, Lu F (2025) Enzyme immobilization advances: a key to unlocking renewable bioenergy potential. Green Chem 27(37):11289–11311. https://doi.org/10.1039/D5GC03388H
He Y, Li K, Wang J, Xu L, Yan J, Yang M, Yan Y (2022) A novel strategy for biodiesel production by combination of liquid lipase, deep eutectic solvent and ultrasonic-assistance in scaled-up reactor: optimization and kinetics. J Clean Prod 372:133740. https://doi.org/10.1016/j.jclepro.2022.133740
Hidayatullah IM, Soetandar F, Sudiyasa PV, Cognet P, Hermansyah H (2023) Ion exchange resin and entrapped candida rugosa lipase for biodiesel synthesis in the recirculating packed-bed reactor: a performance comparison of heterogeneous catalysts. Energies 16(12):4765. https://doi.org/10.3390/en16124765
Huang C, Li R, Tang W, Zheng Y, Meng X (2022) Improve enzymatic hydrolysis of lignocellulosic biomass by modifying lignin structure via sulfite pretreatment and using lignin blockers. Fermentation 8(10):558. https://doi.org/10.3390/fermentation8100558
Ilić N, Milić M, Beluhan S, Dimitrijević-Branković S (2023) Cellulases: from lignocellulosic biomass to improved production. Energies 16(8):3598. https://doi.org/10.3390/en16083598
Ipsen JØ, Hallas-Møller M, Brander S, Lo Leggio L, Johansen KS (2021) Lytic polysaccharide monooxygenases and other histidine-brace copper proteins: structure, oxygen activation and

biotechnological applications. Biochem Soc Trans 49(1):531–540. https://doi.org/10.1042/BST20201031

Jain R, Nigam H, Kalia S, Chauhan N, Gola D, Tyagi P, Gola S, Arya A (2021) Halophilic fungal communities: current research and future challenges. In: Yadav AN (ed) Recent trends in mycological research: volume 1: agricultural and medical perspective. Springer international publishing, pp 203–218. https://doi.org/10.1007/978-3-030-60659-6_9

Jain P, Pundir RK, Yogi R, Gupta MK (2025) New insights into lignocellulosic feedstock conversion into bioethanol: biotechnological perspectives. In: Kapoor RK, Rajan K (eds) Lignocellulosic biomass and enzymes: fundamentals, emerging technologies and applications. Springer Nature, pp 335–359. https://doi.org/10.1007/978-981-96-3037-0_15

Jin F-J, Hu S, Wang B-T, Jin L (2021) Advances in genetic engineering technology and its application in the industrial fungus *Aspergillus oryz*ae. Front Microbiol 12. https://doi.org/10.3389/fmicb.2021.644404

Kandpal V, Jaswal A, Santibanez Gonzalez EDR, Agarwal N (2024) Environmental impact assessment and sustainable energy transition. In: Kandpal V, Jaswal A, Santibanez Gonzalez EDR, Agarwal N (eds) Sustainable energy transition: circular economy and sustainable financing for environmental, social and governance (ESG) practices, pp 273–288). Springer nature switzerland. https://doi.org/10.1007/978-3-031-52943-6_9

Karimi F, Mazaheri D, Saei Moghaddam M, Mataei Moghaddam A, Sanati AL, Orooji Y (2021) Solid-state fermentation as an alternative technology for cost-effective production of bioethanol as useful renewable energy: a review. Biomass Convers Biorefinery. https://doi.org/10.1007/s13399-021-01875-2

Katarina M, Ilić N, Milić M (2024) The realm of smart biomass degrading enzymes in low-carbon fuels and chemicals production. In: Chandel AK (ed) Biorefinery and industry 4.0: empowering sustainability, pp 113–179). Springer nature switzerland. https://doi.org/10.1007/978-3-031-51601-6_5

Khan MF (2025) Fungi for sustainable pharmaceutical remediation: enzymatic innovations, challenges, and applications—a review. Processes 13(4):1034. https://doi.org/10.3390/pr13041034

Konan D, Ndao A, Koffi E, Elkoun S, Robert M, Rodrigue D, Adjallé K (2024) Biodecomposition with *Phanerochaete chrysosporium*: a review. AIMS Microbiol 10(4):1068–1101. https://doi.org/10.3934/microbiol.2024046

Kong W, Wang S, Zhang X, Fu X, Zhang W (2023) Evaluation of biological pretreatment of wormwood rod reies with white rot fungi for preparation of porous carbon. J fungi 9(1):43. https://doi.org/10.3390/jof9010043

Kongkuah M, Alessa N (2025) Renewable energy and carbon intensity: global evidence from 184 countries (2000–2020). Energies 18(13):3236. https://doi.org/10.3390/en18133236

Kumar Chaurasia P, Lata Bharati S, Singh S, Madhavan Sivalingam A, Shankar S, Mani A (2025) Fungal pretreatment methods for organic wastes: advances and challenges in biomass valorization. RSC Sustain 3(3):1234–1266. https://doi.org/10.1039/D4SU00582A

Kumar A, Verma V, Dubey VK, Srivastava A, Garg SK, Singh VP, Arora PK (2023) Industrial applications of fungal lipases: a review. Front Microbiol 14. https://doi.org/10.3389/fmicb.2023.1142536

Kumar A, Singh A, Sharma VK, Goel A, Kumar A (2024) The upsurge of lytic polysaccharide monooxygenases in biomass deconstruction: characteristic functions and sustainable applications. FEBS J 291(23):5081–5101

Lee D, Pan JH, Kim D, Heo W, Shin EC, Kim YJ et al (2024) Mycoproteins and their health-promoting properties: fusarium species and beyond. Compr Rev Food Sci Food Saf 23(3):e13365

Leonel LV, Arruda PV, Chandel AK, Felipe MGA, Sene L (2021) Kluyveromyces marxianus: a potential biocatalyst of renewable chemicals and lignocellulosic ethanol production. Crit Rev Biotechnol 41(8):1131–1152. https://doi.org/10.1080/07388551.2021.1917505

Li N, Li J, Chen Y, Shen Y, Wei D, Wang W (2023) Mechanism of Zn^{2+} regulation of cellulase production in *Trichoderma reesei* Rut-C30. Biotechnol Biofuels Bioprod 16(1):73. https://doi.org/10.1186/s13068-023-02323-1

Li J, Peng C, Mao A, Zhong M, Hu Z (2024) An overview of microbial enzymatic approaches for pectin degradation. Int J Biol Macromol 254:127804. https://doi.org/10.1016/j.ijbiomac.2023.127804

Li Q, Zhang C, Li J, Du G, Li Z, Zhou J, Zhang G (2025) Characterization of *Aspergillus oryzae* mutant and its application in heterologous lipase expression. Synthet Syst Biotechnol 10(2):365–372. https://doi.org/10.1016/j.synbio.2024.12.003

Liu W, Si Q, Sun L, Wang Z, Liu M, Du S, Ge G, Jia Y (2023) Effects of cellulase and xylanase addition on fermentation quality, aerobic stability, and bacteria composition of low water-soluble carbohydrates oat silage. Fermentation 9(7):638. https://doi.org/10.3390/fermentation9070638

Long B, Zhang F, Dai SY, Foston M, Tang YJ, Yuan JS (2025) Engineering strategies to optimize lignocellulosic biorefineries. Nat Rev Bioeng 3(3):230–244. https://doi.org/10.1038/s44222-024-00247-5

Lv X, Qiao Z, Chen C, Hua J, Zhou C (2025) Exploration of multi-source lignocellulose-degrading microbial resources and bioaugmentation strategies: implications for rumen efficiency. Animals 15(13):1920. https://doi.org/10.3390/ani15131920

Madeira-Jr, JV, Gombert AK (2018) Towards high-temperature fuel ethanol production using Kluyveromyces marxianus: on the search for plug-in strains for the Brazilian sugarcane-based biorefinery. Biomass and Bioenergy 119:217–228

Mahato N, Sharma K, Sinha M, Dhyani A, Pathak B, Jang H, Park S, Pashikanti S, Cho S (2021) Biotransformation of citrus waste-I: production of biofuel and valuable compounds by fermentation. Processes 9(2):220. https://doi.org/10.3390/pr9020220

Mariño MA, Fulaz S, Tasic L (2021) Magnetic nanomaterials as biocatalyst carriers for biomass processing: immobilization strategies, reusability, and applications. Magnetochemistry 7(10):133. https://doi.org/10.3390/magnetochemistry7100133

Mehta A, Guleria S, Sharma R, Gupta R (2021) 6—The lipases and their applications with emphasis on food industry. In: Ray RC (ed) Microbial biotechnology in food and health. Academic Press, pp 143–164. https://doi.org/10.1016/B978-0-12-819813-1.00006-2

Milessi TS, Sandri JP, Arruda PV, Esteves TD, Pinheiro LP, Kumar V, Chandel AK (2025) Role of non-genetically modified or native pentose fermenting microorganisms in establishing viable lignocellulosic biorefineries in the Brazilian context. Crit Rev Biotechnol 45(6):1287–1305. https://doi.org/10.1080/07388551.2025.2452628

Mohamed H, Shah AM, Song Y (2021) Conversion of agro-industrial wastes into value-added products. In: Conversion of renewable biomass into bioproducts, vol 1392. American Chemical Society, pp 197–217. https://doi.org/10.1021/bk-2021-1392.ch010

Mohedano MT, Konzock O, Chen Y (2022) Strategies to increase tolerance and robustness of industrial microorganisms. Synthet Syst Biotechnol 7(1):533–540. https://doi.org/10.1016/j.synbio.2021.12.009

Mondal S, Halder SK, Mondal KC (2022) Tailoring in fungi for next generation cellulase production with special reference to CRISPR/CAS system. Syst Microbiol Biomanuf 2(1):113–129. https://doi.org/10.1007/s43393-021-00045-9

Mwaheb MA, El-Aziz BMA, Abd-Elhalim BT, El-Kassim NA, Radwan TEE (2025) Study of different cultivated plants rhizosphere soil fungi-mediated pectinase: insights into production, optimization, purification, biocompatibility, and application. Microb Ecol 87(1):165. https://doi.org/10.1007/s00248-024-02474-0

Naik B, Kumar V, Rizwanuddin S, Chauhan M, Gupta AK, Rustagi S, Kumar V, Gupta S (2023) Agro-industrial waste: a cost-effective and eco-friendly substrate to produce amylase. Food Prod Process Nutr 5(1):30. https://doi.org/10.1186/s43014-023-00143-2

Najim AA, Radeef AY, al-Doori I, Jabbar ZH (2024) Immobilization: the promising technique to protect and increase the efficiency of microorganisms to remove contaminants. J Chem Technol Biotechnol 99(8):1707–1733

Nanda S, Dalai AK (2023) Biomass to bioenergy: modern technological strategies for biorefineries. Elsevier

Nargotra P, Sharma V, Lee Y-C, Tsai Y-H, Liu Y-C, Shieh C-J, Tsai M-L, Dong C-D, Kuo C-H (2023) Microbial lignocellulolytic enzymes for the effective valorization of lignocellulosic biomass: a review. Catalysts 13(1):83. https://doi.org/10.3390/catal13010083

Ndochinwa OG, Wang Q-Y, Amadi OC, Nwagu TN, Nnamchi CI, Okeke ES, Moneke AN (2024) Current status and emerging frontiers in enzyme engineering: an industrial perspective. Heliyon 10(11). https://doi.org/10.1016/j.heliyon.2024.e32673

Ochoa-Chacón A, Martinez A, Poggi-Varaldo HM, Villa-Tanaca L, Ramos-Valdivia AC, Ponce-Noyola T (2022) Xylose metabolism in bioethanol production: saccharomyces cerevisiae vs non-saccharomyces yeasts. Bioenergy Res 15(2):905–923. https://doi.org/10.1007/s12155-021-10340-x

Ojo AO (2023) An overview of lignocellulose and its biotechnological importance in high-value product production. Fermentation 9(11):990. https://doi.org/10.3390/fermentation9110990

Pabbathi NPP, Velidandi A, Tavarna T, Gupta S, Raj RS, Gandam PK, Baadhe RR (2023) Role of metagenomics in prospecting novel endoglucanases, accentuating functional metagenomics approach in second-generation biofuel production: a review. Biomass Convers Biorefinery 13(2):1371–1398. https://doi.org/10.1007/s13399-020-01186-y

Pabbathi NPP, Velidandi A, Pogula S, Gandam PK, Baadhe RR (2025) Advancements in metabolic engineering: enhancing biofuel production through *Escherichia coli* and *Saccharomyces cerevisiae* models. Processes 13(7):2115. https://doi.org/10.3390/pr13072115

Passadis K, Christianides D, Malamis D, Barampouti EM, Mai S (2022) Valorisation of source-separated food waste to bioethanol: pilot-scale demonstration. Biomass Convers Biorefinery 12(10):4599–4609. https://doi.org/10.1007/s13399-022-02732-6

Pimentel PSS-R, de Oliveira JB, Astolfi-Filho S, Pereira N (2021) Enzymatic hydrolysis of lignocellulosic biomass using an optimized enzymatic cocktail prepared from secretomes of filamentous fungi isolated from amazonian biodiversity. Appl Biochem Biotechnol 193(12):3915–3935. https://doi.org/10.1007/s12010-021-03642-5

Ploessl D, Zhao Y, Shao Z (2023) Engineering of non-model eukaryotes for bioenergy and biochemical production. Curr Opin Biotechnol 79:102869. https://doi.org/10.1016/j.copbio.2022.102869

Poddar BJ, Nakhate SP, Gupta RK, Chavan AR, Singh AK, Khardenavis AA, Purohit HJ (2022) A comprehensive review on the pretreatment of lignocellulosic wastes for improved biogas production by anaerobic digestion. Int J Environ Sci Technol 19(4):3429–3456. https://doi.org/10.1007/s13762-021-03248-8

Qin Y, Huang J, Zhou Z, Cui M, Hu H, Yuan M, Yue H, Ma J, Liu G-Q (2025) Genomic and secretomic profiling of *Aspergillus fumigatus* XC6 reveals an expanded lignocellulolytic repertoire and substrate-specific expression patterns for biomass degradation. Ind Crop Prod 233:121469. https://doi.org/10.1016/j.indcrop.2025.121469

Raina N, Chuetor S, Elalami D, Tayibi S, Barakat A (2024) Biomass valorization for bioenergy production: current techniques, challenges, and pathways to solutions for sustainable bioeconomy. Bioenergy Res 17(4):1999–2028. https://doi.org/10.1007/s12155-024-10792-x

Ray RC, Behera SS, Awogbemi O, Sooch BS, Thatoi H, Rath S, Aguilar-Rivera N (2025) Beyond enzymes and organic acids, solid-state fermentation as an alternative for valorizing fruits and vegetable wastes into novel bio-products in a circular economy: a critical review. AIMS Microbiol 11(2):462–500. https://doi.org/10.3934/microbiol.2025021

Ren F, Wu F, Wu X, Bao T, Jie Y, Gao L (2024) Fungal systems for lignocellulose deconstruction: from enzymatic mechanisms to hydrolysis optimization. GCB Bioenergy 16(5):e13130. https://doi.org/10.1111/gcbb.13130

Rothschild-Mancinelli KR (2021) An integrative approach to highthroughput crispr genetic engineering in filamentous fungi. dtu bioengineering

Rout M, Sardar B, Singh PK, Pattnaik R, Mishra S (2022) Utilization of microbial potential for bioethanol production from lignocellulosic waste. In: Biotechnology for zero waste. Wiley, pp 263–282. https://doi.org/10.1002/9783527832064.ch17

Saharan V, Tushir S, Yadav M, Dhaka N, Kapoor RK (2025) Optimization of cellulase enzyme system for lignocellulosic biomass hydrolysis: challenges and future perspectives. In: Kapoor RK, Rajan K (eds) Lignocellulosic biomass and enzymes: fundamentals, emerging technologies and applications. Springer Nature, pp 127–154. https://doi.org/10.1007/978-981-96-3037-0_6

Saye LMG, Navaratna TA, Chong JPJ, O'Malley MA, Theodorou MK, Reilly M (2021) The anaerobic fungi: challenges and opportunities for industrial lignocellulosic biofuel production. Microorganisms 9(4):694. https://doi.org/10.3390/microorganisms9040694

Schwan RF, Joshi VK, Dias DR (2024) Bioreactor Technology in Food Processing. CRC Press

Sethupathy S, Morales GM, Li Y, Wang Y, Jiang J, Sun J, Zhu D (2021) Harnessing microbial wealth for lignocellulose biomass valorization through secretomics: a review. Biotechnol Biofuels 14(1):154. https://doi.org/10.1186/s13068-021-02006-9

Shankar A, Saini S, Sharma KK (2024) Fungal-integrated second-generation lignocellulosic biorefinery: utilization of agricultural biomass for co-production of lignocellulolytic enzymes, mushroom, fungal polysaccharides, and bioethanol. Biomass Convers Biorefinery 14(1):1117–1131. https://doi.org/10.1007/s13399-022-02969-1

Sharif S, Shah AH, Fariq A, Jannat S, Rasheed S, Yasmin A (2023) Optimization of amylase production using response surface methodology from newly isolated thermophilic bacteria. Heliyon 9(1). https://doi.org/10.1016/j.heliyon.2023.e12901

Sharma N, Kapoor N, Kohli SK, Sharma P, Kour J, Devi K, Sharma A, Kaur R, Singh AP, Bhardwaj R (2022) Kitchen and agri waste as renewable, clean and alternative bioenergy resource. In: Agricultural and Kitchen Waste. CRC Press

Sharma S, Tsai M-L, Sharma V, Sun P-P, Nargotra P, Bajaj BK, Chen C-W, Dong C-D (2023) Environment friendly pretreatment approaches for the bioconversion of lignocellulosic biomass into biofuels and value-added products. Environments 10(1):6. https://doi.org/10.3390/environments10010006

Shi K, Liu Y, Chen P, Li Y (2021) Contribution of lignin peroxidase, manganese peroxidase, and laccase in lignite degradation by mixed white-rot fungi. Waste Biomass Valoriz 12(7):3753–3763. https://doi.org/10.1007/s12649-020-01275-z

Shrestha S, Rahman MS, Qin W (2021) New insights in pectinase production development and industrial applications. Appl Microbiol Biotechnol 105(24):9069–9087. https://doi.org/10.1007/s00253-021-11705-0

Shrivastava A, Sharma RK (2023) Unraveling non-conventional *Yeast pichi*a: an emerging lignocellulosic ethanologenic and exoelectrogenic yeast. Bioenergy Res 16(3):1318–1334. https://doi.org/10.1007/s12155-023-10609-3

Shukla A, Kumar D, Girdhar M, Kumar A, Goyal A, Malik T, Mohan A (2023) Strategies of pretreatment of feedstocks for optimized bioethanol production: distinct and integrated approaches. Biotechnol Biofuels Bioprod 16(1):44. https://doi.org/10.1186/s13068-023-02295-2

Singh S (2021) Energy crisis and climate change. In: Energy. Wiley, Ltd, pp 1–17. https://doi.org/10.1002/9781119741503.ch1

Singh B, Soni SK, Mathur P, Garg N (2024) Microbial Multienzyme Viz., Pectinase, cellulase and amylase production using fruit and vegetable waste as substrate—a review. Appl Microbiol 4(3):1232–1246. https://doi.org/10.3390/applmicrobiol4030084

Singh R, Gaur A, Soni P, Jain R, Pant G, Kumar D, Kumar G, Shamshuddin S, Mubarak NM, Dehghani MH, Suhas, Ansari K (2025) A review of biofuels and bioenergy production as a sustainable alternative: opportunities, challenges and future perspectives. J Environ Health Sci Eng 23(2):23. https://doi.org/10.1007/s40201-025-00946-0

Sohail M, Barzkar N, Michaud P, Tamadoni Jahromi S, Babich O, Sukhikh S, Das R, Nahavandi R (2022) Cellulolytic and xylanolytic enzymes from yeasts: properties and industrial applications. Molecules 27(12):3783. https://doi.org/10.3390/molecules27123783

Sokal S, Palsania P, Kaushik G (2022) Bioremediation and functional metagenomics: advances, challenges, and opportunities. In: Kumar V, Thakur IS (eds) Omics insights in environmental bioremediation. Springer Nature, pp 3–36. https://doi.org/10.1007/978-981-19-4320-1_1

Sperandio GB, Filho EXF (2021) An overview of *Trichoderma reesei* co-cultures for the production of lignocellulolytic enzymes. Appl Microbiol Biotechnol 105(8):3019–3025. https://doi.org/10.1007/s00253-021-11261-7

Thakur R (2025) Bioenergy production from algae and microbial systems. In: Algal bioengineering and microbial synergy to green remediation. Springer Nature, pp 231–256. https://doi.org/10.1007/978-981-96-8054-2_10

Torres-Farradá G, Thijs S, Rineau F, Guerra G, Vangronsveld J (2024) White rot fungi as tools for the bioremediation of xenobiotics: a review. J Fungi 10(3):167. https://doi.org/10.3390/jof10030167

Ullah M, Liu P, Xie S, Sun S (2022) Recent advancements and challenges in lignin valorization: green routes towards sustainable bioproducts. Molecules 27(18):6055. https://doi.org/10.3390/molecules27186055

Vasileiadou A (2024) From organic wastes to bioenergy, biofuels, and value-added products for urban sustainability and circular economy: a review. Urban Sci 8(3):121. https://doi.org/10.3390/urbansci8030121

Vasylyshyn R, Ruchala J, Dmytruk K, Sibirny A (2025) Recent progress in engineering yeast producers of cellulosic ethanol. FEMS Yeast Res 25:foaf035. https://doi.org/10.1093/femsyr/foaf035

Vieira RIM, da Peixoto AS, Monclaro AV, Ricart CAO, Filho EXF, Miller RNG, Gomes TG (2025) Fungal coculture: unlocking the potential for efficient bioconversion of lignocellulosic biomass. J Fungi 11(6):458. https://doi.org/10.3390/jof11060458

Wan Mohtar WHM, Wan-Mohtar WAAQI, Zahuri AA, Ibrahim MF, Show P-L, Ilham Z, Jamaludin AA, Abdul Patah MF, Ahmad Usuldin SR, Rowan N (2022) Role of ascomycete and basidiomycete fungi in meeting established and emerging sustainability opportunities: a review. Bioengineered 13(7–12):14903–14935. https://doi.org/10.1080/21655979.2023.2184785

Wei H, Wang Q, Zhang R, Liu M, Zhang W (2023) Efficient biodiesel production from waste cooking oil by fast co-immobilization of lipases from *Aspergillus oryzae* and *Rhizomucor miehei* in magnetic chitosan microcapsules. Process Biochem 125:171–180. https://doi.org/10.1016/j.procbio.2022.12.025

Wilken SE (2020) Developing a systems biology framework to engineer anaerobic gut fungi [UC Santa Barbara]. https://escholarship.org/uc/item/9t3785b5

Wu Y, Jameel A, Xing X-H, Zhang C (2022) Advanced strategies and tools to facilitate and streamline microbial adaptive laboratory evolution. Trends Biotechnol 40(1):38–59. https://doi.org/10.1016/j.tibtech.2021.04.002

Wunderlich G, Bull M, Ross T, Rose M, Chapman B (2023) Understanding the microbial fibre degrading communities & processes in the equine gut. Anim Microbiome 5(1):3. https://doi.org/10.1186/s42523-022-00224-6

Xia S, Lin J, Sayanjali S, Shen C, Cheong L-Z (2024) Lipase-catalyzed production of biodiesel: a critical review on feedstock, enzyme carrier and process factors. Biofuels Bioprod Biorefin 18(1):291–309. https://doi.org/10.1002/bbb.2561

Xiang X, Zhang Y, Wei W, Zhou G (2025) Latest advances in research methods for high-yield cellulase production. Sustain Energy Fuels 9(7):1617–1632. https://doi.org/10.1039/D4SE01553C

Xu S, Gao M, Peng Z, Sui K, Li Y, Li C (2022) Upcycling from chitin-waste biomass into bioethanol and mushroom via solid-state fermentation with *Pleurotus ostreatus*. Fuel 326:125061. https://doi.org/10.1016/j.fuel.2022.125061

Xue Y, Shen R, Li Y, Sun Z, Sun X, Li F, Li X, Cheng Y, Zhu W (2022) Anaerobic fungi isolated from bactrian camel rumen contents have strong lignocellulosic bioconversion potential. Front Microbiol 13. https://doi.org/10.3389/fmicb.2022.888964

Yadav J, Marwah H, Kumar C (2025) Synthetic biology and metabolic engineering paving the way for sustainable next-gen biofuels: a comprehensive review. Energy Adv. https://doi.org/10.1039/D5YA00118H

Yan S, Xu Y, Yu X-W (2021b) Rational engineering of xylanase hyper-producing system in Trichoderma reesei for efficient biomass degradation. Biotechnol Biofuels 14(1):90. https://doi.org/10.1186/s13068-021-01943-9

Yang X, Zhang Y, Pang H, Yuan S, Wang X, Hu Z, Zhou Q, He Y, Yan Y, Xu L (2021) Codisplay of *Rhizopus oryzae* and *Candida rugosa* lipases for biodiesel production. Catalysts 11(4):421. https://doi.org/10.3390/catal11040421

Yanieva OD (2022) Pentose-fermenting yeasts in nature: ecology, biodiversity, and applications. Microbiol J/Mikrobiol Z 84(5)

Yovchevska L, Gocheva Y, Stoyancheva G, Miteva-Staleva J, Dishliyska V, Abrashev R, Stamenova T, Angelova M, Krumova E (2025) Halophilic fungi—features and potential applications. Microorganisms 13(1):175. https://doi.org/10.3390/microorganisms13010175

Yu D, Zhang X, Wang T, Geng H, Wang L, Jiang L, Elfalleh W (2021) Immobilized *Candida Antarctica* lipase B (CALB) on functionalized MCM-41: stability and catalysis of transesterification of soybean oil and phytosterol. Food Biosci 40:100906. https://doi.org/10.1016/j.fbio.2021.100906

Zanuso E, Gomes DG, Ruiz HA, Teixeira JA, Domingues L (2021) Enzyme immobilization as a strategy towards efficient and sustainable lignocellulosic biomass conversion into chemicals and biofuels: current status and perspectives. Sustain Energy Fuels 5(17):4233–4247. https://doi.org/10.1039/D1SE00747E

Zhang C (2022) Development of an efficient c-to-t base-editing system and its application to cellulase transcription factor precise engineering in thermophilic fungus Myceliophthora thermophila. American Society for Microbiology

Zhang H, Feng M, Fang Y, Wu Y, Liu Y, Zhao Y, Xu J (2023a) Recent advancements in encapsulation of chitosan-based enzymes and their applications in food industry. Crit Rev Food Sci Nutr 63(32):11044–11062. https://doi.org/10.1080/10408398.2022.2086851

Zhang Y, Sun T, Wu T, Li J, Hu D, Liu D, Li J, Tian C (2023b) Consolidated bioprocessing for bioethanol production by metabolically engineered cellulolytic fungus *Myceliophthora thermophila*. Metab Eng 78:192–199. https://doi.org/10.1016/j.ymben.2023.06.009

Zhao M, Chen J, Pan X, Zabed HM, Arsalan A, Qi X (2025) Advances in pectinase engineering for food bioprocessing: novel sources, mechanisms, and optimization strategies. J Agric Food Chem 73(37):23078–23097. https://doi.org/10.1021/acs.jafc.5c06547

Zhu X, Cheng S, Fang Z, Zhen G, Lu X, Liu H, Qi J, Zhou Z, Zhang X, Wu Z (2025) Studies on the treatment of anaerobically digested sludge by white-rot fungi: evaluation of the effect of *Phanerochaete chrysosporium* and *Trametes versicolor*. Microb Cell Factories 24(1):23. https://doi.org/10.1186/s12934-024-02611-x

Chapter 4
Bioengineering of Fungal Biomass for Bioethanol Production: Recent Progress and Challenges

Prakram Singh Chauhan and Gaurav Bhardwaj

Abstract Bioethanol utilization is a global move toward the production of renewable energy and the lessening of fossil fuel dependence. In this chapter, we summarize progress in the nutritional bioengineering of fungal biomass and the corresponding challenges that must be mitigated for industrial bioengineering of fungal biomass to be implemented at a global scale. Targeted genetic modifications, specifically to metabolic pathways of fungi, have been the focus of the last 10 years of research. The use of Clustered Regularly Interspaced Short Palindromic Repeats (CRISPR)-Cas, RNAi, and transcriptional engineering has helped to understand and improve traits such as carbohydrate storage, stress tolerance, and substrate utilization effectiveness. Solid-state fermentation and consolidated bioprocessing technologies are very promising and even more so in combination, since they integrate multiple fermentation steps, lowering the energy input plus the demand for exogenous enzymes. The combination of the physicochemical factors and legislative constraints to field release of genetically altered fungi has a long and complex history, especially in countries with more stringent biosafety legislation. Priorities in advancement will include developing strong fungal chassis that can withstand industrial stressors, enhance pentose transport systems, and eliminate expensive pretreatment requirements. The use of synthetic biology, adaptive and computational modeling will be crucial in engineering advanced strains of microorganisms for commercial bioethanol production.

Keywords Lignocellulose degradation · Fungal bioethanol production · Metabolic engineering · *Saccharomyces cerevisiae* · Filamentous fungi · Synthetic biology

---

P. S. Chauhan (✉) · G. Bhardwaj
Chitkara College of Pharmacy, Chitkara University, Rajpura, Punjab, India
e-mail: prakram.chauhan@chitkara.edu.in

M. Tripathi et al. (eds.), *Recent Advancements in Fungal Biotechnology Volume 1*, Fungal Biology, https://doi.org/10.1007/978-981-92-0339-0_4

4.1 Introduction

The ability of bioethanol to mitigate greenhouse gas emissions and promote energy independence and sustainability has placed it at the forefront of biofuel research and commercialization. Conventional methods for making bioethanol rely on food sources, such as sugarcane and corn. This has spurred studies on agricultural sustainable sources of biomass, low-yielding, and low-input systems. The unique biological features of fungi, such as rapid growth, natural hydrolytic enzyme secretion, carbon assimilation from diverse substrates, and high amounts of intracellular carbohydrates, allow for dual purposes as a catalyst for fermentative bioethanol production and as a biomass source. Filamentous fungi such as *Trichoderma reesei*, *Aspergillus niger*, and *Neurospora crassa* are known for their capacity to degrade lignocellulosic materials because of their secretion of cellulases, hemicellulases, and other accessory enzymes (Sivagurunathan et al. 2022; Arshad et al. 2025). On the other hand, yeast species, especially *Saccharomyces cerevisiae*, serve as the predominant microorganism for ethanol fermentation due to their hardiness, high ethanol titer, and long record of safe use in the industry. But the natural fermentation pathway of yeasts is characterized by a frowned-upon trait; the inability to utilize multiple carbon (especially pentose sugars) substrates, which, in turn, has prompted metabolic engineering aimed at improved substrate utilization (Troiano and Studer 2025). By the adoption of omics tools, including genomics, transcriptomics, proteomics, and metabolomics with these tools, researchers are able to identify and characterize key points of the networks that are important in the regulatory mechanisms of sugar. These advanced tools enable computational models to predict the behavior of engineered Fusarium strains under a defined set of fermentation conditions and inform targeted engineering (Dadheech et al. 2018; Baite et al. 2026). The integration of both empirical and simulated data has benefited the construction of Fusarium strains that have a high wild-type performance for the concurrent conversion of biomass to ethanol. The major cost comes from pre-treatment, since plant cell walls are resistant to enzymatic breakdown. At the same time, the improved efficiencies of new bioprocesses have reduced the costs of the production of ethanol at commercial scales. Fortunately, advancements in pre-treatment technologies have helped to greatly improve sugar yields with less formation of chemical byproducts that inhibit further reactions. These new technologies include the use of ionic liquids, biologically pre-treating with white-rot fungi, mild chemical additions in steam explosions, and custom enzymatic cocktails (Fig. 4.1) (Gladysheva 2025). Beyond pre-treatments, the new process of consolidated bioprocessing (CBP), which focuses on the integration of enzyme production and saccharification with the fermentation step into one step, leads to cost effective. Even with the improvements, many challenges remain. During the fermentative process, it is common for fungal strains to experience high-solid and high-ethanol stress, which leads to growth inhibition and reduced yields. To make the problem worse, crossing the cell membranes with specific pentose sugars (xylose and arabinose) is often the biggest bottleneck. In addition, compounds found in lignocellulosic hydrolysates, such as

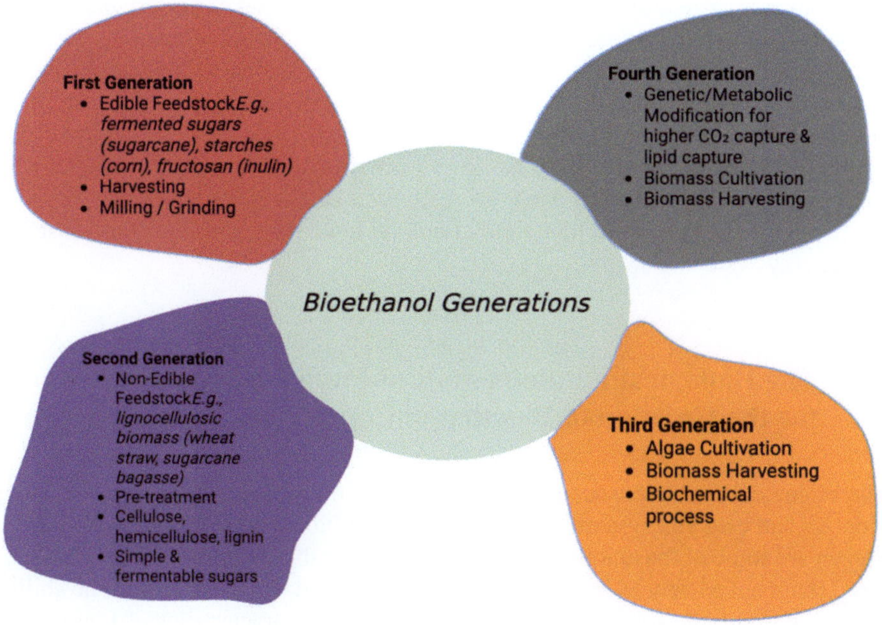

Fig. 4.1 A description of bioethanol's substrates and sequential processes

furans, phenolics, and weak acids, are known to inhibit the fermentative capabilities and viability of fungi. Precision in the integration of metabolic engineering, adaptive evolution, and process optimization is needed to improve the system and overcome the remaining challenges. Although engineered fungal strains often display great potential in laboratory fermenters, industrial bioreactors pose challenges such as temp changes, lack of oxygen, and shear force that tend to show different behaviors. The stability of engineered fungal strains (for commercial purposes) must be assessed in order to determine economic potential, and industrial bioreactors represent a big part of that (Cao et al. 2026). Also, the Dual Use Concerns and Regulations targeting (Genetically Engineered and Modified (GEM)) fungi are still under development and are implemented differently in different regions of the world, which contributes to the limited potential of engineered fungal systems.

For these reasons, this chapter aims to summarize the progress made and setbacks regarding the engineering of fungi for the purpose of biomass bioethanol production. The chapter will introduce the biological characteristics of the fungi that dictate the conversion of biomass to bioethanol (Prakash et al. 2018; de los Santos Mondragón et al. 2023; Benali et al. 2026). These are cell walls and their arrays, streamlined metabolism, and innate polymer-degrading abilities (natural enzymatic activities). Advances in bioprocess technologies that integrate fermentations and biorefineries are also discussed. The chapter also covers the challenges associated with industrial scaling of GEM, the transversal and intertwined nature of environmental regulations, and the closure of bioprocesses. The last part of the chapter discusses the future GEM fungi and the main fungal chassis to be engineered with synthetic biology, which are also provided, along with the multi-omics approach to strain construction and decentralized bioethanol production. It is designed to assist and guide researchers, practitioners, and people in the industrial field toward the commercialization and the sustainable long-term use of biofuels. Biomass, particularly fungal biomass, can become the foundation for ongoing development and interdisciplinary collaboration toward an innovative and sustainable energy system.

4.2 The Biological Fundamentals of Fungal Biomass for the Creation of Bioethanol Concise

The biological characteristics of fungi help explain the potential of fungi for both biomass and catalytic systems in the production of bioethanol (Ntunka et al. 2025). Shown are some of the unique physiological and metabolic characteristics that help fungi stand out compared to other microorganisms used in bioprocesses. The ability to grow on multiple substrates and the physiological characteristics of fungi, such as the synthesis of hydrolytic enzymes and the storage of carbohydrates, make fungi such an appealing candidate in biorefinery systems. This section on fermentation and bioethanol production deals with the physiological characteristics of fungi, such as cell structure, metabolic flexibility, and robust enzyme secretion.

4.2.1 Fungal Diversity and Growth Characteristics

Fungi kingdom is one of the largest taxonomic groups and includes a wide variety of organisms, for example, yeasts, filamentous fungi, and macrofungi. Yeasts, such as *Saccharomyces cerevisiae* and *Pichia stipitis,* and filamentous fungi, such as *Trichoderma reesei, Aspergillus niger,* and *Rhizopus oryzae,* are among the most used organisms for ethanol fermentation and biomass degradation and enzyme production, respectively. The ecological fitness of fungi kingdom is seen in the different morphological characters of the fungi. Yeast is a unicellular organism and reproduces by budding, which make them suitable for high-density fermentation (Onyema et al. 2023). In contrast, filamentous fungi form large mats of hyphal networks called mycelia, which, in addition to their ability to secrete high amounts of extracellular enzymes, help in the colonization of solid substrates. Fungi are also known for their ability to utilize a variety of carbon substrates such as cellulose, hemicellulose, pectin, starch, simple sugars, and lipids, which is a very important trait. This metabolic trait also allows the cultivation of fungi on cheap substrates such as agricultural residuals and byproducts of the industry. In addition, many fungi are able to store carbohydrates, such as glycogen, mannan, and trehalose, which are important for the fuel (i.e., bioethanol) production from the biomass (Chamoli et al. 2025; Khatami et al. 2026).

4.2.2 Cell Wall Architecture and Its Role in Biomass Processing

Fungi possess a chitin, β-glucans, and polysaccharides cell wall that is a dynamic and complex structure. These vary in composition and, together with polysaccharides and mannoproteins, fasciae their mechanical strength, resilience in the environment, and the secretion of enzymes. Filamentous fungi have porous cell walls. These walls facilitate the diffusion of hydrolytic enzymes to lignocellulosic substrates. This feature is especially important in CBP systems, wherein there is a need for simultaneous secretion of enzymes and substrate degradation (Pérez-Contreras et al. 2024).

4.2.3 Metabolic Competence and Carbon Utilization Pathways

A significant constraint in conventional ethanol production has been the inability of yeasts to assimilate pentose sugars such as arabinose and xylose. However, a number of filamentous fungi can metabolize pentoses through the use of both the oxidoreductase and the isomerase pathways. This metabolic property has been the motivation behind transforming *S. cerevisiae* and other yeasts to assimilate pentoses and enhance ethanol production from lignocellulosic hydrolysates (Kobayashi et al. 2018; Chen et al. 2026). Furthermore, fungi exhibit distinct redox balancing strategies that preserve metabolic functions even in hypoxic conditions. Certain species, for instance, *Rhizopus* spp., can make an anaerobic–aerobic metabolic switch to fermentative ethanol production, organic acid accumulation, or aerobic respiration depending on the prevailing conditions. This metabolic nimbleness is an asset for engineering high-yielding, stress-resistant industrial strains.

4.2.4 Hydrolytic Enzyme Production and Secretion Systems

A characteristic of filamentous fungi is their ability to degrade lignocellulosic biomass. They synthesize a diverse range of extracellular enzymes such as cellulases, pectinases, xylanases, lignin-modifying enzymes, and accessory proteins, including lytic polysaccharide mono oxygenases (LPMOs) (Chamoli et al. 2025). *T. reesei* is the most prolific cellulase producer known, making it a leading organism in industrial enzyme production. The secretion of hydrolytic enzymes is controlled by intricate signaling pathways that take into account the nutrients and substrate composition. When fungi encounter lignocellulosic substrates, they initiate the transcription of specific regulator transcription factors that control the expression of the genes in the enzyme clusters. This coordinated response leads to the secretion of both cellulolytic and hemicellulolytic enzymes. Once researchers studied and gained insights into these mechanisms, they were able to increase enzyme secretion through the selective modification of particular promoter regions, engineering transcription factors, and genome editing (de los Santos Mondragón et al. 2023; Kerkaert and Huberman 2023). The ability to secrete enzymes is determined by the porosity of the cellular wall, the efficiency of endomembrane trafficking, the capability of the cell to post-translationally modify proteins (Knychala et al. 2024), and by the systems that control these mechanisms. With some optimization of these mechanisms, the secretion of enzymes into the medium could be further increased, thereby improving the economic viability of the production of bioethanol from lignocellulosic substrates via fungi. In the wild, fungi face several environmental pressures, including sudden temperature changes, differing osmotic conditions, and oxidative stress. There are also toxic compounds that are produced during biomass pretreatment. The ability of fungi to tolerate these environmental stressors is crucial for the industry to use them for fermentation. Similarly, yeasts such as *S. cerevisiae* possess mechanisms that allow them to tolerate the presence of high concentrations of ethanol. These mechanisms include changes to the lipid composition of the membrane, induction of heat shock proteins, and the production of detoxifying enzymes. However, strains used in the production of lignocellulosic bioethanol must also tolerate the presence of inhibitors such as furfural, hydroxymethylfurfural (HMF), and acetic acid. In addition, during the fermentation, fungi secrete catalases, superoxide dismutase, and glutathione-dependent reductases, which neutralize the damaging reactive oxidative species generated during high-density fermentations. Tolerance to oxidative stress, either through enhancing pre-existing traits through genetic engineering or through adaptive evolution, has been shown to improve ethanol yields. Equally important, the thermotolerant fungi are less susceptible to contamination, operate at higher temperatures, and improve reaction kinetics. *Kluyveromyces marxianus* has natural thermotolerance and is being used more frequently for ethanol production.

4.2.5 *Metabolic Engineering of Fungal Strains for Enhanced Bioethanol Production*

With metabolic engineering, conversion of lignocellulose to bioethanol through fermentation using fungi has become more feasible (van Aalst et al. 2022). Although there is growing interest in using fungi to meet the demands of the industry, which include higher yields and productivities during fermentation along with enhanced tolerances to various stressors, fungi themselves still need to be optimized for these traits (Hamden et al. 2025). To meet these demands, extensive efforts have been directed at redesigning the metabolic networks of fungi, engineering the sugar uptake and utilization pathways, and enhancing the overall tolerance of the organism to the fermentation conditions. In this case, metabolic engineering has been directed to optimization of the desired production pathways, enhancing tolerance to various metabolic and oxidative stressors, improving the utilization of sugars, and engineering fungi using synthetic biology approaches for catalyzing better bioethanol production (Table 4.1).

Table 4.1 Key strategies in metabolic engineering of fungi for enhanced bioethanol production

S. No.	Engineering strategy	Description
1.	Redesigning metabolic networks	Involves restructuring native metabolic pathways to increase flux toward ethanol production, minimize byproduct formation, and improve overall carbon efficiency during lignocellulosic fermentation.
2.	Optimization of sugar uptake pathways	Enhances the transport and assimilation of hexose and pentose sugars (glucose, xylose, and arabinose) by engineering high-affinity sugar transporters and regulatory mechanisms.
3.	Improving sugar utilization pathways	Introduces or modifies metabolic routes enabling fungi to efficiently metabolize pentose sugars, which are abundant in lignocellulosic biomass but poorly utilized by native strains.
4.	Enhancing tolerance to fermentation stressors	Focuses on improving resistance to ethanol toxicity, osmotic stress, inhibitors (furfural, HMF, and phenolics) generated during biomass pretreatment, and oxidative stress encountered during high-productivity fermentation.
5.	Synthetic biology-based engineering	Applies genome editing tools, modular genetic parts, and CRISPR-based approaches to construct robust fungal strains tailored for higher ethanol yields and industrial-scale performance.
6.	Boosting co-factor balance and redox homeostasis	Reengineers Nicotinamide adenine dinucleotide hydrogen (NADH)/Nicotinamide adenine dinucleotide $(NAD)^+$ and Nicotinamide Adenine Dinucleotide Phosphate Hydrogen (NADPH)/Nicotinamide Adenine Dinucleotide Phosphate $(NADP)^+$ pathways to support efficient fermentation, reduce metabolic bottlenecks, and improve tolerance to oxidative imbalances.
7.	Minimizing byproduct formation	Targets pathways responsible for glycerol, organic acids, and other side metabolites, ensuring maximum carbon flux toward ethanol formation.
8.	Integration of lignocellulosic biomass degradation traits	Introduces cellulases, hemicellulases, and accessory enzymes to enable CBP, improving biomass hydrolysis and ethanol synthesis in a single organism.

Reviewing productive pathways of primary carbon metabolism improves ethanol yield. The primary metabolic pathways of all fungi determine how efficiently sugars are converted into ethanol. Engineering the primary pathways to improve ethanol yields requires enhancing the glycolytic flux, balancing the redox cofactors, and reducing the carbon drain from competing pathways (Nielsen and Keasling 2016). Among all ethanol-fermenting microorganisms, fungi, and especially the yeasts, *Saccharomyces cerevisiae*, glycolytic metabolism is the most rapid. The most effective way to improve ethanol yields is to increase the expression of the glycolytic rate-limiting enzymes, hexokinase and phosphofructokinase, and pyruvate decarboxylase (van Aalst et al. 2022). Furthermore, the conversion of ethanol and pyruvate is directly mediated by alcohol dehydrogenase, and, hence, the expression of this enzyme should be increased to improve fermentation efficiency to produce ethanol. These same metabolic flux manipulations have also been performed in filamentous fungi, such as increased ethanol yields by genetically engineered *Rhizopus oryzae* to improve the expression of pyruvate decarboxylase. Concurrently, there are a variety of organic acids, glycerol, or biomass that can be produced from fungal metabolism and, therefore, fungi often divert their metabolism to such byproducts. It is critical to redirect these byproducts. One way to achieve this is by deletion or suppression of genes such as GPD1 in yeast, which is responsible for the biosynthesis of glycerol; however, this comes with trade-offs for redox balance and homeostasis. Within the species of Aspergillus, metabolic flux is frequently diverted into oxalic acid, citric acid, and other such metabolites. These pathways can be partially inhibited to divert more carbon into ethanol and are termed transcriptional regulation and enzyme attenuation (de los Santos Mondragón et al. 2023). Maintaining the intracellular balance of NADH/NAD^+ homeostasis is critical in fungi engineered for pentose fermentation, as redox imbalance occurs due to alterations between endogenous pathways and metabolic routes mismatch. Such imbalances can be mitigated by overexpression of transhydrogenases, engineering other oxidoreductases, and replacing native xylose with cofactor-balanced variants of xylitol dehydrogenase (XDH). These imbalances can be resolved more smoothly through pentose and ethanol production, streamlining these processes to enhance carbon flux overall.

4.2.6 Engineering for Efficient Utilization of Pentose Sugars

Bioethanol production from lignocellulosic biomasses has two drawbacks: one is that *Saccharomyces cerevisiae* and other fungi are not able to metabolize sugars such as xylose and arabinose, and the other is that although filamentous fungi can metabolize, the yield is very low (Huang et al. 2024). To overcome this limitation, metabolic engineering works on two approaches of heterologous pathways, primarily pathways involving oxidoreductase, which include xylose reductase (XR) and XDH, or the pathway involving isomerase, which includes xylose isomerase (XI). The heterologous expression of fungal XR-XDH systems from *Scheffersomyces stipites* has been shown to increase the ability of *S. cerevisiae* to utilize xylose. On the other hand, the XI systems from bacteria have been the choice in many instances due to improved redox balance. In parallel, pentose sugar uptake is significantly slower than that of glucose. It can be enhanced by engineering high-affinity xylose transporters, such as through directed evolution and protein engineering, which has enhanced the rate of pentose uptake. Enhanced sugar transport has been achieved through the use of the fungal transporters Gxf1 from *Candida intermedia* and XylHp from *Aspergillus niger* (de los Santos Mondragón et al. 2023). Furthermore, optimization has addressed downstream metabolic bottlenecks through carbon flux balancing via overexpression of TKL1 and TAL1, cofactor specificity of engineering of XR and XDH, and adaptive evolution for laboratory to improve the yield of ethanol from hydrolysates rich in pentose.

4.2.7 The Engineering of Fungal Cell Factories to Improve the Production of Specific Enzymes

The filamentous fungi used in the enzyme-catalyzed lignocellulose degradation process are critical. The engineering of filamentous fungi focuses on the following: increased enzyme production, stability, and expanded ranges of substrate types that the enzymes can degrade (Borkertas et al. 2025). The traditional approach includes overexpression of key cellulase encoding genes: cbh1 (cellobiohydrolase I), cel7A and cel7B (endoglucanases), and xyn2 (xylanase). Strong promoters such as cbh1 and eno1 in *T. reesei* have been used to drive high levels of protein production. The modern CRISPR-Cas9 approach, which facilitates the overexpression of several genes at once as well as the engineering of various promoters, leads to significant improvements in the enzyme titers. The modification of biosynthetic pathways for the purpose of secretion engineering, while often beneficial, is not always necessary. Some of the following approaches, in isolation and in combinations, will likely yield the desired results: removal of proteases that degrade the target enzymes, overexpression of chaperones and protein-folding factors, engineering the pathways that are responsible for vesicle trafficking in the endomembrane system, or stemming and modifying N-glycosylation in order to glycosylate enzymes that are responsive to inactivation. Several of these approaches are reported to have increased the secretion of enzymes in various fungal strains by two to three times. LPMOs are known to convert glycosidic bonds into oxidized metabolites, which are synergistic with the actions of several cellulases. It has been reported that engineering the expression of LMPOs and or the addition of bacteria-cloned LPMOs to fungal host systems is useful for the enhancement of the degradation of cellulosic substrates, which, in due course, alleviates the need for excess addition of exogenous enzymes in large-scale fermenter systems (Guo et al. 2022).

4.3 The Engineering of Stress Tolerance to Attain Industrial Robustness

The routine fermentation of industrially important fungi subjects the organism to significant combined stress. Engineering fungi for stress tolerance is, therefore, an imperative endeavor. The major stresses include thermal, ethanol, osmotic, and mechanical shear. Methods to achieve engineering ethanol tolerance include: modification of membrane sterol composition to raise fluidity, overexpression of heat shock proteins, such as Hsp70 and Hsp104, and the engineering of efflux pumps to export the toxic compound. Adaptive laboratory evolution has also been successful in generating strains that withstand ethanol levels above 16% v/v. Some of the products of biomass pretreatment include furfural, HMF, acetic acid, and phenolic compounds. The fermentation process can be made more effective by engineering fungi to detoxify these inhibitors, for example, by protein overexpression of some of the following: reductases, dehydrogenases, and transporters. Yeast furfural reduction is enhanced by ADH6 overexpression (Tian et al. 2025). The ability of fungi to tolerate heat (thermotolerance) diminishes the chances of contamination while enabling simultaneous saccharification and fermentation (SSCF) at elevated temperatures. Engineering strategies include modifying the composition of membrane fatty acids, overexpression of thermostable enzymes, and implementation of heat shock factor regulatory circuits, mainly *K. marxianus* is engineered for thermotolerance.

4.4 Systems Biology and Omics-Guided Metabolic Engineering

For the past few years, the engineering of fungi has greatly benefited from the recent advances in synthetic biology in the designing, controlling, and optimizing of biological systems. CRISPR-Cas genome editing, a rapid and accurate modification of fungi, has been made possible by the CRISPR systems. Some of the applications are as follows: gene deletions and insertions, multiplexing of pathway genes during editing, and comprehensive rewiring via engineering of the whole genome (Yadav et al. 2025). As a result, transcription is modulated by CRISPRi and CRISPRa systems, and this is done without the alteration of any DNA sequence. The pathway refactoring and modular engineering by the redesigning of entire metabolic pathways in synthetic biology is an emotional fusion of modular promoter libraries, synthetic terminators, multi-gene expression cassettes, and scaffold proteins to cluster enzymes. All these tools ensure balanced flux and reduced metabolic burden. Moreover, biosensors that are responsive to metabolic intermediates permit dynamic regulation of pathways. Fungal biosensors have been designed for NADH, xylose, and ethanol concentrations. Dynamic regulation yields improvements by metabolic overload avoidance (Raheja et al. 2026).

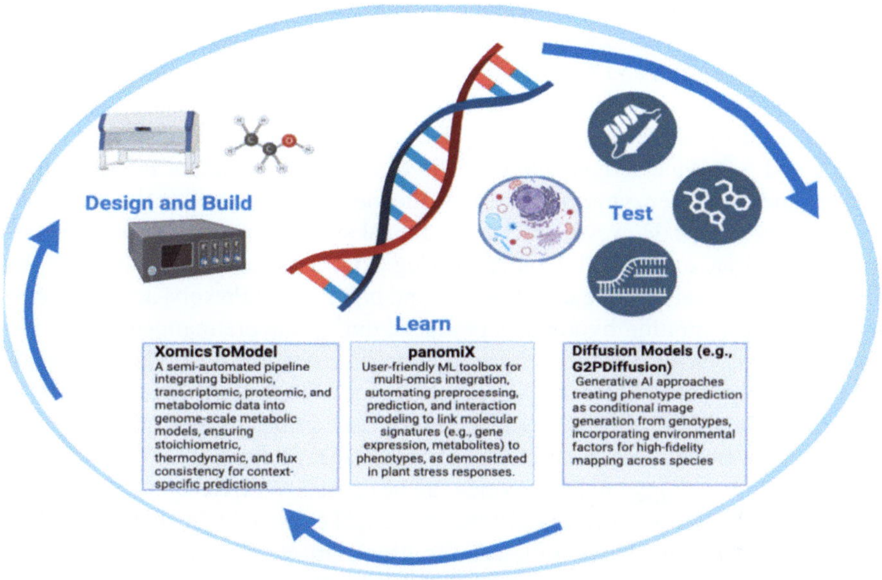

Fig. 4.2 Overview of computational strategies in the learn phase. Omics datasets in test could be interpreted utilizing a range of computational methodologies

Omics technologies have been a game-changer in fungal engineering by enabling data-driven design. The analysis of transcriptomes helps to reveal the genes that are turned on in stress, fermentation, and the secretion of an enzyme. Comparative genomics helps to see the natural diversity of different fungal strains in metabolic pathways. After that, in proteomics and metabolomics, proteomics helps identify the key enzymes and the bottlenecks in the pathways (Yoshikawa et al. 2012). The analysis of metabolomics helps map the intracellular fluxes and the redox imbalances. This information helps in the redesign of the strain (Fig. 4.2). In addition, GEMs provide a distribution of metabolic flows to estimate the target for engineering to redesign. These modes help in rational design strategies and lower the costs of experimentation.

4.5 Challenges and Constraints in Fungal-Based Bioethanol Production

Production of fungal bioethanol is complicated by challenges in fungal metabolic engineering, bioprocessing, and enzyme engineering, and expanding bioprocessing to the industrial level is economically inefficient at the present time. There are biological, processed, economic, and regulatory challenges (Prasanthan et al. 2025). Understanding these challenges is critical to guiding future efforts to improve the stability, effectiveness, and scalability of fungal bioethanol systems. Performance of a fungal system operating at a bioethanol production frontier is constrained by factors articulated in the following subsections, and these are couched in the available economic/engineering and regulatory context, which are limiting the implementation of fungal biotechnologies. The biological behavior of the fungi shows the greatest impact on the effectiveness of the bioprocess. The fungi need to industrially tolerate and withstand fermentation substrates and stresses and, at the same time, produce and secrete enzymes, sugar in the medium, and metabolize the final products of fermentation. However, the natural fungal systems of the pretreatment, process integration, and parallel bioprocessing, current process engineering, and control systems available are unoptimized, and process yield/results are disappointing. There are inhibiting byproducts produced during the pretreatment of the biomass, including furfural, HMF, per acetic acid, and a range of phenolic compounds, and these compounds reduce the process efficiency of fungal growth and controlled systems by inhibiting enzyme production and secretion, fermentation substrate uptake, and overall process control (Yuan et al. 2026). Disrupting the redox balance and the inhibition of glycolytic enzymes can be caused by furfural and HMF. Weak acids, especially acetic acids, can diffuse into fungal cells, reducing the intracellular pH and the energy available in the cell. Transport processes can be impaired, and membrane damage can occur due to phenolic compounds. Even engineered strains of *S. cerevisiae* and *T. reesei* show limited tolerance when the concentrations of the inhibiting compounds exceed the threshold levels (Tian et al. 2025). Added costs and complexities also arise from the process due to added detoxification strategies: ethanol toxicity and cell membrane stress. Increased concentrations of ethanol can compromise fungal membrane integrity. Ethanol accumulation causes increased membrane fluidity, leakage of intracellular metabolites, misfolding of membrane proteins, and a reduction in Adenosine Triphosphate (ATP) synthesis. Particularly filamentous fungi also have limited tolerance of ethanol, especially when compared to traditional yeast systems. This also further limits their direct use in biosynthetic production (CBP) and SSCF configurations. Enzyme secretion is also, in filamentous fungi, highly regulated and relies on complex nutrient-sensing pathways. The regulation is also multifactorial, leading to difficulties in achieving sustained control of enzyme expression in industrial scenarios. This is further complicated by the fact that the secretion of enzymes also competes with the growth of the cell for available resources (Table 4.2).

4.5.1 Storage of Carbohydrates

There are also some fungi that store carbohydrates such as glycogen, trehalose, and mannan. Under hype, fungi could also divert some of the carbon toward storage meshing the yield that could be gotten from the ethanol. Balancing the engineering of these pathways is a delicate task, as it could easily compromise the stress tolerance of the fungi. With the carbs that are available and that can be utilized in an efficient manner, the yield of ethanol that can be produced is also high. However,

Table 4.2 Factors influencing the complex regulation of enzyme secretion in filamentous fungi

S. No.	Regulatory factor	Description
1.	Nutrient-sensing pathways	Enzyme secretion depends on nutrient availability; fungi use sophisticated sensing mechanisms to detect carbon, nitrogen, and environmental cues, which activate or suppress secretion pathways.
2.	Multifactorial regulatory networks	Regulation involves multiple transcription factors, signaling molecules, and pathway interactions, making consistent, long-term control of enzyme expression difficult in industrial conditions.
3.	Competition for cellular resources	Enzyme secretion requires significant energy and metabolic resources, which directly compete with fungal growth needs, often reducing efficiency during large-scale fermentation.
4.	Environmental and stress responses	External conditions such as pH, temperature, oxidative stress, and the presence of inhibitors influence enzyme secretion patterns, adding another layer of complexity.
5.	Industrial process limitations	High cell density, limited nutrients, and fluctuating fermentation conditions in industrial setups often disrupt the natural regulatory balance, making sustained enzyme production challenging.

even with the innovations in engineering, there are still many bottlenecks in the metabolism. Although the heterologous pathways integration has improved pentose assimilation. However, challenges persist due to low affinity of the transporters with xylose and arabinose, which leads to the formation of Xylitol with the imbalances of the cofactors in the pathways of XR and XDH, and limited PPP capacity under high pentose conditions. These limitations are pronounced in lignocellulosic biomass, where the main element to be formed is also pentose and not glucose, and hence the bottlenecks are washed out from the overuse of pentose (Isikgor and Becer 2015). In parallel, like all fungi, many would also rather have respiration than fermentation, and that is especially the case with the filamentous species. As a result, there would be a lower output of ethanol. Additionally, even though the engineered enzyme cocktails are in use, the hydrolysis of enzymes remains slower than the desired outcome due to lignin-mediated cellulose shielding, poor adsorption of the enzyme, which is inadequate, and reduced enzyme stability under an industrial pH or temperature (Agrawal et al. 2022). Ultimately, these factors would be the ones to have a negative effect on what is seen in the fermentation synchrony in the processes of SSF and SSCF.

4.5.2 Process-Level Challenges

Although biological systems can be further tuned, fungal processes must still function within the parameters of an industrialized setting. These include viscosity, sterility, thermal regulation, mixing, oxygen transfer, contamination, and heat. Filamentous fungi produce mycelia, which, in turn, augment the viscosity of the broth, leading to insufficient mixing, diminished oxygen transfer, uneven distribution of nutrients, and dense mycelial aggregation, thereby complicating reactor scale-up and reducing productivity (Sekoai et al. 2026). These conditions aid metabolic inefficiencies by promoting competing metabolic pathways and increasing fermenter downtime from operational instability and cleaning. Aeration proves vital for the production of enzymes, propagation of biomass, and biosynthesis of cofactors. However, as the scale of the reactor increases, the transfer of oxygen decreases. In large reactors, meeting the high oxygen demands of the early fungal growth phase can be very difficult (Cerrone and O'Connor 2025). Furthermore, fungal fermentation is exothermic, but in industrial settings, large-scale reactors lose heat via cooling systems that must be employed to keep the temperature within an optimal range. Heat spikes can lead to a suppression of the fungal metabolism, in turn limiting the quantity of enzymes produced. In addition, under certain conditions, filamentous fungi grow more slowly than bacteria and are thus susceptible to bacterial contamination during industrial fermentations (Khatami et al. 2026). These consequences lead to less bioethanol generation, competing pathways of metabolism, and maintaining sterility in environments loaded with biomass, which is costly and resource-heavy, leading to reduced bioethanol yields. Absolutely engineered fungal systems face potential economic issues, and even more so, when fungal systems are compared to cost-effective food-based ethanol systems and fossil fuels. The major cost holder is the pretreatment of biomass, which can comprise as much as 30–40% of the total production cost, particularly when chemical pretreatment processes that use acids, alkalis, or solvents are used, as these processes aid to expenses linked with neutralization or downstream conditioning. Enzyme production costs further dominate SHF, SSF, and SSCF processes, which come from the enzymes. Enzyme production is resource and energy-costly, even when using engineered fungi. Therefore, uncertainty of feedstock variability in biomass from agriculture can be of various qualities, primarily as a result of differing seasons and heterogeneous composition, which directly affects sugar yields, fermentation consistency, and reactor performance stability. Moreover, lignocellulosic biorefineries require massive investment because of the specialized instruments needed for infrastructure, including pretreatment units, hydrolysis reactors, high-viscosity fermenters, and waste treatment systems. It collectively diminishes the economic viability, particularly in developing economies.

4.5.3 Scaling Up Challenges

The transition from small-scale laboratory culturing of fungi (1–10 L) to larger scales (pilot scales of 100–1000 L and full industrial scales of 10,000–200,000 L) is met with many challenges. Fungal strains that perform well under laboratory conditions often suffer decline in larger scales, due to altered hydrodynamic and physiochemical environments, including changes in mixing pattern, shear force, oxygen gradation, and insufficient heat transfer. All these conditions lead to the "scale shock" phenomenon, and this could be a deal breaker when it comes to bioprocess scaling (Cerrone and O'Connor 2025). Larger-scale bioreactors tend to develop spatial gradients in dissolved oxygen, pH, Temp, and nutrient availability. These gradients elicit metabolic heterogeneity within the pure fungal culture and thus lead to inconsistent production of ethanol. In addition, shear stress within the bioreactor and fungal culture needs to be balanced. Mixing is important to maintain culture homogeneity, but excessive shear damages fungal biomass and reduces enzyme secretion, which, ultimately, lowers the overall productivity of the culture. Therefore, the trade-off between shear stress and mixing device is a great engineering challenge (Ribeiro et al. 2020).

4.5.4 Regulatory and Biosafety Challenges

The bioethanol production system using fungal biomass is highly reliant on the incorporation of genetically modified microorganisms (GMMs). This then leads to various regulatory constraints in their use, which is why the global use is highly limited. The approval of modified microorganisms has many restrictions in many jurisdictions due to concerns related to release in the environment, horizontal gene transfer, and unintended impacts in the environment, resulting in lengthy, complex, and expensive approval processes, resulting in a disincentive for industrial investment. In parallel, facilities utilizing GMM must incorporate significant additional containment measures. Containment results in more expensive infrastructures and more complex operational systems. Beyond regulatory compliance, public perception and societal acceptance of GM technologies play a critical role in shaping biofuel policy frameworks, with prevailing concerns typically focusing on safety, land use, and the environment. But, on the contrary, unfavorable public perceptions of GM technology itself have the potential to drive policy in the biofuels domain. Although bioethanol is designed to be a low-carbon fuel, its sustainability is reliant on the feedstock grown, the process energy intensity used, and the emissions released over the full cycle. Although bioethanol is regarded as a low-carbon fuel, energy-intensive operations, particularly biological pretreatment and the subsequent distillation, result in a net positive carbon impact of bioethanol (Villadiego-del Villar et al. 2026). Large volumes of water are needed for the fungal fermentation, waste streams containing soluble inhibitors, fragments of lignin, and spent enzymes that necessitate effective treatment and disposal strategies. Furthermore, fungi utilize agricultural residues; the large-scale cultivation of fungi will likely require additional feedstock, which raises land use concerns that must be carefully managed to ensure long-term sustainability.

4.6 Future Perspectives and Technological Opportunities in Fungal Bioethanol Systems

Over the course of the past decade, fostering the bioengineering of fungal biomass for bioethanol production has become bioengineering's singular focus, integrating advanced molecular and synthetic biology, metabolic engineering, and bioprocess optimization (Villadiego). Although there are challenges, the most recent learnings from both the biotechnology and bioethanol production systems indicate the beginnings of a pathway for scaling the production of bioethanol from fungal biomass toward industrial competitiveness (Pereira et al. 2026). This chapter outlines emerging perspectives and future directions to strategically position fungal biomass biotechnology at the center of the production of renewable fuels. Opportunities are described within the fields of biological engineering, process engineering, the optimization of sustainability, and within the bioeconomy systems. Assisted machine learning for pathway optimization can amplify the speed of strain design (Costello and Martin 2018). Stress-resistant fungal platforms withstand the following: high levels of ethanol and thermotolerance variations like cellulases will allow SSF and improved kinetics reactions at temperatures above 45 °C, and osmotic pressure. Also, actionable approaches can derive from dynamically engineering a stress response, incorporating robust pathways for cell wall biosynthesis, and systematically genome remodeling to enhance tolerance. With the precision of artificial intelligence and CRISPR systems such as Cas12 and CasΦ, along with base editors, identification of the specific target genes and regulatory elements that can improve the phenotype of the target fungus can be discerned from extensive omics datasets (Dudeja et al. 2025). Engineering stable co-culture dynamics includes synthetically designed communication circuits, quorum-sensing-based population control, nutritional autonomous partitioning through auxotrophs, and stable co-cultures, which minimize the risk of contamination and provide enhanced performance in long-term fermentations. In addition, moving ahead with nanobiotechnology can enable immobilization of bioactive enzymes on nanoparticles, improved reusability of the enzymes, maintained catalytic flow, and protection against thermal denaturation. This can drastically cut down the cost of enzymes. High cell density continuous and semi-continuous fermentation models incorporate cell immobilization, continuous bioreactor functioning, and recycling of biomass, substantially improving volumetric productivity (Guo et al. 2022; Cheng et al. 2023). These strategies enhance the productivity of the system while reducing downtime. The deployment of membrane bioreactors and in situ ethanol removal product recovery techniques, such as pervaporation, gas stripping, or membrane filtration, leads to higher levels of ethanol production, less toxicity to fungal cells, and lower costs for distillation.

Through the use of membrane contactors, ethanol concentrations may be kept below inhibitory levels—in parallel, advanced bioreactor designs, including rotating packed-bed reactor, air-lift hybrid bioreactor, and 3D-printed reactor internals for better mixing and improved hydrodynamics. Collectively, all of these allow for greater oxygen transfer and lower energy use, which provides a scalable and

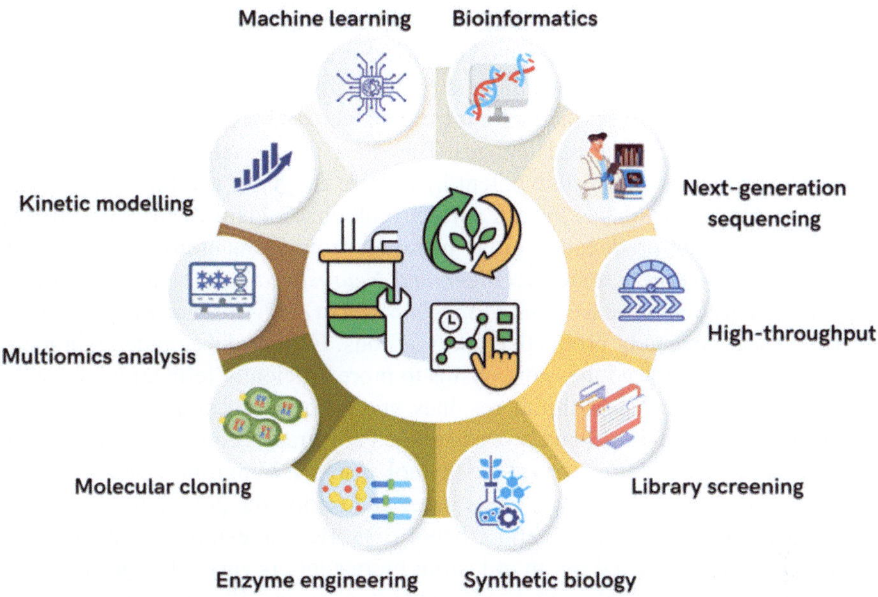

Fig. 4.3 Merging computational, molecular, and bioprocessing methods for high-value bioproduct synthesis is a critical technological pillar for sustaining precision fermentation

cost-effective solution for bioethanol production. Economic viability cannot be achieved solely from the production of bioethanol (Fig. 4.3). Future biorefineries are predicted to operate on integrated systems that create multiple value-added co-products. Bioethanol systems using fungi are expected to incorporate complete engineered fungal consortia, adaptive control of metabolic pathways in real time, next-generation sequencing, biorefineries with no emissions, and multiple income sources. The continuing innovative improvements in these systems will likely give fungi a primary role in supporting existing solar, wind, and green hydrogen systems in achieving global sustainable energy targets (Delhiwala et al. 2025).

4.7 Conclusion

The fungal biomass bioengineering toward the production of bioethanol, fostering the growth of a greener and more sustainable renewable global energy structure, has great potential. In the last 10 years, developments in fungal systems and their industrial applications have been furthered through advances in metabolic engineering, synthetic biology, enzyme development, and bioprocess optimization. In assessing the biological bases of the production of bioethanol from fungi, the engineering approaches that focus on improving fungal capabilities, the process engineering advances that allow the effective conversion of lignocellulosic materials, and the constraints that have been preventing their full commercialization have been analyzed. Collectively, these have attained the insights that not only are fungal biotechnologies complex, but their potential and promise for the sustainable production of renewable fuels are also significant for the organisms that produce bioethanol. Fungi present the greatest ease and advantages. They have the capacity to decompose lignocellulosic biomass and secrete potent hydrolytic enzymes, and their adaptability to varying conditions makes them a perfect biological organism for biorefineries. Fungal enzymes are produced in great quantities from filamentous fungi, relatives of *Trichoderma reesei*, *Aspergillus niger*, and *Neurospora crassa*, and also for the fermentation of ethanol, yeasts such as *Saccharomyces cerevisiae* and *Kluyveromyces marxianus* are extremely productive and vigorous. Further developments in metabolic engineering allowed these organisms to more efficiently ferment mixed sugar streams derived from agricultural residues and other lignocellulosic feedstocks. However, several biological and process-related constraints continue to impede the commercialization of fungal-based bioethanol systems. Fungi must handle several stresses, such as inhibitory compounds generated during the pretreatment, osmotic pressure, temperature changes, and ethanol toxicity. Achieving consistent output during bioreactions carried out in industrial bioreactors is very difficult. Similarly, process-related issues, such as high substrate viscosity, insufficient oxygen transfer, possibilities of contamination, and non-uniformity of the reactor, are significant obstacles to the scale-up of industrial fermenters from the laboratory. All these factors lead to a higher cost of production and slower rates of fermentation, as well as reduced overall stability of the process. Fungal bioethanol production is influenced by advanced technologies and engineering innovations. Advances in CRISPR genome editing, machine learning, metabolic engineering, and multi-omics are allowing the development of fungal chassis for industrial use (Yoshikawa et al. 2012; Dudeja et al. 2025). In synthetic ecology, co-cultures of microorganisms are being used to increase the efficiency of processes by allowing the microorganisms to split tasks (Troiano and Studer 2025). New approaches in enzyme engineering are creating systems that are more efficient and specifically hydrolyze substrates, resulting in lower enzyme use, greater conversion efficiency, and higher reaction rates. Equally important are the advances in continuous fermentation bioreactor design, membrane fermentation, and in situ product removal, which are able to further reduce production costs. The impact of bioethanol systems on the economy is also a reflection of a biorefinery's co-products. Integrated biorefineries are designed to operate on a zero-waste basis and use fungi as the main platform to produce a range of value-added co-products, including bioethanol and bioplastics, organic acids, biogas, nutraceuticals, and lignin-derived aromatics. The development of biorefineries is key to the use of fungi in biofuels, as it will reduce costs, improve economic viability, and provide biofuels with a more robust supply chain. Other crucial aspects influencing the future of biotech are sustainability opportunities. Fungal biotechnology provides systems rooted in climate goals and sustainability. Ethanol systems derived from the bio economy are chiefly used in the decarbonization and climate mitigation of biological wastes, including agricultural residues (Jayakumar et al. 2023), forestry, municipal organic wastes, and non-food biomass.

Community-level decentralized biorefineries diminish transportation emissions and bolster local economies. They enhance flexibility and resource allocation across biorefineries, lowering their vulnerability to supply chain instabilities. There are considerable improvements to be made, though, in the environmental sustainability of fungal bioethanol production, in water recycling, waste disposal, feedstock stream logistics, and life-cycle assessment techniques and processes (Bhardwaj et al. 2025). Regulatory and societal aspects are equally important. Public biosafety and risk management and the use of genetically engineered fungal strains must be balanced. Escaped engineered microorganisms cannot be controlled, and borderless engineered microorganisms raise serious public concerns. Resource rationalization and public environmental advocacy are poor when designed systems fail to exploit rational resources. Safe environmental advocacy, sustainability, and public access to metric systems build societal support. There are many challenges and opportunities involved in developing and scaling fungal-based bioethanol production technologies. Future research must focus on fungal strain improvement and functional characterizations using advanced genetic tools. In addition, research on systems biology, enzymatic catalysis, and synthetic ecology must be integrated to develop and streamline production technologies that harness fungi to bio-manufacture biofuels and biochemicals. While industrial adoption of fungal-based bioethanol technologies has not yet been realized, interdisciplinary research and technological innovation are continuing to construct biofuel systems with fungi. If integrated into a biorefinery that includes other bioprocessing technologies, fibrous fungal biomass has the potential to be a vital contributor in any comprehensive strategy designed to move society toward a low-carbon economy.

References

van Aalst AC, de Valk SC, van Gulik WM, Jansen ML, Pronk JT, Mans R (2022) Pathway engineering strategies for improved product yield in yeast-based industrial ethanol production. Synth Syst Biotechnol 7(1):554–566. https://doi.org/10.1016/j.synbio.2021.12.010

Agrawal R, Bhadana B, Chauhan PS, Adsul M, Kumar R, Gupta RP, Ramakumar SSV (2022) Understanding the effects of low enzyme dosage and high solid loading on the enzyme inhibition and strategies to improve hydrolysis yields of pilot scale pretreated rice straw. Fuel 327:25114. https://doi.org/10.1016/j.fuel.2022.125114

Arshad M, Arshad I, Aslam H et al (2025) Microbial bioproducts: current advances, industrial applications, and future perspectives. J Umm Al-Qura Univ Appl Sci 11:545–560. https://doi.org/10.1007/s43994-025-00247-0

Baite HN, Dhara S, Purkait MK (2026) Emerging pre-treatment technologies for lignocellulosic biomass valorization: advances, challenges, and future perspectives. In: Sharma M, Kumar V, Tabatabaei M, Tripathi M, Gupta VK (eds) Bioproducts from lignocellulosic biomass. Woodhead Publishing, pp 139–169. https://doi.org/10.1016/B978-0-443-21604-6.00001-2

Benali M, Mehdaoui I, Assouguem A, Singh J, Lahrach N, Toyir J (2026) Sustainable valorization of lignocellulosic biomass: recent advances in physicochemical pretreatment methods and the potential of deep eutectic solvents. CABI Rev. https://doi.org/10.1079/cabireviews.2026.0002

Bhardwaj AK, Thakur B, Tripathi SK, Turabayev ME, Aljulaih A, Kharalamova TS, Kulinich SA (2025) Recent advances in valorizing agricultural waste: a sustainable approach. Waste Biomass Valori. https://doi.org/10.1007/s12649-025-03419-5

Borkertas S, Viskelis J, Viskelis P, Streimikyte P, Gasiunaite U, Urbonaviciene D (2025) Fungal biomass fermentation: valorizing the food industry's waste. Fermentation 11(6):351

Cao L, Chen Y, Mashad HME, Pan Z, Zhang R (2026) Techno-economic analysis of fungal biomass production from agricultural byproducts: exploring a low-cost solution for myco-food production. Food Bioprocess Technol 19:112. https://doi.org/10.1007/s11947-025-04202-z

Cerrone F, O'Connor KE (2025) Cultivation of filamentous fungi in airlift bioreactors: advantages and disadvantages. Appl Microbiol Biotechnol 109(1):41. https://doi.org/10.1007/s00253-025-13422-4

Chamoli S, Chamoli AS, Karki K, Kumar R, Kumar V, Kumar P (2025) Recent advances in sustainable strategies for development of innovative nanobiocatalysts using immobilized β-glucosidase for industrial applications. Crit Rev Biotechnol 45:1772–1791. https://doi.org/10.1080/07388551.2025.2517714

Chen T, Zhu X, Liao Q (2026) Resolving the "thick-wall challenge" in *Haematococcus pluvialis*: from metabolic reprogramming to clinical translation. Microorganisms 14:253. https://doi.org/10.3390/microorganisms14010253

Cheng Y, Bi X, Xu Y, Liu Y, Li J, Du G, Lv X, Liu L (2023) Machine learning for metabolic pathway optimization: a review. Comput Struct Biotechnol J 21:2381–2393. https://doi.org/10.1016/j.csbj.2023.03.045

Costello Z, Martin HG (2018) A machine learning approach to predict metabolic pathway dynamics from time-series multiomics data. NPJ Syst Biol Appl 4(1):1–4

Dadheech T, Shah R, Pandit R, Hinsu A, Chauhan PS, Jakhesara S, Kunjadiya A, Rank D, Joshi C (2018) Cloning, molecular modeling and characterization of acidic cellulase from buffalo rumen and its applicability in saccharification of lignocellulosic biomass. Int J Biol Macromol 113:73–81. https://doi.org/10.1016/j.ijbiomac.2018.02.100

Delhiwala Y, Kosuru SMY, Koorla PB, Mekala M (2025) A review on bioethanol production: first generation to fourth generation. Chem Prod Process Model. https://doi.org/10.1515/cppm-2025-0045

Dudeja C, Mishra A, Ali A, Singh PP, Jaiswal AK (2025) Microbial genome editing with CRISPR–Cas9: recent advances and emerging applications across sectors. Fermentation 11(7):410. https://doi.org/10.3390/fermentation11070410

Gladysheva EK (2025) Liquid hot water and steam explosion pretreatment methods for cellulosic raw materials: a review. Polymers 17(13):1783. https://doi.org/10.3390/polym17131783

Guo X, An Y, Liu F, Lu F, Wang B (2022) Lytic polysaccharide monooxygenase–a new driving force for lignocellulosic biomass degradation. Bioresour Technol 362:127803

Hamden Z, El-Ghoul Y, Alminderej FM, Majdoub H (2025) Biomass valorization toward sustainable bioethanol production: a critical review of feedstocks and techno-economic aspects. Waste Biomass Valor. https://doi.org/10.1007/s12649-025-03339-4

Huang M, Jin Z, Ni H, Zhang P, Li H, Liu J, Weng C, Jiang Z (2024) Engineering the xylose metabolism of Saccharomyces cerevisiae for ethanol and single cell protein bioconversion. Biomass Bioenergy 190:107372. https://doi.org/10.1016/j.biombioe.2024.107372

Isikgor FH, Becer CR (2015) Lignocellulosic biomass: a source of sustainable platform chemicals. Polym Chem 6:4497–4559. https://doi.org/10.1039/C5PY00263J

Jayakumar M, Gindaba GT, Gebeyehu KB, Periyasamy S, Jabesa A, Baskar G, John BI, Pugazhendhi A (2023) Bioethanol production from agricultural residues as lignocellulosic biomass feedstock's waste valorization approach: a comprehensive review. Sci Total Environ 879:163158

Kerkaert JD, Huberman LB (2023) Regulation of nutrient utilization in filamentous fungi. Appl Microbiol Biotechnol 107(19):5873–5898. https://doi.org/10.1007/s00253-023-12680-4

Khatami K, Qazanfarzadeh Z, Jiménez-Quero A (2026) Fungal fermentation: the blueprint for transforming industrial side streams and residues. Bioresour Technol 4:133426. https://doi.org/10.1016/j.biortech.2025.133426

Knychala MM, Boing LA, Ienczak JL, Trichez D, Stambuk BU (2024) Precision fermentation as an alternative to animal protein, a review. Fermentation 10:6–315. https://doi.org/10.3390/fermentation10060315

Kobayashi Y, Sahara T, Ohgiya S, Kamagata Y, Fujimori KE (2018) Systematic optimization of gene expression of pentose phosphate pathway enhances ethanol production from a glucose/xylose mixed medium in a recombinant Saccharomyces cerevisiae. AMB Express 8(1):139

de los Santos Mondragón AI, Barragan BA, Sánchez UR, Calleja CA, Millán-Chiu BE, Loske AM, Lim MA (2023) Metabolic engineering of Aspergillus Niger to enhance production of ethanol. Biotechnol Appl Biochem 70(3):1176–1188

Nielsen J, Keasling J (2016) Engineering cellular metabolism. Cell 164:1185–1197

Ntunka MG, Khumalo SM, Makhathini TP, Mtsweni S, Tshibangu MM, Bwapwa JK (2025) Valorization of lignocellulosic biomass to biofuel: a systematic review. Chem Eng 9:58. https://doi.org/10.1016/j.fuel.2021.122678

Onyema VO, Amadi OC, Moneke AN, Agu RC (2023) A brief review: Saccharomyces cerevisiae biodiversity potential and promising cell factories for exploitation in biotechnology and industry processes–west African natural yeasts contribution. Food Chem Adv 2:100162. https://doi.org/10.1016/j.focha.2022.100162

Pereira BD, Silva G, Soares L, Aguiar L, Vasconcelos A, Pessoa V, Pozzan R, Serra J, Campos C, Chevreuil L (2026) Precision fermentation as a frontier in biofuel production advances, challenges, and integration into biorefineries biofuels production. 12:1–35. https://doi.org/10.3390/fermentation12010035

Pérez-Contreras S, Avalos-de la Cruz DA, Lizardi-Jiménez MA, Herrera-Corredor JA, Baltazar-Bernal O, Hernández-Martínez R (2024) Production of ligninolytic and cellulolytic fungal enzymes for agro-industrial waste valorization: trends and applicability. Catalysts 15(1):30

Prakash H, Chauhan PS, General T, Sharma AK (2018) Development of eco-friendly process for the production of bioethanol from banana peel using inhouse developed thermo-alkali-stable depolymerizing enzymes. Bioprocess Biosyst Eng 41(7):1003–1016. https://doi.org/10.1007/s00449-018-1930-3

Prasanthan A, Prasad RK, Balu M (2025) Recent advancements and challenges in the lignocellulosic and algal biomass-based bioethanol production: a review. Biofuels 16:499–518. https://doi.org/10.1080/17597269.2024.2432148

Raheja Y, Gaur VK, Ainala SK, Kumar A, Srivastava JK, Chadha BK (2026) Environmental systems engineering for lignocellulosic biorefineries using AI and genome-scale models. Bioenergy Res 19:24. https://doi.org/10.1007/s12155-026-10965-w

Ribeiro RA, Bourbon-Melo N, Sá-Correia I (2020) The cell wall and the response and tolerance to stresses of biotechnological relevance in yeasts. Front Microbiol 13:953479. https://doi.org/10.3389/fmicb.2022.953479

Sekoai PT, Ezeokoli OE, Mufhandu H, Habimana O (2026) Achieving a circular bioeconomy through fungal biorefineries for single-cell protein production: advances, breakthroughs, and future directions. Biofuels Bioprod Biorefin. https://doi.org/10.1002/bbb.70121

Sivagurunathan S, Raj T, Chauhan PS, Kumari P, Satlewal A, Gupta RP, Kumar R (2022) High-titer lactic acid production from pilot-scale pretreated non-detoxified rice straw hydrolysate at high-solid loading. Biochem Eng J 187:108668. https://doi.org/10.1016/j.bej.2022.108668

Tian L, Qi T, Zhang F, Tran VG, Yuan J, Wang Y, He N, Cao M (2025) Synthetic biology approaches to improve tolerance of inhibitors in lignocellulosic hydrolysates. Biotechnol Adv 78:108477. https://doi.org/10.1016/j.biotechadv.2024.108477

Troiano DT, Studer MH (2025) Microbial consortia for the conversion of biomass into fuels and chemicals. Nat Commun 16(1):6712

Villadiego-del Villar AE, Padilla-Barcasnegra DP, Rojas-Pérez LC (2026) Comparative bioethanol production from sugarcane, cassava, and yam: a biorefinery and economic analysis in Colombia. Biomass Conv Bioref 16:122. https://doi.org/10.1007/s13399-025-07012-7

Yadav J, Marwah H, Kumar C (2025) Synthetic biology and metabolic engineering paving the way for sustainable next-gen biofuels: a comprehensive review. Energy Adv 4:1209–1228. https://doi.org/10.1039/D5YA00118H

Yoshikawa K, Furusawa C, Hirasawa T, Shimizu H (2012) Design of superior cell factories based on systems wide omics analysis. In Syst Metab Eng 57–81. https://doi.org/10.1007/978-94-007-4534-6_3. https://en.metabolic-engineering.jp/publication/detail/5099

Yuan R, Wu Q, Li J et al (2026) Research progress on white-rot fungi pretreatment of lignocellulose based on bibliometric analysis. Syst Microbiol Biom 6:39. https://doi.org/10.1007/s43393-026-00435-x

Chapter 5
Recent Developments in Fungal Biotechnology for Producing Liquid Biofuels: Challenges and Prospects

Atul Kumar Verma, Santosh Kumar Mishra, Mohit Kumar, Rupali Rupasmita, and Narendra Kumar

Abstract The rapid rise in global energy demand and the environmental consequences of fossil fuel dependence have intensified the research for sustainable alternatives. Among various renewal options, fungal biotechnology has emerged as a promising platform for liquid biofuel production due to the metabolic versatility of fungi and their ability to convert diverse feedstock into fuels. A wide range of hydrolysis enzymes are produced and secreted by fungal species, which easily convert lignocellulosic residues into fermentable sugars. The obtained sugars can be converted into bioethanol, biodiesel, and biohydrogen. Oleaginous fungi, capable of accumulating high lipid content, provide unique advantages for biodiesel production, while engineered yeast strains are increasingly optimized for tolerance, productivity, and expanded substrate utilization. A few of the species are deeply examined for their industrial use, such as *Saccharomyces cerevisiae*, *Yarrowia lipolytica,* and *Trichoderma ressei*. Recent advances, such as CRISPR-based genome editing, adapting laboratory evolution, and system biology, have sped up the creation of fungal strains and the use of bioprocessing technologies. However, there are still a lot of challenges to tackle, such as the high cost of pretreatment, the fact that enzymes are not always stable, the fact that lignocellulosic hydrolysate has compounds that stop enzymes from working, and the fact that large-scale operations are not very efficient. Also, competition with food waste feedstocks and economic constraints make it impossible for widespread commercialization to materialize. Even with these issues, putting together fungal biotechnology with waste valorization, biorefinery ideas, and a system with no waste might be a huge step toward a circular bioeconomy. Fungi-based liquid biofuel production could be highly significant for

A. K. Verma · S. K. Mishra · M. Kumar (✉) · R. Rupasmita
Fermentation Laboratory, Department of Life Science, School of Bio Science & Technology, Sharda University, Greater Noida, Uttar Pradesh, India
e-mail: mohit.kumar12@sharda.ac.in

N. Kumar
Department of Biotechnology, Noida Institute of Engineering and Technology, Greater Noida, Uttar Pradesh, India

M. Tripathi et al. (eds.), *Recent Advancements in Fungal Biotechnology Volume 1*, Fungal Biology, https://doi.org/10.1007/978-981-92-0339-0_5

gaining energy security systems. This is because there are more innovative ideas and rules that aid.

Keywords Liquid biofuel · Valorization · Fungal biotechnology · Sustainable · Feedstock

5.1 Introduction

The increase in industrial operations and the growth of the human population are both making the world need more energy. This condition has led to a number of problems that need to be fixed, such as damage to the environment, running out of non-renewable fuels such as coal, natural gas, and oil, and not having enough electricity. The main goal is to make sure that the environment is safe and to stop using chemicals in the process of manufacturing biofuel. To meet energy needs and deal with these problems, it is important to use sustainable methods. These problems are very important to fix, which is why we need to look into and enhance many types of renewable energy, such as solar, wind, hydropower, and notably biomass (from organic matter) (Uzoejinwa et al. 2018).

Biofuels serve as a renewable energy source that holds significant potential for advancement in this sector, especially in the area of transportation. In the upcoming years, the demand for biofuels within transportation infrastructure is anticipated to rise. The main factor contributing to this trend is the growing blending rates and the surging popularity of electric automobiles in nations such as India, Indonesia, and Brazil (Khaleel et al. 2023). More importantly, segments of the transportation sector that rely heavily on petroleum and natural gas and present challenges for decarbonization, such as air transport and maritime commerce, are increasingly lagging behind (Jäger-Waldau 2024). Biofuels are the fuels generated from solid biomass utilizing a range of chemical and biological processes. A few of the most common bio-based fuels are biodiesel, bioethanol, syngas, and biogas. In China, the European Union, the United States, and Brazil, biofuels have primarily been used instead of gasoline and diesel for transportation. Different percentages of bioethanol and biodiesel can be added to normal fuels for transportation. "E5" stands for a blend of 95% gasoline and 5% bio-based ethanol. "B5" implies that the fuel is 95% traditional diesel and 5% bio-based diesel (Lee and Lavoie 2013).

Biomass is estimated to account for 15–50% of worldwide primary power consumption by 2050 (Kumar et al. 2010). Lignin-based biomass serves as one of the largest viable energy sources globally (Moreira 2006), and various alternative energy sources exist globally, which can serve as substitutes for fossil fuels. Fungi, algae, specific bacteria, and plants play an important role in the making of biofuels. At present, three types of bio-based fuels are commonly in practice. This is mostly due to advanced technology, reduced production costs, and an abundance of feedstocks, including food crops, lignocellulosic biomass (LCB), agricultural residues, and microalgae. Additionally, established techniques for modification, such as fermentation and transesterification, alongside appropriate legislation and infrastructure, including regulations for biofuels blending, have facilitated their widespread commercialization (Demirbas 2011; Nigam and Singh 2011). First-generation or conventional biofuels face criticism due to the food versus fuel difficulty (Mukhtar et al. 2022; Padder et al. 2024; Traverso et al. 2020). However, innovative bio-based fuels from the new generation present a significant opportunity, as their potential feedstocks do not interfere with food production. This highlights their significance in mitigating carbon emissions within the transportation sector. Three categories of new generation biofuels include algae-based, genetically modified algae-based, and lignocellulosic bio-based fuels.

Brazil (from sugarcane), China and the United States (from corn), India (from sugarcane molasses), and the European Union (from sugar beet and wheat) produce the largest amount of bio-based fuel in the world, the world's most common bio-based fuel (Li et al. 2019). Less common bioethanol feedstocks include sweet sorghum, sweet potato in China, and cassava in China and Southeast Asia. Many nations that are still developing, specifically in Sub-Saharan Africa and Latin America, are advocating for the advancement of ethanol production from sugarcane (Gasparatos et al. 2012).

5.2 Classification of Biofuels

There are three primary categories of biofuels: solid, liquid, and gaseous. Despite their long-standing presence, individuals continue to utilize solid biofuels such as wood, charcoal, and briquettes for cooking and heating purposes. Liquid fuels, such as ethanol, butanol, and biodiesel, surpass fossil fuels due to their versatility, higher energy density per unit, and compatibility with current transportation infrastructures. Gaseous biofuels, such as biogas, syngas, and biohydrogen, enhance combustion efficiency and can be derived from various organic waste, hence promoting the establishment of a circular bioeconomy. These biofuels differ in the resources utilized, the processes employed for their transformation, their advantages and disadvantages, and their pricing. Table 5.1 delineates their manufacturing processes, applications, advantages, and the challenges they encounter.

Making biofuels from lignin-rich bio-waste involves three primary phases: pretreatment, hydrolysis, and fermentation. Notably, the pretreatment phase is the most expensive and serves as a limiting factor in the process (Satari et al. 2018). By raising the percentage of fermentable carbohydrates after enzymatic saccharification, suitable pretreatment techniques improve the entire process (Maurya et al. 2015). Figure 5.1 shows different types of pretreatment methods: physical, chemical, biological, and physico-chemical. In the pretreatment process, LCBs release various reducing sugars, including arabinose, galactose, fructose, and mannose. Fermentation is facilitated by microorganisms. In the production of bioenergy, sugar serves as a substrate. *Saccharomyces cerevisiae* has long been recognized as the most effective microbe for the conversion of sugars into ethanol (Uday et al. 2016). Biofuels offer

Table 5.1 Comparison between solid, liquid, and gas bio-based fuels

Feature	Liquid biofuels	Solid biofuels	Gas biofuels	References
Definition	Liquid fuels derived from biomass that can be mixed with traditional fuels or utilized in machines	Solid fuels derived from biomass, primarily intended for gasification or combustion	Fuels are gases obtained from anaerobic digestion, gasification, or fermentation of biomass	Demirbas (2011), McKendry (2002), Kapoor et al. (2022)
Examples	Butanol, ethanol, green diesel, biodiesel, and bio-oil	Briquettes, pellets, charcoal, wood, and agricultural byproducts (straw and husk)	Biogas ($CH_4 + CO_2$), Syngas ($H_2 + C) + CO_2$), Biohydrogen (H_2)	McKendry (2002), Kapoor et al. (2022)
Feedstock	Sugarcane, corn, LCB, vegetable oils, animal fats, and algae	Molasses, pellets, charcoal, wood, and crop residues (straw and husk)	Animal manure, sewage sludge, municipal solid waste, crop residues, LCB	Demirbas (2011), Kapoor et al. (2022)
Production methods	Fermentation (ethanol, butanol), transesterification (biodiesel), and pyrolysis/liquefaction (bio-oil)	Direct utilization of wood involves carbonization (charcoal), densification (pellets/ briquettes), and torrefaction	Anaerobic digestion (biogas), Gasification (syngas), Dark fermentation/ photobiological processes (biohydrogen)	Demirbas (2011), Kapoor et al. (2022)
Energy density	High (ethanol: ~27 MJ/kg, Biodiesel: ~37 MJ/ kg)	Lower (wood: ~15–20 MJ/kg, pellets: ~17–19 MJ/kg)	High (bigas: 20–25 MJ/m ^{3}, H_2: 120 MJ/kg)	Demirbas (2011), Kapoor et al. (2022)
Storage and transport	Easy to transport via pipelines or tankers, compatible with the current infrastructure	Bulky, necessitating dehydration and densification, higher transport and storage costs per unit of energy	Requires pressurization/ liquefaction, more complex handling than liquids	McKendry (2002), Kapoor et al. (2022)
Applications	Power generation, heating, and the transportation sector (including automobiles, trucks, and aviation blends)	Industrial furnaces, power plants (co-firing with coal), cooking, and heating	Electricity generation, cool gas, heating, transport fuel (an upgrade)	Demirbas (2011), McKendry (2002), Kapoor et al. (2022)

(continued)

Table 5.1 (continued)

Feature	Liquid biofuels	Solid biofuels	Gas biofuels	References
Environmental impact	Biodiesel may marginally increase NOx, while cleaner combustion reduces greenhouse gas	Fundamentally carbon-neutral, yet inefficient combustion leads to indoor air pollution and particulate particles	Cleaner combustion, reduced greenhouse gases, but methane leakage risk during storage	Demirbas (2011), McKendry (2002), Kapoor et al. (2022)
Commercial status	Extensively marketed (ethanol, biodiesel) with governmental blending requirements	Prevalent in rural and emerging areas, the use of pellets and briquettes is increasing	Cooking gas, electricity generation, heating, and upgraded bigas as transport fuel	Demirbas (2011), McKendry (2002), Kapoor et al. (2022)

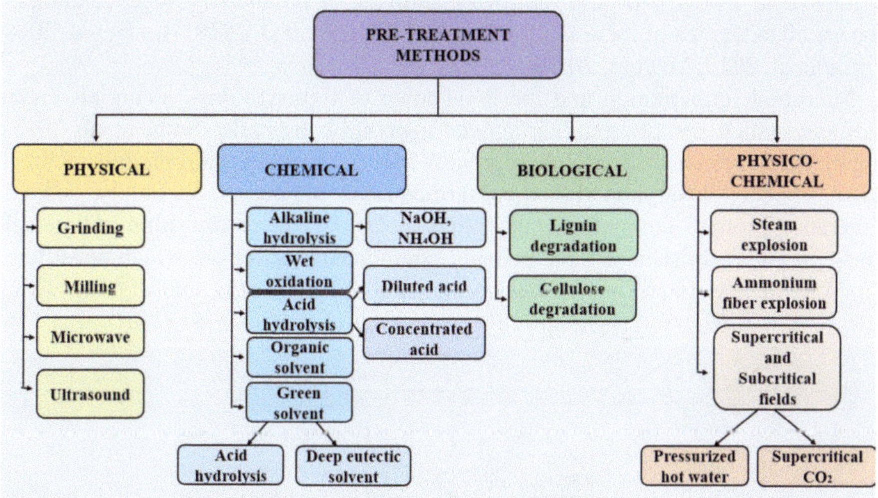

Fig. 5.1 Summary of different pretreatment methods for LCB

several advantages, including their status as a renewable resource and their reduced emission of harmful chemicals during combustion (Dragone et al. 2010; Razzak et al. 2013; Surriya et al. 2014; Voloshin et al. 2016).

5.3 Production of Liquid Biofuels

Liquid biofuels primarily include biodiesel and bioethanol (with emerging options like biobutanol and bio methanol). Both can be used as direct substitutes for conventional biofuels, but biodiesel blends can typically be used in higher proportions without major engine modifications. Their blend level is denoted by a prefix E for ethanol and B for biodiesel, followed by the percentage indicating the maximum volume share of the biofuel. Researchers have focused more on lignocellulosic non-food agricultural residues for the production of liquid biofuels as they are considered economical feedstock for second-generation bio-based fuels (Sims et al. 2010). LCB, owing to its plentiful nature and adaptability, has surpassed all other organic materials in terms of utilization. The transformation of lignocellulose into liquid biofuel involves significant expenses, which poses a challenge to the widespread production of biofuels derived from lignin-rich biomass. To make ethanol from LCB, you need to know a lot about and be good at a number of different processes (Alonso et al. 2010). Bio-based petrol, ethanol, diesel, butanol, hydrogen, and methane are all examples of these kinds of biofuels (Ambat et al. 2018; Huzir et al. 2018; Prussi et al. 2019; Yu et al. 2019).

Microbial fermentation and the breakdown of lignocellulosic materials create biofuels, which are often called second-generation biofuels (Saini et al. 2015). Figure 5.2 represents the complete process that starts when microorganisms break down lignocellulosic material, then fermentation happens, and finally, second-generation biofuels are made (Manochio et al. 2017). The composition of lignocellulose biomass consists of three primary components: cellulose, which constitutes 30–50% of the total; hemicellulose, making up 15–35%; and lignin, which

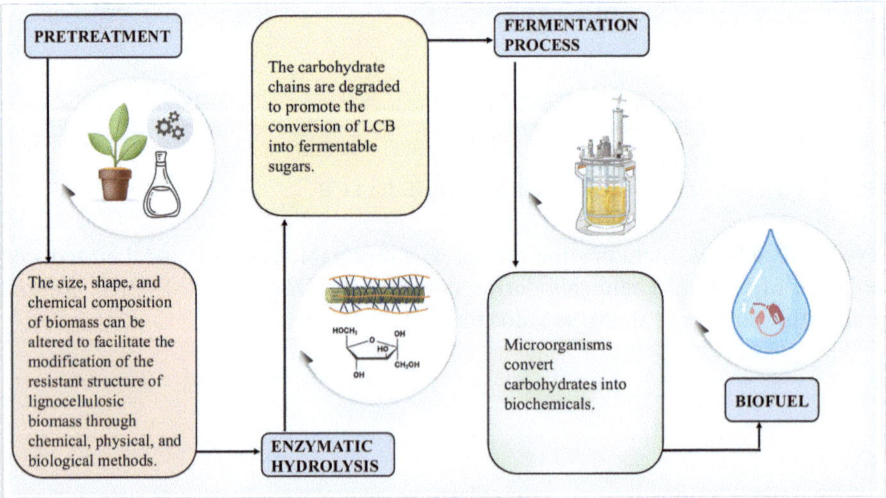

Fig. 5.2 Schematic presentation of various stages involved in liquid biofuel production using lignocellulosic material

comprises 10–20% of the overall biomass (Alonso et al. 2017; Alvira et al. 2010). Bacteria and fungi are widely explored as microbial platforms for sustainable biofuel sustainable biofuel production, though their roles differ significantly in efficacy and metabolic capabilities; bacteria such as *Clostridium acetobutylicum* and engineered *Escherichia coli* are well known for their rapid growth and genetic manipulability, enabling the production of bioethanol, butanol, hydrogen and other biofuels through metabolic engineering (Lee et al. 2008; Peralta-Yahya et al. 2012). However, filamentous fungi, such as *Trichoderma ressei* and *Aspergillus niger,* and oleaginous yeasts, such as *Yarrowia lipolytica* and *Rhodotorula* spp., naturally secrete a large amount of carbohydrate-active enzymes such as cellulase, hemicellulase, and ligninase, enabling efficient hydrolysis of lignocellulosic feedstock (Kuhad et al. 2011; Ljungdahl and Eriksson 1985).

5.4 Fungus Over Bacteria for Liquid Biofuel Production

The majority of contemporary commercial lignocellulolytic formulations predominantly utilize fungal hemicellulases and cellulases (Hu et al. 2011). The investigation concerns prospective advancements in fungal enzymes or the feasibility of developing novel mixtures that incorporate glycosyl hydrolases produced by bacteria and fungi. It is widely acknowledged that fungi produce a significant quantity of cellulases (Bhattacharya et al. 2015). Fungi are increasingly recognized as superior to bacteria in biofuel production, mainly because of their extraordinary enzymatic potential and adaptability to industrial conditions. Unlike bacteria, many fungi secrete large quantities of extracellular enzymes, especially cellulase, hemicellulase, and lignin-degrading oxidase, which are essential for breaking down LCB into fermentable sugars. For instance, *Trichoderma ressei* is one of the best producers of cellulases, and *Aspergillus niger* is one of the best producers of hemicellulases. This gives fungus a big advantage when it comes to turning agricultural residues and woody biomass into ethanol and other biofuels (Dashtban et al. 2010). The main benefit is that fungi may go through solid material with their filamentous hyphae. This structure lets them break down a lot more complicated compounds than bacteria, which can only break down soluble substrates. This ability immediately improves the efficiency of second-generation biofuel processes, which mostly use non-food lignocellulosic feedstock (Saha 2003; Sen et al. 2024). Fungi are also better at dealing with stress from industry. They may still be there and keep their metabolism running, even when the pH is low, the oxygen levels vary, or fermentation inhibitors like acetic acid or furfural are present. These inhibitors often make microorganisms less effective. This strength leads to higher ethanol yields in large-scale fermentations, which lowers both costs and process instability (Hafez et al. 2025; Sánchez and Cardona 2008). The employment of fungi in consolidated bioprocessing, where the same organism does both enzymatic hydrolysis and fermentation, is not the most promising advancement. Some types of fungi, such as genetically modified yeasts, can do both of these jobs as well, which makes the biofuel production process easier and less reliant on expensive outside enzyme supplements (Galbe and Zacchi 2012). Fungi offer a more efficient pathway to high-yield biofuel production than bacteria by directing greater carbon flux toward ethanol and reducing the creation of undesirable byproducts such as organic acids (Lynd et al. 2002).

5.5 Fungal Biofuels

Production of bioethanol from fungi gained researchers' interest, which focused on converting LCB, which was generally constrained by the high cost of enzymes. In the 1980s, various species, such as *Fusarium oxysporium* and *Neurospora crassa*, were identified as effective microbial activators capable of producing C_2H_5OH straight from cellulose, eliminating the necessity for outer enzymes (Christakopoulos et al. 1989; Jouzani and Taherzadeh 2015).

(a) *Ethanol Production*

Ethanol is an extensively researched fungal biofuel. It is primarily produced via glycolysis followed by alcoholic fermentation. *Saccharomyces cerevisiae* is the standard yeast utilized for glucose fermentation. Nonetheless, some fungi, such as *Candida shehatae*, *Pachysolen tannophilus,* and *Scheffersomyces stipitis*, possess the capability to metabolize pentose sugars such as xylose, rendering them suitable for the production of lignocellulosic bioethanol. Table 5.2 presents a compilation of the principal fungus types, their preferred substrates, and the metabolic processes implicated in ethanol production.

(b) *Biodiesel Production*

Oleaginous fungi are crucial for making biodiesel because they can store a lot of lipids. The fungi direct carbon flow into the malonyl-CoA and Acetyl-CoA pathways, which make triacylglycerol (TAG). This TAG might eventually be changed into fatty acid methyl esters so that bio-based diesel can be made. *Mortierella isabellina*, *Cryptococcus curvatus,* and *Yarrowia lipolytica* are some of the most well-known species. In the context of biodiesel, Table 5.3 shows the main types of fungi, how they get their food, and how they store lipids.

(c) *Butanol Production*

Butanol exhibits enhanced effectiveness as an alcohol compared to ethanol. This material offers several benefits, such as improved energy density and reduced hygroscopicity. In fungi, the synthesis of butanol primarily occurs via pathways derived from amino acids. *Saccharomyces cerevisiae* synthesizes isobutanol from valine through the Ehrlich pathway. The process of genetic engineering enables the production of n-butanol from threonine and citramalate, in addition to the metabolism of glycine. *Pichia pastoris* primarily utilizes the valine-dependent Ehrlich pathway, which often results in ethanol formation as a byproduct. Table 5.4 outlines

Table 5.2 Production of ethanol

Fungal species	Substrate	Pathway	Features	References
Saccharomyces cerevisiae	Glucose, sucrose, starch hydrolysates, molasses	Glycolysis→Pyruvate→Acetaldehyde (Pyruvate decarboxylase) →Ethanol (Alcohol dehydrogenase)	It is widely used in the brewing and bioethanol industries	Walker and Stewart (2016)
Scheffersomyces stipitis	Xylose (from lignocellulosic hydrolysates)	Xylose→Xylose reductase→Xylitol dehydrogenase→Xylulose-5-phosphate→Pentose phosphate pthaway→Ethanol	Naturally ferments pentose sugars (xylose, arabinose), unlike *S. cerevisiae*	Liu et al. (2012)
Kluyveromyces marxianus	Glucose, lactose (whey), inulin, plant hydrolysates	Same glycolysis →Pyruvate →Acetaldehyde →Ethanol	Thermotolerant (can grow up to 45–50 °C), making it suitable for industrial bioethanol at high temperatures	Sengupta et al. (2022)

Table 5.3 Production of biodiesel

Fungal species	Substrate	Pathway	Features	References
Yarrowia lipolytica	Glucose, glycerol, vegetable oils, and industrial waste oils	Kennedy pathway: Glucose→Acetyl-CoA→Malonyl-CoA→Fatty Acids→ **TAGs**→Single Cell oils (**SCO**) biodiesel	Engineered strains show high lipid and biodiesel precursor yields	Abghari et al. (2014), Wei et al. (2017)
Aspergillus carneus	Agricultural residues, glucose, starch hydrolysates	Lipid biosynthesis pathway: Carbohydrates→Acetyl-CoA→Fatty Acids→ TAGs→Biodiesel	Produces ~36% lipid content, rich in C16-C18 fatty acids	Ibrahim et al. (2023)
Penicillium citrinum	Cellulose, glucose, agricultural biomass	Lipid accumulation pathway: Sugars > Acetyl-CoA > Fatty Acids > TAGs >Biodiesel	–	Bardhan et al. (2019)
Mucor circinelloides	Glucose, xylose, and agricultural residues	Sugars→Acetyl-CoA→Fatty Acids→ TAGs→Single Cell Oils→Biodiesel	Strong lipid producer, model organism for biodiesel	Zheng et al. (2012)

Table 5.4 Production of butanol

Fungal species	Substrate	Pathway	Features	References
Saccharomyces cerevisiae	Glucose, amino acids (valine, glycine), threonine	Isobutanol pathway Valine (Ehrlich pathway): Valine 2-ketoisovalerate → →Isobutyraldehyde →Isobutanol	Native isobutanol production through valine biosynthesis, engineering introduces multiple heterologous pathways	Branduardi et al. (2013), Shi et al. (2016), Wess et al. (2019)
Pichia pastoris	Glucose (rich medium)	Isobutanol via Valine (Ehrlich pathway): Glucose →Pyruvate →Valine →2-Ketoisovalerate →Isobutyraldehyde →Isobutanol	It shows improving isobutanol yield with additional pathway gene copies, though with co-product ethanol production	Siripong et al. (2018)
Yarrowia lipolytica	Glucose, glycerol	Fatty acid metabolism → Acyl-CoA reduction → Butyryl-CoA →Butanol dehydrogenase	–	Xu et al. (2016)

different fungal species, the substrates they prefer, and the metabolic pathways utilized in the synthesis of butanol.

5.6 Current State and Prospects of Advanced Liquid Biofuels

Worldwide, the most common approach to obtaining bio-based fuel is the first-generation production. Brazil, the United States, Indonesia, China, and the European Union dominate production, accounting for 85% of the overall output. The European Union leads in the production of advanced biofuels, with Indonesia, China, Brazil, and the United States following in succession. The European Union currently derives 31% of its energy from advanced biofuels, a result of a renewable energy policy designed to achieve an emission-free transportation sector (Mizik 2025a). The updated version of the Energy Tax Directive is anticipated to offer additional support by providing tax incentives for advanced biofuels (Mizik et al. 2025). China has the second-highest percentage of advanced bio-based fuels, which is 26%. This is mostly owing to the ongoing decline in subsidies for the production of traditional bio-based fuels (Usmani et al. 2023). The United States ranks third in terms of value, with advanced biofuel accounting for 11% of its total production. The average percentage for Brazil is 7%. The United States and Brazil both have regulations that encourage the use of renewable energy. Brazil has set up national institutions, including the Institute of Alcohol and Sugar, to help the biofuel business (Sajid et al. 2021).

The high expenses of making advanced biofuels are the biggest problem that keeps them from being widely available. Investment expenses are high, especially in the field of thermochemical conversion, where economies of scale are quite important. The efficiencies of conversion processes tend to be relatively low (Mizik and Gyarmati 2022). The production cost of one unit of this exceeds that of fossil fuels. Specifically, fossil fuels are priced at EUR 8–14/gigajoule, while bio-based feedstock ranges from EUR 17–44/gigajoule, and useless raw material falls between EUR 13–29/gigajoule (Brown et al. 2020). Biofuels are usually mixed with traditional fuels; therefore, there is no variation in the cost between the two. However, high-level mixes can only be less expensive if they are backed by distinct policy initiatives. For example, the price of biofuels might be adjusted to account for their lower energy density. This highlights that one of the key aspects of next-generation bio-based production is enhancing productivity and sharing information. In the medium run, commercialization alone might lower production costs by 10–40%. We used the Organisation for Economic Co-operation and Development (OECD) Data Explorer to figure out those numbers because we do not have similar data for 2024 (data retrieved from the OECD Data Explorer). OECD Data Explorer—the updated dataset indicates an increase of 2.6% in global biofuel production. This phenomenon can be attributed to the increases observed in the global context (5%),

Indonesia (6.8%), Brazil (1.7%), and the United States (4.5%). Conversely, China and the European Union experienced reductions of 2.4% and 3.3%, respectively.

It is important to understand that energy policies have a big effect on the growth of biofuels, especially when fossil fuels are cheaper. Combining mandates raises output and improves research and development in the field, which makes production processes more efficient. RED 2 (Renewable Energy Directives) is a good example. It encouraged advanced biofuels by using a double-counting system, but it ended just 3 years later when RED 3 came out (Mizik 2025b). European countries have made strong promises to fight climate change, so it is very important for member states to use renewable energy sources (Bozsik et al. 2023). It costs more to start making sophisticated bio-based fuels, especially those that are third and fourth generations, than it does to start making first-generation fuels. Therefore, tailored research and development incentives at every stage of the production process are crucial for improving efficiency and cultivating a more positive public perspective on bio-based fuels (Moshood et al. 2021).

5.7 Recent Advances in Fungal Biofuel Systems

Recent progress in genetic engineering has led to the creation of a number of tools and methods that make it easier to produce new strains of fungi. CRISPR-Cas9 is a very accurate way to modify genomes. The methods work best when used as ribonucleoprotein (RNP) complexes or in multiplexes or in multiplex mode. This makes it easy and quick to knock out or add genes. For example, it is used in *Saccharomyces cerevisiae* and *Yarrowia lipolytica* to boost lipid accumulation for bio-based diesel and ethanol yields (Cao et al. 2023; Shen et al. 2024). Base editing approaches enable exact modification to individual nucleotides, exemplified by designed *S. cerevisiae* strains with improved ethanol tolerance through the modification of alcohol dehydrogenase genes. At the same time, prime editing offers a way to make precise changes and replacements without changing the double-helix structure (Leal et al. 2024). Moreover, the application of transcriptional regulation via CRISPRa or CRISPRi, employing a nuclease-deactivated Cas9, has been used in *Aspergillus niger* and *Trichoderma ressei* to boost cellulase and xylanase expression, thereby improving LCB hydrolysis for bio-based ethanol production. This enhances the modification of DNA sequences. Initiatives aimed at refining targeted integrations have concentrated on optimizing homology-directed repair, especially through the removal of Ku70/Ku80 and the enhancement of donor template design (Ma et al. 2025).

Concurrently, progress in distribution and transformation technologies has transpired. Agrobacterium-mediated transformation provides benefits for several filamentous fungi. For example, Monascus species has been utilized to augment fatty acid production pertinent to bio-based diesel precursors. The editing process is enhanced through protoplast-based techniques utilizing electroporation or polyethylene glycol (PEG)-mediated release of RNPs (Olzhausen et al. 2021). Episomal

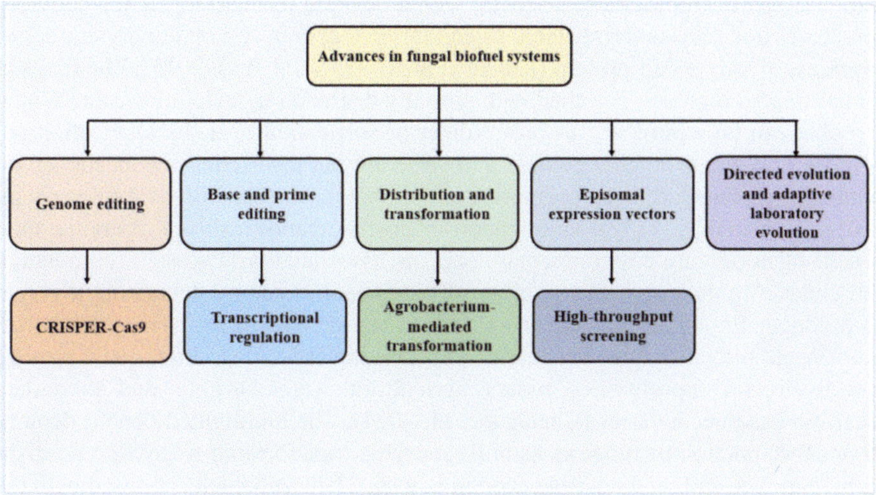

Fig. 5.3 Advances in genetic modulation for fungal biofuel systems

expression vectors, such as AMA1 plasmids, enable temporary expression, allowing for rapid prototyping without the need for permanent genomic integration. High-throughput droplet microfluidics enables the simultaneous screening of millions of fungal mutations, for example, identifying *Y. lipolytica* mutants with superior fatty acid production for bio-based diesel. Identifying beneficial mutations that enhance tolerance and productivity necessitates the application of directed evolution methods and adaptive laboratory evolution (Garg et al. 2025; Lee and Kim 2020; Vladisavljević 2024). Figure 5.3 summarizes the advancements in fungal biofuel systems facilitated by genetic tools, transformation procedures, and evolutionary processes.

5.8 Fungi as a Zero-Waste System

Residues generated after bio-based ethanol production, commonly termed as lignocellulosic distillation stillage or spent biomass, represent an abundant yet underutilized resource within the bioenergy sector (Mathew et al. 2018). Despite the extraction of fermentable sugars during ethanol fermentation, these residues remain rich in unconverted cellulose, hemicellulose, lignin, proteins, and micronutrients (Adeoye et al. 2023). Rather than being disposed of as waste, fungal biotechnology offers an effective strategy for their valorization within a zero-waste and circular bioeconomy framework. Edible mushrooms such as *Pleurotus ostreatus* and *Volvariella volvacea* can directly utilize this spent biomass as a growth substrate, producing protein-rich fruiting bodies with substantial nutritional and economic values (Ali 2021; Silva et al. 2024). In parallel, filamentous fungi like Aspergillus and Trichoderma species secrete hydrolytic and oxidative enzymes, including cellulase, hemicellulase, and laccases, which degrade recalcitrant lignin–cellulose complexes and release fermentable sugars for further conversion into organic acids, biogases, or single-cell protein (Kuhad et al. 1997; Saini et al. 2015). The residual fungal-treated biomass, enriched with partially degraded lignocellulosic and fungal mycelia, can be repurposed as biofertilizer or soil conditioner. Its application not only improves soil organic content and structure but also reduces dependency on synthetic fertilizers, thereby promoting archeological sustainability (Martínez et al. 2005). Additionally, fungal valorization of post—ethanol residues offers the dual benefit of mitigating environmental concerns associated with waste accumulation and enhancing the economic viability of bioethanol plants by generating multiple co-products. For small-scale and rural industries, the integration of mushroom cultivation and fungal enzyme production into existing ethanol facilities could provide new livelihood opportunities, reduce agricultural waste burning, and strengthen local bio-based economies (Chauhan et al. 2024). The highlighted benefit demonstrates the function of fungi as natural recyclers, transforming waste into nourishment, enriched soil, and possible economic gain. Their approach is praised with the zero-waste framework depicted in Fig. 5.4 and further promotes a circular and renewable bioeconomy.

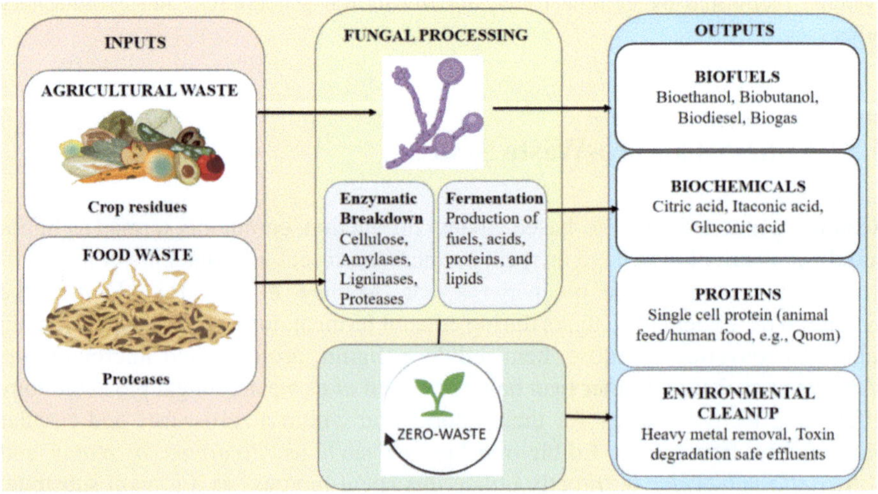

Fig. 5.4 Using agricultural and lignocellulosic waste for the production of biofuel while generating zero waste

5.9 Conclusion

Recent advances in fungal biotechnology have given a strong push to liquid biofuel production and made the process more sustainable. Modified fungal strains can turn lignocellulosic waste and other secondary feedstock into ethanol, biodiesel, and advanced hydrocarbon fuels more efficiently than before. Work on genetics and metabolic pathways of species such as *Saccharomyces cerevisiae*, *Yarrowia lipolytica,* and *Trichoderma ressei* has helped improve substrate use, strengthen tolerance to process stress, and raise overall productivity. Through efforts, researchers have developed fungal strains capable of producing fuels that closely match the qualities of conventional options. Even with this progress, several hurdles remain. Pretreatment methods are still costly, fungal fermentation is not always efficient, and many enzymes struggle to break down complex biomass. Competition with food-based feedstocks, the formation of inhibitory compounds during pretreatment, and the limited stress tolerance of fungi add further complications at large scales. To move forward, biorefineries need better process design and more reliable ways to use non-food biomass and waste material. Current research is exploring new fungal species, co-culture strategies, and integrated biorefineries models, all of which point toward a promising future for the field. Gene editing tools such as CRISPR, along with multi-omics approaches, are expected to unlock new metabolic routes and expand the capabilities of fungal systems. Progress will rely on collaboration across engineering, computational modeling, and microbiology. With steady investment and innovation research, fungal biotechnology could become a key contributor to low-carbon energy production, supporting both environmental health and long-term energy security.

References

Abghari A, Chen S, Paul Christopher L, Dakota S, Kumar Debnath B, di Torino P, Yu-Shen Cheng I (2014) Yarrowia lipolytica as an Oleaginous cell factory platform for production of fatty acid-based biofuel and bioproducts. Frontiersin.Org 2(JUN). https://doi.org/10.3389/FENRG.2014.00021/FULL

Adeoye AO, Lawal OS, Quadri RO, Malomo D, Aliyu MT, Dang GE, Emojevu EO, Maikato MJ, Yahaya MG, Omonije OO, Edidem VG, Abubakar YK, Offor OF, Sochima EH, Peter BE, Hikon BN (2023) Sustainable energy via thermochemical and biochemical conversion of biomass wastes for biofuel production. Springer, Part F647:245–306. https://doi.org/10.1007/978-981-99-2150-8_11

Ali S (2021) Effect of substrate enrichment on contamination and productivity of Oyster mushroom *(Pleurotus ostreatus)*. http://archive.saulibrary.edu.bd:8080/handle/123456789/3732

Alonso DM, Bond JQ, Dumesic JA (2010) Catalytic conversion of biomass to biofuels. Green Chem 12(9):1493–1513. https://doi.org/10.1039/C004654J

Alonso DM, Hakim SH, Zhou S, Won W, Hosseinaei O, Tao J, Garcia-Negron V, Motagamwala AH, Mellmer MA, Huang K, Houtman CJ, Labbé N, Harper DP, Maravelias CT, Runge T, Dumesic JA (2017) Increasing the revenue from lignocellulosic biomass: maximizing feedstock utilization. Science.OrgDM Alonso, SH Hakim, S Zhou, W Won, O Hosseinaei, J Tao, V Garcia-NegronScience Advances, 2017•science.Org. 3(5). https://doi.org/10.1126/SCIADV.1603301

Alvira P, Tomás-Pejó E, Ballesteros M, Negro MJ (2010) Pretreatment technologies for an efficient bioethanol production process based on enzymatic hydrolysis: a review. Bioresour Technol 101(13):4851–4861. https://doi.org/10.1016/j.biortech.2009.11.093

Ambat I, Srivastava V, Sillanpää M (2018) Recent advancement in biodiesel production methodologies using various feedstock: a review. Renew Sust Energ Rev 90:356–369. https://doi.org/10.1016/j.rser.2018.03.069

Bardhan P, Gohain M, Daimary N, Kishor S, Chattopadhyay P, Gupta K, Chaliha C, Kalita E, Deka D, Mandal M (2019) Microbial lipids from cellulolytic oleaginous fungus Penicillium citrinum PKB20 as a potential feedstock for biodiesel production. Springer 69(11):1135–1146. https://doi.org/10.1007/S13213-019-01494-3

Bhattacharya AS, Bhattacharya A, Pletschke BI (2015) Synergism of fungal and bacterial cellulases and hemicellulases: a novel perspective for enhanced bio-ethanol production. Biotechnol Lett 37(6):1117–1129. https://doi.org/10.1007/S10529-015-1779-3/FIGURES/1

Bozsik N, Magda R, Bozsik N (2023) Analysis of primary energy consumption, for the European Union member states. Acta Polytechnica Hungarica 20(10):2023–2089. https://oda.uni-obuda.hu/bitstream/handle/20.500.14044/25428/acta_23_vol_20_issue_10_Bozsik_Magda_Bozsik_.pdf?sequence=1

Branduardi P, Longo V, Berterame NM, Rossi G, Porro D (2013) A novel pathway to produce butanol and isobutanol in Saccharomyces cerevisiae. SpringerP Branduardi, V Longo, NM Berterame, G Rossi, D PorroBiotechnology for Biofuels, 2013•Springer. 6(1). https://doi.org/10.1186/1754-6834-6-68

Brown A, Waldheim L, Landälv I, Saddler J, Ebadian M, Mcmillan JD, Bonomi A, Klein B (2020) Advanced biofuels–potential for cost reduction. Ieabioenergy.ComA Brown, L Waldheim, I Landälv, J Saddler, M Ebadian, JD McMillan, A Bonomi, B KleinIEA Bioenergy, 2020•ieabioenergy.Com. Retrieved August 19, 2025, from https://www.ieabioenergy.com/wp-content/uploads/2020/02/IEABioenergy-2pSummary-CostAdvancedBiofuels.pdf

Cao M, Xiao H, Jiao S, University, T, Kar CB, Mishra S, Yang L, Zhang Q, Zhang Y, Chen S (2023) Application progress of CRISPR/Cas9 genome-editing technology in edible fungi. Frontiersin.OrgY Zhang, S Chen, L Yang, Q ZhangFrontiers in Microbiology, 2023•frontiersin.Org. 14. https://doi.org/10.3389/FMICB.2023.1169884/FULL

Chauhan A, Negi S, Kaur R, Chauhan M, Dhanai R, Negi PS (2024) Cultivation of Oyster mushroom could be a viable option for doubling the farmer's income - an overview. Int J Econ Plants 11(Nov, 4):387–393. https://doi.org/10.23910/2/2024.5725a

Christakopoulos P, Macris BJ, Kekos D (1989) Direct fermentation of cellulose to ethanol by Fusarium oxysporum. Enzym Microb Technol 11(4):236–239. https://doi.org/10.1016/0141-0229(89)90098-7

Dashtban M, Maki M, Leung KT, Mao C, Qin W (2010) Cellulase activities in biomass conversion: measurement methods and comparison. Taylor & Francis 30(4):302–309. https://doi.org/10.3109/07388551.2010.490938

Demirbas A (2011) Competitive liquid biofuels from biomass. Appl Energy 88(1):17–28. https://doi.org/10.1016/j.apenergy.2010.07.016

Dragone G, Fernandes BD, Vicente AA, Teixeira JA (2010) Third generation biofuels from microalgae. Appl Microbiol Microb Biotechnol 2:897–907. https://repositorium.sdum.uminho.pt/handle/1822/16807

Galbe M, Zacchi G (2012) Pretreatment: the key to efficient utilization of lignocellulosic materials. Biomass Bioenergy 46:70–78. https://doi.org/10.1016/J.BIOMBIOE.2012.03.026

Garg S, Kim M, Romero-Suarez D (2025) Current advancements in fungal engineering technologies for Sustainable Development Goals. Trends Microbiol 33(3):285–301. https://doi.org/10.1016/j.tim.2024.11.001

Gasparatos A, Lee L, Maltitz G Von, Mathai M (2012) Biofuels in Africa: impacts on ecosystem services, biodiversity and human well-being. https://researchspace.csir.co.za/bitstream/10204/6327/1/Gasparatos_2012.pdf

Hafez RM, Abdel-Rahman TMA, Yahia MM, El-Gabry KIM, Wahab YMMA (2025) Biochemical and cytological studies of Typha domingensis used for bioethanol production. Biomass Convers Biorefinery 15(10):15361–15384. https://doi.org/10.1007/S13399-024-06229-2/FIGURES/19

Hu J, Arantes V, Saddler JN (2011) The enhancement of enzymatic hydrolysis of lignocellulosic substrates by the addition of accessory enzymes such as xylanase: is it an additive or synergistic effect? Springer 4:36. https://doi.org/10.1186/1754-6834-4-36

Huzir NM, Aziz MMA, Ismail SB, Abdullah B, Mahmood NAN, Umor NA, Syed Muhammad SAF, ad. (2018) Agro-industrial waste to biobutanol production: eco-friendly biofuels for next generation. Renew Sust Energ Rev 94:476–485. https://doi.org/10.1016/J.RSER.2018.06.036

Ibrahim AG, Baazeem A, Al-Zaban MI, Fawzy MA, Hassan SHA, Koutb M (2023) Sustainable biodiesel production from a New Oleaginous fungus, Aspergillus carneus strain OQ275240: biomass and lipid production optimization using box–Behnken design. Sustainability 15(8):6836. https://doi.org/10.3390/su15086836

Jäger-Waldau A (2024) Snapshot of photovoltaics – February 2024. EPJ Photovolt 15:21. https://doi.org/10.1051/EPJPV/2024018

Jouzani GS, Taherzadeh MJ (2015) Advances in consolidated bioprocessing systems for bioethanol and butanol production from biomass: a comprehensive review. Biofuel Res J 2(1):152–195. https://doi.org/10.18331/BRJ2015.2.1.4

Kapoor R, production, M. R.-G. nano solution for bioenergy, 2022, undefined (2022) Recent advances in conversion of agricultural waste to biofuel by nanoparticles. SpringerRT Kapoor, M RafatullahGreen Nano Solution for Bioenergy Production Enhancement, 2022•Springer:245–273. https://doi.org/10.1007/978-981-16-9356-4_10

Khaleel M, Yusupov Z, Ahmed A, Alsharif A, Nassar Y, El-Khozondar H (2023) Towards sustainable renewable energy. Springer 59(4):557–567. https://doi.org/10.3103/S0003701X23600704

Kuhad RC, Singh A, Eriksson KE (1997) Microorganisms and enzymes involved in the degradation of plant fiber cell walls. SpringerRC Kuhad, A Singh, KEL ErikssonBiotechnology in the Pulp and Paper Industry, 2006•Springer. 57:45–125. https://doi.org/10.1007/BFB0102072

Kuhad RC, Gupta R, Singh A (2011) Microbial cellulases and their industrial applications. Wiley Online Lib 2011(1). https://doi.org/10.4061/2011/280696

Kumar A, Kumar K, Kaushik N, Sharma S, Mishra S (2010) Renewable energy in India: current status and future potentials. Renew Sust Energ Rev 14(8):2434–2442. https://doi.org/10.1016/j.rser.2010.04.003

Leal K, Rojas E, Madariaga D, Contreras MJ, Nuñez-Montero K, Barrientos L, Goméz-Espinoza O, Iturrieta-González I (2024) Unlocking fungal potential: the CRISPR-Cas system as a strategy for secondary metabolite discovery. J Fungi 10(11):748. https://doi.org/10.3390/jof10110748

Lee S, Kim P (2020) Current status and applications of adaptive laboratory evolution in industrial microorganisms. J Microbiol Biotechnol 30(6):793–803. https://doi.org/10.4014/jmb.2003.03072

Lee RA, Lavoie JM (2013) From first- to third-generation biofuels: challenges of producing a commodity from a biomass of increasing complexity. Anim Front 3(2):6–11. https://doi.org/10.2527/AF.2013-0010

Lee SY, Park JH, Jang SH, Nielsen LK, Kim J, Jung KS (2008) Fermentative butanol production by Clostridia. Wiley Online LibrarySY Lee, JH Park, SH Jang, LK Nielsen, J Kim, KS JungBiotechnology and Bioengineering, 2008•Wiley Online Library 101(2):209–228. https://doi.org/10.1002/BIT.22003

Li Y, Chiu YH, Wang L, Liu YC, Chiu CR (2019) A comparative study of different energy efficiency of OECD and non-OECD countries. Trop Conserv Sci 12. https://doi.org/10.1177/1940082919837441/ASSET/CD76F95B-9B3D-493A-AD5A-A4ECC4373B8A/ASSETS/IMAGES/LARGE/10.1177_1940082919837441-FIG6.JPG

Liu T, Zou W, Liu L, Chen J (2012) A constraint-based model of Scheffersomyces stipitis for improved ethanol production. Springer 5. https://doi.org/10.1186/1754-6834-5-72

Ljungdahl LG, Eriksson K-E (1985) Ecology of microbial cellulose degradation, pp 237–299. https://doi.org/10.1007/978-1-4615-9412-3_6

Lynd LR, Weimer PJ, van Zyl WH, Pretorius IS (2002) Microbial cellulose utilization: fundamentals and biotechnology. Microbiol Mol Biol Rev 66(3):506–577. https://doi.org/10.1128/MMBR.66.3.506-577.2002/ASSET/6F6A28CF-2C07-40BC-B7A7-BAF0F705A33E/ASSETS/GRAPHIC/MR0320014111T.JPEG

Ma B, Li Y, Wang T, Li D, Jia S (2025) Advances in CRISPR/Cas9-based gene editing in filamentous fungi. J Fungi 11(5):350. https://doi.org/10.3390/jof11050350

Manochio C, Andrade BR, Rodriguez RP, Moraes BS (2017) Ethanol from biomass: a comparative overview. Renew Sust Energ Rev 80:743–755. https://doi.org/10.1016/J.RSER.2017.05.063

Martínez Á, Speranza M, Ruiz-Dueñas F, Ferreira P (2005) Biodegradation of lignocellulosics: microbial, chemical, and enzymatic aspects of the fungal attack of lignin. https://scielo.isciii.es/pdf/im/v8n3/07%20Martinez.pdf

Mathew AK, Abraham A, Mallapureddy KK, Sukumaran RK (2018) Lignocellulosic biorefinery wastes, or resources? In: Waste biorefinery: potential and perspectives, pp 267–297. https://doi.org/10.1016/B978-0-444-63992-9.00009-4

Maurya DP, Singla A, Negi S (2015) An overview of key pretreatment processes for biological conversion of lignocellulosic biomass to bioethanol. 3 Biotech 5(5):597–609. https://doi.org/10.1007/S13205-015-0279-4/TABLES/1

McKendry P (2002) Energy production from biomass (part 1): overview of biomass. Bioresour Technol 83(1):37–46. https://doi.org/10.1016/S0960-8524(01)00118-3

Mizik T (2025a) European Union guidelines for the production of different generations of biofuels. In: Biofuels and sustainability. Elsevier, pp 205–219. https://doi.org/10.1016/B978-0-443-21433-2.00006-2

Mizik T (2025b) European Union guidelines for the production of different generations of biofuels. In: Biofuels and sustainability: life cycle assessments, system biology, policies, and emerging technologies, pp 205–219. https://doi.org/10.1016/B978-0-443-21433-2.00006-2

Mizik T, Gyarmati G (2022) Three pillars of advanced biofuels' sustainability. Fuels 3(4):607–626. https://doi.org/10.3390/FUELS3040037

Mizik T, Igbeghe CB, Deák Z (2025) Production efficiency of advanced liquid biofuels: prospects and challenges. Energies 18(4):1008. https://doi.org/10.3390/en18041008

Moreira JR (2006) Global biomass energy potential. Mitig Adapt Strateg Glob Chang 11(2):313–333. https://doi.org/10.1007/S11027-005-9003-8/METRICS

Moshood TD, Nawanir G, Mahmud F (2021) Microalgae biofuels production: a systematic review on socioeconomic prospects of microalgae biofuels and policy implications. Environ Challeng 5:100207. https://doi.org/10.1016/j.envc.2021.100207

Mukhtar A, Saqib S, Lin H, Hassan Shah MU, Ullah S, Younas M, Rezakazemi M, Ibrahim M, Mahmood A, Asif S, Bokhari A (2022) Current status and challenges in the heterogeneous catalysis for biodiesel production. Renew Sust Energ Rev 157:112012. https://doi.org/10.1016/j.rser.2021.112012

Nigam PS, Singh A (2011) Production of liquid biofuels from renewable resources. Prog Energy Combust Sci 37(1):52–68. https://doi.org/10.1016/j.pecs.2010.01.003

Olzhausen J, Grigat M, Seifert L, Ulbricht T, Schüller HJ (2021) Increased biosynthesis of acetyl-CoA in the yeast Saccharomyces cerevisiae by overexpression of a deregulated pantothenate kinase gene and engineering of the. SpringerJ Olzhausen, M Grigat, L Seifert, T Ulbricht, HJ SchüllerApplied Microbiology and Biotechnology, 2021•Springer 105(19):7321–7337. https://doi.org/10.1007/S00253-021-11523-4

Padder SA, Khan R, Rather RA (2024) Biofuel generations: new insights into challenges and opportunities in their microbe-derived industrial production. Biomass Bioenergy 185:107220. https://doi.org/10.1016/j.biombioe.2024.107220

Peralta-Yahya PP, Zhang F, del Cardayre SB, Keasling JD (2012) Microbial engineering for the production of advanced biofuels. Nature 488(7411):320–328. https://doi.org/10.1038/nature11478

Prussi M, Padella M, Conton M, Postma ED, Lonza L (2019) Review of technologies for biomethane production and assessment of Eu transport share in 2030. J Clean Prod 222:565–572. https://doi.org/10.1016/j.jclepro.2019.02.271

Razzak SA, Hossain MM, Lucky RA, Bassi AS, de Lasa H (2013) Integrated CO2 capture, wastewater treatment and biofuel production by microalgae culturing—a review. Renew Sust Energ Rev 27:622–653. https://doi.org/10.1016/j.rser.2013.05.063

Saha BC (2003) Hemicellulose bioconversion. J Ind Microbiol Biotechnol 30(5):279–291. https://doi.org/10.1007/S10295-003-0049-X

Saini JK, Saini R, Tewari L (2015) Lignocellulosic agriculture wastes as biomass feedstocks for second-generation bioethanol production: concepts and recent developments. SpringerJK Saini, R Saini, L Tewari3 Biotech, 2015•Springer 5(4):337–353. https://doi.org/10.1007/S13205-014-0246-5

Sajid Z, da Silva M, Danial S (2021) Historical analysis of the role of governance systems in the sustainable development of biofuels in Brazil and The United States of America (USA). Sustainability 13(12):6881. https://doi.org/10.3390/su13126881

Sánchez ÓJ, Cardona CA (2008) Trends in biotechnological production of fuel ethanol from different feedstocks. Bioresour Technol 99(13):5270–5295. https://doi.org/10.1016/J.BIORTECH.2007.11.013

Satari B, Karimi K, Kumar R (2018) Cellulose solvent-based pretreatment for enhanced second-generation biofuel production: a review. Sustain Energy Fuels 3(1):11–62. https://doi.org/10.1039/C8SE00287H

Sen S, Malakar C, Banik B, Devi NB, Pakshirajan K (2024) Drop-in biofuel production using fungi, pp 183–219. https://doi.org/10.1007/978-3-031-61637-2_7

Sengupta P, Mohan R, Wheeldon I, Kisailus D, Wyman CE, Cai CM (2022) Prospects of thermotolerant Kluyveromyces marxianus for high solids ethanol fermentation of lignocellulosic biomass. SpringerP Sengupta, R Mohan, I Wheeldon, D Kisailus, CE Wyman, CM CaiBiotechnology for Biofuels and Bioproducts, 2022•Springer. 15(1). https://doi.org/10.1186/S13068-022-02232-9

Shen Q, Ruan H, Zhang H, Wu T, Zhu K, Han W, Dong R, Ming T, Qi H, Zhang Y (2024) Utilization of CRISPR-Cas genome editing technology in filamentous fungi: function and advancement potentiality. Frontiersin.Org 15. https://doi.org/10.3389/FMICB.2024.1375120/FULL

Shi S, Si T, Liu Z, Zhang H, Ang EL, Zhao H (2016) Metabolic engineering of a synergistic pathway for n-butanol production in Saccharomyces cerevisiae. Sci Rep 6(1):25675. https://doi.org/10.1038/srep25675

Silva M, Ramos AC, Lidon FJ, Reboredo FH, Gonçalves EM (2024) Pre- and postharvest strategies for Pleurotus ostreatus mushroom in a circular economy approach. Foods 13(10):1464. https://doi.org/10.3390/foods13101464

Sims REH, Mabee W, Saddler JN, Taylor M (2010) An overview of second generation biofuel technologies. Bioresour Technol 101(6):1570–1580. https://doi.org/10.1016/j.biortech.2009.11.046

Siripong W, Wolf P, Kusumoputri TP, Downes JJ, Kocharin K, Tanapongpipat S, Runguphan W (2018) Metabolic engineering of Pichia pastoris for production of isobutanol and isobutyl acetate. SpringerW Siripong, P Wolf, TP Kusumoputri, JJ Downes, K Kocharin, S TanapongpipatBiotechnology for Biofuels, 2018•Springer 11(1):1. https://doi.org/10.1186/S13068-017-1003-X

Surriya O, Saleem SS, Waqar K, Gul Kazi A, Öztürk M (2014) Bio-fuels: a blessing in disguise. SpringerO Surriya, SS Saleem, K Waqar, A Gul Kazi, M ÖztürkPhytoremediation for Green Energy, 2014•Springer:11–54. https://doi.org/10.1007/978-94-007-7887-0_2

Traverso L, Colangeli M, Morese M, Pulighe G, Branca G (2020) Opportunities and constraints for implementation of cellulosic ethanol value chains in Europe. Biomass Bioenergy 141:105692. https://doi.org/10.1016/j.biombioe.2020.105692

Uday USP, Choudhury P, Bandyopadhyay TK, Bhunia B (2016) Classification, mode of action and production strategy of xylanase and its application for biofuel production from water hyacinth. Int J Biol Macromol 82:1041–1054. https://doi.org/10.1016/J.IJBIOMAC.2015.10.086

Usmani RA, Mohammad AS, Ansari SS (2023) Comprehensive biofuel policy analysis framework: a novel approach evaluating the policy influences. Renew Sust Energ Rev 183:113403. https://doi.org/10.1016/j.rser.2023.113403

Uzoejinwa BB, He X, Wang S, El-Fatah Abomohra A, Hu Y, Wang Q (2018) Co-pyrolysis of biomass and waste plastics as a thermochemical conversion technology for high-grade biofuel production: recent progress and future directions elsewhere worldwide. Energy Convers Manag 163:468–492. https://doi.org/10.1016/J.ENCONMAN.2018.02.004

Vladisavljević G (2024) Droplet microfluidics for high-throughput screening and directed evolution of biomolecules. Micromachines 15(8):971. https://doi.org/10.3390/mi15080971

Voloshin RA, Rodionova MV, Zharmukhamedov SK, Nejat Veziroglu T, Allakhverdiev SI (2016) Review: biofuel production from plant and algal biomass. Int J Hydrog Energy 41(39):17257–17273. https://doi.org/10.1016/J.IJHYDENE.2016.07.084

Walker G, Stewart G (2016) Saccharomyces cerevisiae in the production of fermented beverages. Beverages 2(4):30. https://doi.org/10.3390/beverages2040030

Wei S, Jian X, Chen J, Zhang C, Hua Q (2017) Reconstruction of genome-scale metabolic model of Yarrowia lipolytica and its application in overproduction of triacylglycerol. SpringerS Wei, X Jian, J Chen, C Zhang, Q HuaBioresources and Bioprocessing, 2017•Springer 4(1):51. https://doi.org/10.1186/S40643-017-0180-6

Wess J, Brinek M, Boles E (2019) Improving isobutanol production with the yeast Saccharomyces cerevisiae by successively blocking competing metabolic pathways as well as ethanol and glycerol. SpringerJ Wess, M Brinek, E BolesBiotechnology for Biofuels, 2019•Springer 12(1). https://doi.org/10.1186/S13068-019-1486-8

Xu P, Qiao K, Ahn WS, Stephanopoulos G (2016) Engineering Yarrowia lipolytica as a platform for synthesis of drop-in transportation fuels and oleochemicals. Pnas.OrgP Xu, K Qiao, WS Ahn, G StephanopoulosProceedings of the National Academy of Sciences, 2016•pnas.Org 113(39):10848–10853. https://doi.org/10.1073/PNAS.1607295113

Yu Q, Liu R, Li K, Ma R (2019) A review of crop straw pretreatment methods for biogas production by anaerobic digestion in China. Renew Sust Energ Rev 107:51–58. https://doi.org/10.1016/J.RSER.2019.02.020

Zheng Y, Yu X, Zeng J, Chen S (2012) Feasibility of filamentous fungi for biofuel production using hydrolysate from dilute sulfuric acid pretreatment of wheat straw. Springer 5. https://doi.org/10.1186/1754-6834-5-50

Chapter 6
Recent Developments in Fungal Biotechnology for Biodiesel Production: Application Potential, Challenges, and Perspectives

Ayush Thakur, Vijay Kumar, Padma Sharma, Megha Sethi, Indu Bhushan Prasher, Sachitra Kumar Ratha, and Rajesh Jeewon

Abstract The growing concern with sustainable energy sources has driven extensive research in biodiesel production by microorganisms, and fungal biotechnology is one of the most promising directions. The oleaginous fungi and yeasts have the ability to store lipids and convert cheap byproduct materials, including plant industry wastes, agricultural wastes, and lignocellulosic biomass, into desirable lipids. The recent innovations in strain optimization, genetic engineering, and fermentation methodologies have increased the production of biodiesels, hence increasing the economic and operational feasibility of the production. Fungal lipases provide a biocatalytic path to transesterification, eliminating the need to use chemical pro-

A. Thakur · M. Sethi · I. B. Prasher (✉)
Department of Botany, Panjab University, Chandigarh, India

V. Kumar (✉)
Department of Botany, Panjab University, Chandigarh, India

Patanjali Herbal Research Department, Patanjali Research Foundation, Haridwar, Uttarakhand, India
P. Sharma
Department of Ecology, Environment, and Remote Sensing, Paryavaran Bhawan, Gladni, Jammu and Kashmir, India

S. K. Ratha
Algology Laboratory, CSIR-National Botanical Research Institute, Lucknow, Uttar Pradesh, India

Academy of Scientific and Innovative Research, Ghaziabad, Uttar Pradesh, India

R. Jeewon
Department of Health Sciences, Faculty of Medicine and Health Sciences, University of Mauritius, Réduit, Mauritius

Honghe Center for Mountain Futures, Kunming Institute of Botany, Chinese Academy of Sciences, Honghe County, Yunnan, People's Republic of China

School of Pharmacy, Zunyi Medical University, Zunyi, Guizhou, China

M. Tripathi et al. (eds.), *Recent Advancements in Fungal Biotechnology Volume 1*, Fungal Biology, https://doi.org/10.1007/978-981-92-0339-0_6

cesses. However, the issue of scalability, affordability, and process optimization continues to pose a major challenge to industrial implementation. This chapter discusses the current trends, significant applications, technological developments, and the ongoing challenges of producing fungal biodiesel. Moreover, future possibilities such as the CRISPR-based strain engineering, embedded biorefinery methods, and governing policy infrastructure that might stimulate the commercialization of fungal biodiesel are also discussed. Overall, the chapter gives a comprehensive overview of fungal biotechnology in the future biofuel economy.

Keywords Fungi · Bioprospecting · Oleaginous fungi · Fuel · Lignocellulosic biomass · Biorefinery methods

6.1 Introduction

The widespread use of non-renewable fossil fuels and the destruction of the environment have aroused interest in the research of sustainable renewable energy (Amin et al. 2022). The alternative sources of fossil fuels put more emphasis on eco-friendliness, affordability, renewability, sustainability, and widespread accessibility. Biodiesels are mono-alkyl esters of long-chain fatty acids, which are formed by the catalyzed interaction of vegetable oils or animal fats with alcohol. Biodiesel is a good substitute for a more conventional fossil fuel because it is low-carbon, non-toxic, and exhibits better lubricity (John et al. 2022). It is produced through the process of transesterification of fatty acids obtained from various feedstocks, such as microalgae, cyanobacteria, yeasts, fungi, bacteria, and plant-seed oils (Ibrahim et al. 2023).

The global demand for biofuels has risen, with the United States and Brazil constituting nearly 85% of the world's bioethanol production, totaling 15.8 billion and 8.6 billion gallons, respectively (Karp et al. 2021). The European Union and China produced 1.5 million and 1.1 million gallons, respectively (Ramachandraiah et al. 2021). India has developed legislation to increase the percentage of ethanol in gasoline to 10% with a target of 20% (Ramachandraiah et al. 2021). Biodiesel comprises fatty acid esters and short-chain monoalcohols that can be used in diesel engines as either fuels or as various mixes with mineral diesel. The biodiesel that is industrially produced consists of soybean, rape, and palm oil that have been prepared into the ester and transester forms. Minor contributions are the waste, such as the cooking oil and animal fats. The rapid increase in the production of biodiesel in the last few years makes it clear that the agricultural land left to produce food will be used to meet the future need of biofuels at the cost of lowered prices and supply of edible oils (Azócar et al. 2010).

Microbial-derived oils (MDOs) are an attractive alternative to saponifiable plant crop and animal waste lipids (Bento et al. 2021). Microorganisms such as microalgae, bacteria, and fungi produce lipidic oils, which are less prone to seasonal and climatic changes and often lead to higher productivity (Ochsenreither et al. 2016;

Patel et al. 2020). Fungi have been at the center stage when it comes to biotechnological utilization in biorefineries due to their outstanding biocatalytic properties, as they have become the preferred choice to harness the full potential of a vast variety of substrates to make a broad range of products (de Souza et al. 2025). Many fungal strains have been identified as potential alternative sources of triacylglycerols, exhibiting physicochemical properties similar to those of traditional vegetable oils.

However, obstacles such as the shortage of agricultural land and sustainable raw materials still exist. Conventional biodiesel production methods are based on vegetable oils and animal fats, which, in turn, pose a competition to the food industry (Zheng et al. 2012). MDO is considered a good substitute for vegetable oil and animal fat due to its high production capacity and low land demand. In this regard, microbial oils offer an attractive alternative for better economic and environmental sustainability.

Studies have shown that fungi have oleaginous properties in glucose and sucrose-supplemented media. Fungal biomass is an effective source of lipid extraction, and the microbial lipid can be subsequently transformed into fatty acid methyl esters (FAMEs) or biodiesel by employing cost-effective methodologies. They can be grown on waste or low-quality biomass as a substrate in large-scale bioreactors with high efficiency. This chapter provides insights into fungal biotechnology in biodiesel production.

6.2 Fungal Strains in Biodiesel Production

Fungi can be utilized as highly efficient biofactories to produce biodiesel by using two principle approaches: first, oleaginous fungal strains are grown on inexpensive organic biomass, often agricultural or industrial waste, to accumulate large intracellular stores of neutral lipids—triacylglycerides, which are utilized in the biodiesel production, and second with filamentous fungi to utilize their ability to secrete hydrolytic enzymes, especially extracellular lipases, which can act as efficient and sustainable biocatalysts to facilitate the transesterification reaction. Table 6.1 lists the different fungi strains used to produce lipid and lipase.

6.3 Fungal Biotechnology for Biodiesel Feedstock

Fungal species can be derived from various sources, including soil, industrial waste, wastewater, ocean, plants, agro-industrial byproducts, and lignocellulosic materials (Li et al. 2023; Norphanphoun et al. 2018; Gomdola et al. 2024). Fungal cultures are isolated using standard methods (Senanayake et al. 2020). The different types of fungi have the ability to metabolize a wide range of sugars, glycerol, and lignocellulosic compounds (Hyde et al. 2018). The fungal biodiesel generation process is usually carried out by inoculation and growth of a suitable fungal strain on a

Table 6.1 List of fungal and yeast strains with respect to carbon source-dependent lipid accumulation for biofuel prospects

Fungal strain	Carbon source	% Lipid content (w/w)	References
Aspergillus carneus OQ275240	Glucose and yeast extract	36.20%	Ibrahim et al. (2023)
Aspergillus niger	Glucose and xylose	12.0%	Hasni et al. (2023)
A. niger	Glucose	9.6%	Zheng et al. (2012)
A. niger	Xylose	8.0%	Zheng et al. (2012)
A. niger MH079049.1 and *A. niger* MH078571.1	Tween 80 (for lipase production)	Not reported—fungi were used as lipase producers and biocatalysts for transesterifying plant oils into biodiesel.	Al-Anazi et al. (2024)
A. niger AS-101	Pure cellulose	14.9%	Singh (1991)
Aspergillus oryzae	Glucose and xylose	11.9%	Hasni et al. (2023)
A. oryzae	Glucose	27.0%	Ahmad et al. (2015)
A. oryzae	Xylose	21.0%	Ahmad et al. (2015)
A. oryzae	EFB hydrolysate (xylose-rich)	14.0%	Ahmad et al. (2016)
Aureobasidium melanogenum	Inulin	66.2%	Li et al. (2018)
Cutaneotrichosporon oleaginosus	4-Hydroxybenzoic acid	44.9%	Yaguchi et al. (2020)
C. oleaginosus	p-Coumarate	48.2%	Yaguchi et al. (2020)
C. oleaginosus	Resorcinol	45.6%	Yaguchi et al. (2020)
C. oleaginosus	Phenol	29.2%	Yaguchi et al. (2020)
Gilbertella persicaria DSR1	Glucose/KandR (Kendrick and Ratledge) medium	34.5%	Shah et al. (2021)
Lichtheimia corymbifera	Watermelon peel waste	39.56%	Hashem et al. (2020a)
Lipomyces starkeyi	Xylose	85.1%	Juanssilfero et al. (2018)
L. starkeyi	Cellobiose and xylose	52%	Gong et al. (2012)
L. starkeyi	Sugarcane bagasse hydrolysate	27.3%	Xavier et al. (2017)

(continued)

Table 6.1 (continued)

Fungal strain	Carbon source	% Lipid content (w/w)	References
Mucor circinelloides	Vinasse and molasses	25%	Reis et al. (2020)
M. circinelloides IBT-83	Corn steep solids, plant oil, glucose	60%	Szczęsna-Antczak et al. (2018)
M. circinelloides Q531	Dried lignocellulosic materials	28.8%	Qiao et al. (2018)
M. circinelloides URM 4182	Sucrose-based feedstocks (brown sugar, sugarcane juice, and sugarcane syrup)	39.22%, 29.15%, and 20.03%, respectively	Bento et al. (2019)
M. circinelloides URM 4182	Sugarcane bagasse	25%	Carvalho et al. (2019)
M. circinelloides URM 4182	Sugarcane molasses	26%	Bento et al. (2020)
M. circinelloides URM 4182	Cheese whey	22.5%	Braz et al. (2020)
M. circinelloides URM 4182	Sugarcane molasses	29%	Bento et al. (2021)
M. circinelloides URM 4182	Glucose	43%	Carvalho et al. (2018)
M. circinelloides ZSKP	Glucose	21.4%	Zininga et al. (2019)
M. griseocyanus (as *M. circinelloides*)	Sugarcane vinasse	23%	Rodrigues Reis et al. (2019)
M. indicus	Glucose	21%	Shafiei Alavijeh et al. (2020)
M. lusitanicus ATCC 1216B (as *M. circinelloides* f. *lusitanicus* ATCC 1216B)	Hydrolyzed whey permeates	32%	Chan et al. (2020)
M. racemosus	Glucose and xylose	22.6%	Assawah et al. (2020)
M. racemosus AH1	Starch	32.4%	Hashem et al. (2023)
Rhizopus oryzae	Glucose	34.8%	Zheng et al. (2012)
R. oryzae	Xylose	20.0%	Zheng et al. (2012)
R. stolonifer	Glucose and xylose	37.5%	Assawah et al. (2020)
Rhizopus sp.	Glucose and xylose	17.0%	Hasni et al. (2023)

(continued)

Table 6.1 (continued)

Fungal strain	Carbon source	% Lipid content (w/w)	References
Rhodotorula toruloides	Glucose	61.1%	Zhang et al. (2016b)
R. toruloides	Xylose	43.4%	Zhang et al. (2016b)
R. toruloides	Glucose	75.6%	Zhang et al. (2016a)
R. toruloides	Glucose and xylose	59%	Fei et al. (2016)
R. toruloides	Levoglucosan	35.6%	Lian et al. (2013)
R. toruloides	Sugarcane juice and urea	45%	Soccol et al. (2017)
R. toruloides	Corn stover hydrolysate	60.8%	Sànchez I Nogué et al. (2018)
R. toruloides	Glycerol	37.7%	Xu et al. (2017a)
Saccharomyces cerevisiae	Glucose	25.4%	Ferreira et al. (2018)
Syncephalastrum racemosum	Watermelon peel waste	29.4%	Hashem et al. (2020b)
Trichosporon coremiiforme	Glucose, xylose, and arabinose	37.8%	Huang et al. (2013)
Trichosporon cutaneum	Inulin	33.6%	Wang et al. (2015)
Yarrowia lipolytica	Glucose	81.4%	Xu et al. (2017b)
Y. lipolytica	Glucose	67%	Gong et al. (2012)
Y. lipolytica	Glucose	>70%	Xu et al. (2016)
Y. lipolytica	Glucose	66.8%	Qiao et al. (2017)
Y. lipolytica	Glucose	77%	Friedlander et al. (2016)
Y. lipolytica	Xylose	–	Li and Alper (2016)
Y. lipolytica	Cellobiose	–	Guo et al. (2015)
Y. lipolytica	Galactose	34%	Lazar et al. (2015)
Y. lipolytica	Molasses	>30%	Gajdoš et al. (2015)
Y. lipolytica	Fructose	49.8%	Hapeta et al. (2017)
Y. lipolytica	Starch	–	Ledesma-Amaro et al. (2015)

suitable substrate. The mycelium is then isolated by filtration using a muslin cloth after incubating. The visualization of intracellular lipid stores is achieved through staining with Sudan Black B and Nile red and is divided into lipophilic and non-lipophilic. When solubilized in lipid globules, Sudan Black B gives a dark-brown color, which can be viewed under a light microscope (Thakur et al. 1989). The Nile Red gives a golden yellow to deep red fluorescence, and is observed using a fluorescent microscope. (Thakur et al. 1989). The presence of lipid inclusions in the populations of fungal cells leads to the evaluation of the oleaginous possibilities. The mycelial biomass is then dried to constant weight, and then lipids are extracted by using known methods such as the Bligh-Dyer and Folch methods (Iverson et al. 2001). A fungal strain can be called oleaginous when the proportion of lipid mass is more than 20% of the dry biomass. The next process is the transesterification of triacylglycerols to form FAMEs, which are an ideal biodiesel precursor. FAMEs are produced by the met-alkylation of triacylglycerols with methanol using acid or base catalysts (Ejikeme et al. 2010). The composition of FAME is characterized using gas chromatography-mass spectrometry (GC-MS). The oleaginous fungi that produce stearic and oleic acids are an excellent biodiesel feedstock source since the resulting FAMEs have a higher energy density and better oxidative stability and lend to easier production of biodiesels in industrial processes (Meng et al. 2009). Figure 6.1 illustrates the overall workflow.

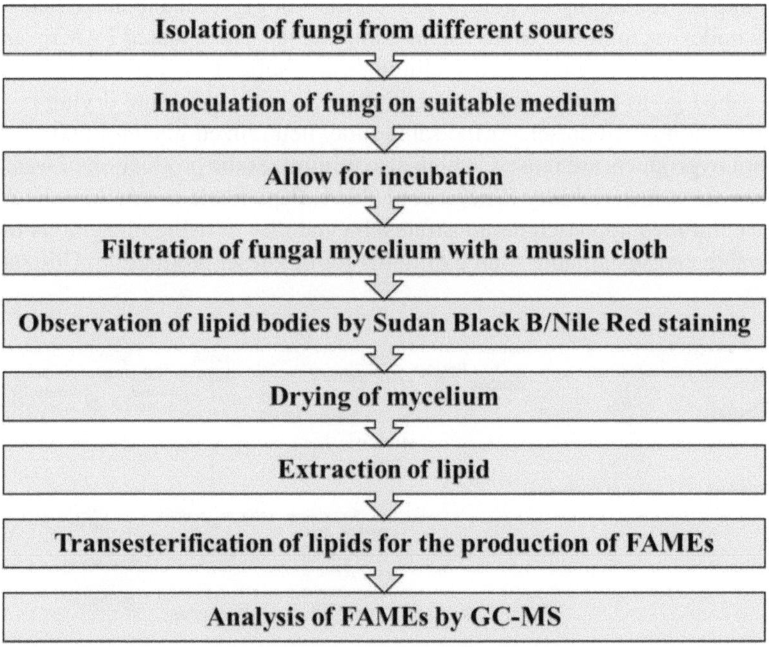

Fig. 6.1 Sequential steps involved in producing biodiesel by oleaginous fungi

6.3.1 Biodiesel Production from Agricultural Waste Using Oleaginous Fungi

Oleaginous fungi are filamentous fungi that can store lipids up to 90% of their dry cell biomass. They are known to have the ability to produce useful compounds, especially long-chain polyunsaturated fatty acids. In turn, these fungi have been deemed as a key concern in the framework of biorefinery since they can effectively utilize a wide range of substrates and convert these substrates into lipids that can be used to produce high-quality bio-diesel (Subhash and Mohan 2014). Lipid extraction processing in oleaginous fungi to produce bio-diesel comprises a stepwise chain of activities that aim at maximizing the characteristics of yield and also increasing the sustainability of the process (Fig. 6.2). First, agro-industrial waste products are used as substrates; they are also free and cost-effective sources of carbon and are eco-friendly (Ratledge and Wynn 2002). The growth of these fungi leads to the accumulation of large amounts of intracellular lipids in the biomass (Ageitos et al. 2011). After harvesting the biomass, transesterification is performed in a reactor in which the lipid-rich material is reacted with the methanol and a suitable catalyst to produce the FAMEs, which form the main constituents of biodiesel (Meng et al. 2009).

After transesterification, the mixture is transferred to the primary settling unit (Settler I), which facilitates the separation of FAMEs, methanol, and other byproducts. Methanol obtained during the process is then recycled to enhance overall efficiency, and the remaining mixture, which consists of glycerol and unreacted FAME, is dried under vacuum to extract useful components. The isolated FAMEs are then subjected to a washing procedure to eliminate any impurities. The FAMEs that have been washed go to a secondary settler (Settler II), and additional drying results in the purification of biodiesel. At the same time, neutralized glycerol and recovered methanol byproducts are reused, which also minimizes the production of wastes and increases economic viability (Sitepu et al. 2014; Papanikolaou and Aggelis 2011).

This combined approach demonstrates the viability of using oleaginous fungi as a renewable and sustainable source of lipids for biodiesel production. Utilization of

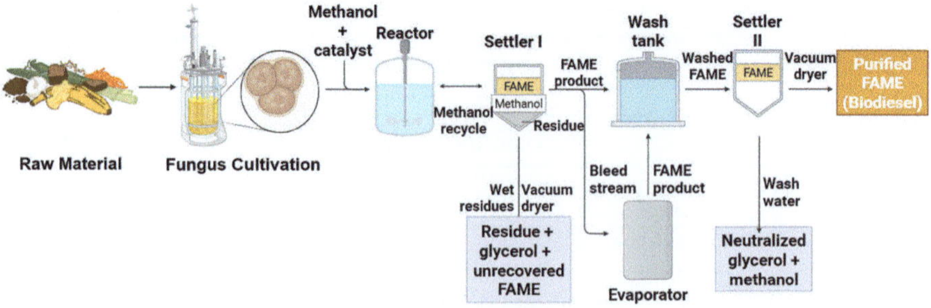

Fig. 6.2 Schematic process of lipid extraction from oleaginous fungi utilizing waste materials as feedstock for biodiesel production

waste-based feedstocks contributes not only to the reduction of the production costs but also to the concepts of a circular bioeconomy and waste valorization, which offers an environmentally friendly alternative to more standardized systems of biodiesel production (Subramaniam et al. 2010; Li et al. 2008; Papanikolaou and Aggelis 2019).

6.3.2 A Single-Step Biodiesel Production Using Fungal Biomass

A one-step method has been established in order to transform lipids in dried microbial biomass to FAME to eliminate the need for a prior lipid extraction step as well as the subsequent generation of biodiesel using isolated microbial oil (Fig. 6.3) (Vicente et al. 2009). The process combines lipid extraction, acid-catalyzed transesterification of the extracted saponifiable lipids, and acid-catalyzed esterification of the extracted free fatty acids in one step; it was initially recommended due to its significant reduction of both time and dissolution of the extracted lipids and solvent in analytical applications.

Similar procedures that do not require the lipid extraction step have been reported in various substrates, including marine tissues, yeast, fungi, bacteria, and microalgae (Bautista et al. 2012). In this context, the use of the technique on microorganisms can be particularly beneficial, since in the case of microorganisms, the solvent extraction method of obtaining lipids is typically used instead of the relatively inexpensive physical extraction methods used when it comes to extracting oilseeds (Bautista et al. 2012).

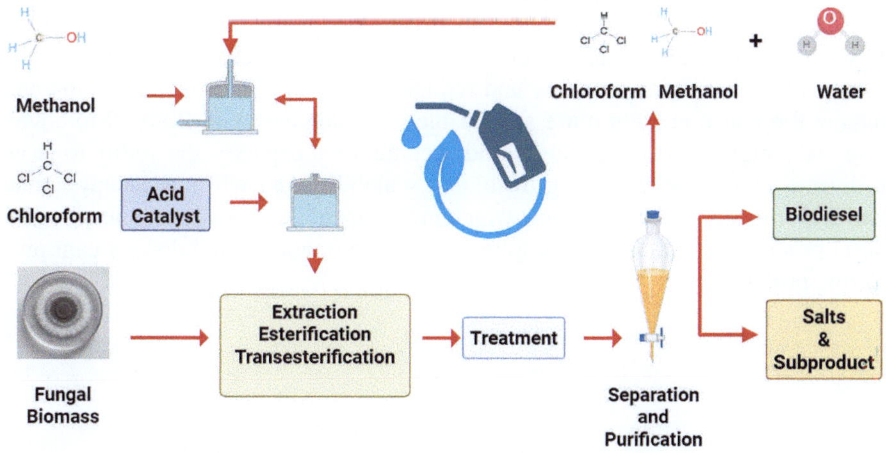

Fig. 6.3 A single-step biodiesel production using fungal biomass. (Adapted from Bautista et al. (2012))

Direct transformation of microbial lipid extracts to biodiesel has produced a greater purity and productivity than the traditional two-step process (Vicente et al. 2009; Bautista et al. 2012). In addition, waste production is also reduced by combining the uses of lipid extraction and acid-catalyzed transesterification/esterification of the extract lipids in a single process.

Oleaginous microorganisms, including yeast and filamentous fungi, can store large amounts of lipid in the cell, often more than 70% of dry biomass, especially when they grow under metabolic stress. Solid-state fermentation (SSF) is an advantageous approach for synthesizing fungal lipase, with filamentous fungi and yeasts frequently exhibiting superior productivities relative to submerged fermentation (Ramos Sanchez 2015).

Latest studies show that the non-sterile SSF of palm biomass waste using the *Aspergillus tubingensis* TSIP9 can produce oils with desirable biodiesel fuel properties (Intasit et al. 2020). Additionally, *Rhodotorula paludigena* CM3333 in pilot-scale fermentation using agro-industrial byproducts such as crude glycerol, cassava wastewater, and sugarcane bagasse hydrolysate has been effectively utilized as a source of carbon in lipid production by fungi (Sriphuttha et al. 2023). The lipids produced through these fungal biomasses can be isolated and can be transesterified to produce biodiesel (Chintagunta et al. 2021). Anaerobic digestion coupled with fungal fermentation, such as oleaginous fungi, can convert residual carbon sources and nutrients in digestate into biofuel precursor lipids and thus offer a water-free and energy-positive process. Moreover, a number of fungi, such as *Gliocladium roseum,* possess a robust lipid-synthesizing metabolism that facilitates the conversion of plant-derived carbon into diesel-like hydrocarbons, including alkanes, alkyl esters, and fatty acids (Strobel et al. 2008). The synthesis of hydrocarbons is significantly increased in conditions of microaerophilicity and the use of cellulose- or plant-based substrates (Strobel et al. 2008). This biocatalytic process takes advantage of the fact that fungi can metabolize low-cost raw materials and agricultural byproducts. Oleaginous organisms can utilize low-cost carbon feeds, including agricultural leftovers, lignocellulosic matter, crude glycerol, and other waste materials, which enables the exploitation of inexpensive feedstocks and encourages the concepts of the circular economy and reducing the overall cost of production and making the entire process more sustainable (Chintagunta et al. 2021). Moreover, fungi have high growth rates, high lipid accumulation capacity, the ability to grow in diverse environments, easy growth, and scalability and will not occupy arable land and compete with food crops, which makes them an appealing choice for biodiesel production in industries despite ongoing optimization and downstream processing problems.

6.4 Applications of Fungal Biotechnology in Biodiesel Production

The development of biotechnology has led to the genetic and metabolic characterization of fungi and, therefore, facilitated the identification of characteristics that can be used to maximize the production of biodiesel (Fig. 6.4). Genome editing and synthetic biology can also be used to enhance lipid biosynthetic pathways, which can increase oil production efficiency (Hussain et al. 2020; Xin et al. 2022; de Souza et al. 2025). There are also prospects in genetic and metabolic engineering to enhance cultivation modalities that can be used in large-scale biodiesel production by fungi. In an effort to curb the growing need for biodiesel and in an endeavor to increase their sustainability in multiple industries, there is an urgent need to understand the metabolic pathways and mode of movement of lipids.

The development of technologies has a central role in the production of biodiesel with the help of fungi. Enzymatic hydrolysis is a useful process that converts lignocellulosic biomass to fermentable sugars, which increases the availability of substrates by fungal fermentation (Carvalho et al. 2019). The method enhances the rate of conversion and prevents the production of counteractive byproducts. Recent biorefinery research has increasingly focused on the systematic identification and optimization of cellulolytic and hemicellulolytic enzyme systems, as enzymatic hydrolysis represents a major technical and economic constraint in second-generation biofuel production (Da Silva et al. 2024). Lignocellulosic feedstocks necessitate physicochemical pretreatment and subsequent enzymatic saccharification to release fermentable sugars; nevertheless, the efficacy of this conversion is significantly influenced by enzyme composition, operating pH, temperature, and

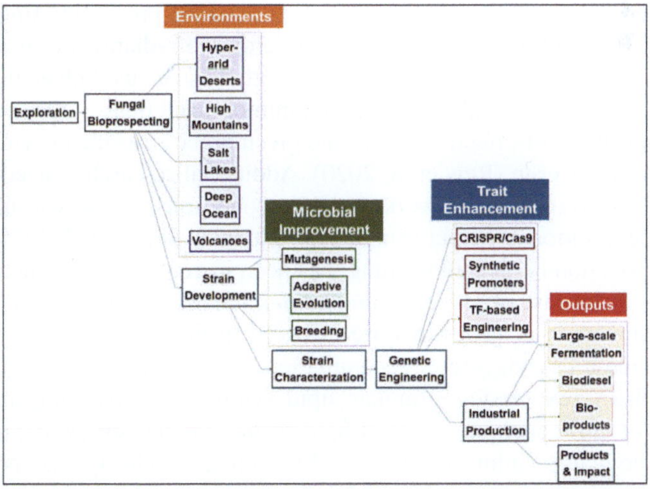

Fig. 6.4 Bioprospecting and biotechnological utilization of fungi for industrial production of biodiesel

reaction duration (Da Silva et al. 2024). Research on bamboo culm cellulose unequivocally shows that the strategic manipulation of hydrolysis parameters and enzyme formulations can significantly improve glucose release, achieving hydrolysis efficiencies close to total cellulose conversion under ideal conditions (Da Silva et al. 2024). Thus, the optimization of enzymatic saccharification is considered a crucial factor for increasing sugar yields, decreasing enzyme-related process expenses, and improving the overall technical and economic viability of lignocellulosic biorefineries.

Improvements in cultivation conditions have been significantly achieved due to bioreactor design, cultivation strategies, and improvements in reactor systems (Carvalho et al. 2019; Rodrigues Reis et al. 2019; Mhlongo et al. 2021). Mass and oxygen exchange in fixed-bed bioreactors is enhanced, leading to increased productivity (de Souza et al. 2025). Automated control and monitoring systems allow the accurate control of cultivation parameters, which is vital when increasing the scale of production from the laboratory to the industrial one (Mhlongo et al. 2021).

Although there has been development in the large-scale production of microbial lipids, there are various hurdles that still hinder commercial application as a precursor of biodiesel. The various types of cultivation methods, including fed-batch, continuous, and two-stage processes, are being optimized in order to attain high yields of lipids at lower costs (Karamerou and Webb 2019). Triglycerides have been shown to be effectively extracted in pressurized stainless-steel reactors, allowing the concomitant esterification and transesterification of triglycerides with ethanol acting as both a solvent and an acid acceptor. This has been significantly enhanced by the addition of heterogeneous catalysts, such as H_3PMo/Al_2O_3 (Reis et al. 2020; Bento et al. 2021).

One of the fields of innovation is the transesterification techniques, which transform lipids into biodiesel. The use of transesterification with the help of microwaves has increased the reaction efficiency and decreased the processing time (Carvalho et al. 2019; Bento et al. 2021). It utilizes microwave radiations to heat reactants, increasing the rate of lipid transformations to ethyl esters and enhancing economic viability. The use of biocatalysts, including immobilized lipases, may also increase the efficiency and cost-effectiveness of the given process, as the enzymes are reusable and cheaper to run (Reis et al. 2020). Additional research is needed to determine the biocatalysts that can work at milder conditions, and this would make the overall processes more efficient (Singh et al. 2020; de Souza et al. 2025).

Lipid production in oleaginous fungi has been increased using genetic engineering, such as in the case of *Mucor circinelloides* and *Mucor lusitanicus* (Wang et al. 2021). Manipulation of genes that are involved in lipid metabolism has been targeted to increase triacylglycerol accumulation. As an example, malate transporter mt and sodit overexpression stimulate lipid synthesis by diverting malate to the production of citric acid, an essential lipid-forming precursor (Wang et al. 2021). Similarly, the manipulation of fatty-acid biosynthetic pathways has promoted the synthesis of individual fatty acids, which have prospective uses in fungal biotechnology (Zhang et al. 2022).

Recent advances have been beneficial to the field, especially CRISPR/Cas9-mediated genome editing being the most successful in enabling targeted genetic editing to silence or silence lipid production-related genes. Next-generation sequencing and metagenomic techniques have been further used to determine new genetic targets, thus making it possible to develop more specific means of producing lipids (Shanmugam et al. 2019). Such approaches enhance lipid production by sustainable use of resources and reducing external outputs. These advanced technologies, when combined with genetic engineering, are defining the future of biotechnological use of biofuels (Bento et al. 2021). New approaches, including synthetic and systems biology, have the potential to provide more insight into the production of biodiesel through fungal systems. Genetic manipulation that optimizes metabolic processes can result in lipid-producing strains that are also resistant to adverse environmental factors, hence increasing the efficiency of production and the responsiveness to market needs.

6.5 Challenges in Fungal Biodiesel Production

Fungi have the potential to produce biodiesel; however, there are still many obstacles. The major problem is in the recalcitrance of the lignocellulosic substrates that hampers the release of fermentable sugars because of the complex nature of plant biomass due to the presence of cellulose, hemicellulose, and lignin. Polymerization of this biomass is characteristically technologically complex and high-energy-driven. The result of this transformation is the formation of inhibitory compounds, that is, hydroxymethyl-furfural and furfural, which could impede fungal growth and fermentation activity (Carvalho et al. 2019; Hussain et al. 2020; Mhlongo et al. 2021). There are other complications, caused by the heterogeneity of the fungi; some taxa can generate significant amounts of lipid, and others are not well-performing in the same conditions (Hashem et al. 2020a; Fazili et al. 2022).

Genetic and phenotypic characterization is crucial for identifying and utilizing the most productive fungal and yeast strains for biofuel applications (de Souza et al. 2025). However, species identification based solely on morphological characteristics is often unreliable due to the vast diversity and plasticity of fungal forms.

Phenotypic methods of characterization are often not able to differentiate closely related taxa due to the lack of molecular data (Kurtzman et al. 2015). Further, many organisms, such as industrial yeasts, have a high intraspecific genetic variation, leading to the appearance of phenotypically different variants with different metabolic potentials (Onyema et al. 2023). The observations highlight the importance of combining genotypic and phenotypic studies in order to taxonomically define the boundaries and to select the best strains. The experimental studies carried out with *Rhizopus oryzae* B97 have established that the accumulation of lipids and fatty-acid productivity are significantly affected by the pH, temperature, nutrient composition, and the environmental conditions, thus showing that the lack of standardized cultivation parameters prevents the effective achievement of high biodiesel yields (Ullah et al. 2025). In line with this, literature on oleaginous fungi convincingly shows that the factors influencing cultivation have a profound control over the cellular metabolism (Mhlongo et al. 2021).

The absence of standardized protocols that can be universally applicable in laboratories and on a production scale undermines the quality of data comparability, hampers the scaling of a process, and is a major setback in industrial implementation. Despite the promising biocatalytic properties, their growth is often associated with contamination issues that negatively impact the yield, purity, and reproducibility of the desired bioactive molecules.

Although interest in using fungi as potential biocatalysts to convert lignocellulosic biomass into biofuels and biodiesel biorefineries is increasing, their broader deployment is hindered by biological and process-engineering issues. Anaerobic and filamentous fungi have a high sensitivity to oxygen levels, nutrient composition, and retention time, and large-scale cultivation is prone to contamination and biological instability (Saye et al. 2021). Further production of fatty-acid and lipid-derived biofuels is significantly influenced by the type of substrate, enzyme formulation, pH, temperature, and fermentation strategy. Even slight deviation of the optimal conditions of hydrolysis significantly decreases the accessibility of fermentable sugars, so there are no standardized, universally applicable production guidelines in fungal or non-fungal microbial systems (Da Silva et al. 2024). There is a significant economic challenge posed by downstream recovery, as lignocellulosic biorefineries do not produce fuel-relevant chemicals as easily because the fermentation matrices chemically become complex (Da Silva et al. 2024). Overall, detailed bioprospecting and strain identification are necessary because anaerobic and lignocellulose-degrading fungi are poorly studied, yet have incredibly diverse cellulosome systems and carbohydrate-active enzyme systems that are not well characterized. Therefore, new taxa and enzymatic systems should be discovered to enhance the resilience and productivity, as well as the industrial feasibility of fungal biofuel systems (Saye et al. 2021).

There is a lack of large-scale infrastructure for bioconversion, which turns out to be a limiting factor. Even though biodiesels are gaining popularity, there is a lack of processing plants in various regions, similar to fossil fuels. This shortcoming can be addressed by means of bidirectional integration of biodiesel production with parallel bioprocessing activities, but it requires enormous investment and support (Mhlongo et al. 2021; Saye et al. 2021). The policies and regulations that govern the industry of biofuels often languish behind the needs of the industry, favoring the conventional energy sources. The progress in this field is valuable only by advancing policies that support the development of second- and third-generation biofuels and their application (Parsons et al. 2018).

Finally, there is the continuing challenge of production costs. Fungi have the potential to convert waste into a raw material; however, the costs of processing and conversion are still high. The cost of fungi-based biodiesel production has to be made commercially viable and competitive through research to maximize the processes and minimize operational costs in the biofuels sector.

6.6 Future Perspectives

The utilization of fungi in biodiesel production is promising, and there are various recent studies in different fields. In order to maximize microbial oil production, there is an urgent need to discover and characterize fungal strains that have high lipid-producing ability. Cellular and molecular biology of fungi provides information on lipid deposition and resistance to the inhibitors, which can help to generate the optimized strain (Wang et al. 2021). These organisms are likely to be used in synthetic biotechnology to boost the production of biodiesel. Genetic engineering has the capacity to enhance the metabolic function, allowing fungi to accumulate lipids even within poor environmental conditions (Zhang et al. 2021; Wang et al. 2021). The gene-editing methods like CRISPR can play a significant role in this process by producing strains that are more productive of lipids and also have better survival under unfavorable environments (Shanmugam et al. 2019; Hussain et al. 2020).

Furthermore, various opportunities regarding large-scale lipid production are introduced by the implementation of continuous and cascade fermentation regimes. A more efficient and cost-effective system can be provided as a continuous process that combines various phases of different processes (Mhlongo et al. 2021). Real-time monitoring technologies also optimize cultivation processes, thus streamlining the conditions for lipid biosynthesis (Carvalho et al. 2019; Mhlongo et al. 2021).

Production of fungal biodiesel will require interdisciplinary cooperation of scientific, engineering, and industrial aspects. The interface between academia, research, and business can help speed up technology transfer and enable scale-up. The knowledge sharing across disciplines can result in new methods that can meet the market needs. Finally, biodiesel based on fungi will rely on the level of awareness and governmental support. Sustainable biodiesel can be used to support a low-carbon economy by adopting biodiesel, which can be facilitated by the use of policies that stimulate the exploration and implementation of sustainable biodiesel.

6.7 Conclusion

Fungal biotechnology is significant in the development of biodiesel production as a sustainable, feasible, and eco-friendly alternative to global energy demands. The oleaginous fungi and yeast can accumulate high lipid content, often using inexpensive raw materials such as agricultural residues and lignocellulosic biomass; hence, they solve the twofold task of waste reuse and renewable fuel production. In addition, the fungal lipases provide an adequate biocatalytic substitute to the traditional chemical transesterification procedures, improving the quality and producing sustainable biodiesel. The improvement of fungal strains, metabolic engineering, and new fermentation technologies have enhanced the yield of lipids and efficiency of the processes to a great extent. However, despite such promising advances, various challenges, such as scalability, downstream processing, and economic viability, still limit large-scale implementation. Recent advances in novel biorefinery strategies, combined with modern genome-editing tools, like CRISPR-Cas systems, have significant potential to overcome these barriers. Moreover, integrating fungal biodiesel manufacturing in circular bioeconomy models would be the best way to ensure maximum use of resources and reduced environmental footprints.

In the future, interdisciplinary collaboration between microbiology, metabolomics, and policy support will be necessary to take the fungal biodiesel out of laboratory research and into industrial reality. Infrastructure, regulatory, and sustainable supply chain strategic investments significantly facilitate commercialization. Altogether, fungal biotechnology is one of the cornerstones of the future of renewable energy, as it has the potential to decrease the use of fossil fuels and promote global energy security to achieve a new climate-aware world.

References

Ageitos JM, Vallejo JA, Veiga-Crespo P, Villa TG (2011) Oily yeasts as oleaginous cell factories. Appl Microbiol Biotechnol 90(4):1219–1227. https://doi.org/10.1007/s00253-011-3200-z

Ahmad FB, Zhang Z, Doherty WOS, O'Hara IM (2015) A multi-criteria analysis approach for ranking and selection of microorganisms for the production of oils for biodiesel production. Bioresour Technol 190(264):273. https://doi.org/10.1016/j.biortech.2015.04.083

Ahmad FB, Zhang Z, Doherty WO, O'Hara IM (2016) Evaluation of oil production from oil palm empty fruit bunch by oleaginous microorganisms. Biofuels Bioprod Biorefin 10(4):378–392. https://doi.org/10.1002/bbb.1645

Al-Anazi NA, Alabdalall AH, Alsoufi MH, Al-Ghamdi A, Aldakheel FA (2024) Promising abilities of fungal lipases of *Aspergilli* strains in the production of biodiesel from plant oil wastes. Energies 17(2):381. https://doi.org/10.3390/en17020381

Amin M, Shah HH, Fareed AG, Khan WU, Chung E, Zia A, Rahman Farooqi ZU, Lee C (2022) Hydrogen production through renewable and non-renewable energy processes and their impact on climate change. Int J Hydrog Energy 47(77):33112–33134. https://doi.org/10.1016/j.ijhydene.2022.07.172

Assawah S, Gadidy M, Allah K, Badry A (2020) Evaluation of the oleaginous Rhizopus stolonifer capability for lipid production using agricultural wastes as cheap nutritional sources for biodiesel production. Egypt J Exp Biol (Bot) 16(1):35. https://doi.org/10.5455/egyjebb.20200228103519

Azócar L, Ciudad G, Heipieper HJ, Navia R (2010) Biotechnological processes for biodiesel production using alternative oils. Appl Microbiol Biotechnol 88(3):621–636. https://doi.org/10.1007/s00253-010-2804-z

Bautista LF, Vicente G, Garre V (2012) Biodiesel from microbial oil. In: Luque R, Melero JA (eds) Advances in biodiesel production. Elsevier, Amsterdam, pp 179–120

Bento HBS, Carvalho AKF, Reis CER, De Castro HF (2019) Microbial biodiesel production: from sucrose-based carbon sources to alkyl esters via enzymatic transesterification. Process Saf Environ Prot 121:349–356. https://doi.org/10.1016/j.psep.2018.11.010

Bento HBS, Carvalho AKF, Reis CER, De Castro HF (2020) Single cell oil production and modification for fuel and food applications: assessing the potential of sugarcane molasses as culture medium for filamentous fungus. Ind Crop Prod 145:112141. https://doi.org/10.1016/j.indcrop.2020.112141

Bento HBS, Reis CER, Cunha PG, Carvalho AKF, De Castro HF (2021) One-pot fungal biomass-to-biodiesel process: influence of the molar ratio and the concentration of acid heterogenous catalyst on reaction yield and costs. Fuel 300:120968. https://doi.org/10.1016/j.fuel.2021.120968

Braz CA, Carvalho AKF, Bento HBS, Reis CER, De Castro HF (2020) Production of value-added microbial metabolites: oleaginous fungus as a tool for valorization of dairy byproducts. Bioenergy Res 13(3):963–973. https://doi.org/10.1007/s12155-020-10121-y

Carvalho AKF, Bento HBS, Rivaldi JD, De Castro HF (2018) Direct transesterification of Mucor circinelloides biomass for biodiesel production: effect of carbon sources on the accumulation of fungal lipids and biofuel properties. Fuel 234:789–796. https://doi.org/10.1016/j.fuel.2018.07.029

Carvalho AKF, Bento HBS, Reis CER, De Castro HF (2019) Sustainable enzymatic approaches in a fungal lipid biorefinery based in sugarcane bagasse hydrolysate as carbon source. Bioresour Technol 276:269–275. https://doi.org/10.1016/j.biortech.2018.12.118

Chan LG, Dias FFG, Saarni A, Cohen J, Block D, Taha AY, De Moura Bell JMLN (2020) Scaling up the bioconversion of cheese whey permeate into fungal oil by *Mucor circinelloides*. J Americ Oil Chem Soc 97(7):703–716. https://doi.org/10.1002/aocs.12372

Chintagunta AD, Zuccaro G, Kumar M, Kumar SPJ, Garlapati VK, Postemsky PD, Kumar NSS, Chandel AK, Simal-Gandara J (2021) Biodiesel production from lignocellulosic biomass using oleaginous microbes: prospects for integrated biofuel production. Front Microbiol 12:658284. https://doi.org/10.3389/fmicb.2021.658284

Da Silva MF, Flaibam B, De Mélo AHF, Sampaio U, Clerici MTPS, Goldbeck R (2024) Optimization of enzymatic hydrolysis of cellulose extracted from bamboo culm for bioethanol production by Saccharomyces cerevisiae modified via CRISPR/Cas9. Food Res Int 192:114768. https://doi.org/10.1016/j.foodres.2024.114768

de Souza CAF, de Medeiros EV, Duda GP, Menezes RSC, Antonino ACD, Pacheco Filho JGDA, Araujo ASF (2025) Unlocking the biodiesel potential of Mucorales fungi: emerging trends in sustainable biofuel production. Microbe 7:100314. https://doi.org/10.1016/j.microb.2025.100314

Ejikeme PM, Anyaogu ID, Ejikeme CL, Nwafor NP, Egbuonu CAC, Ukogu K, Ibemesi JA (2010) Catalysis in biodiesel production by transesterification processes-an insight. J Chem 7(4):1120–1132. https://doi.org/10.1155/2010/689051

Fazili ABA, Shah AM, Zan X, Naz T, Nosheen S, Nazir Y, Ullah S, Zhang H, Song Y (2022) *Mucor circinelloides*: a model organism for oleaginous fungi and its potential applications in bioactive lipid production. Microb Cell Factories 21(1):29. https://doi.org/10.1186/s12934-022-01758-9

Fei Q, O'Brien M, Nelson R, Chen X, Lowell A, Dowe N (2016) Enhanced lipid production by *Rhodosporidium toruloides* using different fed-batch feeding strategies with lignocellulosic hydrolysate as the sole carbon source. Biotechnol Biofuels 9(1):130. https://doi.org/10.1186/s13068-016-0542-x

Ferreira R, Teixeira PG, Gossing M, David F, Siewers V, Nielsen J (2018) Metabolic engineering of *Saccharomyces cerevisiae* for overproduction of triacylglycerols. Metab Eng Commun 6:22–27. https://doi.org/10.1016/j.meteno.2018.01.002

Friedlander J, Tsakraklides V, Kamineni A, Greenhagen EH, Consiglio AL, MacEwen K, Crabtree DV, Afshar J, Nugent RL, Hamilton MA, Joe Shaw A, South CR, Stephanopoulos G, Brevnova EE (2016) Engineering of a high lipid producing *Yarrowia lipolytica* strain. Biotechnol Biofuels 9(1):77. https://doi.org/10.1186/s13068-016-0492-3

Gajdoš P, Nicaud J-M, Rossignol T, Čertík M (2015) Single cell oil production on molasses by *Yarrowia lipolytica* strains overexpressing DGA2 in multicopy. Appl Microbiol Biotechnol 99(19):8065–8074. https://doi.org/10.1007/s00253-015-6733-8

Gomdola D, Jeewon R, Gentekaki E, Jayawardena RS, Hyde KD, Alotibi F, Wang Y (2024) Assigning Xiuguozhangia (genus incertae sedis) to Pseudoberkleasmiaceae (Dothideomycetes, Pleosporomycetidae, Pleosporales) and introducing X. broussonetiae as a novel species. MycoKeys 111:181–210. https://doi.org/10.3897/mycokeys.111.136202

Gong Z, Wang Q, Shen H, Hu C, Jin G, Zhao ZK (2012) Co-fermentation of cellobiose and xylose by Lipomyces starkeyi for lipid production. Bioresour Technol 117:20–24. https://doi.org/10.1016/j.biortech.2012.04.063

Guo Z, Duquesne S, Bozonnet S et al (2015) Development of cellobiose-degrading ability in Yarrowia lipolytica strain by overexpression of endogenous genes. Biotechnol Biofuels 4(8):109. https://doi.org/10.1186/s13068-015-0289-9

Hapeta P, Rakicka M, Dulermo R, Gamboa-Meléndez H, Cruz-Le Coq A-M, Nicaud J-M, Lazar Z (2017) Transforming sugars into fat – lipid biosynthesis using different sugars in *Yarrowia lipolytica*: lipid biosynthesis using different sugars in *Yarrowia lipolytica*. Yeast 34(7):293–304. https://doi.org/10.1002/yea.3232

Hashem AH, Hasanin MS, Khalil AMA, Suleiman WB (2020a) Eco-green conversion of watermelon peels to single cell oils using a unique oleaginous fungus: *Lichtheimia corymbifera* AH13. Waste Biomass Valor 11(11):5721–5732. https://doi.org/10.1007/s12649-019-00850-3

Hashem AH, Suleiman WB, Abu-elreesh G, Shehabeldine AM, Khalil AMA (2020b) Sustainable lipid production from oleaginous fungus *Syncephalastrum racemosum* using synthetic and watermelon peel waste media. Bioresour Technol Rep 12:100569. https://doi.org/10.1016/j.biteb.2020.100569

Hashem AH, Abu-Elreesh G, El-Sheikh HH, Suleiman WB (2023) Isolation, identification, and statistical optimization of a psychrotolerant *Mucor racemosus* for sustainable lipid production. Biomass Conv Bioref 13(4):3415–3426. https://doi.org/10.1007/s13399-022-02390-8

Hasni MH, Ahmad FB, Athoillah AZ (2023) The production of microbial biodiesel from cellulose-derived fungal lipid via consolidated bioprocessing. Environ Technol Innov 30:103123. https://doi.org/10.1016/j.eti.2023.103123

Huang C, Chen X, Xiong L, Yang X, Chen X, Ma L, Chen Y (2013) Microbial oil production from corncob acid hydrolysate by oleaginous yeast *Trichosporon coremiiforme*. Biomass Bioenergy 49:273–278. https://doi.org/10.1016/j.biombioe.2012.12.023

Hussain SA, Garcia A, Khan MAK, Nosheen S, Zhang Y, Koffas MAG, Garre V, Lee SC, Song Y (2020) Increased accumulation of medium-chain fatty acids by dynamic degradation of long-chain fatty acids in *Mucor circinelloides*. Genes 11(8):890. https://doi.org/10.3390/genes11080890

Hyde KD, Norphanphoun C, Chen J, Dissanayake AJ, Doilom M, Hongsanan S, Jayawardena RS, Jeewon R et al (2018) Thailand's amazing diversity: up to 96% of fungi in northern Thailand may be novel. Fungal Divers 93:215–239

Ibrahim AG, Baazeem A, Al-Zaban MI, Fawzy MA, Hassan SHA, Koutb M (2023) Sustainable biodiesel production from a new oleaginous fungus, *Aspergillus carneus* strain OQ275240: biomass and lipid production optimization using Box–Behnken design. Sustainability 15(8):6836. https://doi.org/10.3390/su15086836

Intasit R, Cheirsilp B, Louhasakul Y, Boonsawang P, Chaiprapat S, Yeesang J (2020) Valorization of palm biomass wastes for biodiesel feedstock and clean solid biofuel through non-sterile repeated solid-state fermentation. Bioresour Technol 298:122551. https://doi.org/10.1016/j.biortech.2019.122551

Iverson SJ, Lang SLC, Cooper MH (2001) Comparison of the bligh and dyer and folch methods for total lipid determination in a broad range of marine tissue. Lipids 36(11):1283–1287. https://doi.org/10.1007/s11745-001-0843-0

John CB, Raja SA, Deepanraj B, Ong HC (2022) Palm stearin biodiesel: preparation, characterization using spectrometric techniques and the assessment of fuel properties. Biomass Conv Bioref 12(5):1679–1693. https://doi.org/10.1007/s13399-020-01265-0

Juanssilfero AB, Kahar P, Amza RL, Miyamoto N, Otsuka H, Matsumoto H, Kihira C, Thontowi A, Yopi, Ogino C, Prasetya B, Kondo A (2018) Effect of inoculum size on single-cell oil production from glucose and xylose using oleaginous yeast *Lipomyces starkeyi*. J Biosci Bioeng 125(6):695–702. https://doi.org/10.1016/j.jbiosc.2017.12.020

Karamerou EE, Webb C (2019) Cultivation modes for microbial oil production using oleaginous yeasts – a review. Biochem Eng J 151:107322. https://doi.org/10.1016/j.bej.2019.107322

Karp SG, Medina JDC, Letti LAJ, Woiciechowski AL, De Carvalho JC, Schmitt CC, De Oliveira Penha R, Kumlehn GS, Soccol CR (2021) Bioeconomy and biofuels: the case of sugarcane ethanol in Brazil. Biofuels Bioprod Biorefin 15(3):899–912. https://doi.org/10.1002/bbb.2195

Kurtzman CP, Mateo RQ, Kolecka A, Theelen B, Robert V, Boekhout T (2015) Advances in yeast systematics and phylogeny and their use as predictors of biotechnologically important metabolic pathways. FEMS Yeast Res 15(6):fov050. https://doi.org/10.1093/femsyr/fov050

Lazar Z, Gamboa-Meléndez H, Le Coq A-MC, Neuvéglise C, Nicaud J-M (2015) Awakening the endogenous Leloir pathway for efficient galactose utilization by *Yarrowia lipolytica.* Biotechnol Biofuels 8(1):185. https://doi.org/10.1186/s13068-015-0370-4

Ledesma-Amaro R, Dulermo T, Nicaud JM (2015) Engineering *Yarrowia lipolytica* to produce biodiesel from raw starch. Biotechnol Biofuels 8(1):148. https://doi.org/10.1186/s13068-015-0335-7

Li H, Alper HS (2016) Enabling xylose utilization in *Yarrowia lipolytica* for lipid production. Biotechnol J 11(9):1230–1240. https://doi.org/10.1002/biot.201600210

Li Q, Du W, Liu D (2008) Perspectives of microbial oils for biodiesel production. Appl Microbiol Biotechnol 80(5):749–756. https://doi.org/10.1007/s00253-008-1625-9

Li Y-F, Jiang H, Hu Z, Liu G-L, Chi Z-M, Chi Z (2018) Overexpression of an inulinase gene in an oleaginous yeast, *Aureobasidium melanogenum* P10, for efficient lipid production from inulin. Microb Physiol 28(4):190–200. https://doi.org/10.1159/000493139

Li J, Jiang H, Jeewon R, Hongsanan S, Bhat DJ et al (2023) *Alternaria*: update on species limits, evolution, multi-locus phylogeny, and classification. Stud Fungi 8:1. https://doi.org/10.48130/SIF-2023-0001

Lian J, Garcia-Perez M, Chen S (2013) Fermentation of levoglucosan with oleaginous yeasts for lipid production. Bioresour Technol 133:183–189. https://doi.org/10.1016/j.biortech.2013.01.031

Meng X, Yang J, Xu X, Zhang L, Nie Q, Xian M (2009) Biodiesel production from oleaginous microorganisms. Renew Energy 34(1):1–5. https://doi.org/10.1016/j.renene.2008.04.014

Mhlongo SI, Ezeokoli OT, Roopnarain A, Ndaba B, Sekoai PT, Habimana O, Pohl CH (2021) The potential of single-cell oils derived from filamentous fungi as alternative feedstock sources for biodiesel production. Front Microbiol 12:637381. https://doi.org/10.3389/fmicb.2021.637381

Norphanphoun C, Raspé O, Jeewon R, Wen T-C, Hyde KD (2018) Morphological and phylogenetic characterisation of novel *Cytospora* species associated with mangroves. MycoKeys 38:93–120. https://doi.org/10.3897/mycokeys.38.28011

Ochsenreither K, Glück C, Stressler T, Fischer L, Syldatk C (2016) Production strategies and applications of microbial single cell oils. Front Microbiol 7:1539. https://doi.org/10.3389/fmicb.2016.01539

Onyema VO, Amadi OC, Moneke AN, Agu RC (2023) A brief review: Saccharomyces cerevisiae biodiversity potential and promising cell factories for exploitation in biotechnology and industry processes – West African natural yeasts contribution. Food Chem Adv 2:100162. https://doi.org/10.1016/j.focha.2022.100162

Papanikolaou S, Aggelis G (2011) Lipids of oleaginous yeasts. Part I: biochemistry of single cell oil production. Eur J Lipid Sci Technol 113(8):1031–1051. https://doi.org/10.1002/ejlt.201100014

Papanikolaou S, Aggelis G (2019) Sources of microbial oils with emphasis to *Mortierella* (*Umbelopsis*) *isabellina* fungus. World J Microbiol Biotechnol 35(4):63. https://doi.org/10.1007/s11274-019-2629-1

Parsons S, Chuck CJ, McManus MC (2018) Microbial lipids: progress in life cycle assessment (LCA) and future outlook of heterotrophic algae and yeast-derived oils. J Clean Prod 172:661–672. https://doi.org/10.1016/j.jclepro.2017.10.014

Patel A, Karageorgou D, Rova E, Katapodis P, Rova U, Christakopoulos P, Matsakas L (2020) An overview of potential oleaginous microorganisms and their role in biodiesel and omega-3 fatty acid-based industries. Microorganisms 8(3):434. https://doi.org/10.3390/microorganisms8030434

Qiao K, Wasylenko TM, Zhou K, Xu P, Stephanopoulos G (2017) Lipid production in Yarrowia lipolytica is maximized by engineering cytosolic redox metabolism. Nat Biotechnol 35(2):173–177. https://doi.org/10.1038/nbt.3763

Qiao W, Tao J, Luo Y, Tang T, Miao J, Yang Q (2018) Microbial oil production from solid-state fermentation by a newly isolated oleaginous fungus, *Mucor circinelloides* Q531 from mulberry branches. R Soc Open Sci 5(11):180551. https://doi.org/10.1098/rsos.180551

Ramachandraiah A, Kaushik J, Gowda L, Chigadannavar PS, Krishnappa R, Venkatesh S, Sivadas S, Judith S, Ramesh UM, Quadri Z, Dey P, Chinthala M, Gola D, Chauhan N, Bharti RK (2021) Prospect of biofuel production by fungus. In: Gupta VK, Pandey A, Sharma GD, Tuohy MG (eds) Fungi bio-prospects in sustainable agriculture, environment and nano-technology. Elsevier, Amsterdam, pp 285–313

Ramos Sanchez LB (2015) Fungal lipase production by solid-state fermentation. J Bioprocess Biotech 5(2):1000203. https://doi.org/10.4172/2155-9821.1000203

Ratledge C, Wynn JP (2002) The biochemistry and molecular biology of lipid accumulation in oleaginous microorganisms. Adv Appl Microbiol 51(1):51. https://doi.org/10.1016/S0065-2164(02)51000-5

Reis CER, Valle GF, Bento HBS, Carvalho AKF, Alves TM, de Castro HF (2020) Sugarcane byproducts within the biodiesel production chain: vinasse and molasses as feedstock for oleaginous fungi and conversion to ethyl esters. Fuel 277:118064. https://doi.org/10.1016/j.fuel.2020.118064

Rodrigues Reis CE, Furtado Carvalho AK, Bento HBS, de Castro HF (2019) Integration of microbial biodiesel and bioethanol industries through utilization of vinasse as substrate for oleaginous fungi. Bioresour Technol Rep 6:46–53. https://doi.org/10.1016/j.biteb.2018.12.009

Sànchez I Nogué V, Black BA, Kruger JS, Singer CA, Ramirez KJ, Reed ML, Cleveland NS, Singer ER, Yi X, Yeap RY, Linger JG, Beckham GT (2018) Integrated diesel production from lignocellulosic sugars via oleaginous yeast. Green Chem 20(18):4349–4365. https://doi.org/10.1039/C8GC01905C

Saye LMG, Navaratna TA, Chong JPJ, O'Malley MA, Theodorou MK, Reilly M (2021) The anaerobic fungi: challenges and opportunities for industrial lignocellulosic biofuel production. Microorganisms 9(4):694. https://doi.org/10.3390/microorganisms9040694

Senanayake IC, Rathnayaka AR, Marasinghe DS, Calabon MS, Gentekaki E, Lee HB, Hurdeal VG, Pem D, Dissanayake LS, Wijesinghe SN, Bundhun D, Nguyen TT, Goonasekara ID, Abeywickrama PD, Bhunjun CS, Jayawardena RS, Wanasinghe DN, Jeewon R, Bhat DJ, Xiang MM (2020) Morphological approaches in studying fungi: collection, examination, isolation, sporulation and preservation. Mycosphere 11(1):2678–2754. https://doi.org/10.5943/mycosphere/11/1/20

Shafiei Alavijeh R, Karimi K, Van Den Berg C (2020) An integrated and optimized process for cleaner production of ethanol and biodiesel from corn stover by *Mucor indicus*. J Clean Prod 249:119321. https://doi.org/10.1016/j.jclepro.2019.119321

Shah AM, Mohamed H, Zhang Z, Song Y (2021) Isolation, characterization and fatty acid analysis of *Gilbertella persicaria* DSR1: a potential new source of high value single-cell oil. Biomass Bioenerg 151:106156. https://doi.org/10.1016/j.biombioe.2021.106156

Shanmugam K, Ramalingam S, Venkataraman G, Hariharan GN (2019) The CRISPR/Cas9 system for targeted genome engineering in free-living fungi: advances and opportunities for lichenized fungi. Front Microbiol 10:62. https://doi.org/10.3389/fmicb.2019.00062

Singh A (1991) Lipid production by a cellulolytic strain of *Aspergillus niger*. Lett Appl Microbiol 12(5):200–202. https://doi.org/10.1111/j.1472-765X.1991.tb00539.x

Singh N, Dhanya BS, Verma ML (2020) Nano-immobilized biocatalysts and their potential biotechnological applications in bioenergy production. Mater Sci Energy Technol 3:808–824. https://doi.org/10.1016/j.mset.2020.09.006

Sitepu IR, Sestric R, Ignatia L, Levin D, Bruce CG, Boundy-Mills KL (2014) Oleaginous yeasts for biodiesel: current and future trends in biology and production. Biotechnol Adv 32(7):1336–1360. https://doi.org/10.1016/j.biotechadv.2014.08.003

Soccol CR, Dalmas Neto CJ, Soccol VT, Sydney EB, Da Costa ESF, Medeiros ABP, Vandenberghe LPDS (2017) Pilot scale biodiesel production from microbial oil of *Rhodosporidium toruloides* DEBB 5533 using sugarcane juice: performance in diesel engine and preliminary economic study. Bioresour Technol 223:259–268. https://doi.org/10.1016/j.biortech.2016.10.055

Sriphuttha C, Boontawan P, Ketudat-Cairns M, Boontawan A (2023) Fermentation of non-edible lipids using agro-industrial residues (AIRs) for biodiesel production by *Rhodotorula paludigena* CM33. In: 2023 International conference on power and renewable energy engineering (PREE). IEEE, Tokyo, pp 76–81

Strobel GA, Knighton B, Kluck K, Ren Y, Livinghouse T, Griffin M, Spakowicz D, Sears J (2008) The production of myco-diesel hydrocarbons and their derivatives by the endophytic fungus *Gliocladium roseum* (NRRL 50072). Microbiology (Reading) 154(Pt 11):3319–3328. https://doi.org/10.1099/mic.0.2008/022186-0

Subhash GV, Mohan SV (2014) Lipid accumulation for biodiesel production by oleaginous fungus *Aspergillus awamori*: influence of critical factors. Fuel 116:509–515. https://doi.org/10.1016/j.fuel.2013.08.035

Subramaniam R, Dufreche S, Zappi M, Bajpai R (2010) Microbial lipids from renewable resources: production and characterization. J Ind Microbiol Biotechnol 37(12):1271–1287. https://doi.org/10.1007/s10295-010-0895-1

Szczęsna-Antczak M, Struszczyk-Świta K, Rzyska M, Szeląg J, Stańczyk Ł, Antczak T (2018) Oil accumulation and in situ trans/esterification by lipolytic fungal biomass. Bioresour Technol 265:110–118. https://doi.org/10.1016/j.biortech.2018.05.094

Thakur MS, Prapulla SG, Karanth NG (1989) Estimation of intracellular lipids by the measurement of absorbance of yeast cells stained with Sudan Black B. Enzym Microb Technol 11(4):252–254. https://doi.org/10.1016/0141-0229(89)90101-4

Ullah H, Anjum M, Noor B, Qadeer S, Nawaz R, Khalid A, Saleem AR, Kabeer B, Idris AM, Sohail MT, Rao Z, Ullah H, Anjum M, Noor B, Qadeer S, Nawaz R, Khalid A, Saleem AR, Kabeer B, Idris AM, Sohail MT, Rao Z (2025) Sustainable biofuel production from sludge by oleaginous fungi: effect of process variables on lipid accumulation. Catalysts 15(11):1009. https://doi.org/10.3390/catal15111009

Vicente G, Bautista LF, Rodríguez R, Gutiérrez FJ, Sádaba I, Ruiz-Vázquez RM, Torres-Martínez S, Garre V (2009) Biodiesel production from biomass of an oleaginous fungus. Biochem Eng J 48(1):22–27. https://doi.org/10.1016/j.bej.2009.07.014

Wang J, Zhang H, Bao J (2015) Characterization of inulin hydrolyzing enzyme(s) in oleaginous yeast *Trichosporon cutaneum* in consolidated bioprocessing of microbial lipid fermentation. Appl Biochem Biotechnol 177(5):1083–1098. https://doi.org/10.1007/s12010-015-1798-5

Wang X, Mohamed H, Bao Y, Wu C, Shi W, Song Y, Yang J (2021) Heterologous expression of two malate transporters from an oleaginous fungus *Mucor circinelloides* improved the lipid accumulation in *Mucor lusitanicus*. Front Microbiol 12:774825. https://doi.org/10.3389/fmicb.2021.774825

Xavier MCA, Coradini ALV, Deckmann AC, Franco TT (2017) Lipid production from hemicellulose hydrolysate and acetic acid by *Lipomyces starkeyi* and the ability of yeast to metabolize inhibitors. Biochem Eng J 118:11–19. https://doi.org/10.1016/j.bej.2016.11.007

Xin F, Dang W, Chang Y, Wang R, Yuan H, Xie Z, Zhang C, Li S, Mohamed H, Zhang H, Song Y (2022) Transcriptomic analysis revealed the differences in lipid accumulation between spores and mycelia of *Mucor circinelloides* WJ11 under solid-state fermentation. Fermentation 8(12):667. https://doi.org/10.3390/fermentation8120667

Xu P, Qiao K, Ahn WS, Stephanopoulos G (2016) Engineering *Yarrowia lipolytica* as a platform for synthesis of drop-in transportation fuels and oleochemicals. Proc Natl Acad Sci USA 113(39):10848–10853. https://doi.org/10.1073/pnas.1607295113

Xu J, Zhao X, Du W, Liu D (2017a) Bioconversion of glycerol into lipids by *Rhodosporidium toruloides* in a two-stage process and characterization of lipid properties. Eng Life Sci 17(3):303–313. https://doi.org/10.1002/elsc.201600062

Xu P, Qiao K, Stephanopoulos G (2017b) Engineering oxidative stress defense pathways to build a robust lipid production platform in *Yarrowia lipolytica*. Biotechnol Bioeng 114(7):1521–1530. https://doi.org/10.1002/bit.26285

Yaguchi A, Franaszek N, O'Neill K, Lee S, Sitepu I, Boundy-Mills K, Blenner M (2020) Identification of oleaginous yeasts that metabolize aromatic compounds. J Ind Microbiol Biotechnol 47(9–10):801–813. https://doi.org/10.1007/s10295-020-02269-5

Zhang S, Ito M, Skerker JM, Arkin AP, Rao CV (2016a) Metabolic engineering of the oleaginous yeast *Rhodosporidium toruloides* IFO0880 for lipid overproduction during high-density fermentation. Appl Microbiol Biotechnol 100(21):9393–9405. https://doi.org/10.1007/s00253-016-7815-y

Zhang S, Skerker JM, Rutter CD, Maurer MJ, Arkin AP, Rao CV (2016b) Engineering *Rhodosporidium toruloides* for increased lipid production. Biotechnol Bioeng 113(5):1056–1066. https://doi.org/10.1002/bit.25864

Zhang Y, Liu Q, Li P, Wang Y, Li S, Gao M, Song Y (2021) Enhanced lipid production by addition of malic acid in fermentation of recombinant *Mucor circinelloides* Mc-MT-2. Sci Rep 11(1):12674. https://doi.org/10.1038/s41598-021-92324-7

Zhang X-Y, Li B, Huang B-C, Wang F-B, Zhang Y-Q, Zhao S-G, Li M, Wang H-Y, Yu X-J, Liu X-Y, Jiang J, Wang Z-P (2022) Production, biosynthesis, and commercial applications of fatty acids from oleaginous fungi. Front Nutr 9:873657. https://doi.org/10.3389/fnut.2022.873657

Zheng Y, Yu X, Zeng J, Chen S (2012) Feasibility of filamentous fungi for biofuel production using hydrolysate from dilute sulfuric acid pretreatment of wheat straw. Biotechnol Biofuels 5(1):50. https://doi.org/10.1186/1754-6834-5-50

Zininga JT, Puri AK, Govender A, Singh S, Permaul K (2019) Concomitant production of chitosan and lipids from a newly isolated *Mucor circinelloides* ZSKP for biodiesel production. Bioresour Technol 272:545–551. https://doi.org/10.1016/j.biortech.2018.10.035

Chapter 7
Fungal Biotechnology for Biogas Production: Applications, Challenges, and Perspectives

Urmila Gupta Phutela, Isha, Yashika Aggarwal, and Tanisha Gupta

Abstract Fungal biotechnology presents an efficient and green route for energy generation through the valorization of complex organic substrates. Global demand for renewable energy has increased in recent decades due to the rate of urbanization and industrialization. Agro-industrial byproducts and lignocellulosic biomass present significant challenges to preserving environmental sustainability. Innovative fungal applications and contemporary biotechnological technologies are needed to manage such material. This chapter offers several methods for improving biogas production in pretreatment, enzymatic hydrolysis, and microbial consortia design that are based on fungal techniques. It is evident that the efficiency of fungal species lies in their exceptionally developed enzymatic machinery capable of depolymerizing complex plant polymers into fermentable intermediates. The continuous discovery of new fungal strains supplemented by technological advancement increases their potential in biogas technology. Another important aspect of the integration of fungal formulation in anaerobic digestion processes is that the utilization of residues becomes more feasible and economically viable. Besides, supplementation of fungal metabolites and enzyme cocktails contributed to improved methane yield and process stability. Advanced fungal bioprocesses may help in attaining much better sustainability and productivity for the future bioenergy system.

Keywords Biogas · Enzymatic hydrolysis · Fungal biotechnology · Pretreatment

U. G. Phutela (✉)
Department of Microbiology, Punjab Agricultural University, Ludhiana, Punjab, India

Department of Renewable Energy Engineering, Punjab Agricultural University, Ludhiana, Punjab, India
e-mail: urmilphutela@pau.edu

Isha · Y. Aggarwal
Department of Microbiology, Punjab Agricultural University, Ludhiana, Punjab, India

T. Gupta
Department of Plant Pathology, Punjab Agricultural University, Ludhiana, Punjab, India

M. Tripathi et al. (eds.), *Recent Advancements in Fungal Biotechnology Volume 1*, Fungal Biology, https://doi.org/10.1007/978-981-92-0339-0_7

7.1 Introduction

Global greenhouse gas emissions have significantly increased due to the expansion of industrial activities; in 2021, the manufacturing sector accounted for around 34% of carbon emissions and contributed around 20 gigatons of CO_2 annually (Dhakal et al. 2022). Many nations have set aggressive net-zero or carbon neutrality goals to address the growing effects of climate change, with a significant emphasis on lowering energy use and carbon footprints in industrial sectors. Biogas and other renewable energy sources provide a practical means to turn biological waste into clean energy while promoting the ideas of the circular economy. In addition to lowering reliance on fossil fuels, this bioenergy source helps achieve sustainability objectives, stabilize energy consumption, and minimize methane emissions through the incorporation of biogas. As a result, in the midst of the global energy crisis, biogas technology plays a crucial role in sustainable energy transitions (Harirchi et al. 2023).

The world's energy issue is still a major problem, especially in emerging nations where the energy supply is still dominated by fossil fuels such as coal, oil, and gas. Ensuring energy security, fostering economic progress, and safeguarding the environment are interrelated issues for these nations. However, forecasts show that fossil fuel reserves will run out within the next century, necessitating the immediate adoption of alternative energy sources.

In order to reduce greenhouse gas emissions and slow down climate change, international accords such as Agenda 21 and the Kyoto Protocol have sparked a move toward renewable and low-carbon energy. Biogas is unique among renewable energy sources since it can produce clean energy from biological waste, promoting sustainable development. Heating, lighting, and the creation of biofuel are a few of the industrial and home uses for biogas (Sahota et al. 2018).

Methane and carbon dioxide make up the majority of the biogas produced by anaerobic digestion (AD), a biological process in which microorganisms break down organic materials in the absence of oxygen. Different microbial populations help each of the four steps of the process, which are hydrolysis, acidogenesis, acetogenesis, and methanogenesis. Long retention durations, sensitivity to operating conditions (such as temperature), and partial breakdown of some compounds are common problems with conventional AD systems. Microbial activity and the stability of biogas generation can be disturbed by the accumulation of volatile fatty acids (VAFs) and inhibition by toxic substances such as ammonia and heavy metals. Additionally, the complicated structure of lignocellulose biomass limits the efficacy of conventional AD in breaking it down, making feedstock diversification difficult. Research into process optimization and the use of microbial consortia, particularly fungi, to increase resilience and efficiency, is prompted by these limitations, which restrict the scalability and economic viability of standard AD.

Over two-thirds of the world's population makes their living primarily from agriculture, which provides food for everyone. For almost 82% of the world's population, smallholder agriculture and allied industries are their primary source of income (Sarkar et al. 2020). Access to contemporary energy services is still a major obstacle, though; in 2012, over 836 million Indians did not have access to modern electricity (Kabeyi and Oludolapo 2020). Sustainable food and energy production depends on agricultural technologies that improve revenue while conserving resources.

7.2 Bioenergy Systems: Fungal Biotechnology Importance

As fungi offer highly effective biological tools for transforming lignocellulosic biomass and organic wastes into usable energy carriers, including bioethanol, biogas, and bioelectricity, fungal biotechnology is essential to contemporary bioenergy systems. Fungi lower process energy demand and chemical inputs through their potent and varied enzyme systems, making bioenergy production more sustainable and in line with the objectives of the circular bioeconomy (Deshpande and Dhakephalkar 2025). Fungi serve as both fermentation hosts and enzyme producers in integrated biorefineries, allowing the same fungal platforms to saccharify biomass and transform sugars into bioethanol, organic acids, lipids, and value-added biochemicals that enhance process economics. Such fungal bioprocessing is a key component of the circular bioeconomy, where agro-industrial leftovers are converted into fuels and bioproducts rather than being wasted, according to conceptual and policy-oriented articles. This establishes fungal biotechnology as a crucial technique for achieving energy sector decarbonization and sustainable development goals (Lange 2017).

Recent research on fungal bio-electricity shows that fungi can be included in electrochemical systems and microbial fuel cells that both treat contaminated effluents and produce electricity, going beyond traditional fuels. By combining the removal of pollutants with the production of renewable energy, these systems extend the use of fungal biotechnology from biomass conversion to more comprehensive environmental bioenergy solutions. These developments collectively demonstrate that fungi are essential biological engines powering low-carbon, sustainable bioenergy systems rather than merely supporting organisms (Umar et al. 2024).

With an emphasis on agricultural and organic residues as essential feedstock, this chapter critically analyzes how the global energy crisis and climate mitigation goals are fuelling interest in biogas as a flexible, low-carbon energy carrier within sustainable energy carrier within sustainable energy transitions. The traditional AD process is described, with a focus on operating procedures and common limitations such as feedstock, heterogeneity, process instability, and suboptimal methane outputs.

Methane (CH_4) and carbon dioxide (CO_2) make up the majority of the renewable energy resource produced by the complex biochemical process of biogas production, which involves the anaerobic process of biogas production, which involves the anaerobic decomposition of organic materials through a succession of microbial-mediated steps.

As shown in Fig. 7.1, the process starts with hydrolysis, where hydrolytic bacteria release enzymes to convert complex polymers such as carbohydrates, lipids, and

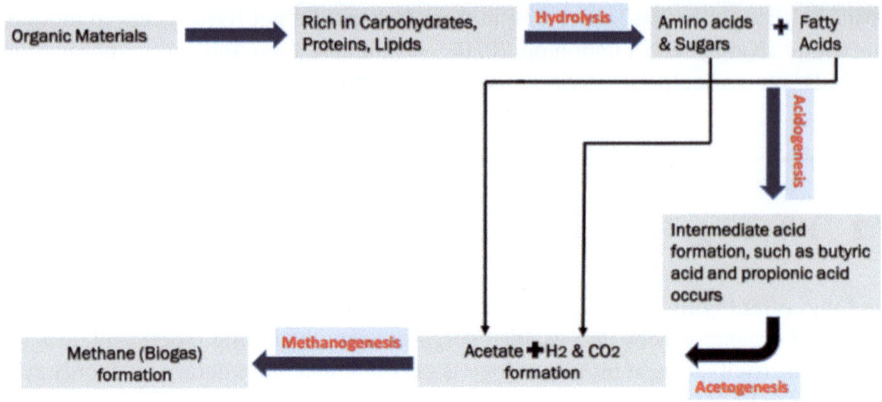

Fig. 7.1 Diagrammatic representation of the biochemical process involved in biogas production

proteins into simpler, soluble monomers such as sugars, fatty acids, and amino acids (Kabeyi and Olanrewaju 2022). Due to the recalcitrance of lignocellulose substrates, this stage is particularly important because it frequently affects the overall pace of digestion. After that, acidogenic bacteria ferment these monomers into VAFs, alcohols, hydrogen, ammonia, and CO_2, producing an acidic environment that nurtures the subsequent stages. During acetogenesis, syntrophic acetogenic bacteria convert VFAs and alcohols into mainly acetate, hydrogen, and CO_2 in tightly coupled reactions with hydrogen-consuming microbes to maintain low hydrogen partial pressures necessary for thermodynamically favorable transformations. Methanogenesis is the terminal phase in which methanogenic bacteria convert acetate, hydrogen, and CO_2 into methane, completing the energy-rich biogas formation. The methane content typically ranges between 50% and 70%, making biogas a valuable bioenergy source.

Technical, biological, and operational methods are used to stabilize the biogas generation process. By dissolving intricate lignocellulose structures, pretreatment techniques such as heat, chemical, and ultrasonic approaches have shown effectiveness in speeding up hydrolysis and significantly raising methane production. Furthermore, syntrophic interactions are strengthened, and microbial constriction is stimulated by process additives such as charcoal, catalysts, or trace nutrients. Microbial and gas production rates are optimized by maintaining critical factors, including temperature (mesophilic or thermophilic regimes), hydraulic retention time, and pH (preferably 6.5–8.0). The increasing adaptability and sustainability of biogas systems are demonstrated by creative improvements such as combining biogas production with hydroponic farming, steam methane reforming for hydrogen synthesis, and waste heat recovery for electricity. Together, these developments help biogas become a key technology in the world's sustainable energy transition, supporting waste management, renewable energy supply, and greenhouse gas mitigation objectives (Kabeyi and Olanrewaju 2020).

7.3 Biogas Production and Applications in the Sustainable Energy Transition

By determining microbial accessibility, process kinetics, stability, and overall methane production efficiency, feedstock variables have a significant impact on biogas yield in AD. These characteristics include biochemical composition, physical structure, nutrient stoichiometry, moisture dynamics, and inhibitory factors, all of which interact to determine specific biogas yield (0.2–0.8 m^3/kg VS) and biochemical methane potential (BMP, usually 200–500 m L CH_4/g VS), as thoroughly examined in Kulichkova et al. (2020). Optimal feedstocks strike a balance between process resilience and degradability, frequently necessitating co-digestion or pretreatment to reduce constraints.

7.3.1 Biochemical Composition and Degradability

The organic fraction, quantified as volatile solids (VS, ideally >80% of total solids/TS), directly correlates with biogas potential: carbohydrates yield ~0.38 m^3 CH_4/kg VS, proteins ~0.6 m^3/kg VS, and lipids up to 1.0 m^3/kg VS due to their high hydrogen-to-carbon ratio. Lignocellulose components dominate plant feedstocks—cellulose (40–50%) and hemicellulose (20–30%) are fermentable, but lignin (5–25%) acts as a recalcitrant barrier, reducing hydrolysis rates by 50–70% in straws (e.g., wheat straw BMP 200–300 mL CH_4/g vs. maize silage 350–400 mL/g VS). Soluble sugars/starch in energy crops accelerate acidogenesis, while high-lipid wastes (e.g., slaughterhouse offal) boost yields but risk long-chain fatty acids (LCFAs) inhibition above 1–2 g/L.

7.3.2 Nutrient Balance and C/N Ratio

Microbial consortia require a carbon-to-nitrogen (C/N) ratio of 20–30:1 for balanced growth; imbalances disrupt syntrophy. Low C/N (<20:1, e.g., crop residues) starves methanogens of nitrogen, slowing acetogenesis. Co-digestion (e.g., crop residues) starves methanogens of nitrogen, slowing acetogenesis. Co-digestion (e.g., straw + manure) optimizes this, enhancing yields by 20–40%; trace elements (Fe, Ni, and Co for methanogens) and phosphorus/sulphur must align to avoid deficiencies.

Table 7.1 Process parameters impacting biogas production

Characteristic	Optimal range	Positive impact	Negative impact	Mitigation
VS/TS ratio	>80%	High biogas potential	Low VS (manure <70%)	Select lipid/ carbohydrate-rich crops
C/N ratio	20–30:1	Microbial balance	High (>40:1, straw): N-limitation	Co-digestion with N-rich wastes
Lignin content	<15%	Rapid hydrolysis	>20% (wheat straw): 50% yield loss	Pretreatment (alkaline/steam)
Moisture	80–95%	Stability/ pumpability	<60% TS: dilution needed	Add process water
Inhibitors	<Thresholds (NH_3 <1.5 g/L)	No disruption	LCFA >2 g/L (lipid waste): inhibition	Dose control/ acclimation

7.3.3 Physical and Structural Properties

Particle size (<10–20 mm) and fiber structure govern hydrolysis accessibility—coarse lignocellulose increases retention times (30–60 days) and lowers yields, mitigated by pretreatments (steam explosion: +50% BMP; alkaline: +20–30%). Moisture contents (80–95%) ensure pumpability and dilution without nutrient washout; overly dry feedstocks (<60% TS) demand water addition, while hypersaline (>5% salts) or high-ash (>20%, e.g., rice straw silica) materials reduce VS digestibility by 10–30%. Seasonal harvesting affects composition: late-cut grasses accumulate lignin, dropping BMP from 350 to 3250 mL CH_4/g VS.

7.3.4 Inhibitors, Toxicity, and Process Interactions

Inhibitory compounds—ammonia, sulphides (H_2S >100 mg/L), phenolics, heavy metals, and antibiotics—selectively target methanogens, causing process crashes; for example, tannins in leaves reduce yields 15–25%. pH sensitivity (optimal 6.8–7.2) amplifies VFA toxicity during acidogenesis overload. As in Table 7.1, the characteristics interact holistically—for example, high-lignin residues need C/N balanced co-digestion and pretreatment for viability—underscoring feedstock analysis (BMP tests, VS/C/N profiling) as essential for 20–50% yield gains.

7.4 Role of Fungi in Biomass Degradation

Fungi play central roles in degrading lignocellulosic biomass for biogas, through diverse ecophysiological groups and powerful enzyme systems that act in both aerobic pretreatment and AD environments. Their ecological diversity (aerobic, anaerobic, and facultative) and highly specialized enzyme systems make them central to biomass pretreatment and enhanced hydrolysis in biogas processes (Andlar et al. 2018).

7.4.1 Biogas-Relevant Fungal Groups

Biogas-relevant fungi can be broadly grouped into the following (Hess et al. 2020):

- Aerobic filamentous fungi (mainly Ascomycota and Basidiomycota) such as white-rot, brown-rot, and soft-rot wood decayers and soil saprotrophs, which are used primarily in solid-state pretreatment or enzyme production under oxygenated conditions;
- Obligate anaerobic fungi in the phylum *Neocallimastigomycota*, native to rumen and hindgut environments, which thrive strictly in anoxic conditions and colonize plant fibers inside animals or digesters;
- Facultative fungi and yeast-like forms that tolerate microaerophilic or fluctuating redox conditions and can contribute to early biomass colonization or produce lignocellulolytic enzymes in processes that are not strictly anaerobic throughout.

7.4.2 Enzymatic Mechanisms in Lignocellulose Breakdown

Fungi degrade lignocellulose using a coordinated suite of hydrolytic and oxidative enzymes that act on cellulose, hemicellulose, and lignin (Kumar and Chandra 2020):

- Cellulases include endoglucanases, cellobiohydrolases (exoglucanases), and β-glucosidases; they hydrolyze crystalline and glucose, which can be fermented.
- Hemicellulases such as xylanases, mannanases, β-xylosidases, and accessory debranching enzymes (arabinosidases and glucronidases) attack heteropoysaccharides such as xylan and mannan, opening up the matrix around cellulose microfibrils.
- Lignin-modifying enzymes include lignin peroxidases (LiP), which cleave non-phenolic lignin structures via high-redox potential oxidations, and manganese peroxidases (MnP), which oxidize Mn (II) to Mn (III), generating diffusible oxidants that attack phenolic lignin units.
- Laccases, multicopper oxidases, catalyze one-electron oxidation of phenolic lignin units and various aromatic compounds using oxygen as a terminal electron acceptor, often working synergistically with LiP/MnP and low-molecular-weight mediators.

White-rot, brown-rot, and soft-rot fungi are among the most important biological tools for pre-treating lignocellulosic biomass, because they selectively or non-selectively alter lignin and polysaccharides, making substrates more amenable to enzymatic hydrolysis and AD. White-rot fungi such as *Phanerochaete chrysosporium, Ceriporiopsis subvermispora, Pleurotus ostreatus,* and *Trametes versicolor* secrete powerful lignin-modifying enzymes (laccases, LiP, and manganese peroxidases) that preferentially remove lignin and increase cellulose accessibility; several studies report 20–40% higher methane yields after white-rot pretreatment of wheat straw, corn stover, or woody biomass compared with untreated controls. Brown-rot fungi (e.g., *Gloeophyllum trabeum*) mainly depolymerize carbohydrates via *non-enzymatic* Fenton chemistry and subsequent hydrolases, leaving a modified lignin skeleton that can still improve porosity and enzyme penetration, while soft-rot fungi are important in high-moisture or soil-live environments where they slowly erode cell walls and complement other decay types (Singh and Singh 2014).

Anaerobic fungi of the phylum Neocallimastigomycota are obligate anaerobes native to the rumen and hindgut of herbivores, and they are increasingly recognized as powerful lignocellulose degraders with direct relevance for biogas processes. Their invasive rhizoids and broad fibrolytic enzyme systems enable them to attack intact plant tissue more aggressively than most rumen bacteria, making them attractive candidates for enhancing hydrolysis in digesters and for biomass pretreatment (Griffith et al. 2010).

7.4.3 Fungal Pretreatment for Enhanced Biogas Production

An increasingly significant biological technique for enhancing the breakdown of lignocellulosic biomass prior to AD is fungal pretreatment. Fungal techniques, in contrast to conventional thermal or chemical methods, selectively modify lignin, hemicellulose, and cellulose structures by utilizing the natural metabolic and enzymatic capabilities of filamentous fungi. Fungi release enzymes that begin to degrade lignin, reduce fiber crystallinity, and loosen the lignocellulosic matrix when they are allowed to grow on biomass with regulated moisture, air, and temperature (Kumar et al. 2025). This change improves the access of carbohydrates to anaerobic microbes, speeding up hydrolysis and increasing methane production (Kainthola et al. 2021). As a biological and environmentally friendly technology, fungal pretreatment prevents the creation of harmful compounds that usually occur in acid or alkali processes (Dar et al. 2019). It is much more energy efficient, operates under normal conditions, and is capable of breaking down a large number of agro-residues. However, it may turn out to be a slow process depending on the fungus used. The prolonged growth period of the fungus may also consume fermentable carbohydrates. To address these issues, novel methods such as co-culture of fungus populations, fungus immobilized reactors, or fungus improvement methods using adaptive evolution or genome editing could be added to pretreatment strategies (Salazar-Cerezo et al. 2023). In totality, fungus pretreatment is a selectively, sustainably, and economically viable process to produce a methane-rich gas, that is, biogas, from lignocellulosic biomass (Sahil et al. 2023).

7.4.4 Delignification and Structural Modification of Substrates

Biological delignification is also of interest due to its specificity to the substrate, lack of environmental hazards, and ability to eliminate lignin without damaging carbohydrate fractions. The white-rot, brown-rot, and soft-rot fungi secrete various ligninases such as laccases, LiP, manganese peroxidases, and auxiliary oxidoreductases. The action of these enzymes leads to the oxidative breakage of lignin–carbohydrate linkages in the lignin–carbohydrate complexes. This leads to the opening up of ether linkages, increased porosity, reduction in fiber rigidity, and accessibility of cellulose microfibrils to hydrolysis (Chaurasia et al. 2025). The direct and indirect enzymatic pathways are involved in fungal biomass degradation process, highlighting their hierarchical and cooperative roles in biomass conversion (Fig. 7.2). Substrate-specific enzymes are involved in direct pathways: cellulases hydrolyze β-1,4-glycosidic linkages in cellulose, hemicellulases (xylanases, mannanases, and esterases) cleave hemicellulose backbones and side chains, and ligninolytic enzymes (laccases, LiP, and manganese peroxidase) oxidatively depolymerize lignin. Esterases eliminate acetyl and phenolic cross-links, oxidative delignification breaks lignin–carbohydrate complexes, and radical-mediated reactions lessen cellulose crystallinity and steric hindrance, improving enzyme accessibility (Ren et al. 2024). Conversely, the main driving force for the modification of the biomass struction is the indirect pathway, which is concerned with the fragmentation of direct bonds through the action of delignification, which attacks the ligno-carbohydrate bonds of lignin, carbohydrate complexes, the action of the esterase, which breaks acetyl and

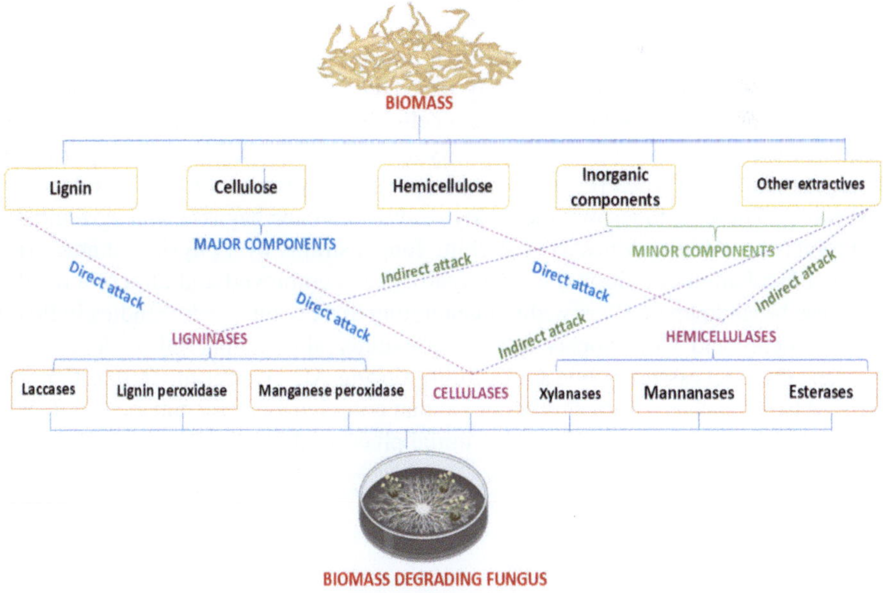

Fig. 7.2 Fungal biomass components modification pathways

phenolic bonds of the cross-links, and the radical reaction, which attacks the crystallinity of the celluloses, hence improving the accessibility of the enzymes (Yu et al. 2025). These modifications occur before the direct modification and are indicative of the fungal biomass degradation process, which is a non-linear process where the structure in the early stages determines the level of polysaccharide degradation in the fermentation processes for the production of biofuels (Balk et al. 2023).

Although single fungus technology is predominantly used for biomass delignification, microbial consortia of fungi and bacteria have also shown promising biological agents for structural transformations. These consortia possess a combination of enzymatic assets, where fungal lignolytic enzymes are involved in the initial stages of delignification, and bacterial cellulases and hemicellulases act on crystalline as well as amorphous polysaccharides (Kaur et al. 2023). Studies related to rice straw hydrolysis indicate that microbial consortia-treated substrates demonstrate enhanced endoglucanase, exoglucanase, and β-glucosidase activities. This results in reduced recalcitrance and faster hydrolysis (Sahil et al. 2023).

7.5 Fungal Technology in Biomass Hydrolysis

Following delignification, fungal-assisted hydrolysis improves the conversion of lignocellulosic biomass into fermentable sugars. Fungi such as *Phanerochaete chrysosporium*, *Trichoderma reesei*, *Coriolus versicolor*, and *Trametes* spp. are able to secrete different types of enzyme systems involving cellulases, xylanases, β-glucosidases, and carbohydrate-active accessory enzymes. The secreted enzymes depolymerize cellulose and hemicellulose, thereby converting them into oligosaccharides. These are efficient and effective enzymes on lignin–carbohydrate complexes. These enzymes work efficiently to reduce crystallinity and help in opening up the structural lignin cellulose and hemicellulose matrix for subsequent biodegradation by methanogenic microorganisms (Chaurasia et al. 2025). Fungal hydrolysis is largely employed when residues contain physical barriers such as silica deposits and waxes. These residues include paddy straw, sugarcane bagasse, and agricultural and food processing residues. By matching fungal strains with physical characteristics and incubation conditions, fungal hydrolysis is employed and effective for the immense breakdown of fibers without generating toxic compounds. Studies indicate that Trametes and Coprinopsis species efficiently alter holocellulose levels and increase their accessibility and digestibility during AD (Dar et al. 2019). Fungal-assisted hydrolysis is a critical component that complements and enhances holocellulose changes commenced during the initial phase of delignification.

7.5.1 Fungal Pretreatment and Its Effect on Microbial Digestion

The fungal-assisted hydrolysis enhances not only the process of structural transformations in lignocellulosic materials but also has large synergy effects on microbial AD in particular. Fungi that induce oxidative delignification and partial hydrolysis provide a substrate environment that is more suitable for colonization by anaerobic hydrolytic, acidogenic, and methanogenic populations (Vieira et al. 2025). Because aerobic fungi and anaerobic microorganisms have complementary metabolic powers, they work together to overcome the inherent resistance of agricultural and food-based wastes in the following ways:

(i) The fungal pretreatment begins the essential lignin oxidation reactions via laccase, LiP, and manganese peroxidase. These enzymes produce reactive species, including phenoxy radicals and manganese complexes (Mn^{3+}), to cleave the ether bonds and lignin–carbohydrate complexes. The white-rot fungi, including *Phanerochaete chrysosporium* and *Trametes versicolor*, undergo lignin oxidation reactions, making biomass structure accessible to anaerobic bacteria in a manner not otherwise possible (Hou et al. 2020). Additionally, the fungal pretreatment begins the carbohydrate depolymerization reactions by releasing cellulases, xylanases, and β-glucosidases. This results in a reduction of crystallinity of the cellulose structure and the transformation of the biomass polysaccharides into oligosaccharides. These intermediate products include cellodextrins, xylo-oligossaccharides, and simple sugars, which are direct nutrients for anaerobic hydrolyzing and fermenting bacteria (Parmar et al. 2024). The microbial part nership closely collaborates in the biomass fractionation process and confirms the direct contributions of fungi, including *Aspergillus* and *Trichoderma* species, in the provision of crucial hydrolytic enzymes for bacterial efforts in the efficient microbial conversion of modified lignocellulose into fermentable products (Sahil et al. 2023).

(ii) The produced fungal intermediates, in turn, interact with microbial processes, in particular, acetogenesis, syntrophic oxidation, and methanogenesis, in the existing microbial pathways after being introduced into the anaerobic bioreactor. The use of fungal-assisted treatments reduces the activation energy for hydrolytic bacteria, for instance, in *Clostridium* spp., since they rely on easier access to carbohydrates. The use of soluble aromatic moieties produced in the fungal oxidation of lignin is also beneficial in acetoclastic bacteria in reactions such as the Wood-Ljungdahl process, also known as the reductive acetyl-CoA pathway, yielding acetate, CO_2, and H_2, which serve as substrates for methanogens. Recent metagenomics analyses clearly indicate that hydrogenotrophic methanogens, Methanoculleus and Methanobacterium, grow more favorably in pretreated materials in a way that clearly demonstrates biochemical complementarities between fungal oxidation reactions and microbial reduction processes (Manyi-Loh et al. 2019).

The cumulative effect of these integrated biochemical reactions brought about by fungal enzyme stimulants brings in a more rapid hydrolysis phase, improved fermentation stability, enhanced syntrophic interactions, and streamlined biomass conversion to methane production. This synergistic relationship reinforces fungal symbiosis with microbes involved in AD and increases the importance of the pretreatment process using fungi as the strategic biochemical process in the microbial breakdown of lignocellulosic materials.

7.5.2 Fungal Pretreatment Advantage

Fungal pretreatment is highly beneficial for biomass biogas-producing biorefineries. First, due to the high selectivity and ability to selectively oxidize lignin during fungal action should leave the cellulose and hemicellulose untouched, contrary to the toxic byproducts generated by traditional chemical treatments (Nadir et al. 2019; Jyani et al. 2024). Second, the processes work under natural conditions and without the addition of any chemicals, which reduces the environmental footprint due to reduced energy consumption (Garg et al. 2025). Finally, the technology has the ability to work well even in decentralized settings where the fungi Trichoderma and Aspergillus can simply be raised from agricultural waste (Kaur et al. 2017).

Additionally, fungal-treated substrates improve microbial uptake through improvements in porosity, reduced hardness, and the production of soluble compounds that can easily be fed directly into acetogenic and methanogenic pathways (Hu et al. 2024). Compared to other physical or chemical techniques, fungal treatments are distinct for involving structural modification, detoxification, and hydrolysis processes all in one biological treatment, which are core to sustainable biogas production processes.

7.5.3 Fungal Enzymes in Biogas Enhancement

Fungal biotechnology plays an important role in increasing biogas generation by utilizing a varied spectrum of lignocellulolytic enzymes capable of breaking down the structural complexity of biomass (Table 7.2). The fungal enzymes generally responsible are ligninolytic, cellulolytic, or hemicellulolytic, with each contributing uniquely to the pretreatment and biodegradability of lignocellulosic substrates. During AD, their coordinated action determines how well material is broken down and metabolized (Vasco-Correa and Shah 2019).

(i) *Ligninolytic enzymes'* primary role is to undertake oxidation of lignin, the most refractive component of plant cell walls, via ligninolytic enzymes. White-rot fungi such as *Aspergillus niger*, *Pleurotus florida*, and *Pleurotus ostreatus* have great applications in this aspect, and their application in pretreatment has

Table 7.2 Major classes of fungal enzymes and their roles in biogas enhancement

Enzyme class	Representative enzymes	Main substrate/ target	Primary role in biogas enhancement	Key references
Ligninolytic enzymes	Laccase, LiP, MnP	Lignin, lignin–carbohydrate complexes	White-rot fungi	Pérez-Contreras et al. (2024)
Cellulolytic enzymes	Endoglucanase, Exoglucanase, β-glucosidase	Cellulose: amorphous and crystalline regions	Phylum *Neocallimastigomycota*, white-, brown-, and soft-rot fungi	Vinzelj et al. (2020) Sahil et al. (2023)
Hemicellulolytic enzymes	Xylanase, Mannanase, Arabinofuranosidase	Hemicellulose (xylans, mannans, arabinans)	Aspergillus species, *Trichoderma* spp.	Gruening de Mattos et al. (2025)
Auxiliary redox enzymes	Lytic polysaccharide monooxygenases, oxidases, dehydrogenases	Crystalline cellulose, lignin-linked regions	*Phanerochaete chrysosporium*, *Trametes versicolor*, *Aspergillus* spp., *Streptomyces viridosporus*	Contato and Conte-Junior (2025)

been proven to improve the digestibility of the substrate to a great extent (Pérez-Contreras et al. 2024).

(ii) *Cellulolytic enzymes* act sequentially to hydrolyze cellulose into glucose units. Fungal strains screened using biogas slurry or agricultural residues often exhibit strong cellulase activity, demonstrating their utility in accelerating hydrolysis, a rate-limiting step in AD (Sahil et al. 2023). Their action increases the availability of fermentable sugars, thereby supporting more efficient microbial conversion pathways (Sahni and Phutela 2012).

(iii) *Hemicellulolytic enzymes* complement this process by degrading hemicellulose and hence weakening the biomass structure of plants even further. The presence of these three groups of enzymes solves the issue of mass transfer limitations, hence improving biodegradation (Østby and Várnai 2023).

7.5.4 Enzymatic Cocktails for Hydrolysis and Saccharification

The enzymatic cocktails are being widely adopted as helpful and effective in biotechnology for improving the hydrolysis and saccharification efficiency of lignocellulosic feedstocks before AD. These usually consist of mixtures of cellulases, hemicellulases, lignin-active enzymes, and auxiliary oxidoreductases that work together in degrading the whole complex plant cell wall structure. By incorporating enzymes such as endoglucanases, cellobiohydrolases, β-glucosidase, xylanases, mannanases, and side-chain cleaving esterases, the cocktails target both linear and branched polysaccharides. This increases the release of fermentable sugars and shortens the time needed for biomass solubilization (Pérez-Contreras et al. 2024).

Recently, enzyme mixtures generated by fungal bacterial consortia during bioenzyme production from organic waste peels in facultative anaerobic fermentation have emerged as an efficient and sustainable pretreatment method for lignocellulosic biomass before biogas production. These naturally derived enzyme complexes provide several benefits over purified commercial enzymes. They are cost-effective, made from waste materials, diverse in enzyme functions, and well-suited for mixed-substrate environments (Kaur et al. 2026). They are especially useful for low-cost decentralized biogas systems and circular bioeconomy applications due to their complex enzymatic profiles, which frequently comprise ligninolytic oxidases, cellulolytic hydrolases, hemicellulases, proteases, pectinases, and other enzymes (Phutela et al. 2025).

In addition, several studies have addressed the development of substrate-specific enzyme cocktails that match the biomass composition, such as rice straw, sugarcane bagasse, maize stover, or microalgal lignocellulosic blends, by varying the ratio of cellulolytic to hemicellulolytic enzymes. Fungal secretomes from *Trichoderma, Aspergillus, Penicillium*, and white-rot basidiomycetes act as natural models and are commonly augmented with specific activities to rectify deficiencies, such as low β-glucosidase activity or low accessibility of xylanases (Kaur et al. 2017). Biodigested slurry rich in enzymes and fungal broths can act as cheaper, naturally available enzymatic cocktails. Such preparations, when prepared from agricultural residues and biogas slurry, possessed high cellulase, xylanase, and accessory enzyme activities that augmented substrate solubilization and enhanced methane yields in farm-scale digesters (Garg et al. 2025).

Overall, whether as purified, semi-purified enzymes or fungal–bacterial consortia, enzymatic cocktails are fundamental in the reduction of biomass resistance and increased rate of hydrolysis, thereby improving overall AD efficiency. Their multiplicity of approaches, low environmental impacts, and complementarity with fungal biotechnology allow them to constitute an integral part of up-to-date methods for pretreatment in biogas production.

7.5.5 Recombinant and Engineered Fungal Enzymes in Biogas Processes

The low activity at the digester temperatures, narrow pH ranges, and low expression levels are some of the limitations that are increasingly being addressed by recombinant DNA technology and protein engineering. Recent advances involve mutagenesis and directed evolution of cellulases and xylanases for improved thermostability, catalytic efficiency, and resistance to end-product inhibition (Singh et al. 2024). Bifunctional and multifunctional enzymes, combining β-1,3/1,4-glucanase with laccase or cellulase with xylanase in one polypeptide, have also exhibited enhanced depolymerization efficiency. They also reduce the amount of enzyme added in total, thereby benefiting the biogas plants, which are cost-sensitive. Production of fungal enzymes in alternative hosts such as *Pichia pastoris*, *Aspergillus* spp., or in engineered Trichoderma strains enables higher yields and better control over glycosylation, secretion, and cofactor supply (Goudarzi et al. 2025). In AD, this allows for the design of enzyme sets optimized for conditions relevant to digesters, such as 35–55 °C, near-neutral pH, and the presence of ammonia and VAFs. For instance, the activity of lignin in the case of engineered laccases and peroxidases may be tuned for moderate temperatures, thus aiding in the delignification and detoxification of phenolics known to inhibit methanogens (Singh et al. 2024).

From the perspective of fungal biotechnology, recombinant expression is combined with traditional techniques of strain improvement, such as random mutagenesis and adaptive laboratory evolution, to allow for the development of next-generation production strains able to secrete complex lignocellulolytic repertoires in powerful production titers.

7.5.6 Immobilization and Reuse of Fungal Enzymes for Process Intensification

Furthermore, immobilized fungal enzymes on solid carriers have been identified as an important process intensification technique in biogas production. This is because immobilized enzymes have been known to be more resistant to shear stress and inhibitors compared to free enzymes. Additionally, immobilized enzymes can be reused several times to pretreat biomass with considerably lower costs compared to biogas produced (Hassan et al. 2025).

Laccase is the most deeply investigated enzyme in terms of immobilized preparations and applications, especially by immobilized fungal laccases on a support material such as silica, biochar, or alginate, as well as synthetic polymers, used for oxidative delignification and detoxification treatments of pretreated lignocellulosic hydrolysates, which reduced levels of phenolic and furanic compounds before AD (Paraschiv et al. 2022). Besides laccase, immobilized cellulases and hemicellulases are currently investigated as packed-bed or fluidized-bed reactors as a continuously operating pretreatment reactor preceding the digester. These reactors enable a large enzyme–substrate contact area without enzyme loss and enable separate optimization of reactor conditions, independent from digestion conditions (such as a slightly higher temperature to speed up enzymatic reactions). Co-immobilized mixtures of different enzymes (such as endoglucanase, β-glucosidase, and xylanase) improve synergistic effects and minimize limitations by mass transfer in the lignocellulosic substrate (Garg et al. 2026).

AD-scale biorefineries will specifically benefit from immobilized fungal enzymes in the following three ways: (i) cost minimization due to the potential for use and longevity, (ii) resilience to variation in quality of input substrate, and (iii) modularity to improve retrofit capability of enzymatic pretreatment plants to gas plants. The development of immobilization methods and materials is expected to increase in influence in the likely shift toward AD-scale gas plants incorporating fungal pretreatment and enzyme mixes (Hassan et al. 2025).

7.6 Co-culture and Consortium Engineering

7.6.1 Fungal–Bacterial–Archaeal Consortia and Syntrophic Interactions

A new research avenue of engineering co-cultures involving mixed fungal, bacterial, and archaeal species has proven useful in overcoming the shortcomings of single-cell pretreatment bioreactors and optimizing processes in AD bioprocesses. Fungi are known as "front-end decomposers," initiating the step of oxidative delignification and hydrolysis as the first step of degradation. Bacteria further degrade soluble carbohydrates into volatile compounds, which are followed by Archaea in the final step of methanogenesis (Pérez-Contreras et al. 2024).

Fungal pretreatment fosters a syntrophic lifestyle based on cellodextrin, xylo-oligossacharide, and partially oxidized aromatic compound secretion. Hydrolyzing and fermenting bacteria rapidly produce acetate, hydrogen, and formate, which are central intermediates of the methane metabolic pathway. Hydrogen-requiring archaea, such as Methanoculleus, Methanobacterium, and Methanospirillum, reduced the hydrogen partial pressure, thus facilitating the operation of hydrogen-consuming reactions that benefit the bacteria. The syntrophy between bacteria and archaea maintains a thoroughly integrated metabolic pathway, where the role of a fungus to enhance bacterial fermentation rates and hydrogen-consumption by archaea maintains the metabolic pathway (Kapoor et al. 2025).

Co-culture engineering is directed at the introduction of complementary organisms, as well as the fixation of the growing syntrophic relationships, facilitating methane conversion. Techniques such as the co-immobilization of fungi and bacteria, staged bioreactors, and adaptive co-cultures have been increasingly adopted to promote interspecies electron transfer and shock resistance (Sanglier 2014).

7.6.2 Mixed-Culture Fermentation Systems and Metabolic Cross-Feeding

The mixed-culture fermentation processes represent the "bridge" in the metabolic pathway from fungal pretreatment to methanogenic processes. The processes make use of the wide range of microbial communities involved in the following microbial processes: cellulolysis, saccharolysis, proteolysis, acetogenesis, and syntrophs, each carrying out their own specific biochemical reaction. The benefit of the use of mixed-culture microorganisms lies in their ability to efficiently function irrespective of changes in substrates, toxins, and so on (Lee 2022).

Cross-feeding by metabolic complementation plays a key role in this aspect of functioning. Fungal enzymatic hydrolysis products such as glucose, xylose, arabinose, and acetate esters are utilized by fermentative bacteria to produce organic acids, hydrogen, and carbon dioxide gas. These byproducts are then used by acetogenic bacteria/methanogenic archaea. Cross-feeding occurs in addition to providing substrates; bacteria can produce vitamins, factors, and/or cofactors that promote fungal metabolic activity, and the latter produce detoxifying enzymes like oxidases that decrease phenolic/lignin-derived compounds acting as bacterial and archaeal inhibitors (Sanglier 2014).

In mixed-culture fermentations, anaerobic fungi that belong to the phylum Neocallimastigomycota are also included. These anaerobic fungi have strong cellulase and hemicellulase activities. They are anaerobic. Their inclusion in mixed cultures raises the level of fiber hydrolysis and prevents VFA inhibition. When added to white-rot fungi that are used in aerobic treatment, they form an ideal transition process for the degradation of biomass, from oxidation to fermentation (Vinzelj et al. 2020).

7.7 Systems Biology Approaches for Optimizing Mixed Consortia

Recent advances in systems biology have transformed the way mixed microbial consortia are designed, monitored, and optimized for biogas production. Multi-omics tools, such as metagenomics, metatranscriptomics, proteomics, and metabolomics, allow researchers to decode community structure, functional gene expression, and metabolite fluxes in unprecedented detail. These insights enable targeted manipulation of consortia to enhance hydrolysis, promote syntrophy, and improve methane production (Guerrero et al. 2025).

Engineering strategies informed by systems biology include the following (Asemoloye et al. 2021):

- *Selective enrichment* of ligninolytic fungi or cellulolytic bacteria based on predicted pathway bottlenecks;
- *Manipulation of reactor conditions* such as pH, loading rate, and trace minerals helps to favor key syntrophic or methanogenic taxa;
- *Metabolic augmentation*, a process by which missing pathways are added by supplementing with particular cofactors or using engineered fungal strains;
- *Bioaugmentation* is done by using customized microbial/fungal blends, which can improve stability or restore performance following inhibition events.

Ultimately, fungal biotechnology systems are positioned as highly optimized, self-regulating engines for next-generation biogas production because of systems biology's ability to facilitate the shift from empirical, trial-and-error methods to predictive, design-driven consortia engineering.

7.8 Process Optimization and Bioreactor Design

Efficient biogas generation through fungal biotechnology is dependent on combining reactor configuration, environmental control, and computational optimization. Recent reviews revealed that combining process engineering with advanced data-driven models significantly enhances methane yield and reactor stability (Reza et al. 2024; Chong et al. 2023). Therefore, the following subsections summarize critical aspects of reactor design and process optimization that underpin fungal-based biogas systems.

7.8.1 Reactor Configurations for Fungal-Based Digestion

Reactor configuration plays an important role in defining the efficiency of fungal-mediated AD systems. Among commonly adopted configurations, batch and continuous stirred tank reactors (CSTRs) govern laboratory and pilot setups due to their controllable mixing and ease of monitoring (Zhang et al. 2024). However, fungal biomass is often sensitive to shear stress in CSTRs, which may compromise enzymatic activity and fungal integrity.

Upflow anaerobic sludge blanket (UASB) reactors provide an alternative with improved biomass retention, yet they require adjustment of fungal consortia to granule formation. Solid-state reactors are increasingly explored for lignocellulosic feedstocks, particularly in co-cultivation systems comprising Trichoderma or Aspergillus species with cellulolytic bacteria (Kostopoulou et al. 2023). The structural stability of fungal mycelia in such systems enhances substrate–enzyme contact along with the hydrolysis rates.

Recent reactor innovations integrate membrane bioreactors (MBRs) with hybrid CSTR–UASB systems to retain fungal biomass at low redox potential. Nonetheless, maintaining true anaerobicity while sustaining fungal metabolism is still technically demanding. Large-scale deployment is still limited by factors such as operational stability, mixing uniformity, and the avoidance of biomass washout.

7.8.2 Parameters Influencing Fungal Survival and Activity

Table 7.3 shows that the survival and activity of fungi under anaerobic or microaerophilic conditions depend on pH, temperature, substrate composition, and nutrient balance. Most fungal strains involved in biogas production are mesophilic, with optimal ranges of 30–37 °C and pH 6.0–7.5, though some thermotolerant *Aspergillus* and *Mucor* species tolerate higher values (Chong et al. 2023).

Another key factor is the C/N ratio of the substrate, since too much carbon results in limiting the nitrogen availability, whereas nitrogen-rich substrates lead to ammonia accumulation, which can hamper methanogenesis. Furthermore, the accumulation of lignin-derived phenolics and VFAs can weaken fungal membranes and decrease the enzyme efficiency. Maintaining redox balance is particularly challenging because fungi need trace amounts of oxygen to produce sterols and unsaturated fatty acids. Controlled micro-aeration strategies have been proposed to provide minimal oxygen flow without interfering with AD (Reza et al. 2024). These subtle modifications require precise monitoring of the bioreactor, usually with a dissolved oxygen probe and online VAF sensors.

7.8.3 Scale-Up Challenges in Fungal Biogas Systems

Scaling fungal biogas systems from laboratory to pilot or industrial scales remains a persistent challenge. When compared to their bacterial counterparts, fungal systems exhibit excellent hydrolytic potential but relatively slow growth rates, which results in low overall volumetric productivity (Zhang et al. 2024). Further scale-up is complicated by other factors such as biomass washout, difficulty in maintaining consistent fungal populations, and uneven substrate colonization. Furthermore, the heterogeneity of lignocellulosic feedstocks impacts the reactor mixing and mass transfer. Techniques such as carrier-based immobilization, biofilm reactors, and granule support matrices have been studied to stabilize fungal biomass and increase solids retention. However, these approaches introduce new design constraints, including clogging, oxygen gradients, and cost escalation. Competition between

Table 7.3 Key operational parameters affecting fungal performance in the AD systems

Parameter	Optimal range	Limitation	References
Temperature	30–37 °C	Deviation reduces enzyme efficiency	Chong et al. (2023)
pH	6.0–7.5	Extreme pH inactivates fungal enzymes	Zhang et al. (2024)
C/N ratio	20–30	Imbalance causes nutrient limitation or ammonia inhibition	Kostopoulou et al. (2023)
Lignin content	<20% of TS	The presence of phenolics inhibits growth	Reza et al. (2024)
Redox potential	–300 to –400 mV	Required for stable anaerobic conditions	Chen et al. (2022)

fungal and bacterial communities also results in operational instability. Co-culturing necessitates fine control over substrate loading rates, hydraulic retention times, and inoculum ratios. According to Kostopoulou et al. (2023), maintaining microbial equilibrium under changing organic loading rates remains one of the most important challenges to industrial operation.

7.8.4 Process Modeling, Kinetic Studies, and Optimization Using RSM or ANN

For achieving predictable performance in fungal-based digestion, advanced modeling and optimization tools have become essential. Response surface methodology (RSM) is widely applied for assessing interactions among numerous process factors, such as temperature, pH, C/N ratio, and hydraulic retention time, providing a statistically guided optimization path (Reza et al. 2024). Corresponding to this, artificial neural networks (ANNs) and adaptive neurofuzzy inference systems (ANFIS) are gaining attention for their capacity to handle nonlinear and complex datasets inherent to microbial bioprocesses (Chen et al. 2022). By comparing RSM, ANN, and ANFIS models for predicting biogas yield, it was found that hybrid ANN-RSM frameworks resulted in the highest predictive accuracy with minimal experimental replications (Chong et al. 2023). Combining kinetic modeling with these computational techniques helps in the easier identification of rate-limiting steps in hydrolysis and methanogenesis-like processes. Moreover, digital twin concepts and real-time control algorithms are evolving for large-scale process stability and fault prediction. Such tools not only speed up optimization but also provide trails for operational scalability, addressing one of the most enduring issues in fungal biotechnology for biogas production. Overall, optimizing fungal biogas processes requires a multifaceted approach including reactor configuration, physiological control, and computational modeling. The integration of experimental insights with data-driven modeling (RSM and ANN) provides a promising pathway toward fungal-based biogas systems with high efficiency, scalability, and resilience. Future research should focus on adaptive reactor designs, resilient consortia engineering, and dynamic control systems to guarantee constant performance under actual operational conditions.

7.8.5 Challenges and Limitations

Despite significant advances in the understanding of fungal physiology and their metabolic role in lignocellulosic conversion, several technical, biological, and operational limitations continue to prevent their actual integration into biogas production systems. Although anaerobic fungi have unique enzymatic capabilities, they still remain sensitive to environmental fluctuations, process inhibitors, and microbial competition. Moreover, issues such as process stability and scalability also create delays in industrial applications. Therefore, the following sections describe the major obstacles affecting fungal-based AD.

7.8.6 Maintaining Fungal Activity Under Strict Anaerobic Conditions

Maintaining active fungal metabolism under strictly anaerobic conditions is a major challenge for utilizing fungal biotechnology for biogas production. Anaerobic fungi from the phylum *Neocallimastigomycota* are obligate anaerobes having hydrogenosomes in place of mitochondria (Saye et al. 2021). These organelles not only enable hydrogen production and energy conservation without oxygen, but they also make the fungi extremely sensitive to even a little oxygen availability. Any contact with air or aberrations in redox potential can result in irreversible metabolic inhibition, cell lysis, and community collapse. According to a study conducted by Gruninger et al. (2014), it was found that anaerobic fungi are highly specialized to work in rumen-like environments, which makes their cultivation in artificial reactors difficult. Maintaining consistent temperature, pH, and nutrient conditions equivalent to their natural habitat is important for obtaining desired activity. The dual life cycle of anaerobic fungi involves flagellated zoospores and rhizoidal thalli, both of which require tightly regulated anaerobic microhabitats, as reported by Hess et al. (2020). Interruptions during these growth stages result in sporulation failure and reduced biomass yield. From an engineering point of view, creating a long-term stable inoculum of anaerobic fungi in digesters is still challenging. Fungal biomass often washes out in continuous stirred systems because of its slow growth and weak cell aggregation. Therefore, maintaining strict anaerobiosis through the means of immobilization matrices, closed-loop inoculum recycling could help in stabilizing fungal populations. Moreover, mixed cultures aiming at combining anaerobic fungi with syntrophic bacteria and methanogenic archaea are also gaining attention as a feasible means to maintain activity, along with confirming continuous gas production.

7.8.7 Process Inhibition Due to Lignin-Derived Compounds or Metabolites

Lignocellulosic feedstocks such as agricultural residues, forestry waste, and energy grasses are present in abundance and renewable in nature, yet their high lignin content has extreme limitations in terms of biochemical constraints. During pretreatment and hydrolysis, lignin depolymerization produces inhibitory aromatic compounds that restrict fungal growth and enzymatic activity (Sun et al. 2024). The common inhibitors include vanillin, syringaldehyde, ferulic acid, and 5-hydroxymethylfurfural, which inhibit cellular respiration and protein synthesis-like processes. Along with these functions, these substances also result in causing oxidative stress that impairs fungal enzymes important for lignocellulose degradation. Khan and Ahring (2019) reported that lignin-derived intermediates accumulate in digesters, causing toxicity and metabolic arrest of the microbial populations. Fungi, although much more tolerant compared to bacteria, also have a reduced level of enzymatic secretion when high phenolic concentration is present. In biogas systems, this results in decreased substrate conversion efficiency and methane yields. Strategies such as mild alkali or oxidative pretreatment can reduce these effects, but it is very unlikely to eliminate inhibitory residues. Fungal pretreatment has thus emerged as an option to minimize lignin interference. White-rot fungi, for instance, secrete oxidative enzymes such as LiP and manganese peroxidases capable of selectively removing lignin (Hatakka and Hammel 2010). However, incomplete degradation can still produce toxic byproducts that diffuse into anaerobic digesters. Therefore, maintaining a balance between pretreatment severity, retention time, and co-substrate blending is crucial to minimizing the buildup of inhibitory metabolites. Understanding fungal–lignin interactions at the molecular level will be key to optimizing pretreatment compatibility with AD processes.

7.8.8 Low Growth Rate and Competition with Bacteria

A significant obstacle in integrating fungi into AD systems is their very low growth rate compared with bacteria. Anaerobic fungi have longer lag phases, slower biomass accumulation, and less tolerance to environmental stressors. Dollhofer et al. (2015) reported that fungal growth doubling times can exceed 24 h, whereas bacterial consortia multiply within a few hours under similar conditions. This difference results in bacterial dominance in mixed cultures, gradually diminishing the functional contribution of fungi over time. Nevertheless, the enzymatic machinery of fungi provides a critical advantage. Fungal cellulases and hemicellulases can break down complex plant polymers inaccessible to bacterial enzymes. Tian et al. (2025) showed that co-culturing of fungi with either hydrolytic or methanogenic bacteria leads to substrate degradation efficiency improvement through synergistic metabolism. Such cooperation enables fungi to produce fermentable sugars, which bacteria subsequently convert into VAFs and methane. However, this synergy depends on maintaining balanced nutrient flows and spatial separation to avoid direct substrate competition. Li et al. (2021) highlighted that uncontrolled bacterial growth can lead to substrate monopolization, suppressing the fungal metabolism. End-product inhibition, primarily due to VAF accumulation, also results in disrupting fungal enzyme production further. To counter these limitations, reactor configurations such as two-stage digesters or MBRs have been proposed, allowing fungal hydrolysis in the first stage and bacterial methanogenesis in the second. Immobilization and biofilm formation can also enhance retention mechanisms for fungi, improving overall stability of the system.

7.8.9 Operational Stability and Scalability Barriers

Achieving stable operation and scalability in fugal-assisted AD remains one of the major challenges. In full-scale systems, fungal communities after initial adaptation phases tend to decline due to feedstock heterogeneity and environmental stress (Bengelsdorf et al. 2013). Factors such as fluctuating organic loading rates, ammonia accumulation, and temperature shifts can selectively inhibit fungal populations. As a result, maintaining constant community structure and metabolic activity is difficult over long operational periods. Li et al. (2011) discussed that in solid-state AD, limited mass transfer and heterogeneous distribution of substrate can hinder fungal access to nutrients and oxygen-free niches. This physical limitation lowers the overall biodegradation efficiency. In addition to this, scaling up of fungal-based systems introduces difficulties in process control, including maintaining a uniform pH range, redox potential, and moisture content throughout the large reactor volumes. Process instability, foaming, and accumulation of intermediate acids are other common issues that disrupt steady-state conditions in biogas plants. Fungal washout and low population resilience further add up these problems. Emerging solutions such as immobilized fungal carriers, adaptive control algorithms, and co-digestion strategies that buffer metabolic stress are being developed. However, translation of these laboratory-scale developments to industrial digesters remains challenging due to their cost, reactor design constraints, and monitoring issues. Collectively, physiological sensitivity, metabolic inhibition to microbial competition, and operational instability are the present challenges defining fungal biotechnology for biogas production. Anaerobic fungi have numerous biocatalytic potential for lignocellulose deconstruction, but their successful deployment is dependent on improving cultivation stability, tolerance to inhibitors, and reactor compatibility. Advanced reactor engineering, synthetic co-culture design, and omics-based microbial optimization are thought to play a transformative role in surpassing these barriers. Future research integrating fungal ecology with process modeling will be of huge importance to unlock the full potential of fungi in sustainable biogas generation.

7.9 Emerging Tools and Future Perspectives

7.9.1 Metagenomics and Metatranscriptomics for Fungal Community Profiling

Metagenomics and metatranscriptomics are revolutionizing our understanding of fungal community structure and function in AD-related bioprocesses. These high-resolution omics techniques enable parallel profiling of both microbial diversity and metabolic activity, alongside revealing the ecological roles of both dominant and rare fungal taxa. Studies demonstrate that metagenomic sequencing allows identification of lignocellulose-degrading enzymes and anaerobic fungal lineages that interact synergistically with methanogenic archaea to enhance biogas production (Aguiar-Pulido et al. 2016; Schneider et al. 2021). Butowski et al. (2025) reviewed the potential role of metatranscriptomics as a tool to understand the active microbial processes, particularly under dynamic environmental conditions typical of biogas reactors. The combination of DNA with an RNA-based analysis improves the accuracy to efficiently distinguish between the viable fungal species and dormant or dead fungal cells.

Similarly, Auer et al. (2024) revealed the functional relations between fungal guilds and ecosystem processes through metatranscriptomics mapping, correlating expression patterns of genes and substrate cycling. These findings are particularly useful for optimizing feedstock selection and reactor operating parameters. Sharuddin et al. (2022) provided a broader general overview of the evaluation of metatranscriptomics toward productive microbiome analysis, keeping in mind the rising trend of opting for multi-omics approaches combining metabolomics and proteomics. Such integration holds a huge potential in biogas biotechnology, by providing an all-time parallel assessment of fungal metabolic activity and identification of novel enzymes involved in biomass hydrolysis. Further research should highly focus on machine learning-driven multi-omics analytics for understanding fungal community dynamics, accelerating toward the development of a more stable design of resilient biogas microbial consortia.

7.9.2 CRISPR/Cas and Synthetic Biology in Fungal Strain Improvement

The CRISPR/Cas genome-editing toolkit has rapidly revolutionized fungal biotechnology by allowing precise control over metabolic and regulatory pathways relevant to biogas production. Ma et al. (2025) gave a comprehensive review of their application in filamentous fungi, particularly in modulating cellulolytic and ligninolytic enzyme expression for enhancing substrate degradation efficiency. The ease of use and transportability, like factors of CRISPR/Cas9, make it an indispensable platform without which the high-performance fungal strains capable of robust growth under semi-AD conditions cannot be developed. Shen et al. (2024) elaborated on opportunities and technical challenges associated with applying CRISPR systems to non-model fungal species. Their work highlighted the role of optimizing transformation protocols and promoter systems in overcoming species-specific barriers in gene editing. Leal et al. (2024) further proved how CRISPR can be incorporated into synthetic-biology approaches for building modular fungal cell factories that co-express multi-enzyme complexes, enabling faster depolymerization of lignocellulose and co-metabolism of complex feedstocks. A comprehensive review by Shen et al. (2023) gave industrial-scale insights, suggesting that combining CRISPR with systems biology and adaptive laboratory evolution could act as a key to engineered fungal consortia with improved tolerance to inhibitors and fluctuating pH or temperature. The integration of genome editing, automation, and computational design is gaining attention for the development of next-generation fungal biocatalysts in biogas production by shifting the focus from empirical strain selection to predictive strain engineering.

7.9.3 Fungal Biofilms and Immobilization Strategies for Stable Biogas Yield

Fungal biofilms and immobilized cell systems have emerged as sustainable solutions for maintaining process stability and enhancing biogas production. Mohamed Ali et al. (2023) reported that immobilized biofilms significantly enhanced the food waste hydrolysis and other organic wastes, besides functioning as natural protective matrices against mechanical shear and toxic metabolites in the environment. Such immobilization not only enhances enzyme retention but also ensures higher microbial density, in turn, leading to enhanced methane yields and better reactor performance. Saini et al. (2023) provided a comprehensive review of biofilm formation mechanisms, structural composition, and engineered carrier materials, highlighting their role in wastewater and anaerobic treatment systems. Fungal biofilms offer distinct advantages due to their filamentous form and the high level of extracellular polymeric substance production, which helps them to stick to the support surfaces and provide further mechanical strength. In addition to these, Madusanka et al. (2024) discussed mycelium-based composites and immobilization strategies, which are basically an innovative application of fungal mycelia in immobilized biocatalyst systems. Such systems aim at exploiting the inherent rigidity of fungal networks for entrapment or surface attachment, thus creating biocatalytic matrices that can be recycled across multiple digestion cycles. In conclusion, these reviews suggest that optimizing carrier material, biofilm composition, and reactor configuration will determine the development of industrial-scale, biofilm-based AD units.

7.9.4 Circular Bioeconomy and Waste-to-Energy Integration

The fungal biotechnology is being increasingly recognized as one of the major key drivers of the circular bioeconomy. Meyer et al. (2020) demonstrated how fungal processes are capable of transforming waste biomass into bioenergy and bioproducts, thus closing the nutrient and carbon loops. The range of enzyme produce by fungi allows the conversion of agricultural residues and food waste into valuable products, aligning with sustainable resource management goals. Dhiman et al. (2024) emphasized fungal bioprocessing for valorizing lignocellulosic feedstocks and thus integrating these systems into existing agro-industrial chains. Their review showed fungal fermentation as a flexible platform to produce biofuels, organic acids, and enzymes, along with methane, by supporting the concept of zero-waste biorefinery. Wikandari et al. (2022) similarly discussed the importance of filamentous fungi in the advancement of the sustainable circular bioeconomy by highlighting their compatibility with AD and the potential they hold in the co-digestion of industrial effluents. Combined, these studies demonstrate how fungal biotechnology has the potential to bridge the gap between waste management and renewable energy production. Expanding waste-to-energy frameworks with fungal bioprocessing at the same time not only aims at contributing to energy generation but also supports nutrient cycling and greenhouse gas reduction, which are the main indicators of sustainable development.

7.9.5 Policy and Industrial Perspectives for Fungal Biotechnology in Biogas Systems

The technological potential of fungal biotechnology for biogas production is clear. However, large-scale implementation will probably depend on supportive policy frameworks and industry and academia collaboration. Gustafsson et al. (2024) conducted an in-depth investigation of biogas industry trends in the 12-member countries, which highlighted policy heterogeneity, financial incentives, and infrastructure readiness as key factors. The report also acknowledged the rising curiosity in biological enhancement technologies, such as fungal inocula and enzyme supplementation, but pointed out the absence of dedicated regulatory pathways for such innovations. The IEA (2020) report stated that the use of advanced microbial and fungal systems could significantly enhance methane productivity and process resilience. Nevertheless, the implementation of these technologies requires standard quality matrices, safety checks, and the setting of pilot-scale demonstration plants. Harmonized bioenergy policies, green credit mechanisms, and circular-economy targets could speed up the process of getting fungal biogas technologies out of the laboratory and into industrial deployment. Thus, in short, policy foresight should keep pace with technological advances. International collaboration, capacity building, and cross-sectoral funding initiatives will be crucial initiatives to make fungal biotechnology a major contributor in the renewable energy sector.

7.10 Conclusion

Recent developments in fungal biotechnology have shown that fungi have great potential in terms of improving biogas production. Their strong ability to disintegrate complex plant materials, production of potent enzymes, and ability to withstand harsh conditions make them efficient for enhancing the efficiency of the AD process. These developments have enabled researchers to map out how fungi work, their interactions with other microbes, and the exploitation of their enzymatic machinery to yield more methane. In the future, research should focus on developing more robust fungal strains, enhancing enzyme secretion, and assembling highly efficient, productive fungal–bacterial consortia. There is also a great potential for applying fungal pretreatment methods at an industrial scale to reduce processing time and thereby boost biogas output. However, other issues, such as optimization of growth conditions, large-scale cost, and maintaining stable fungal activity, need to be resolved. With continued innovation and closer collaboration between researchers and industry, fungal biotechnology can become a major contributor to cleaner energy production and more sustainable waste management.

References

Aguiar-Pulido V, Huang W, Suarez-Ulloa V, Cickovski T, Mathee K, Narasimhan G (2016) Metagenomics, metatranscriptomics, and metabolomics approaches for microbiome analysis: supplementary issue: bioinformatics methods and applications for big metagenomics data. Evol Bioinforma 12:EBO-S36436

Andlar M, Rezić T, Marđetko N, Kracher D, Ludwig R, Šantek B (2018) Lignocellulose degradation: an overview of fungi and fungal enzymes involved in lignocellulose degradation. Eng Life Sci 18(11):768–778. https://doi.org/10.1002/elsc.201800039

Asemoloye MD, Marchisio MA, Gupta VK, Pecoraro L (2021) Genome-based engineering of ligninolytic enzymes in fungi. Microb Cell Factories 20(1). https://doi.org/10.1186/s12934-021-01510-9

Auer L, Buée M, Fauchery L, Lombard V, Barry KW, Clum A et al (2024) Metatranscriptomics sheds light on the links between the functional traits of fungal guilds and ecological processes in forest soil ecosystems. New Phytol 242(4):1676–1690

Balk M, Sofia P, Neffe AT, Tirelli N (2023) Lignin, the lignification process, and advanced, lignin-based materials. Int J Mol Sci 24(14):11668. https://doi.org/10.3390/ijms241411668

Bengelsdorf FR, Gerischer U, Langer S, Zak M, Kazda M (2013) Stability of a biogas-producing bacterial, archaeal and fungal community degrading food residues. FEMS Microbiol Ecol 84(1):201–212

IEA (2020) Outlook for biogas and biomethane: prospects for organic growth, IEA, Paris. https://www.iea.org/reports/outlook-for-biogas-and-biomethane-prospects-for-organic-growth, Licence: CC BY 4.0

Butowski CF, Dixit Y, Reis MM, Mu C (2025) Metatranscriptomics for understanding the microbiome in food and nutrition science. Metabolites 15(3):185

Chaurasia PK, Bharati SL, Singh S, Sivalingam AM, Shankar S, Mani A (2025) Fungal pretreatment methods for organic wastes: advances and challenges in biomass valorization. RSC Sustain 3(3):1234–1266. https://doi.org/10.1039/d4su00582a

Chen WY, Chan YJ, Lim JW, Liew CS, Mohamad M, Ho CD et al (2022) Artificial neural network (ANN) modelling for biogas production in pre-commercialized integrated anaerobic-aerobic bioreactors (IAAB). Water 14(9):1410

Chong DJS, Chan YJ, Arumugasamy SK, Yazdi SK, Lim JW (2023) Optimisation and performance evaluation of response surface methodology (RSM), artificial neural network (ANN) and adaptive neuro-fuzzy inference system (ANFIS) in the prediction of biogas production from palm oil mill effluent (POME). Energy 266:126449

Contato AG, Conte-Junior CA (2025) Fungal lytic polysaccharide monooxygenases (LPMOs): functional adaptation and biotechnological perspectives. Eng 6(8):177. https://doi.org/10.3390/eng6080177

Dar RA, Yaqoob M, Parmar M, Phutela UG (2019) Biofuels from food processing wastes. Mater Res Found 46:249–288. https://doi.org/10.21741/9781644900116-10

Deshpande P, Dhakephalkar PK (2025) From gut to green energy: the essential role of anaerobic fungi in sustainable biofuel production. Sustain Energy Fuels 9(24). https://doi.org/10.1039/D5SE00240K

Dhakal S, Minx JC, Toth FL, Abdel-Aziz A, Meza MJF, Hubacek K, Jonckheere IGC, Kim YG, Nemet GF, Pachauri S, Tan XC, Wiedmann T (2022) Emissions trends and drivers. In: Shukla PR et al (eds) Climate change 2022: mitigation of climate change. Cambridge University Press, Cambridge

Dhiman S, Kaur P, Narang J, Mukherjee G, Thakur B, Kaur S, Tripathi M (2024) Fungal bioprocessing for circular bioeconomy: exploring lignocellulosic waste valorization. Mycology 15(4):538–563

Dollhofer V, Podmirseg SM, Callaghan TM, Griffith GW, Fliegerová K (2015) Anaerobic fungi and their potential for biogas production. In: Biogas science and technology. Springer, Cham, pp 41–61

Garg D, Goyal N, Syan AS, Gupta U (2025) Advancements in microbial formulation for energy production from waste biomass. In: Biotechnological advancements in biomass to bioenergy biotransformation, pp 211–244. https://doi.org/10.1007/978-981-96-9282-8_10

Garg D, Bala S, Kumari S, Phutela UG (2026) Syngas and biogas production from lignocellulosic waste biomass: cleaning and upgrading technologies. In: Biofuels and bioproducts from lignocellulosic materials, pp 211–229. https://doi.org/10.1016/b978-0-443-21527-8.00010-1

Goudarzi R, Jahanshahi DA, Kavousi A, Ariaeenejad S (2025) Discovery and engineering of bifunctional enzymes for lignocellulose degradation: metagenomic and computational approaches. Biotechnol Rep 48:e00926. https://doi.org/10.1016/j.btre.2025.e00926

Griffith GW, Baker S, Fliegerova K, Liggenstoffer A, Van Der Giezen M, Voigt K, Beakes G (2010) Anaerobic fungi: neocallimastigomycota. IMA Fungus 1(2):181–185

Gruening de Mattos PB, Ramos Neyra LC, Matte Borges Machado C, Porto de Souza Vandenberghe L, Rosas Vega FE, Zanlorenzi Weber M, Soccol CR (2025) Hemicellulolytic enzymes applied in biofuel production. In: Enzymes applied in biofuels production: new technologies and innovation, pp 75–104. https://doi.org/10.1007/978-3-032-02420-6_4

Gruninger RJ, Puniya AK, Callaghan TM, Edwards JE, Youssef N, Dagar SS et al (2014) Anaerobic fungi (phylum Neocallimastigomycota): advances in understanding their taxonomy, life cycle, ecology, role and biotechnological potential. FEMS Microbiol Ecol 90(1):1–17

Guerrero IAC, Santos JTG, Valdez ARÁ, Romo DLV, Escareño MPL, Almanza AYH, Beltrán JUH, Terán FH, Balagurusamy N (2025) Strategies for microbiome management in biogas digesters using biomolecular tools. In: Innovations in the global biogas industry, pp 27–67. https://doi.org/10.1016/b978-0-443-22372-3.00002-9

Gustafsson M, Meneghetti R, Souza Marques F, Trim H, Dong R, Al Saedi T, Rasi S, Thual J, Kornatz P, Wall D, Berntsen C, Saxegaard S, Lyng KA, Nagele HJ, Heaven S, Bywater A (2024) A perspective on the state of the biogas industry from selected member countries. Gustafsson M, Liebetrau J (ed) IEA Bioenergy Task 37, 2024:2

Harirchi S et al (2023) Microbiological insights into anaerobic digestion for biogas production. Renew Sust Energ Rev 153:111825

Hassan ME, Zhu X, de Souza EF, Elnashar MM, Lu F (2025) Enzyme immobilization advances: a key to unlocking renewable bioenergy potential. Green Chem 27(37):11289–11311. https://doi.org/10.1039/d5gc03388h

Hatakka A, Hammel KE (2010) Fungal biodegradation of lignocelluloses. In: Industrial applications. Springer Berlin Heidelberg, Berlin, Heidelberg, pp 319–340

Hess M, Paul SS, Puniya AK, Van der Giezen M, Shaw C, Edwards JE, Fliegerová K (2020) Anaerobic fungi: past, present, and future. Front Microbiol 11:584893

Hou L, Ji D, Dong W, Yuan L, Zhang F, Li Y, Zang L (2020) The synergistic action of electro-Fenton and white-rot fungi in the degradation of lignin. Front Bioeng Biotechnol 8. https://doi.org/10.3389/fbioe.2020.00099

Hu Y, Zhou Z, Shen C (2024) Microbial community acclimation during anaerobic digestion of high-oil food waste. Sci Rep 14(1). https://doi.org/10.1038/s41598-024-77136-9

Jyani M, Palanisamy K, Phutela UG (2024) Characterization and extraction of silica particles from green solvent pretreated paddy straw. Bull Mater Sci 47(3). https://doi.org/10.1007/s12034-024-03266-5

Kabeyi MJB, Oludolapo AO (2020) Characteristics and applications of geothermal wellhead powerplants in electricity generation. SAIIE31 Proceedings, South Africa, pp 31222–31235

Kabeyi MJB, Olanrewaju OA (2022) Biogas production and applications in the sustainable energy transition. J Energy 2022:1–43. https://doi.org/10.1155/2022/8750221

Kainthola J, Podder A, Fechner M, Goel R (2021) An overview of fungal pretreatment processes for anaerobic digestion: applications, bottlenecks and future needs. Bioresour Technol 321:124397. https://doi.org/10.1016/j.biortech.2020.124397

Kapoor A, Tiwari AK, Tripathi SC, Asiri M, Pal DB, Tripathi M (2025) Sustainable production of biofuels from lignocellulosic biomass using microbial applications: status, challenges and prospects. Mol Biotechnol. https://doi.org/10.1007/s12033-025-01479-5

Kaur A, Bhatia S, Phutela U (2017) Partial purification and characterization of cellulolytic enzymes extracted from *Trichoderma reesei* inoculated digested biogas slurry. J Pure Appl Microbiol 11(4):1953–1963. https://doi.org/10.22207/jpam.11.4.37

Kaur N, Kaur A, Gupta Phutela U (2023) Enhanced production of lignocellulolytic enzymes by bacterial culture of Delftia sp. through submerged state fermentation of biodigested slurry and utilization for biogas augmentation using paddy straw. Cellul Chem Technol 57(3–4):305–323

Kaur H, Phutela UG, Aggarwal Y, Palanisamy K, Singh I (2026) Yeast mediated expeditious Citrus limetta (sweet lemon) peels bioenzyme preparation and characterization using response surface methodology. Curr Res Nutr Food Sci 14(1):324–337. https://doi.org/10.12944/CRNFSJ.14.1.22

Khan MU, Ahring BK (2019) Lignin degradation under anaerobic digestion: influence of lignin modifications-a review. Biomass Bioenergy 128:105325

Kostopoulou E, Chioti AG, Tsioni V, Sfetsas T (2023) Microbial dynamics in anaerobic digestion: a review of operational and environmental factors affecting microbiome composition and function. Environmental Earth Sciences

Kulichkova GI, Ivanova TS, Blume YB (2020) Plant feedstocks and their biogas production potentials. Open Agric J 14:219–231

Kumar A, Chandra R (2020) Ligninolytic enzymes and their mechanisms. Heliyon 6(2):e03364

Kumar N, Saini H, Aggarwal NK, Jangra N, Dhiman K, Sangwan I (2025) Biological pretreatment of lignocellulosic biomass as a sustainable option for biofuel production. Next Sustain 5:100133. https://doi.org/10.1016/j.nxsust.2025.100133

Lange L (2017) Fungal enzymes and yeasts for conversion of plant biomass to bioenergy and high-value products. In J Heitman, BJ Howlett, PW Crous, EH Stukenbrock, TY James NAR Gow (eds) The fungal kingdom. ASM Press, pp 1027–1048. https://doi.org/10.1128/9781555819583.ch51

Leal K, Rojas E, Madariaga D, Contreras MJ, Nuñez-Montero K, Barrientos L et al (2024) Unlocking fungal potential: the CRISPR-Cas system as a strategy for secondary metabolite discovery. J Fungi 10(11):748

Lee C (2022) Engineering direct interspecies electron transfer for enhanced methanogenic performance. In: Renewable energy technologies for energy efficient sustainable development, pp 23–59. https://doi.org/10.1007/978-3-030-87633-3_2

Li Y, Park SY, Zhu J (2011) Solid-state anaerobic digestion for methane production from organic waste. Renew Sust Energ Rev 15(1):821–826

Li Y, Meng Z, Xu Y, Shi Q, Ma Y, Aung M et al (2021) Interactions between anaerobic fungi and methanogens in the rumen and their biotechnological potential in biogas production from lignocellulosic materials. Microorganisms 9(1):190

Ma B, Li Y, Wang T, Li D, Jia S (2025) Advances in CRISPR/Cas9-based gene editing in filamentous fungi. J Fungi 11(5):350

Madusanka C, Udayanga D, Nilmini R, Rajapaksha S, Hewawasam C, Manamgoda D, Vasco-Correa J (2024) A review of recent advances in fungal mycelium-based composites. Discov Mater 4(1):13

Manyi-Loh CE, Mamphweli SN, Meyer EL, Okoh AI (2019) Microbial anaerobic digestion: process dynamics and implications from the renewable energy, environmental and agronomy perspectives. Int J Environ Sci Technol 16(7):3913–3934. https://doi.org/10.1007/s13762-019-02380-w

Meyer V, Basenko EY, Benz JP, Braus GH, Caddick MX, Csukai M et al (2020) Growing a circular economy with fungal biotechnology: a white paper. Fungal Biol Biotechnol 7(1):5

Mohamed Ali A, Zahangir MA, Mohamed Abdoul-Latif F, Jami MS, Mohamed J, Ainane T (2023) Hydrolysis of food waste with immobilized biofilm as a pretreatment method for the enhancement of biogas production. Sustainability 15(4):3316

Nadir N, Ismail NL, Hussain AS (2019) Fungal pretreatment of lignocellulosic materials. In: Biomass for bioenergy-recent trends and future challenges. https://doi.org/10.5772/intechopen.84239

Østby H, Várnai A (2023) Hemicellulolytic enzymes in lignocellulose processing. Essays Biochem 67(3):533–550. https://doi.org/10.1042/ebc20220154

Paraschiv G, Ferdes M, Ionescu M, Moiceanu G, Zabava BS, Dinca MN (2022) Laccases—versatile enzymes used to reduce environmental pollution. Energies 15(5):1835. https://doi.org/10.3390/en15051835

Parmar M, Patel SAH, Phutela UG, Dhawan M (2024) Comparative analysis of ligninolytic potential among Pleurotus ostreatus and Fusarium sp. with a special focus on versatile peroxidase. Appl Microbiol 4(3):1348–1361. https://doi.org/10.3390/applmicrobiol4030093

Pérez-Contreras S, Avalos-de la Cruz DA, Lizardi-Jiménez MA, Herrera-Corredor JA, Baltazar-Bernal O, Hernández-Martínez R (2024) Production of ligninolytic and cellulolytic fungal enzymes for agro-industrial waste valorization: trends and applicability. Catalysts 15(1):30. https://doi.org/10.3390/catal15010030

Phutela UG, Arora R, Kumar A, Aggarwal Y (2025) Potential use of mosambi peel-based bio-enzymes for wound management in animals. Indian J Anim Reprod 46(1):61–67. https://doi.org/10.48165/ijar.2025.46.01.10

Ren F, Wu F, Wu X, Bao T, Jie Y, Gao L (2024) Fungal systems for lignocellulose deconstruction: from enzymatic mechanisms to hydrolysis optimization. GCB Bioenergy 16(5). https://doi.org/10.1111/gcbb.13130

Reza A, Chen L, Mao X (2024) Response surface methodology for process optimization in livestock wastewater treatment: a review. Heliyon 10(9):e30326

Sahil S, Karvembu P, Kaur R, Katyal P, Phutela UG (2023) Enhanced biogas production from rice straw through pretreatment with cellulase producing microbial consortium. Energy Nexus 12:100246. https://doi.org/10.1016/j.nexus.2023.100246

Sahni N, Phutela UG (2012) Effect of thermophilic fungus Humicola fuscoatra MTCC 1409 on paddy straw digestibility and biogas production. Int J Sci Res 2(9):357–359. https://doi.org/10.15373/22778179/sep2013/125

Sahota S, Shah G, Ghosh P, Kapoor R, Sengupta S, Singh P, Vijay V, Sahay A, Vijay VK, Thakur IS (2018) Review of trends in biogas upgradation technologies. Bioresour Technol Rep 1:79–88

Saini S, Tewari S, Dwivedi J, Sharma V (2023) Biofilm-mediated wastewater treatment: a comprehensive review. Mater Adv 4(6):1415–1443

Salazar-Cerezo S, de Vries RP, Garrigues S (2023) Strategies for the development of industrial fungal producing strains. J Fungi 9(8):834. https://doi.org/10.3390/jof9080834

Sanglier J-J (2014) Faculty opinions recommendation of co-cultivation—a powerful emerging tool for enhancing the chemical diversity of microorganisms. [dataset]. In: Faculty opinions – post-publication peer review of the biomedical literature. H1 Connect. https://doi.org/10.3410/f.718293800.793491612

Sarkar S, Skalicky M, Hossain A, Brestic M, Saha S, Garai S, Ray K, Brahmachari K (2020) Management of crop residues for sustainability. Sustainability 12:9808

Saye LM, Navaratna TA, Chong JP, O'Malley MA, Theodorou MK, Reilly M (2021) The anaerobic fungi: challenges and opportunities for industrial lignocellulosic biofuel production. Microorganisms 9(4):694

Schneider AN, Sundh J, Sundström G, Richau K, Delhomme N, Grabherr M et al (2021) Comparative fungal community analyses using metatranscriptomics and internal transcribed spacer amplicon sequencing from Norway spruce. Msystems 6(1):10–1128

Sharuddin SS, Ramli N, Yusoff MZM, Muhammad NAN, Ho LS, Maeda T (2022) Advancement of metatranscriptomics towards productive agriculture and sustainable environment: a review. Int J Mol Sci 23(7):3737

Shen JY, Zhao Q, He QL (2023) Application of CRISPR in filamentous fungi and macrofungi: from component function to development potentiality. ACS Synth Biol 12(7):1908–1923

Shen Q, Ruan H, Zhang H, Wu T, Zhu K, Han W et al (2024) Utilization of CRISPR-Cas genome editing technology in filamentous fungi: function and advancement potentiality. Front Microbiol 15:1375120

Singh AP, Singh T (2014) Biotechnological applications of wood-rotting fungi. Biomass Bioenergy 62:198–206

Singh H, Janiyani K, Gangawane A, Pandya S, Jasani S (2024) Engineering cellulolytic fungi for efficient lignocellulosic biomass hydrolysis: advances in mutagenesis, gene editing, and nanotechnology with CRISPR-Cas innovations. Discov Appl Sci 6(12). https://doi.org/10.1007/s42452-024-06405-z

Sun Z, Liu Q, Li Y, Mazarji M, Feng L, Pan J (2024) Deciphering the impact of lignin on anaerobic digestion: focus on inhibition mechanisms and methods for alleviating inhibition. ACS Omega 9(44):44033–44041

Tian Y, Li Y, Zhang H, Huang T, Tian W, Wang Z, Qian J (2025) Synergy between bacteria and fungi contributes to biodegradation and methane production of lignocellulosic anaerobic co-digestion exposing to surfactants. J Environ Manag 373:123579

Umar A, Mubeen M, Ali I, Iftikhar Y, Sohail MA, Sajid A, Kumar A, Solanki MK, Divvela PK, Zhou L (2024) Harnessing fungal bio-electricity. Front Microbiol 14:1291904

Vasco-Correa J, Shah A (2019) Techno-economic bottlenecks of the fungal pretreatment of lignocellulosic biomass. Fermentation 5(2):30. https://doi.org/10.3390/fermentation5020030

Vieira RIM, Peixoto A. da S, Monclaro AV, Ricart CA O, Filho EXF, Miller RNG, Gomes TG (2025) Fungal coculture: unlocking the potential for efficient bioconversion of lignocellulosic biomass. J Fungi 11(6):458. https://doi.org/10.3390/jof11060458

Vinzelj J, Joshi A, Insam H, Podmirseg SM (2020) Employing anaerobic fungi in biogas production: challenges & opportunities. Bioresour Technol 300:122687. https://doi.org/10.1016/j.biortech.2019.122687

Wikandari R, Hasniah N, Taherzadeh MJ (2022) The role of filamentous fungi in advancing the development of a sustainable circular bioeconomy. Bioresour Technol 345:126531

Yu S, Hu Y, Cao T, Zhu Y, Li H, You T, Xu F (2025) Revealing the molecular interactions between lignin and carbohydrates towards Improved Lignocellulose utilization. Prog Polym Sci 102070. https://doi.org/10.1016/j.progpolymsci.2025.102070

Zhang X, Wang Y, Jiao P, Zhang M, Deng Y, Jiang C et al (2024) Microbiome-functionality in anaerobic digesters: a critical review. Water Res 249:120891

Chapter 8
Fungal Application in Bioelectricity Production: Applications, Challenges, and Perspectives

Arpitha Manchan, Anupama Nisarani, Anwesha Patel, Olivia Anthony, and Sourav Bhattacharya

Abstract Fungal bioelectrochemical systems exploit the metabolic versatility of fungi to generate electricity and enable integrated wastewater treatment, biosensing, and resource recovery platforms. This chapter documents fungi-mediated electron transfer, including direct extracellular electron transfer, redox-active enzymes, and the role of endogenous and exogenous mediators in enhancing electricity production. Representative fungal genera reported for bioelectricity applications and degradative capabilities are reviewed alongside reactor configurations that optimize fungal colonization, electrode interfacing, and mass transfer for sustained electricity generation. Wastewater-driven power generation, integrated bioremediation-power platforms, biosensing, and decentralized energy recovery from lignocellulosic residues are mentioned. The chapter addresses principal challenges: inherently low power densities, limited electron transfer kinetics, biofilm management and longevity, scale-up complexities linked to electrode design, and process economics. The chapter evaluates engineering responses such as electrode surface functionalization, tailored electrode architectures, fungal strain selection, consortia design, and metabolic or protein-level engineering to enhance redox flux. Perspectives on integrating fungal fuel cells into circular bioeconomy frameworks are discussed, emphasizing their potential for decentralized energy systems and resource recovery. Future research priorities, including molecular elucidation of fungal electron pathways, development of standardized performance metrics, hybridization with bacterial electroactive partners, and demonstrations at pilot scale to validate long-term stability, are highlighted in this chapter. By depicting current research and technological advances, this chapter positions fungi as key players in next-generation bioelectrochemical systems, offering insights into their untapped potential for clean energy innovation.

A. Manchan · A. Nisarani · A. Patel · S. Bhattacharya (✉)
Department of Microbiology and Botany, School of Sciences, JAIN (Deemed-to-be University), Bangalore, Karnataka, India

O. Anthony
Department of Pharmaceutical Sciences, School of Pharmacy, University of Maryland, College Park, MD, USA

M. Tripathi et al. (eds.), *Recent Advancements in Fungal Biotechnology Volume 1*, Fungal Biology, https://doi.org/10.1007/978-981-92-0339-0_8

Keywords Bioelectricity · Electron mediator · Fungal cells · Microbial fuel cell · Wastewater treatment

8.1 Introduction

The current worldwide energy crisis is characterized by a contradictory increase in demand and a pull toward fossil fuels that are still supplying almost 80% of the total global energy. Energy consumption in 2024 increased by 2.2%, which can be attributed to intense industrial activities and very hot weather conditions, thus resulting in record-high emissions despite an enormous increase in the capacity of clean energy. This scenario brings in a vital climate change issue: the Intergovernmental Panel on Climate Change in 2024 has issued a stern warning that there is a need to cut down the global greenhouse gas emissions by 43% by the year 2030 in order to keep the 1.5 °C pathway open. This change necessitates a complete makeover of energy systems, transforming from the usage of carbon-intensive sources to a portfolio of renewable energy that is more resilient (International Energy Agency 2025).

The transition to bioelectricity brings along the benefit of "dispatchable" and base-load power that is reliable. Bioelectricity, which comes from agricultural and organic waste, behaves just like a stable energy supply, which facilitates the grid's flexibility. The recent reports say that the use of bioenergy is responsible for about half of all the renewable energy consumption today. The deployment of bioelectricity on a wider scale allows countries to convert waste streams into energy and, at the same time, offer a low-carbon solution to the existing thermal power plants. Bioelectricity is the process of converting the chemical energy in organic matter to electricity through microbial metabolism directly. The core of this technology is mostly based on microbial fuel cells (MFCs). The basic concept is that exoelectrogenic bacteria in anaerobic conditions oxidize the organic matter in the anode chamber, thus liberating protons and electrons. The protons move through a selective blockade while electrons go the opposite way and flow through an external circuit to the cathode, where the reaction with the oxygen takes place, producing a continuous current (Singh et al. 2024).

According to Shajid et al. (2025), one of the main benefits of bioelectricity compared to traditional biofuels (such as ethanol or biodiesel) is its ability to be directly converted. It does not depend on the energy-consuming methods of refining or burning and is able to perform well even at low temperatures. In addition, bioelectricity has the special potential of "waste-to-watts," which allows the treatment of wastewater and the generation of power at the same time and without land-use conflicts between "food and fuel." The idea can be traced back to 1911 when Michael Cressé Potter first demonstrated the production of electricity from bacteria. The technology took a major turn in 1999 when Byung Hong Kim invented the mediator-less MFCs that not only dropped the use of pricey and hazardous electron shuttles but also laid down the groundwork for present-day bioelectrochemical systems (BES).

Primarily, due to their superior metabolic versatility, fungi present distinct benefits over bacteria in bioelectrochemical systems. Bacteria, the most common among the electrogenic microorganisms, are limited to simple organic acids, while fungi grow in acidic to alkaline conditions and at low to high temperatures. The latter's resilience transforms the "cell factories" concept, as fungi can operate under industrial or environmental conditions that would normally inhibit bacterial growth. An important advantage of fungi is their capability to break down complex, resistant substances such as lignocellulose and even plastics. Bacteria usually depend on preprocessed sugars, but fungi make use of a powerful army of extracellular enzymes, including laccases and peroxidases. These enzymes gradually decompose the tough lignin and cellulose structures, thus allowing the direct conversion of agricultural waste to electricity without much pretreatment. Besides, fungi have extraordinary extracellular electron transfer (EET) mechanisms. There are redox proteins that are membrane-bound and the vast mycelial hyphae that form natural conductive networks. The advantages of filamentous structures include their huge surface area for interacting with electrodes and the ability to cross the physical divides in the biofilm, thus providing a more stable and efficient electron flux to the anode than nonfilamentous bacteria (Islam et al. 2025).

In this chapter, a detailed guide is given on how to employ fungi for the production of bioelectricity, thus highlighting the technology as a significant remedy for the worldwide energy crisis. It first lays down the basic concepts of bioelectrochemical systems and their ability to generate power directly. The proposal of employing fungi to gain central attention is supported by their metabolic flexibility and exceptional capability to consume a wide range of difficult substrates that, surprisingly, limit the performance of traditional bacterial systems. This chapter subsequently examines the routes of technical mechanisms, covering both direct and mediated electron transfer (MET), and assessing the potential of different species such as white-rot fungi (WRF) and yeasts. Theoretical possibilities have been looked into in the areas of waste-to-energy conversion, bioremediation, and remote biosensing. The chapter also talks about existing technical issues and suggests cutting-edge methods for resolving them, for example, through CRISPR/Cas9 gene editing and new materials, making it easier to incorporate fungal electricity into a sustainable circular bioeconomy.

8.2 Fundamentals of Fungal Bioelectricity Generation

8.2.1 Mechanisms of Fungal Electron Transfer

8.2.1.1 Direct Electron Transfer

In fungal bioelectrochemical systems, direct electron transfer (DET) mainly occurs through the outer cell's specialized redox-active proteins. Cytochromes, which are heme-containing proteins, are the main ones for the electron flow. They draw electrons from the metabolic internal pathways and transport them to the electrode through the sequence of redox reactions. Besides, many different proteins on the cell surface help maintain the stability of the physical contact between the fungal mycelium and the anode. All these living beings together establish a molecular bridge in the form of conductive pili or nanowires (i.e., long and thin filaments), which act like biological cables. They give a mediator-less current generation facilitation directly from the fungal cell wall to the conductive material, even from cells that are not in close contact with the electrode (Araújo Soares et al. 2025).

8.2.1.2 Mediated Electron Transfer

In fungi, MET is based on the use of metabolites that have redox properties and are soluble, which act as electron shuttles between the fungal cell and the anode. This way of doing things is especially advantageous since it allows for the current produced by a larger area of the biofilm, rather than just the area around the electrode surface. Fungi secrete a lot of these endogenous molecules in their natural state to enable EET. The shuttles are reduced during the internal metabolic processes and then move through the extracellular matrix to the electrode, where they give up their electrons and go back to the cell to start the process again.

The indirect transfer is made easier by a wide range of redox mediators. Quinone compounds, especially the perylenequinones like hypocrellins, which are notable for their exceptional reversible redox behavior, are the main fungal metabolites that help with survival in anaerobic environments by means of such behaviors. While phenazines (like pyocyanin) are more common in bacterial systems, they are still used in cooperative fungal–bacterial fuel cells for overall electron flux increase. Additionally, others, such as flavins and manganese (Mn^{3+}) ions produced by WRF, serve as necessary redox slots. By using these carriers, fungal bioelectrochemical systems have the ability to produce more power and utilize broader substrates (Segundo et al. 2024b).

In fungal bioelectrochemical systems, mixed mechanisms of electron transfer cases are a complex, multi-modal strategy that uses fungi's power of DET and MET at the same time. The utilization of both pathways at once results in a process that is not only more efficient but also stronger because it makes use of the advantages of both ways: the direct ones, that is, contact-based, and the indirect, that is, diffusion-based (Umar et al. 2024).

The communication between the two organisms is largely facilitated by the enzyme secretion; one such enzyme is cellobiose dehydrogenase (CDH). CDH is a special type of flavocytochrome because it consists of two different domains: the oxidoreductase domain (DH) that catalyzes the reaction and the mobile cytochrome domain (CYT). The DH domain is responsible for the reduction of soluble mediators (MET), whereas the CYT domain can make direct contact with the electrode surface to support DET. As a result, the fungus is able to generate a continuous current irrespective of its location in relation to the anode (Scheiblbrandner and Ludwig 2020).

Additionally, the vast mycelium networks of filamentous fungi function like "biological wires," which get their power from nature. These networks can literally cover large distances in a biofilm by maintaining direct contact at the hyphal tips and, at the same time, discharging endogenous shuttles such as quinones to transport electrons through the surrounding fluid. This coordination that combines redox-active proteins with secreted metabolites not only greatly increases the power density but also makes the process of breaking down complicated organic pollutants easier in varied environmental conditions. Power density represents the actual energy output of the system (Hunter 2023).

8.2.2 Fungal Species with Bioelectricity Potential

The choice of fungal species for the production of bioelectricity is based on their metabolic capabilities, the range of enzymes they produce, and their tolerance to the electrochemical conditions of the MFC. The possible use of fungi in MFC technology has not been completely ignored, as they nevertheless stand out as "novel cell factories" because of their exceptional characteristics (Moubasher et al. 2024).

8.2.2.1 White-Rot Fungi

WRF, such as *Phanerochaete chrysosporium* and *Trametes versicolor*, can be counted among the best fungal biocatalysts. They are credited largely with their ligninolytic enzymes, which encompass the three main enzyme types, lignin peroxidases (LiP), manganese peroxidases (MnP), and laccases. These enzymes are most active in redox reactions and can, therefore, degrade complicated organic matter and, at the same time, make electrons available for power generation. The case of *T. versicolor* is an example of such a scenario, where it produced a power density of 1200 mW/m^3 through its laccases being very effectively used in the reduction reaction of oxygen taking place at the cathode or the oxidation reaction of substrate happening at the anode.

These species are often applied in the anode to decompose complicated polymers (e.g., wood) into simple sugars that can be oxidized to get electrons, or as biocathodes where their enzymes facilitate oxygen reduction. They can utilize "waste" those bacteria cannot, such as sawdust or industrial dyes, facilitating a circular bioeconomy (Umar et al. 2024).

8.2.2.2 Yeasts

Yeasts are the most thoroughly examined fungal group in bioelectrochemical systems because of their quick growth and their genetics that are well-understood. The most common model is *Saccharomyces cerevisiae*, which has the potential for both direct and MET. However, some other genera, such as *Candida* (e.g., *C. melibiosica*), *Pichia*, and *Hansenula,* have excelled in certain industrial applications. *Hansenula anomala*, in particular, has beaten many pure bacterial cultures by giving off up to 2900 mW/m^3 of power, which is an impressive output. In addition, they are the most preferred yeast species because of their ability to convert various types of raw materials to fermentation products, including the residues of brewing and farming operations, which are otherwise discarded (Segundo et al. 2024b).

8.2.2.3 Other Electroactive Species

Recent research has expanded the catalogue of electrogenic fungi to include *Aspergillus* and *Trichoderma* species. *Aspergillus niger* has been successfully utilized with sugarcane bagasse to generate bioelectricity while simultaneously degrading lignocellulosic waste. Furthermore, species such as *Aspergillus* sp. and *Paecilomyces* have gained attention for their "dual-action" potential-generating current while degrading recalcitrant pollutants such as microplastics and heavy metals. These "emergent exoelectrogens" represent a frontier in sustainable bioremediation, where the chemical energy of a pollutant is directly converted into electrical energy (Sarma et al. 2021). Table 8.1 compares the different fungi and substrates used in bioelectricity generation and the advantages of the process.

8.2.3 Bioelectrochemical Systems Architecture

The MFC's architectural design is what decides the success of the whole process of getting the fungus's metabolic activity translated into bioelectricity. Unlike bacteria, fungi are different in many aspects. However, their unique structure, namely, hyphal growth and large enzyme production, requires special arrangements in order to achieve maximum electron recovery and substrate degradation (Sekrecka-Belniak and Toczyłowska-Mamińska 2018).

8.2.3.1 Fundamental Designs of MFCs

There are two main architectural frameworks for bioelectricity production through fungi, which are single-chamber and dual-chamber designs. Single-chamber MFCs (SCMFCs) are known for their easy and economical design, where the anode and cathode are housed in a single compartment, most of the time, only separated by a membrane or a thin air cathode. SCMFCs are considered for their low internal resistance and inexpensive fabrication. Normally, in fungal systems, SCMFCs adopt a porous air cathode that makes it possible to do oxygen reduction directly without active aeration. This setup is very useful for fungi that are either aerobic or microaerophilic because it permits a gradient of oxygen supply. Nevertheless, one of the

Table 8.1 Summary of fungal biocatalysts, substrate compatibility, and metabolic advantages for bioelectricity generation

Fungal species	Substrate utilized	Peak power output	Primary advantages
T. versicolor	Wood/oak hydrolysate, paper mill effluent	~1200 mW/m³	Robust ability to break down lignin and xenobiotics, high biomass accumulation
S. cerevisiae	Glucose, molasses, domestic wastewater	146.7–218 mW/m²	Well-understood genetics for optimization, high sugar tolerance, easy to cultivate
A. niger	Citric acid waste, starch, potato broth	24–438 mW/m²	Efficiently processes starch and industrial food waste, produces high levels of organic acids
K. marxianus	Whey, lactose, natural organic waste	~850 mW/m³	Operates at higher temperatures, fast growth rate, utilizes a wide range of carbon sources
C. boidinii	Fruit waste (blueberry), glucose	~3155 mW/m²	High power density in specialized waste streams, efficient in complex organic mixtures
P. chrysosporium	Lignocellulosic waste, bagasse effluent	10–50 mW/m²	The "gold standard" for lignin breakdown, can be used as a biocathode or for pretreatment
Hansenula anomala	Glucose, sucrose	2900 mW/m³	Exhibits high electron transfer efficiency compared to many other yeast strains
P. fermentans	Wheat straw hydrolysate	65.1 mW/m²	Effective at converting agricultural residues into electricity with minimal pretreatment

main issues is the oxygen diffusion into the anode area (oxygen crossover), which could either obstruct the anaerobic fungal pathways or cause the substrate to be oxidized directly by oxygen, thus lowering Coulombic efficiency (CE) (the ratio of the actual electrons retrieved as current to the maximum theoretical number of electrons that could be generated from the substrate oxidation). If CE is high, it means that the bacteria are using the metabolic energy mainly for electricity production instead of biomass or heat (Dakal et al. 2025).

Dual-chamber MFCs (DCMFCs) are designed with anodes and cathodes in two separate compartments and a proton exchange membrane (PEM) in between them. This physical separation provides a way to maintain strictly controlled conditions for the anode (such as anaerobic or buffered conditions), essential for certain fungi such as *A. niger* or *S. cerevisiae* if one's interest is in particular intermediates. Nafion 117 is known for its high proton conductivity and is thus the most commonly used PEM. However, its being expensive led to the investigation of low-cost separators such as glass fiber, ceramic membranes, or even natural polymer-based membranes, such as cellulose and agar, which are especially suitable for the "green" approach of mycotechnology. The DCMFCs are designed to produce higher open-circuit voltages than the normal fuel cells since they reduce the oxygen crossover, but they are characterized by a higher internal resistance (Ohmic losses), which is attributed to the membrane used and the longer distance between the electrodes (Jalili et al. 2024).

The selection of electrode materials is essential in the case of fungal biofilm attachment. The carbonaceous materials are the best choice for anodes. Carbon felt, carbon cloth, and graphite brushes offer a large surface area and biocompatibility that is necessary for the invasion of hyphae. Recent studies have been conducted to investigate the graphene and carbon nanotube (CNT) application to alter these substrates for the purpose of increasing the conductivity and the EET rate. Fungal hyphae, unlike bacteria, can circumvolve fibers; thus, the anode's structure becomes the most important determinant of its long-lastingness (Chen et al. 2022). The substances used in cathodes consist of platinum (Pt), which is still the best-performing catalyst for oxygen reduction despite the fact that its price is very high, thereby limiting the application on a larger scale. Activated carbon or metal-organic frameworks are usually used as less expensive alternatives in the research of fungal fuel cells (FFCs). Besides, some designs adopt "biocathodes," which are where fungal enzymes such as laccase are fixed on the cathode to speed up the process of reduction of oxygen, thus making it an entirely fungal system (Costa de Oliveira et al. 2020).

8.2.3.2 Specific Considerations for FFCs

FFCs are very different from the traditional bacterial MFCs, mainly in their operational flexibility. One of the differences is that the substrates can be in different states. Bacteria, however, are very fast exoelectrogens (shuttling electrons faster), but their efficiency is limited to a very short range of substrates. On the other hand, fungi are able to act as "bio-refineries," converting complex lignocellulosic polymers into simpler sugars that can then be oxidized for the generation of electricity. Thus, they are the preferred option when it comes to the treatment of "raw" agricultural waste. In traditional FFCs, the use of liquid-state (submerged) configurations is a classical method where the fungi are cultivated in the nutrient broth. This method is suitable for yeasts, but it can present a difficulty for filamentous fungi as they may form hefty "pellets" that would restrict mass transfer (Sonawane et al. 2022).

Conversely, solid-state FFCs (SSFFCs) make use of the natural living conditions of fungi, which are damp and solid organic material. In an SSFFC, the fungus directly grows on a solid substrate like wood chips or crop residue that is in contact with the anode. This arrangement minimizes the use of water and active mixing, thereby significantly lessening the energy "tax" of the system. However, the transition from milliliter-scale laboratory prototypes to liter-scale reactors is a major challenge. With the increase in volume, the internal resistance rises at an exponential rate. In addition, fungal biofilms may grow too thick, which causes the problems of biofouling and clogging of membranes or electrode pores that hinder nutrient diffusion and electron flow (Umar et al. 2024).

8.2.3.3 Hybrid Systems: Fungal–Bacterial Consortia

One of the most encouraging vistas in the area of fungal bioelectricity is the invention of mixed-culture consortia. Fungi and bacteria are frequently found in symbiotic relationships. In a mixed MFC, fungi are considered the "primary degraders" (among others) using their powerful extracellular enzymes (LiP and cellulases) to create tiny molecules from the degradation of complex organic matter (like lignin or cellulose). Bacteria (such as *Geobacter*, *Pseudomonas*, or *Shewanella*) then take the role of "exoelectrogens," transferring the electrons from these simpler molecules to the anode quickly. This separation of duties allows overcoming the metabolic constraints of the single-species cultures. Research has demonstrated that the power density of the fungi–bacteria consortium can be up to two to three times more than that of solely fungi systems, and there is also a significant increase in chemical oxygen demand (COD) removal. This is especially important for wastewater treatment, as it indicates the effective removal of the organic load due to the metabolic activity of the fungi throughout the treatment process, measured as a percentage. Furthermore, mixed cultures are generally more resistant to changes in the environment (pH and temperature) and to the variations in the substrate (Getu et al. 2025).

8.3 Applications of Fungal Bioelectricity

Fungal bioelectrochemical systems are quite multifunctional as they can be used for environmental clean-up, power production, and sensor applications. They use the distinct fungal metabolic capabilities and hyphal networks to not only effectively degrade the pollutants but also generate electricity. Such eco-friendly biotechnological methods indicate a great potential in the management of agricultural waste and even in remote sensing for solving modern environmental and energy problems (Umar et al. 2024).

8.3.1 Waste-to-Energy Conversion

The conversion of waste to energy through the fungal bioelectrochemical systems is a sustainable way that copes with two problems together, one being waste management and the other being renewable energy production. The fungus-based batteries are able to do so not only by transforming various types of organic wastes, such as agricultural by-products, industrial waste, and municipal solid waste, into power but also by diminishing the amount of the pollutants that are harmful to the environment (Ali et al. 2025).

8.3.1.1 Agricultural Waste Valorization

Fungal bioelectrochemical systems are an eco-friendly solution that allows energy production and agricultural waste management at the same time. Yeasts and filamentous fungi use their wide hyphal networks for electron transfer and, at the same time, degrade organic materials. Lignocellulosic biomass is an enormous renewable resource for these systems, although it is often very hard to break down and needs to undergo physical pretreatment like milling to make it more biodegradable. FFCs not only produce electricity that is stable and can be generated from a wide range of substrates, such as kitchen waste, fruit peels, and so on, but they also reduce the adverse effects of the environment by breaking down the toxins present in the organic waste into less complex and more harmless components (Dakal et al. 2025).

Phanerochaete chrysosporium, a white-rot fungus, has laccases and other enzymes in its arsenal and thus points out how efficient the fungi are in undergoing the lignocellulose degradation process. Many others, such as *Aspergillus* and *Trichoderma* spp., are strong contenders directly for biofuel and electricity production of high quality. Recent experimentation has raised *Aspergillus tubingensis* to the spotlight with its super-active cellulase, especially for working on agricultural residues. Moreover, it has been demonstrated, with applications involving *Pleurotus ostreatus*, that the sugarcane bagasse being processed can be concurrent with the current and voltage being measured, which corroborates the claim that these systems can indeed reduce the COD significantly while they produce power (Segundo et al. 2024b).

8.3.1.2 Industrial Wastewater Treatment

Industrial wastewater treatment through FFCs has a two-fold advantage as it not only cleans up the polluted effluents but also produces bioelectricity at the same time. These systems are very good at treating waste from various sectors such as textile, brewing, and food processing industries. High COD brewery and distillery wastewaters have been successfully treated with fungal consortia such as *Aspergillus* and *Penicillium* spp. resulting in remarkable pollution reduction along with high power densities (Umar et al. 2024). Besides, food processing wastes can also be used as a good agricultural substrate for energy production. Although the textile industry presents a big challenge due to the presence of resistant azo dyes, *P. chrysosporium* has been shown to be quite effective in degrading these dyes by employing complex oxidative and reductive mechanisms. Moreover, sophisticated setups with ionic liquid membranes are a new era of technological achievements as they not only selectively transport protons but also create optimum pH conditions for the microorganisms, which support and enhance the technology's sustainability in managing complex industrial liquid waste streams (Roy et al. 2023).

8.3.1.3 Municipal Solid Waste and Performance Metrics

The organic fraction of municipal solid waste is a very valuable feedstock for fungal bioelectrochemical systems, as it provides together waste volume reduction and energy generation at the same time. Biogenic waste has low costs associated with its use in the electricity generation process, as it usually needs only minimal pretreatment. Latest studies have verified that waste from fermented beverages can result in high voltage and power density. Though performance shows variation according to the operating conditions, production of COD occurs frequently in the range of 70–85% when optimized systems are being used. Successful utilization necessitates meticulous regulation of factors such as electrode material, membrane characteristics, pH, temperature, and organic loading rates (Chatterjee and Venkata Mohan 2022).

8.3.2 Bioremediation with Simultaneous Energy Recovery

Fungal bioelectrochemical systems have the potential to degrade pollutants and, at the same time, produce electrical energy, thus providing a unique dual strategy for environmental remediation. Innate ecological habitats of these microorganisms make bioremediation, implemented through MFCs, a feasible option for the cleaning of polluted areas. They take a share of the power generated; the more power they produce, the more oil they can effectively biodegrade under anaerobic conditions. Besides, through methods such as biosorption and enzymatic transformation, fungi can easily cope with heavy metal pollution. Filamentous genera such as *Aspergillus* and *Penicillium* are able to metabolize organic pollutants for energy; thus, they can degrade complex aliphatic and polycyclic aromatic hydrocarbons (Abbas and Rafatullah 2021).

Fungi would not only majorly degrade the persistent organic pollutants such as gasoline and diesel but also do it by breaking down the complex matter through their extensive enzyme systems. Pilot projects validate the practicality of soil MFCs for in-situ remediation, which is the provision of power continuously. Fungal mycelium, in addition to direct degradation, acts as a transporter that helps the degrading bacteria move in the soil. This combined effort greatly speeds up the elimination of the contaminants. The scientific research supports that these systems can reach very high degradation rates while electricity is generated, thus making it economically viable for pollutants to be cleaned through electric power (Vaksmaa et al. 2023).

8.3.3 Biosensing Applications

FFCs may be viewed as the most, if not the only, promising platforms for biosensing of water quality and environmental contaminants. Their primary benefits are cost-effectiveness, robustness, and the ability to operate with no external power supply.

One of the main applications is the sensing of biological oxygen demand, in which the electroactive biofilm transforms the environmental changes into electrical signals. The recent 3D-printed designs have been successful in assessing the quality of urban wastewater, with the sensitivity of the tests being consistent and comparable to the standard methods. These setups have remarkably long-lasting performance and stability, and some designs have been able to operate for more than 5 years nonstop with just a little maintenance (Salvian et al. 2024).

One of the main applications where the sensors perform toxicity detection is now focusing on heavy metals such as cadmium, which is detected at very low concentrations with quick response times. Floating sensors, which are part of the portable field-deployable versions, are used for monitoring oil spills and freshwater oxygen levels. Such sensors, in comparison with enzymatic alternatives, have a major advantage in terms of durability since they do not require costly enzyme purification processes. Single-chamber arrangements are favored greatly due to their straightforwardness in design. They extensively eliminate the necessity of a constant catholyte supply and simultaneously accommodate both electrodes in a single reaction chamber (Yao et al. 2023).

8.3.4 Remote and Off-Grid Power Generation

Fungal bioelectrochemical systems are capable of powering low-energy devices in very remote areas, where conventional power sources such as batteries and solar panels usually cannot provide enough energy. Dirty solar panels or a complete lack of sunlight place them at a disadvantage, while labor-intensive operations of battery replacements in wide areas are not economically feasible. Soil MFCs, on the other hand, turn out to be a great solution since they continue to produce energy from soil organic carbon. Surprisingly, plant MFCs with *Cyperus papyrus* have been all the while drawing supercapacitors for 900 days, powering environmental sensors. This claims their excellent durability and suitability for long-term off-grid monitoring (Bataillou et al. 2024).

Paper-based systems with *Shewanella putrefaciens* at the core have been able to power wireless sensor nodes that can transmit data to smartphones. Proper power management is essential to avoid any voltage reversal in the connected units, and thus, a customized solution is required for efficient operation. The most efficient way to scale up is to use a modular approach, where multiple small-scale cells are employed, yielding higher power densities than larger volumes. The trials conducted in agricultural fields have demonstrated that these systems make it possible to operate at a power level that is more than sufficient for the sensors, thus greatly cutting down the maintenance of sensors located in hard-to-reach areas (Veerubhotla et al. 2019).

8.3.5 Integration with Biorefinery Concepts

The alliance of fungal bioelectrochemical systems with biorefineries opens up the possibility of producing electricity and other value-added products in one go, which thus increases their economic viability. These environment-friendly setups take in the maximum of the feedstock by putting to use the fungal enzymes such as endoglucanases and xylanases for the complete degradation of the plant biomass. That means the process will not generate much waste while being able to convert a lot of biomass. Moreover, the fermentation of the enzyme along with the fungal cell biomass gives considerable value as this can be changed into useful compounds like heteropolysaccharides and glycoproteins (Troiano et al. 2020).

Fungal biorefineries are fed with cheap lignocellulosic feedstocks from the agricultural and food industry, which can then be turned into biofuels along with the production of high-value items such as organic acids and single-cell proteins. The principle of sequential valorization is followed in this system, where the electricity generated is used for heating the processing of the residues while the residues are further processed. The task of consolidating bioprocessing is made easier as it involves the production of the enzyme together with the fermentation in one reactor using a single bioreactor. The recent hybrid systems further enhance the efficiency of the processes by combining the bioelectrochemical technologies with the enzymatic fuel cells to increase the electron transfer rates and product output (Corona-Martínez et al. 2025). Figure 8.1 illustrates the versatility of bio-engineered fungal hubs across four primary sectors, that is, waste management, bioremediation, biosensing, and energy generation, highlighting specific applications such as pollutant degradation and sustainable power sources.

8.4 Challenges in Fungal Bioelectricity Systems

The commercialization of fungal bioelectrochemical systems still faces a long journey ahead due to major technical, biological, and economic obstacles. Overcoming these different concerns is a prerequisite for transferring laboratory research into large-scale industrial applications.

8.4.1 Technical Challenges

The main technical challenge that fungal bioelectrochemical systems face is the low power output and current density that are inherent to them when compared to the bacterial ones. The studies state that the FFCs obtain power output that is markedly lower than in the case of bacterial or mixed-culture systems. This is due to the slower substrate oxidation and the lower electron transfer rates among the factors.

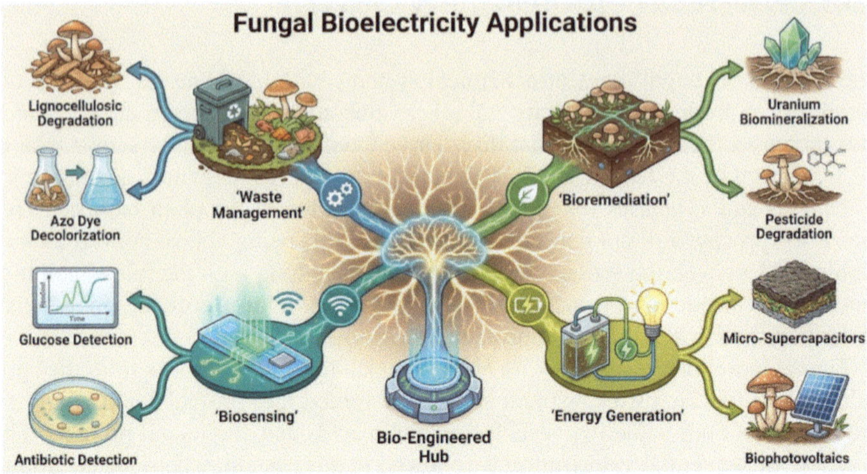

Fig. 8.1 Schematic representation of fungal bioelectricity applications

To this day, the mentioned limitations are worsened by the high internal resistance of the system. The ohmic resistance in the electrodes and the electrolytes combines with the activation and concentration polarization and results in energy generation being micro-inefficient. Therefore, the amount of electricity produced is very often not enough for practical applications because the system fails to find a way around these fundamental metabolic and physical barriers to efficient operation (Wang et al. 2022).

Another main obstacle is that electron transfer is inefficient, which is mainly a consequence of the fungal cell wall's physical structure. The barrier that is this thick and complex consists of chitin and polysaccharides, which hinder the direct interaction of intracellular and extracellular metabolic processes by means of electrodes. Most fungi, unlike special bacteria such as *Geobacter* spp., which have conductive pili, do not possess any machinery for DET at all. The very low electroactivity in this case makes it necessary for the system to use MET and hence requires redox-active compounds. Nonetheless, these mediators also bring in their own inefficiencies and potential toxicity problems that make it even harder to stabilize and maintain the bioelectrochemical system's performance (Thapa et al. 2022).

In the arena of operational feasibility, the slow growth rates of the most industrially relevant fungi present a major hindrance. The initiation of the cultivation of functional biofilms on electrode surfaces can last for several days or even weeks, resulting in high production costs and the potential for contamination by rapidly multiplying microorganisms. Also, the ways in which fungal biofilms develop are quite different from those in bacterial systems. Fungal hyphae possess distinct growth patterns and an extracellular matrix that sometimes counteract the ideal situation for electrode colonization. Such biological barriers elongate the commissioning period and lessen the overall practicality of installing FFCs in urgent or widespread industrial locations (Roy et al. 2022).

On top of that, the management of environmental parameters introduces a complex set of challenges. The majority of the electroactive fungi are aerobic, which creates a problem for the anodic chamber that usually requires anaerobic conditions for optimal efficiency. The oxygen supplied for the cathode can mix with that at the anode, which can lead the fungi to take in the O_2 for respiration instead of transferring electrons to the electrode. This leads to significant power loss (Mahesh and Mahidhara 2025). Besides that, the fungal systems are very much affected by pH and temperature changes. Continuous running of the system may cause the accumulation of metabolic by-products that change the pH of the system to levels outside the narrow optimal zones needed by fungi. The resulting instability reduces the metabolic activity and, consequently, the electron transfer kinetics, making the systems less reliable in the long run (Zhang et al. 2025).

8.4.2 Material and Design Challenges

One of the primary hurdles in the development of implantable electrodes is their biocompatibility. The conductivity of the material must be sufficiently high, and at the same time, the surface must be so venous that it harbors wall-resisting fungi yet remains non-toxic to living cells. Among carbon-based materials, brushes and cloths are the ones that excel, but their high price still acts as a barrier to commercialization. In fact, the capital costs for the first generation of these systems can be up to 30 times more than the ones needed for conventional sewage treatment (Borja-Maldonado and López Zavala 2022).

Another challenge that leads to a reduction of the system performance is membrane fouling caused by microbial growth, which, in turn, obstructs proton migration and then lowers power output. Although there are some techniques, such as hydrophobic coatings and polarity reversal, the major issue with them is the short life of the operation because of catalyst poisoning and component degradation. Further, the process of moving these systems from the lab to the industry involves a lot of engineering challenges, particularly where the power output does not scale up proportionately with the size. Thus, the inability to preserve the energy efficiency on large scales, along with the need for frequent maintenance, has already determined the overall economic viability and sustainability of the technology as compromised (Khan et al. 2024).

8.4.3 Biological and Metabolic Challenges

Non-electroactive microorganisms still present a constantly worrisome situation in open systems by taking over the feeding and lowering the efficiency. Usually, such competitors enlarge faster than electroactive fungi and thus create an unstable situation for the process. Besides, the metabolic ability of fungi to adapt mostly leads to the diversion of electrons toward growth by respiration or fermentation rather than electricity generation. Such an inefficiency is compounded by genetic instability in which mutations occurring over a period of time render electron transfer less efficient and change critical biofilm properties (Chong et al. 2025).

Operational challenges become heavier with the presence of complicated substrates like lignocellulosic biomass, which will entail expensive pretreatment to make it less resistant. The removal of toxins, though, greatly improves the activity. However, it makes processing costs increase substantially. The success of the fungi strains relies on matching them precisely to the waste streams, but the wide differences in efficiency of the strains and the adaptation to the changing conditions of the wastewater complicate the large-scale application (Zhao et al. 2021).

8.4.4 Economic and Practical Challenges

The high capital expenses are the main economic obstacle for the commercialization of the fungal bioelectrochemical systems. These systems usually need an initial investment that is about 30 times higher than the traditional wastewater treatment due to the expensive electrodes, catalysts, and separators among the components. Besides construction costs, the energy return on investment is often very low, and the coefficients are often below the level needed for the project to be economically viable. The ratio of energy produced to energy invested is so poor that many projects are deemed immature or exploratory and hence lose their appeal to investors whose focus is on sustainability and financial returns that are already proven (Unuofin et al. 2023).

The market entry battle is further complicated by intense competition from the mature and optimized technologies, such as activated sludge processes that have supply chains in place. The extensive use of these systems would require huge investments in infrastructure and personnel trained in the technology that would extend far beyond just installing a simple reactor unit. The absence of large-scale commercial demonstrations leads to a considerable confidence gap that prevents performance and reliability validation under real conditions. Furthermore, the uncertainty from regulations and the public's fear of the safety and environmental impact make the situation worse since the regulatory agencies often do not have clear-cut guidelines for green-lighting such biotechnological inventions (Tsekouras et al. 2022). Table 8.2 outlines the primary technical and biological hurdles associated with developing fungal bioelectrical systems. It categorizes challenges into seven distinct areas, that is, contamination, electron transfer, growth kinetics, metabolic flux, oxygen management, pH stability, and power output.

Table 8.2 Major technical and biological challenges in fungal bioelectrochemical systems

Challenge category	Specific issue	Impact on system performance	Potential solutions
Contamination	Competition with non-electroactive microbes	Reduced system efficiency, unpredictable performance	Selective enrichment, controlled inoculation, antimicrobial strategies
Electron transfer	Thick fungal cell wall barrier	Reduced current generation, inefficient electron transfer	Genetic modification, mediator addition, cell wall disruption
Growth kinetics	Slow fungal growth rates	Extended start-up time (days to weeks)	Pre-cultured inoculum, growth promoters, temperature optimization
Metabolic flux	Diversion to non-electroactive pathways	Lower electricity generation	Metabolic engineering, substrate optimization, operational parameter control
Oxygen management	Aerobic nature of most fungi	Substrate loss to respiration, reduced CE	Microaerobic conditions, oxygen-limiting strategies
pH stability	Narrow optimal pH ranges	System instability, metabolic disruption	Buffer systems, pH monitoring and control, adaptive strains
Power output	Low current density compared to bacteria	Limited practical applications	Electrode modification, optimized reactor design, mixed cultures

8.5 Strategies for Improvement and Recent Advances

Fungal bioelectricity has witnessed a significant transformation from the era of proof-of-concept experiments to the time of sophisticated engineering methods. Today, researchers are exploring the basic hurdles that have inhibited the development of fungal MFCs. The advancements that have been made come from a variety of fields, such as molecular biology to materials science, and show a keen realization that conquering fungal bioelectricity needs to be tackled from all angles. Essentially, the strategies being discussed here are the crest of the wave of the current scientific inquiries conducted in laboratories around the world, with each breakthrough making us nearer to feasible applications (Islam et al. 2025).

8.5.1 Genetic and Metabolic Engineering

The latest developments in fungus genetic engineering have made it possible to use and develop electroactive capabilities to a great extent by overcoming the previously existing barriers of complicated cell structure. One of the main methods is to make the electron-transferring proteins, such as cytochromes, very active. These proteins are called biological shuttles because they move electrons through the membranes. One of the cool examples is the increasing of cytochrome c in

Saccharomyces cerevisiae, which resulted in a 2.3 times bigger current production. Another effective method is the subtractive approach, where the researchers are knocking out the metabolic paths that are competing for electrons. The metabolic paths, which were the diversions, are now eliminated, and hence, the flux is now directed toward the respiratory chain. This was proven in *Aspergillus niger*, where the cutting down of the fermentative pathway led to better CE by 1.8 times (Wu et al. 2024).

Also, the scientists are trying to use inter-kingdom hybridization by putting the electron transfer machinery of the bacteria, such as *Shewanella* and *Geobacter*, into the fungi. The hybrids of *Pichia pastoris* resulted in cell lines that produced current 3.1 times larger than controls. The adaptation of Cas9 technology has completely changed the situation by making it possible to do precise and stable multiplex gene edits. In *Yarrowia lipolytica*, simultaneously, the genes for transfer were overexpressed, and the pathways that were competing for resources were deleted, resulting in a 4.2-fold power density increase. Finally, synthetic biology is not only about the simple modifications but also about the designing of turning the whole new genetic circuits (Luo et al. 2022).

These synthetic operons, like the ones made in *Aspergillus oryzae,* can make microbes change their gene expression according to the environmental signals. This way, they can be made into smart systems that manage electricity production depending on the food supply. In combination, the advances in protein engineering, metabolic redirection, cross-species gene transfer, and synthetic circuit design are converting fungi into very efficient and highly programmable platforms for the sustainable generation of bioelectricity (Cairns et al. 2021).

8.5.2 Material Science Innovations

The development of material science is directly impacting the fungal MFCs to a great extent by ensuring optimal interface at the point of their convergence with technology. Carbon-based nanomaterials such as graphene and CNTs have turned out to be the best choices for electrodes due to their high conductivity and surface area. As an example, three-dimensional graphene aerogels allowed the formation of thick *Aspergillus niger* biofilms, which produced power densities of 687 mW/m^2, while arrays of vertically aligned CNTs multiplied the power output by 3.4 because they created the electron transfer pathways (Wang and Li 2024).

Metal organic frameworks are another extraordinary idea that comes with the capability to adjust crystalline structures that generate the best growing conditions for fungi, and thus, more than 800 mW/m^2 power densities are obtained. Conductive polymers such as polypyrrole can be made to carry adhesive peptides that will quicken the process of biofilm formation and give stability for a long time. In addition to this, three-dimensional printing now makes it possible to produce custom-made lattice-structured electrodes with the geometry that is perfectly controlled and is optimized for the growth of fungi (Kižys et al. 2025).

Surface modifications, apart from bulk materials, are very important in boosting performance. Boosting the features of the electrodes with chitosan or exercising plasma treatments makes the fungi stick better and lowers the resistance of the interface. The technique of membranes is also getting better with composite membranes that contain graphene oxide, which give the features of great proton conduction and anti-fouling to avert the blockage by biomass. The bio-inspired membranes with the surface patterns being specifically designed also keep the effectiveness intact by not allowing the fouling while letting through a high proton flux (Ramesh et al. 2025).

The combined methods of material use have made it possible for the fuel cell to not only meet but also benefit from the genetic modifications of the fungal strains, thus enhancing the bioelectricity generation through better mass transport, higher electrical connectivity, and more durability in the biocompatibility at the junction of the fungal electrode (Yaqoob et al. 2021).

8.5.3 System Design and Operational Optimization

The use of fungal MFCs nowadays follows the paradigm of basic batch cultures to sophisticated systems that depend not just on high efficiency but also on precise control and integrated designs to bring the best efficiency out of the system. The fed batch operation has proved to be an effective way to hold the power output consistently over a prolonged period, as evidenced by the systems of *Pleurotus ostreatus* that took care of the metabolic inhibition by adding the substrates according to the consumption rates of the fungi (Segundo et al. 2024b).

Hydraulic retention time is another important factor, where a 3–5-day window of the optimal amount will give the maximum removal of the COD along with the highest power density. Stacked configurations are utilized by the researchers to connect multiple units in series to overcome the low voltage limitations, using real-time monitoring and bypass circuits to recurrently check voltage reversal in the weaker units (Castellano-Hinojosa et al. 2025).

Hybrid systems are a substantial ground of development because they combine the strong enzymatic powers of fungi like *Trichoderma reesei* with the dominion in the electron transfer of bacteria, thereby getting power densities that are 2.7 times higher than those of monocultures. In addition, the integration of fuel cells into broader biorefinery concepts, for instance, using spent mushroom substrate for electricity generation, not only improves the economic viability of the technology but also makes it more attractive (Christwardana et al. 2021).

Finally, the advanced automation through fuzzy logic control systems makes it possible to have real-time adjustments of feeding and aeration based on live sensor data, which consequently leads to higher average power outputs and better resilience to environmental disturbances. Having a holistic approach to system management ensures that genetic and material improvements are fully realized in practical and large-scale applications (Islam et al. 2025).

8.5.4 Mediator and Co-factor Enhancement

Even though the theoretical advantage would categorize direct transfer as the best option, in reality, MET using redox-active compounds frequently surpasses it regarding practical power density. Neutral red and methylene blue, which are artificial mediators, move electrons between the fungal cells and the electrodes like shuttles without physical contact. The study on *Aspergillus niger* reported that using optimized concentrations of the mediator resulted in a 4.5 times power output. Nevertheless, the environmental concerns associated with synthetic chemicals have turned the search for natural and endogenous alternatives (Sarma et al. 2021).

Fungi have their own redox-active compounds, such as quinones and flavins, and by adding riboflavin precursors to the cultures, *Candida albicans* was able to produce power densities that are on par with those of the synthetic systems. Furthermore, in situ generation has advanced to such an extent that it is now possible to genetically engineer *Aspergillus oryzae* for the overexpression of phenazine biosynthesis genes from bacteria, consequently forming self-mediating systems that do not need any external additives. Another creative method making use of redox-active nanoparticles has been developed, which comprises the use of gold nanoparticles that are organic molecule modified and mimic the behavior of a mediating substance that is continuously replenished at the fungal electrode interface, thus creating a stable electron transfer hotspot lasting for more than 100 days (Verma and Mishra 2021).

Mediator management precision is vital because using the wrong concentrations can either cause metabolic inhibition or result in wastage. By employing electrochemical methods to analyze cycling kinetics, researchers can coordinate that the generation of the mediator is in proportion to the fungal metabolic rate, which has been proven to increase CE by as much as 35%. These advances in the strategies of synthetic, natural, and nanoparticle-based mediation are essential for the transition of FFCs into energy systems that are both sustainable and of high performance (Pal et al. 2021). Figure 8.2 explains the multidimensional framework of fungal-hybrid MFC technology, encompassing diverse reactor configurations, symbiotic consortia strategies, and their integration into circular bioprocesses and automated monitoring systems.

8.6 Future Perspectives and Research Directions

Fungal bioelectricity is a power source of the future that is still at an interesting crossroads. The basic research has been done where fungi are recognized as electricity producers, the mechanisms are known, and a list of applications is available. However, fungal MFCs are still a long way from being used by mankind. The way ahead calls for a combination of gradual enhancements and remarkable breakthroughs. The thrill of this area is that the advancements are being made at different levels concurrently, for example, from laboratory optimization to initial pilot

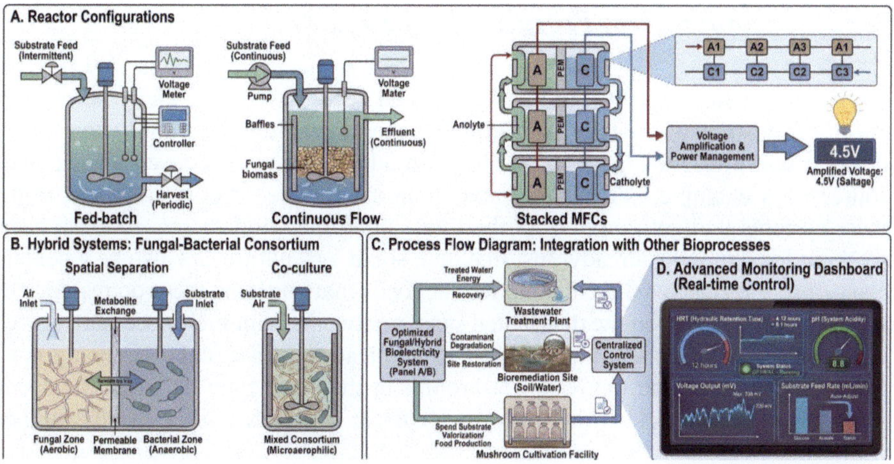

Fig. 8.2 Overview of fungal-hybrid bio-electrochemical systems, showing reactor configurations, microbial consortia, and integrated operational monitoring

experiments, from main mechanism explorations to techno-economic evaluation. The following timeline shows the realistic goals and expectations for the next decades (Umar et al. 2024).

8.6.1 *Short-Term Goals (2–5 Years)*

Bringing fungal MFC power densities to levels where they could compete would still be a major focus. Currently, the output of them ranges from 100 to 800 mW/m^2, while bacterial systems have already surpassed 2000 mW/m^2. The target aimed at 1500–2000 mW/m^2 needs the perfect combination of genetic modification and high-performance materials within 5 years (Luo et al. 2022).

The urgent need for standardized testing protocols must be met as well, which will enable accurate performance benchmarking across various labs. A lack of basic mechanistic understanding exists that this field has to tackle, however, since advanced microscopy has already shown that the microbial interactions of cytochromes and nanowires differ from each other according to species. In order to take the laboratory curiosities to market, pilot-scale demonstrations of 100–1000 L would be required first to pinpoint the mass transfer and biomass management challenges at that scale (Islam et al. 2025).

The larger systems are indispensable for the comprehensive techno-economic evaluations, and the determining factor for capital and operating costs would be conducting. FFCs might be suitable for niche applications like remote sensing or specialized waste treatment, but broader commercial viability is heavily reliant on substantial cost reductions coupled with a better grasp of system lifetimes, regulatory requirements, and so forth (Yek et al. 2022).

8.6.2 Medium-Term Prospects (5–10 Years)

The future of fungal MFCs in the commercial sector will most likely be in very specific industries where their self-sufficient characteristic and the ability to convert difficult waste into energy will be the main advantages over conventional power sources. For example, in environmental monitoring, which is done remotely, using these systems that get power from the forest litter, the wireless sensors have been working for 6 months already. Just like that, smart agriculture can also be a beneficiary of this scenario, where the hidden fuel cells that consume the crop residues to supply the entire season the distributed sensor networks with their necessary power could be the scenario. Another very good application of a circular economy is to produce electricity and cultivate mushrooms at the same time by using the spent substrate as a feedstock, which converts a current wasteland into a revenue stream (Luo et al. 2022).

The combination of stable fungal–bacterial consortia over the next decade will be a crucial factor in the prevention of population instability and the maximization of electron transfer through synergy that has already come up with free energy for the ecosystem in bioreactors supported by super-technical genetic tools like genome-scale CRISPR in species such as *Aspergillus niger*. This will also include metabolic pathways' complete engineering instead of just simple knockouts. Standardization of these cutting-edge synthetic biology techniques will not only help the researchers in the lab and near the end-users of the product but also, at the same time, bring the fungal-based fuel cells from their existing research state to the undoubted reliable energy solutions for both the distributed and industrial applications (Bai et al. 2021).

8.6.3 Long-Term Vision (10+ Years)

In the distant future, fungal bioelectricity is seen as a major player in the circular bioeconomy. It will have turned the urban organic waste into a reliable source of dispersed energy. Urban electricity consumption could be greatly reduced by these systems if they succeeded in reaching the quite ambitious power densities of 3000–5000 mW/m^2. Moreover, this would make landfills unnecessary. Besides simple waste processing, fungal MFCs could find their way into green infrastructure, such as urban parks and green walls, where they could treat graywater and power lighting along the pathways (Ali et al. 2025).

Further, the newfangled studies point out the possibility of carbon capture; that is, fungi would be the ones to produce electricity, but, at the same time, they would be capturing carbon in the form of stable polymers like melanin, which would contribute to the fight against climate change. On top of that, the bioremediation powered by the fungi themselves is the easy way out for cleaning contaminated industrial or agricultural sites, as they would be degrading the pollutants while at the same time providing the energy for the monitoring equipment (Umar et al. 2024).

Perhaps the most revolutionary idea is that of myco-electronic interfaces, which involve living mycelial networks being directly connected to electronic circuits, resulting in the creation of self-regenerating biobatteries or environmental sensors. These speculative but scientifically founded visions portray a future where the interaction between biology and technology becomes so close that highly sophisticated urban ecosystems develop that are both productive and self-healing (Thulasinathan et al. 2021).

8.6.4 Interdisciplinary Research Needs

To tap into the entire possibility of fungal bioelectricity, it will be necessary to have mycologists, electrochemists, and engineers working hand in hand like never before in order to transfer the information from one discipline to the other. The combination of biology and genome-scale metabolic modeling allows scientists to better foresee the effect of genetic changes on power output, thus avoiding the traditional trial-and-error methods that are time-consuming and inefficient. Moreover, machine learning provides the community with very good optimization tools as long as high-quality standardized datasets are established. Last but not least, the conduct of thorough life cycle assessments is integral in proving the environmental advantages of these systems by considering all material inputs and operational costs from construction to disposal (Cucurachi et al. 2018).

8.7 Conclusion

Fungal bioelectricity is an innovative method of using fungi to transfer electrons from their hyphal networks out of the environment, which, in turn, transforms the bioelectricity generated into their food or products and, at the same time, benefits nature through pollutant degradation. Fungi are capable of living and developing unceasingly in the dark by feeding on the organic matter in the soil, and the hyphal networks that they grow help them by providing a very good surface area and conductivity. While the current electricity output is between 60 mWm^{-3} and 86.316 mW/cm^2, the technology is still bound by scalability and operational stability concerns. Due to the inefficiencies in the electron transfer process and the resultant power densities being lower than those of bacterial fuel cells, most of the systems are confined currently to the lab. However, fungal systems have the advantage of being very versatile in terms of the substrates they can use and the degradation process they can employ through enzymes. Hence, they are highly suitable for applications like biosensors or decentralized wastewater treatment, where the dual functional capability of the systems adds significant value.

In order to reach the stage of commercialization, the researchers will have to do the optimization of the electrode materials and use genetic engineering to make the fungi more electroactive. The interdisciplinary collaboration of artificial intelligence and metabolic modeling is crucial for the prediction of optimal performance modifications. The future regulatory framework should be such that it promotes pilot projects and the incorporation of these systems into the existing infrastructure, thus contributing to the circular bioeconomy. At the same time, that fungal bioelectricity does not compete with wind or solar energy at the grid scale, it is a necessary waste-to-energy solution. Through collaboration with the global sustainability targets, this technology opens up an evolutionary route for biotechnological solutions that reconcile economic desires with ecological responsibilities by means of efficient waste valorization.

Acknowledgments The author wishes to extend his sincere gratitude to the management of JAIN (Deemed-to-be University) for providing the research facilities.

Conflict of Interests The authors have no relevant financial or non-financial interests to disclose.

Consent for Publication The authors mutually agreed to submit the book chapter. The book chapter has not been published elsewhere, either wholly, in part, or in any other form. The book chapter has not been submitted to another publisher and will not be published elsewhere.

References

Abbas SZ, Rafatullah M (2021) Recent advances in soil microbial fuel cells for soil contaminants remediation. Chemosphere 272:129691. https://doi.org/10.1016/j.chemosphere.2021.129691

Ali HE, Hemdan BA, El-Naggar ME, El-Liethy MA, Jadhav DA, El-Hendawy HH, Ali M, El-Taweel GE (2025) Harnessing the power of microbial fuel cells as pioneering green technology: advancing sustainable energy and wastewater treatment through innovative nanotechnology. Bioprocess Biosyst Eng 48:343–366. https://doi.org/10.1007/s00449-024-03115-z

Araújo Soares K, Schembek Silva JA, Wang X, Valente Bueno A, Leite Lobo F (2025) Tools for enhancing extracellular electron transfer in bioelectrochemical systems: a review. Fermentation 11(7):381. https://doi.org/10.3390/fermentation11070381

Bai X, Lin T, Liang N et al (2021) Engineering synthetic microbial consortium for efficient conversion of lactate from glucose and xylose to generate electricity. Biochem Eng J 172:108052. https://doi.org/10.1016/j.bej.2021.108052

Bataillou G, Ondel O, Haddour N (2024) 900-Days long term study of plant microbial fuel cells and complete application for powering wireless sensors. J Power Sources 593:233965. https://doi.org/10.1016/j.jpowsour.2023.233965

Borja-Maldonado F, López Zavala MÁ (2022) Contribution of configurations, electrode and membrane materials, electron transfer mechanisms, and cost of components on the current and future development of microbial fuel cells. Heliyon 8(7):e09849. https://doi.org/10.1016/j.heliyon.2022.e09849

Cairns TC, Zheng X, Feurstein C, Zheng P, Sun J, Meyer V (2021) A library of Aspergillus niger chassis strains for morphology engineering connects strain fitness and filamentous growth with submerged macromorphology. Front Bioeng Biotechnol 9:820088. https://doi.org/10.3389/fbioe.2021.820088

Castellano-Hinojosa A, Gallardo-Altamirano MJ, Pozo C, González-Martínez A, González-López J (2025) Inoculum selection and hydraulic retention time impacts in a microbial fuel cell treating saline wastewater. Appl Microbiol Biotechnol 109(1):29. https://doi.org/10.1007/s00253-024-13377-y

Chatterjee S, Venkata Mohan S (2022) Fungal biorefinery for sustainable resource recovery from waste. Bioresour Technol 345:126443. https://doi.org/10.1016/j.biortech.2021.126443

Chen J, Zuo K, Li B, Xia D, Lin L, Liang J, Li X (2022) Embedment of graphene in binder-free fungal hypha-based electrodes for enhanced membrane capacitive deionization. Sep Purif Technol 304:122381. https://doi.org/10.1016/j.seppur.2022.122381

Chong PL, Chuah JH, Chow CO, Ng PK (2025) Plant microbial fuel cells: a comprehensive review of influential factors, innovative configurations, diverse applications, persistent challenges, and promising prospects. Int J Green Energy 22(3):599–648. https://doi.org/10.1080/15435075.2024.2421325

Christwardana M, Joelianingsih J, Yoshi LA (2021) Performance of yeast microbial fuel cell integrated with sugarcane bagasse fermentation for COD reduction and electricity generation. Bull Chem React Eng Catal 16(3):446–458. https://doi.org/10.9767/bcrec.16.3.9739.446-458

Corona-Martínez DA, Martínez-Amador SY, Rodríguez-De la Garza JA, Laredo-Alcalá EI, Pérez-Rodríguez P (2025) Recent advances in scaling up bioelectrochemical systems: a review. Biotech 14(1):8. https://doi.org/10.3390/biotech14010008

Costa de Oliveira MA, D'Epifanio A, Ohnuki H, Mecheri B (2020) Platinum group metal-free catalysts for oxygen reduction reaction: applications in microbial fuel cells. Catalysts 10(5):475. https://doi.org/10.3390/catal10050475

Cucurachi S, van der Giesen C, Guinée J (2018) Ex-ante LCA of emerging technologies. Proc CIRP 69:463–468. https://doi.org/10.1016/j.procir.2017.11.005

Dakal TC, Singh N, Kaur A, Dhillon PK, Bhatankar J, Meena R, Sharma RK, Gadi BR, Sahu BS, Patel A, Singh B, Kumari K (2025) New horizons in microbial fuel cell technology: applications, challenges, and prospects. Biotechnol Biofuels Bioprod 18(1):79. https://doi.org/10.1186/s13068-025-02649-y

Getu AA, Dessie W, Sugira Murekezi J, Sarker MS, Chen G, Hazzan OO, Xiao Y (2025) Microbial synergistic interactions in mixed cultures for improved and sustainable power generation in microbial fuel cells: a review. Sustainability 17(24):10942. https://doi.org/10.3390/su172410942

Hunter P (2023) The fungal grid: fungal communication via electrical signals has inspired the hypothesis of a wood wide web of plants and fungi: fungal communication via electrical signals has inspired the hypothesis of a Wood Wide Web of plants and fungi. EMBO Rep 24(5):e57255. https://doi.org/10.15252/embr.202357255

International Energy Agency (IEA) (2025) Renewables 2025: analysis and forecast to 2030. https://www.iea.org/reports/renewables-2025

Islam MS, Chantal AAK, Islam A, Somtochukwu OS, Poudyal RR, Wesseh CB, Faiyaz OI, Kong X, Wei X (2025) Bioelectricity harvesting from microorganism: review of recent advancements. Front Fungal Biol 6:1739847

Jalili P, Ala A, Nazari P, Jalili B, Ganji DD (2024) A comprehensive review of microbial fuel cells considering materials, methods, structures, and microorganisms. Heliyon 10(3):e25439. https://doi.org/10.1016/j.heliyon.2024.e25439

Khan SS, Amjad M, Shareef H, Larkin S (2024) Review of microbial fuel cell from a techno-economic perspective. Energy Explor Exploit 42(1):373–398. https://doi.org/10.1177/01445987231208510

Kižys K, Pirštelis D, Bružaitė I, Morkvėnaitė I (2025) Polypyrrole-modified *Saccharomyces cerevisiae* used in microbial fuel cell. Biosensors 15(8):519. https://doi.org/10.3390/bios15080519

Luo J, Tian W, Jin H, Yang J, Li J, Wang Y, Shen W, Ren Y, Zhou M (2022) Recent advances in microbial fuel cells: a review on the identification technology, molecular tool and improvement strategy of electricigens. Curr Opin Electrochem 37:101187. https://doi.org/10.1016/j.coelec.2022.101187

Mahesh A, Mahidhara G (2025) Membrane-less microbial fuel cells for energy production-a comprehensive review. Fuel Cells 25(6):e70032. https://doi.org/10.1002/fuce.70032

Moubasher H, Tammam A, Saleh M (2024) Enhancing electricity generation using fungal laccase-based microbial fuel cell. J Microbiol Biotechnol Food Sci 14(2):e9703. https://doi.org/10.55251/jmbfs.9703

Pal M, Shrivastav A, Sharma RK (2021) Electrochemical response of *Saccharomyces cerevisiae* with stainless steel and carbon fiber electrodes. Mater Today Proc 43:2979–2984. https://doi.org/10.1016/j.matpr.2021.01.327

Ramesh P, Gupta R, Koventhan C, Muralitharan G, Lo AY, Huang YJ, Ramasamy S (2025) Recent trends in the use of electrode materials for microbial fuel cells accentuating the potential of photosynthetic cyanobacteria and microalgae. Processes 13(5):1348. https://doi.org/10.3390/pr13051348

Roy AS, Sharma A, Thapa BS, Pandit S, Lahiri D, Nag M, Sarkar T, Pati S, Ray RR, Shariati MA, Wilairatana P, Mubarak MS (2022) Microbiomics for enhancing electron transfer in an electrochemical system. Front Microbiol 13:868220. https://doi.org/10.3389/fmicb.2022.868220

Roy H, Rahman TU, Tasnim N, Arju J, Rafid MM, Islam MR, Pervez MN, Cai Y, Naddeo V, Islam MS (2023) Microbial fuel cell construction features and application for sustainable wastewater treatment. Membranes 13(5):490. https://doi.org/10.3390/membranes13050490

Salvian A, Farkas D, Ramirez-Moreno M, Torruella-Salas D, Berná A, Avignone-Rossa C, Varcoe JR, Esteve-Núñez A, Gadkari S (2024) Resilience of anodic biofilm in microbial fuel cell biosensor for BOD monitoring of urban wastewater. npj Clean Water 7(1):53. https://doi.org/10.1038/s41545-024-00350-5

Sarma H, Bhattacharyya PN, Jadhav DA, Pawar P, Thakare M, Pandit S, Mathuriya AS, Prasad R (2021) Fungal-mediated electrochemical system: prospects, applications and challenges. Curr Res Microb Sci 2:100041. https://doi.org/10.1016/j.crmicr.2021.100041

Scheiblbrandner S, Ludwig R (2020) Cellobiose dehydrogenase: bioelectrochemical insights and applications. Bioelectrochemistry 131:107345. https://doi.org/10.1016/j.bioelechem.2019.107345

Segundo RF, Magaly DLCN, Luis CC, Otiniano NM, Soto-Deza N, Terrones-Rodríguez N (2024b) Reducing plastic waste and generating bioelectricity simultaneously through fuel cells using the fungus *Pleurotus ostreatus*. Sustainability 16(18):7909. https://doi.org/10.3390/su16187909

Sekrecka-Belniak A, Toczyłowska-Mamińska R (2018) Fungi-based microbial fuel cells. Energies 11(10):2827. https://doi.org/10.3390/en11102827

Shajid SR, Mourshed M, Kibria MG, Shabani B (2025) The potential of microbial fuel cells as a dual solution for sustainable wastewater treatment and energy generation: a case study. Energies 18(14):3725. https://doi.org/10.3390/en18143725

Singh P, Srivastava A, Srivastava N, Sinha N, Sharma V, Upadhyay A (2024) Microbial fuel cell as innovative approach for bio-electricity generation: a review. Zašt Mater 65(3):426–439. https://doi.org/10.62638/ZasMat1184

Sonawane JM, Mahadevan R, Pandey A, Greener J (2022) Recent progress in microbial fuel cells using substrates from diverse sources. Heliyon 8(12):e12353. https://doi.org/10.1016/j.heliyon.2022.e12353

Thapa BS, Kim T, Pandit S, Song YE, Afsharian YP, Rahimnejad M, Kim JR, Oh SE (2022) Overview of electroactive microorganisms and electron transfer mechanisms in microbial electrochemistry. Bioresour Technol 347:126579. https://doi.org/10.1016/j.biortech.2021.126579

Thulasinathan B, Jayabalan T, Sethupathi M, Kim W, Muniyasamy S, Sengottuvelan N, Naina Mohamed S, Ponnuchamy K, Arun A (2021) Bioelectricity generation by natural microflora of septic tank wastewater (STWW) and biodegradation of persistent petrogenic pollutants by basidiomycetes fungi: an integrated microbial fuel cell system. J Hazard Mater 412:125228. https://doi.org/10.1016/j.jhazmat.2021.125228

Troiano D, Orsat V, Dumont MJ (2020) Status of filamentous fungi in integrated biorefineries. Renew Sust Energ Rev 117:109472. https://doi.org/10.1016/j.rser.2019.109472

Tsekouras GJ, Deligianni PM, Kanellos FD, Kontargyri VT, Kontaxis PA, Manousakis NM, Elias CN (2022) Microbial fuel cell for wastewater treatment as power plant in smart grids: utopia or reality? Front Energy Res 10:843768. https://doi.org/10.3389/fenrg.2022.843768

Umar A, Mubeen M, Ali I, Iftikhar Y, Sohail MA, Sajid A, Kumar A, Solanki MK, Kumar Divvela P, Zhou L (2024) Harnessing fungal bio-electricity: a promising path to a cleaner environment. Front Microbiol 14:1291904. https://doi.org/10.3389/fmicb.2023.1291904

Unuofin JO, Iwarere SA, Daramol MO (2023) Embracing the future of circular bio-enabled economy: unveiling the prospects of microbial fuel cells in achieving true sustainable energy. Environ Sci Pollut Res 30(39):90547–90573. https://doi.org/10.1007/s11356-023-28717-0

Vaksmaa A, Guerrero-Cruz S, Ghosh P, Zeghal E, Hernando-Morales V, Niemann H (2023) Role of fungi in bioremediation of emerging pollutants. Front Mar Sci 10:1070905. https://doi.org/10.3389/fmars.2023.1070905

Veerubhotla R, Nag S, Das D (2019) Internet of Things temperature sensor powered by bacterial fuel cells on paper. J Power Sources 438:226947. https://doi.org/10.1016/j.jpowsour.2019.226947

Verma M, Mishra V (2021) Recent trends in upgrading the performance of yeast as electrode biocatalyst in microbial fuel cells. Chemosphere 284:131383. https://doi.org/10.1016/j.chemosphere.2021.131383

Wang L, Li M (2024) Improving the performance of microbial fuel cell electrode materials to enhance electricity production. J Energy Biosci 15(4):233–242. https://doi.org/10.5376/jeb.2024.15.0022

Wang J, Ren K, Zhu Y, Huang J, Liu S (2022) A review of recent advances in microbial fuel cells: preparation, operation, and application. Biotech 11(4):44. https://doi.org/10.3390/biotech11040044

Wu N, Wu X, Zhang M, Zhang C, Xu Q (2024) Metabolic engineering of *Aspergillus niger* for accelerated malic acid biosynthesis by improving NADPH availability. Biotechnol J 19(5):e2400014. https://doi.org/10.1002/biot.202400014

Yao H, Xiao J, Tang X (2023) Microbial fuel cell-based organic matter sensors: principles, structures and applications. Bioengineering 10(8):886. https://doi.org/10.3390/bioengineering10080886

Yaqoob AA, Ibrahim MN, Guerrero-Barajas C (2021) Modern trend of anodes in microbial fuel cells (MFCs): an overview. Environ Technol Innov 23:101579. https://doi.org/10.1016/j.eti.2021.101579

Yek PNY, Wan Mahari WA, Kong SH et al (2022) Pilot-scale co-processing of lignocellulosic biomass, algae, shellfish waste via thermochemical approach: recent progress and future directions. Bioresour Technol 347:126687. https://doi.org/10.1016/j.biortech.2022.126687

Zhang R, Dai W, Xiang H, Chen J, Yi T, Li J, Zhang J, Yang Q, Xiao R, Li X (2025) Characteristics and driving factors of power generation performance in microbial fuel cells: an analysis based on the CNKI database. Front Microbiol 16:1620539. https://doi.org/10.3389/fmicb.2025.1620539

Zhao J, Li F, Cao Y, Zhang X, Chen T, Song H, Wang Z (2021) Microbial extracellular electron transfer and strategies for engineering electroactive microorganisms. Biotechnol Adv 53:107682. https://doi.org/10.1016/j.biotechadv.2020.107682

Chapter 9
Biohydrogen Production Using Fungi: Recent Advances and Applications

Atin Kumar and Sharad Sachan

Abstract Sustainable and renewable energy sources have made biohydrogen a potential alternative to fossil fuels because of its high energy output and absence of carbon emissions. Fungal systems are also considered efficient and environmentally friendly solutions to the problem of hydrogen generation, due to their exceptional enzyme activities and ability to withstand various organic substances. This chapter gives a broad profile of fungal biohydrogen generation, which focuses on the mechanisms involved, including dark fermentation, and the benefits of fungal pretreatment of lignocellulosic and organic wastes. Recent developments in solid-state fermentation of fungi, metabolic and genetic engineering, and co-culture with hydrogen-producing bacteria are mentioned as the dominant approaches to improve hydrogen yield. The incorporation of fungal biohydrogen generation into a biorefinery and a circular bioeconomy is also described, presenting opportunities to convert low-value waste streams into high-value bioenergy products. Although significant progress has been made, issues like scalability, process optimization, and economic feasibility are critical to industrial adoption. This chapter suggests future perspectives and research directions, including technological advances, hybrid methods of pretreatment, and optimization of the process, with the help of systems biology and machine learning. Altogether, this chapter highlights the important role of fungi in enhancing sustainable hydrogen generation and a low-carbon energy future.

Keywords Biohydrogen · Circular bioeconomy · Dark fermentation · Fungal pretreatment · Genetic engineering · Hydrogen production · Lignocellulosic biomass · Metabolic pathways · Renewable energy · Waste valorization

A. Kumar (✉)
School of Agriculture, Uttaranchal University, Dehradun, Uttarakhand, India

S. Sachan
Department of Agricultural Economics and Extension, School of Agriculture, Lovely Professional University, Phagwara, Punjab, India
e-mail: sharad.19461@lpu.co.in

M. Tripathi et al. (eds.), *Recent Advancements in Fungal Biotechnology Volume 1*, Fungal Biology, https://doi.org/10.1007/978-981-92-0339-0_9

9.1 Introduction

The annual global hydrogen demand is projected to exceed 500 Mt by 2050, driven by the decarbonization of transport, industry, and power sectors. Among the renewable hydrogen pathways, biological hydrogen (biohydrogen) offers the advantage of coupling waste valorization with low-carbon fuel generation. Unlike thermochemical or electrochemical hydrogen production, biohydrogen production relies on microbial metabolism to convert organic substrates into H_2 under mild conditions (Fan et al. 2024; Ramprakash et al. 2022).

The production of biohydrogen takes advantage of the metabolic potential of a variety of microorganisms, such as bacteria, microalgae, archaea, and fungi, to transform organic materials into hydrogen gas. Though bacteria and microalgae have traditionally played a leading role in research and industrial applications due to their relatively high hydrogen yields and well-understood metabolic pathways, recent years have seen an increasing interest in the individual merits of using fungi (Fan et al. 2024). Fungi, especially anaerobic and lignocellulolytic ones, have a well-developed enzymatic apparatus that can hydrolyze complex vegetal biomass (agricultural residues and food waste) into fermentable sugars. This capability is not only used to convert low-value or waste substances into high-value energy carriers effectively but also corresponds to the ideas of a circular bioeconomy and sustainable use of resources (Manuel et al. 2022; Saha et al. 2021).

The reactions occurring in the background of the production of fungal biohydrogen are diverse and can include dark fermentation, co-culture with other microorganisms, and joint use in combination with complex bioprocessing methods. Fungi can also operate in a wide range of settings and can be extremely strong against the changeability of materials, pH shifts, and inhibitory byproducts, in contrast to bacterial systems (Fan et al. 2024; Saha et al. 2021). This strength is such that it is particularly attractive for use at scale, such as at the industrial level, where the heterogeneity of feedstocks and the reliability of the processes are the key elements.

Emerging trends in the industry have been guided toward maximizing fungal pretreatment processes, optimizing hydrogen production through metabolic engineering, and integrating biorefinery processes that would result in the production of hydrogen and other value-added products. As a case in point, fungal solid-state fermentation (SSF) usage has been shown to be a terrific supplement to the digestibility of lignocellulosic feeds, which consequently boosts the effectiveness of the subsequent stages of fermentation in producing hydrogen (Manuel et al. 2022; Zhao et al. 2014). Also, fungi can be co-cultured with hydrogen-generating bacteria or methanogens to form synergistic interactions that increase total hydrogen production and process resilience (Saha et al. 2021; Panahi et al. 2022).

Even with these encouraging developments, a number of issues still prevail. Hydrogen yields from fungal systems have improved, but in many cases are not as high as those of optimized bacterial processes. In addition, fungal biohydrogen production needs to be assessed for scalability, economic viability, and stability over time, which should be further explored and developed through technology (Fan et al. 2024; Ramprakash et al. 2022; Saha et al. 2021). These problems need to be addressed in the quest to realize the potential of fungi as a key element of the sustainable hydrogen economy. In this chapter, the authors provide the general picture of the available knowledge base, recent technological advances, and practical aspects of the production of fungal biohydrogen, and define the main challenges and prospects of the specified area.

9.2 Fundamentals of Biohydrogen Production

9.2.1 Biological Hydrogen Production Mechanisms

The biological processes that are used to generate biohydrogen include direct and indirect biophotolysis, photofermentation, and dark fermentation. These processes are performed by a number of populations of microorganisms, and they require particular environmental conditions (Fan et al. 2024; Ramprakash et al. 2022; Rittmann and Herwig 2012). The most fascinating aspect of fungal systems is the so-called dark fermentation process that can decompose an extremely large range of organic substances without the need for light (Fan et al. 2024; Manuel et al. 2022). Dark fermentation is the breakdown of organic matter by microorganisms under anaerobic conditions into hydrogen, carbon dioxide, and other organic acids. It is particularly appropriate for the treatment of organic wastes and of lignocellulosic biomass, which is a widespread and renewable feedstock (Ramprakash et al. 2022; Saha et al. 2021).

Photofermentation, in turn, is the use of photosynthetic bacteria or algae to convert organic acids to hydrogen in light conditions. Higher yields of hydrogen may be realized with the help of this process, but it is impractical to use because of the need for constant availability of light and the sensitivity of organisms to oxygen. Biophotolysis is a direct or indirect system that makes use of the photosynthetic apparatus of microalgae or cyanobacteria in order to separate the molecules of water and liberate hydrogen. The rate of biophotolysis is, however, commonly limited by low photon conversion efficiency and oxygen evolution that may inhibit hydrogenase enzymes (Rady et al. 2024; Ahmed et al. 2021).

Various factors affect the selection of the production mechanism, and they include the type of microorganism, availability of substrate, environmental factors, and desired end products. In fungal-assisted biohydrogen systems, dark fermentation is the key hydrogen-generating stage, where fermentable sugars released by fungal pretreatment are converted to H_2 and CO_2 by hydrogen-producing bacteria (Fig. 9.1).

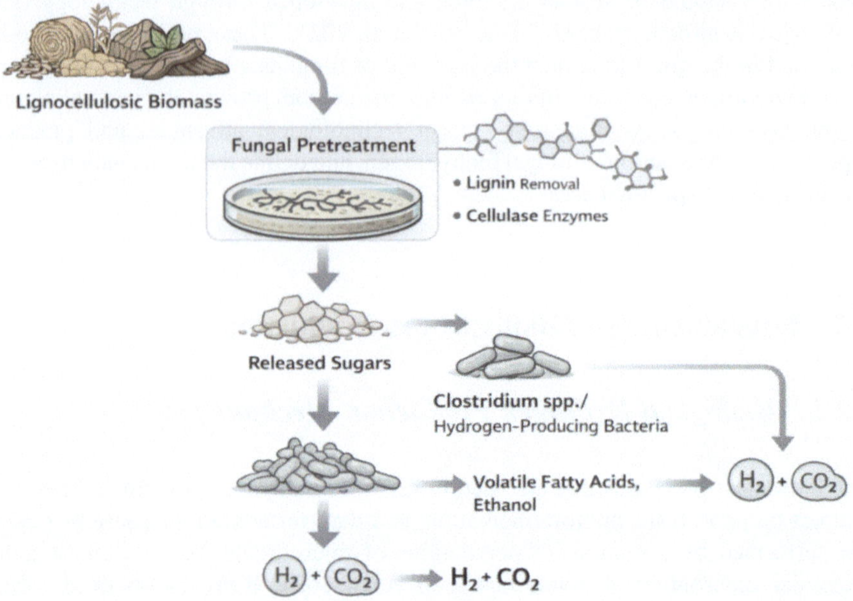

Fig. 9.1 Fungal-assisted fermentative biohydrogen production from lignocellulosic biomass

9.2.2 Microorganisms Involved

Hydrogen can be produced by a number of microorganisms, such as bacteria (e.g., Clostridia, purple non-sulfur bacteria), microalgae (e.g., Chlamydomonas reinhardtii), archaea, and fungi. Although bacteria and microalgae have received a lot of interest, fungi are also receiving interest due to their distinct enzyme functions and durability under various conditions. Fungi, especially representatives of the phyla Ascomycota, Basidiomycota, and Neocallimastigomycota, have numerous carbohydrate-active enzymes (CAZymes) that allow them to degrade lignocellulosic biomass effectively. This arsenal of enzymes enables fungi to enter and degrade complex polysaccharides that are inaccessible, which are commonly recalcitrant to degradation by bacteria. Anaerobic fungi, which are commonly present in the digestive tracts of herbivores, are especially efficient at lignocellulosic biomass degradation because of their production of the most active cellulases and hemicellulases. White-rot fungi, in turn, are characterized by lignin-degrading ability, which assists in the conversion of plant cell wall polysaccharides to fermentable sugars (Saha et al. 2021; Panahi et al. 2022). The variety of fungal species and their metabolic routes have great potential for the creation of efficient and strong biohydrogen production systems.

9.2.3 Comparison of Biological and Non-biological Hydrogen Production

Some of the benefits of biological hydrogen production compared to conventional systems include reduced energy needs, renewable feedstock usage, and fewer greenhouse gas emissions. Other non-biological processes, such as steam methane reforming, coal gasification, and water electrolysis, are the prevailing technologies for producing hydrogen; however, they are linked to high energy consumption and high levels of carbon emissions. Conversely, waste management systems can be combined with biological processes, which will be both a source of energy and a means to clean the environment. But such issues as low hydrogen yields and the scalability of processes should be overcome to render the biological methods cost--effective. There are various factors that affect the effectiveness of biological hydrogen production, and such factors include types of microorganisms, environmental conditions, type of substrate, and the way in which the process is configured. The key to these challenges is in the field of metabolic engineering, design of bioreactors, and integration of the processes that will help achieve the potential of biohydrogen production (Ramprakash et al. 2022; Pachaiappan et al. 2024; Saha et al. 2021).

9.2.4 Factors Affecting Biohydrogen Production

The influences on the efficiency of biohydrogen production are varied but could include temperature, pH, the concentration of the substrate, the type of inoculum, and the presence of inhibitory substances (Fan et al. 2024; Ramprakash et al. 2022; Manuel et al. 2022). The best conditions are different depending on the microorganism and pathway of production. An example of this is that thermophilic strains are more efficient in the conversion of substrates and mesophilic ones in the aspect of volumetric productivity (Rittmann and Herwig 2012). The feedstock is also very crucial, as different feedstocks contain different amounts of hydrogen depending on composition and digestibility (Saha et al. 2021). Fungal solid-state fermentation could potentially be employed as a pretreatment for the process to improve the level of fermentable sugars and hydrogen production (Manuel et al. 2022; Zhao et al. 2014). There are also possibilities of competing microorganisms that may affect the production of hydrogen since they have the capacity to divert substrates out of hydrogen-producing pathways. The processes, including design of experiments (DOE) techniques and strategies, have to be optimized in order to find the optimal conditions for biohydrogen production and maximize the yields (Rittmann and Herwig 2012; Manuel et al. 2022).

9.3 Fungi in Biohydrogen Production

9.3.1 Types of Fungi Relevant to Hydrogen Production

Some of the fungi used to synthesize biohydrogen include anaerobic fungi (e.g., Neocallimastigomycota), white-rot fungi (e.g., Phanerochaete chrysosporium), and other lignocellulolytic fungi. They can hydrolyze complex lignocellulosic biomass into fermentable sugars, which are then converted into hydrogen by hydrogen--producing bacteria. Specifically, a pool of anaerobic fungi, which are abundant in the gut of herbivores, are highly efficient in the degradation of lignocellulosic biomass, as they produce highly active cellulases and hemicellulases. The white-rot fungi have, in turn, been associated with the ability to deconstruct lignin and release fermentable sugars, which are trapped within the walls of the plant cells (Saha et al. 2021; Panahi et al. 2022). The successful growth of fungal species and the routes followed by the fungi to generate biohydrogen offer the possibility of developing potent and efficient systems of biohydrogen generation. Recent reports have also shown that a number of promising systems of fungal pretreatment, which favor biohydrogen production, such as Aspergillus tubingensis, Meyerozyma caribbica, and Trametes versicolor, were discovered (Manuel et al. 2022; Saha et al. 2021). These fungi are very efficient in the hydrolysis of lignocellulosic materials and also produce high hydrogen yields from dark fermentation.

9.3.2 Unique Enzymatic Systems and Metabolic Pathways

Carbohydrate-active enzymes (CAZymes) and cellulosomes are abundantly present in fungi, and they help them readily degrade lignocellulosic biomass. These enzyme systems are often more effective than those of bacteria, particularly for the breakdown of recalcitrant materials. Mediating fungi are used to produce hydrogen through bacterial fermentation of sugars released during fungal biomass hydrolysis in the dark (Fig. 9.1), and glycolysis and mixed-acid fermentation are the pathways supporting H_2 evolution.

The occurrence of fermentation in the presence of hydrogenase and formate dehydrogenase enzymes is critical in the production of hydrogen gas. Secondary metabolites are also produced by fungi and may affect the effectiveness of the hydrogen production process and microbial community stability (Saha et al. 2021; Panahi et al. 2022). The development of omics technologies, including genomics, transcriptomics, and proteomics, is offering a novel understanding of the metabolic networks and regulatory processes that drive fungal biohydrogen generation.

9.3.3 Advantages and Limitations

The general benefits of fungi as a biohydrogen source are their versatile nature in terms of substrate selection, environmental resilience, and the possibility of being used together with other bioprocesses (Saha et al. 2021; Panahi et al. 2022). Fungi can grow in various conditions, such as extreme pH and temperature conditions, and therefore can be used in industry. Moreover, the fermentation of lignocellulosic biomass by fungi can also exploit the fermentable sugars in further processes by increasing their availability (Manuel et al. 2022; Zhao et al. 2014).

Nonetheless, drawbacks are reduced hydrogen yields relative to certain bacterial systems, the requirement for additional optimization of metabolic routes, and problems with the scalability of the process and its economic viability (Ramprakash et al. 2022). The existence of competing microorganisms, including lactic acid bacteria, can also influence the hydrogen production by diverting substrates away from hydrogen-producing pathways (Manuel et al. 2022). To solve these problems, metabolic engineering, process integration, and bioreactor design will need improvements (Saha et al. 2021; Panahi et al. 2022).

9.4 Substrates and Pretreatment Strategies

9.4.1 Lignocellulosic Biomass and Other Organic Wastes

Agricultural waste, food waste, and forestry byproducts are forms of lignocellulosic biomass, which is a renewable and abundant feedstock to produce biohydrogen. The recalcitrance of lignocellulose, however, necessitates effective pretreatment to make it more digestible. Fungi are particularly well-suited to the pretreatment of lignocellulosic feeds, as they express enzymes that break down lignin, e.g., laccases, peroxidases, and cellulases. These enzymes break down the hardened structure of the plant cell wall and liberate fermentable sugars, which can be utilized by hydrogen--producing microorganisms (Ali et al. 2024; Kovács et al. 2022; Sheng et al. 2018; Zhi and Wang 2014; Zhou and Li 2016).

The research undertaken in recent times has revealed the utility of various kinds of fungi for the pretreatment of different types of lignocellulosic biomass. An example is that it is proven that wheat straw, corn stover, and willow chips can be degraded by Penicillium aurantiogriseum, Trichoderma reesei, and other filamentous fungi, which results in anaerobic degradation and biogas generation (Kovács et al. 2022). The reason behind this is that the hyphae penetrate deeply into the substrate, as demonstrated by a scanning electron microscope, which augments the hydrolysis mechanism and yield of fermentable sugars. In particular, it has been established that corn stover can be a highly effective source of biomass to take advantage of during both biomass degradation and biogas production, which explains why substrate selection could turn out to be the main determinant of biohydrogen production (Kovács et al. 2022).

Food waste is another good potential biohydrogen source. It has been shown that food waste can be fermented in fungal solid-state fermentation (SSF) by Aspergillus tubingensis and Meyerozyma caribbica strains to liberate significant amounts of soluble carbohydrates, which can be further converted to hydrogen by dark fermentation (Manuel et al. 2022). Nevertheless, competition for substrates can occur because of the presence of lactic acid bacteria during SSF, which decreases the yields of hydrogen. This brings attention to the fact that microbial communities should be properly controlled in terms of pretreatment and fermentation processes (Manuel et al. 2022).

9.4.2 *Fungal Solid-State Fermentation and Enzymatic Hydrolysis*

Fungal solid-state fermentation (SSF) is one of the most common techniques of pretreatment that exploits the enzymatic potential of fungi to decompose lignocellulosic materials. This causes not only the enhancement of the availability of fermentable sugars but also eliminates the necessity of harsh chemical treatment. SSF refers to the growth of fungi on wet solid materials without free-flowing water, which resembles the natural environment of most lignocellulolytic fungi. Direct conversion of the obtained hydrolysates into hydrogen through dark fermentation can be done (Manuel et al. 2022; Kovács et al. 2022; Sheng et al. 2018; Zhi and Wang 2014).

SSF is affected by a variety of factors, among them being the species of fungi, the type of substrate to be used, the moisture content, and the incubation period. As an example, edible fungi, like Gymnopus contrarius, were applied to pretreat rice straw, which increased the yield of hydrogen by 74% relative to untreated straw. Microscopic and spectroscopic analyses revealed that the pretreatment activity resulted in the total removal of lignin and the structural disorder of the rice straw. Notably, the pretreated rice straw was nutrient-rich to support the activity of the hydrogen-producing bacteria, which did not require any extra supplements in the fermentation medium (Sheng et al. 2018).

Equally, pseudo-white-rot fungi, such as Phanerochaete chrysosporium, have been used to pretreat wheat straw, where significant lignin removal and better hydrogen production are achieved by means of simultaneous saccharification and fermentation (SSF) with Clostridium perfringens (Zhi and Wang 2014). Fungal pretreatment followed by enzymatic hydrolysis and fermentation is an exciting approach to the ultimate goal of biohydrogen production using lignocellulosic biomass.

9.4.3 *Nanomaterial-Assisted and Hybrid Pretreatment Methods*

New developments have investigated nanomaterials, including magnetite nanoparticles, to improve the thermal and pH stability of fungal enzymes, thereby improving the effectiveness of biomass hydrolysis and subsequent hydrogen generation. Also, hybrid pretreatment systems that involve fungal and chemical or physical treatment are being examined due to the potential to generate as much hydrogen as possible (Ali et al. 2024; Hashemi et al. 2022).

As an example, the fusion of fungal pretreatment with the Fenton reaction, where iron ions are used together with hydrogen peroxide to form reactive oxygen species, has been used to delignify steam-exploded birch wood. This method enhanced the methane production by up to 63% relative to the controls that were not treated, and this was further enhanced by adding enzymatic hydrolysis. Pleurotus ostreatus and Lentinula edodes, used as pretreatments or as a mixture with the Fenton reaction, produced up to 25% delignification and had a much higher stimulating effect on biogas production (Hashemi et al. 2022).

These observations suggest that with the help of an integrated pretreatment approach, the recalcitrance of lignocellulosic biomass can be overcome, and the overall efficiency of biohydrogen production can be increased. Nevertheless, these approaches need to be economically viable and scalable, which requires additional research, especially regarding large-scale industrial implementation (Ali et al. 2024).

9.5 Recent Advances in Fungal Biohydrogen Production

9.5.1 *Genetic and Metabolic Engineering*

Genetic and metabolic engineering have been used to alter the strains of fungi to promote the production of hydrogen, express more enzymes, and improve the use of substrates. Genome editing with CRISPR/Cas9 to enhance hydrogen production and overexpression of essential hydrogenase genes, as well as optimization of metabolic pathways, has been utilized to boost yields of hydrogen and process strength (Ali et al. 2024; Fan et al. 2024). Although most of the studies that have been conducted so far have concerned bacterial systems, there is increasing interest in applying these techniques to fungi. As an illustration, optimization of enzyme production in Aspergillus fumigatus by the Taguchi approach involved a drastic rise in soluble carbohydrate production from polysaccharidic wastes and a 56% increase in yield of biohythane in a two-stage procedure (Basak et al. 2019). This proves that there is a possibility of achieving optimization of genetic processes to make the fungi more efficient in the production of biohydrogen.

9.5.2 Co-cultivation and Microbial Consortia

Fungal–bacterial consortia play a critical role in biohydrogen systems by coupling biomass deconstruction with hydrogen fermentation. Fungi hydrolyze lignocellulose into soluble sugars, while hydrogenogenic bacteria, such as Clostridium spp., convert these sugars into H_2, creating a syntrophic production platform. Fungi can be co-cultivated with hydrogen-producing bacteria or methanogens, and synergistic interactions lead to an increase in the efficiency of the overall process and production of hydrogen. These consortia can also enhance the breakdown of complex substances and the stability of the process. As an illustration, applying a fungal consortium of Trichoderma viride and Phanerochaete chrysosporium to pretreat oat straw led to an increase in hydrogen yield by 165% above that of untreated straw and was more effective than pretreatments using individual species (Zhou and Li 2016). Scanning electron microscopy was used to show that lignin was extensively degraded along with cell wall breakdown, both of which allowed the fermentable sugars to be released and enhanced the production of hydrogen. On the same note, the co-culture of white-rot fungi and Clostridium species was found to boost the production of hydrogen through the liberation of fermentable sugars and the alleviation of the formation of inhibitory byproducts (Zhi and Wang 2014). These results highlight the importance of microbial community engineering for the improvement of biohydrogen production processes.

9.5.3 Bioreactor Designs and Process Optimization

New bioreactor models, such as hybrid and sequential fermentation, have been created in order to optimize the environment where fungi can grow to generate hydrogen. The critical process parameters that affect the maximization of yields are temperature, pH, substrate concentration, and hydraulic retention time. The use of immobilized cell systems, membrane bioreactors, and continuous flow reactors has been explored for use in increasing process stability and scalability (Ali et al. 2024; Fan et al. 2024; Show et al. 2012).

As an example, fungal pretreatment coupled with dark fermentation as a part of a consolidated bioprocessing strategy has been demonstrated to enhance hydrogen productivity and decrease the complexity of the process (Zhi and Wang 2014). The use of design of experiments (DOE) approaches, like response surface methodology, may be useful in finding the best process settings and optimizing the production of hydrogen (Kovács et al. 2022; Rady et al. 2024).

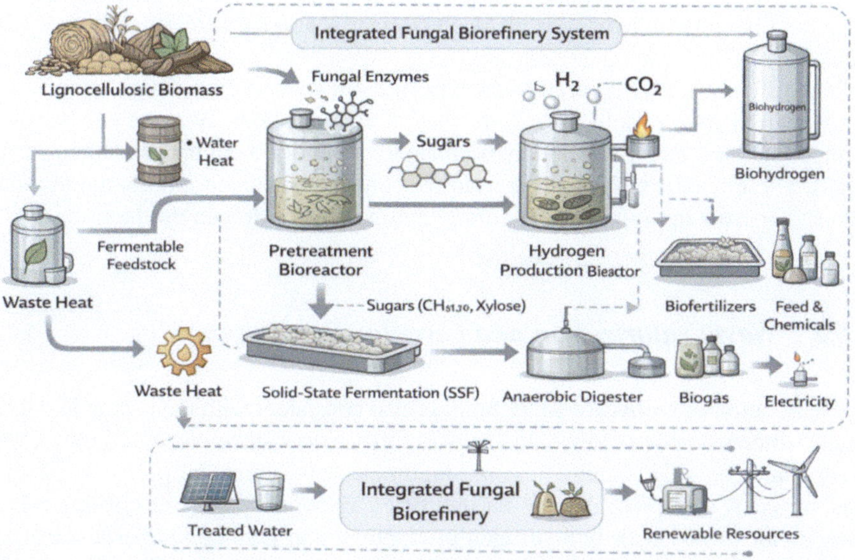

Fig. 9.2 Integrated fungal biorefinery system for sustainable biohydrogen production

9.5.4 Integration with Other Bioprocesses

Biohydrogen production should be coupled with other bioproducts, e.g., biodiesel production, so that the overall energy is recovered and the economies are viable. As an example, oleaginous yeasts can be grown using spent media of hydrogen fermentation to produce lipids, forming a closed-loop biorefinery system. Such an approach provides not only better resource use but also less waste and a decrease in the impact on the environment (Ali et al. 2024; Fan et al. 2024).

Another strategy that can enhance the sustainability and profitability of biohydrogen production is the development of integrated fungal biorefinery systems that combine hydrogen production with biogas, biofuels, and other value-added products (Fig. 9.2).

9.6 Applications and Case Studies

9.6.1 Industrial and Pilot-Scale Applications

A number of pilot-scale experiments have proven that fungal biohydrogen production using different waste sources, including food waste, agricultural residues, and brewery sludge, is feasible. These applications point to the possibility of large-scale implementation and valorization of waste. To illustrate this point, fungal pretreatment of cornstalk in consolidated bioprocessing has been demonstrated to substantially enhance hydrogen yields in comparison to the level of substrates (Zhi and Wang 2014).

In a different study, the pretreatment of rice straw with edible fungi led to a 74% increase in hydrogen yield, which proves that lignocellulosic biomass can be transformed into cost-effective and environmentally friendly biohydrogen (Sheng et al. 2018). These results highlight the need for substrate selection, pretreatment methods, and process optimization in scaling up fungal biohydrogen production.

9.6.2 Waste Valorization and Circular Bioeconomy

The generation of biohydrogen by fungi is also consistent with the principles of the circular bioeconomy, as low-value or waste materials can be converted into useful energy carriers. This strategy not only minimizes environmental pollution but also generates new sources of income for industries. Waste management systems, which must be integrated with biohydrogen production, can help create sustainable and resilient urban infrastructure (Ali et al. 2024; Fan et al. 2024; Bundhoo et al. 2015).

A case in point is food waste, which is used as a substrate for fungal SSF and further hydrogen generation to offer both energy and waste management (Manuel et al. 2022). The pretreatment of agricultural residues and other lignocellulosic wastes using fungi and fermentation to produce bioethanol enhances the establishment of closed-loop biorefinery systems and the efficiency of resources (Ali et al. 2024; Fan et al. 2024). The generation of biohydrogen through fungal-assisted biorefineries enables circular bioeconomy pathways by converting agricultural and food wastes into hydrogen, biogas, and biofertilizers within an integrated production loop (Fig. 9.2).

9.6.3 Environmental and Economic Impacts

Fungal biohydrogen production has the following environmental advantages: it lowers greenhouse gas emissions, reduces fossil fuel dependence, and enhances waste disposal. Economic studies indicate that integrated biorefinery strategies have the potential to improve the profitability and sustainability of biohydrogen production. Nevertheless, issues with process scalability, feedstock, and acceptance in the market also need to be solved to achieve the maximum potential of this technology (Ali et al. 2024; Fan et al. 2024; Bundhoo et al. 2015).

The cost and accessibility of the substrates and the effectiveness of the pretreatment and fermentation processes also define one of the key aspects in the economic feasibility of fungal biohydrogen production. The renewable energy market requires fungal biohydrogen production to be more competitive, which is possible only through the enhancement of process integration, metabolic engineering, and bioreactor design (Ali et al. 2024; Fan et al. 2024; Bundhoo et al. 2015).

9.7 Challenges in Fungal Biohydrogen Production

9.7.1 Technical Barriers

Although the production of fungal biohydrogen has a bright future, there exist a number of technical issues that inhibit the large-scale use of the alternative and its commercial attractiveness. The recalcitrance of lignocellulosic biomass is one of the major barriers. The intricate composition of plant cell walls—cellulose, hemicellulose, and lignin—prevents microorganisms from gaining easy access to fermentable sugars (Ali et al. 2024; Soares et al. 2019). Although fungal pretreatment has the capacity to enhance digestibility, the procedure is sometimes time-consuming and can fail to achieve full delignification, thereby restricting the overall yield of hydrogen.

Optimization of fermentation conditions is another technical issue. Temperature, pH, substrate concentration, and hydraulic retention time (HRT) are crucial factors that affect hydrogen production (Soares et al. 2019; Rittmann and Herwig 2012). An example of this is that mesophilic temperatures (around 37 °C), neutral pH, and an HRT of about 72 h are generally reported as having the most favorable level of dark fermentation, yet these variables can vary depending on the fungal strain and substrate involved (Soares et al. 2019). It can be hard to attain and sustain these conditions at scale, particularly with heterogeneous waste streams.

The problem of inhibitory byproducts is also a major concern. Organic acid and ethanol buildup, and other metabolites that may accumulate during fermentation, may suppress the activity of hydrogenase and decrease hydrogen yields (Manuel et al. 2022; Soares et al. 2019). Competing microorganisms, such as lactic acid bacteria, may also divert substrates from hydrogen-producing pathways, and this may be seen in the fermentations of food waste, where lactic acid bacteria outcompete hydrogen producers because of faster growth rates (Manuel et al. 2022).

9.7.2 Economic and Scale-Up Barriers

Economically, the pretreatment of substrates is still a significant impediment. Although fungal pretreatment is environmentally friendly compared to chemical treatment, it is slow and can be costly to execute since it involves cultivating fungi on a large scale, which is an added cost (Ali et al. 2024). The production of fungal biohydrogen is not yet fully scalable because it requires specific bioreactors and control systems to ensure optimal growth and fermentation conditions (Soares et al. 2019; Brar et al. 2022).

Other economic issues are process integration and downstream processing. Separation and purification of hydrogen in the fermentation gases requires efficient procedures, which might be energy-consuming and costly in real applications. Moreover, the majority of the available studies have been carried out at the laboratory level, and no techno-economic analyses or pilot-scale demonstrations have been done to confirm the possibility of large-scale implementation (Soares et al. 2019; Ali et al. 2024).

9.7.3 Environmental and Regulatory Barriers

Environmental challenges include the management of residual biomass and effluents generated during the fermentation process. While the use of waste substrates aligns with circular economy principles, the disposal or further treatment of fermentation residues must be addressed to minimize the environmental impact (Ali et al. 2024). Regulatory barriers may also arise, particularly concerning the use of genetically modified organisms (GMOs) in biohydrogen production and the integration of new technologies into the existing waste management and energy infrastructure (Brar et al. 2022).

9.8 Future Perspectives and Research Directions

9.8.1 Technological Innovations

The next round of research must be on improving the pretreatment technology to further expand on the digestibility of the lignocellulosic biomass. Fungal enzymes have demonstrated the prospect of increased stability and activity through the addition of nanomaterials, including magnetite nanoparticles, resulting in increased sugar and hydrogen production (Srivastava et al. 2021). Hybrid pretreatment strategies that incorporate biological, chemical, and physical processes will likely continue to be developed to make the processes more efficient and less expensive (Ali et al. 2024).

Another potential line of research is metabolic and genetic engineering of fungi. Through the overexpression of several important hydrogenase genes or even the incorporation of new metabolic networks, it is possible to raise the yields of hydrogen and broaden the selection of substrates that can be successfully utilized (Brar et al. 2022). Synthetic biology and systems biology methods are capable of further understanding the regulatory systems that govern hydrogen production and of developing new targets that can be enhanced in strains (Ramprakash et al. 2022).

9.8.2 Process Integration and Scale-Up

The economic viability and sustainability of the process can be increased by the development of integrated biorefinery systems that can produce biohydrogen along with other value-added products, including biogas, bioethanol, and organic acids

(Ali et al. 2024; Brar et al. 2022). Waste management systems can be integrated with these processes, and the use of continuous or semi-continuous bioreactors can contribute to better scalability and operational stability of the processes (Soares et al. 2019; Ali et al. 2024).

Process modeling and machine learning are new technologies to be used in optimizing the production of biohydrogen. Machine learning algorithms can be used to predict the performance of the process, detect the bottlenecks, and design more efficient bioprocesses by analyzing large amounts of data on operational parameters and microbial community dynamics (Pandey et al. 2022).

9.8.3 Environmental and Economic Assessments

Techno-economic and life-cycle analyses should be carried out to determine the sustainability and commercial feasibility of fungal biohydrogen generation. Such evaluations would take into account the costs and benefits of various pretreatment systems, the environmental impact of obtaining the sources and disposing of the wastes, and how they can be integrated with the existing energy and waste systems (Ali et al. 2024; Brar et al. 2022).

Regulatory frameworks and policy support will also be important in enabling the biohydrogen technologies to be adopted. Barriers to commercialization can be removed by incentives to produce renewable energy and supporting pilot-scale demonstrations, as well as by giving a clear signal regarding the use of GMOs and waste-derived substrates (Brar et al. 2022).

9.9 Compilation and Comparison Tables

A summary of representative fungal species, substrates used,pretreatment processes, and their reported effects on hydrogenproduction is presented in Table 9.1.

The major differences in substrate utilization, hydrogen yield, environmental tolerance, and industrial applications among fungi, bacteria and algae are summarized in Table 9.2.

The major challenges associated with fungal biohydrogen production and potential technological solutions are summarized in Table 9.3.

9.10 Conclusions

The use of fungi to produce biohydrogen is a promising and environmentally friendly method for producing renewable hydrogen. Recent developments in fungal pretreatment, process integration, and bioreactor design have greatly enhanced the

Table 9.1 Fungal species, substrates, and hydrogen yields

Fungal species	Substrate	Pretreatment/ Process	Qualitative label	Key findings/ Notes	Citations
Aspergillus tubingensis	Food waste	Solid-state fermentation	Moderate enhancement	Best hydrolyzer among the tested fungi	Manuel et al. (2022)
Piromyces rhizinflata	Corn silage, cattail	Anaerobic bioaugmentation	Reported stimulation	Improved digestibility, initial H_2/CH_4 rate increase	Soares et al. (2019)
Pleurotus ostreatus	Brewery sludge	Fungal pretreatment + DF	High enhancement	25% higher yield than DF alone	Ali et al. (2024)
Gymnopus contrarius	Rice straw	Edible fungal pretreatment	High enhancement	74% higher than untreated	Ramprakash et al. (2022)

Source: Compiled from Manuel et al. (2022), Soares et al. (2019), Ali et al. (2024), and Ramprakash et al. (2022)

Table 9.2 Comparison of fungi, bacteria, and algae in biohydrogen production

Attribute	Fungi	Bacteria (Clostridia)	Algae (Chlamydomonas)	Citations
Substrate utilization	Lignocellulosic, complex	Simple sugars, some complex	Simple sugars, water splitting	Fan et al. (2024) and Soares et al. (2019)
Pretreatment requirement	Lower (due to enzymes)	Higher	Moderate	Manuel et al. (2022) and Ali et al. (2024)
Hydrogen yield	Moderate	High (optimized)	Variable	Fan et al. (2024) and Soares et al. (2019)
Environmental tolerance	Wide pH, temp range	Moderate	Sensitive to light, nutrients	Fan et al. (2024) and Soares et al. (2019)
Industrial application	Emerging	Established	Experimental	Ali et al. (2024) and Brar et al. (2022)

Source: Compiled from Fan et al. (2024), Soares et al. (2019), Manuel et al. (2022), Ali et al. (2024), and Brar et al. (2022)

effectiveness and usability of these systems. Fungi, with their well-developed enzymatic parameters and the capacity to break down complex lignocellulosic materials, provide a special opportunity to valorize agricultural residues, food waste, and other organic substances.

Although these advances have been made, there are a number of challenges. The recalcitrance of lignocellulosic biomass, the development of effective and scalable pretreatment processes, and the optimization of fermentation conditions are current topics of study. The development of fungal biohydrogen production and integration

Table 9.3 Challenges and solutions in fungal biohydrogen production

Challenge	Solution/Approach	Citations
Low hydrogen yield	Metabolic engineering, consortia	Soares et al. (2019) and Ali et al. (2024)
Substrate competition	Process optimization, co-cultures	Manuel et al. (2022) and Ali et al. (2024)
Scale-up and stability	Bioreactor design, process control	Soares et al. (2019) and Brar et al. (2022)
Economic feasibility	Integrated processes, waste valorization	Ali et al. (2024) and Brar et al. (2022)
Environmental impact	Use of waste streams, green pretreatment	Manuel et al. (2022) and Ali et al. (2024)

Source: Compiled from Soares et al. (2019), Manuel et al. (2022), Ali et al. (2024), and Brar et al. (2022)

into the renewable energy environment will require some serious economic and environmental evaluations and policy facilitation to succeed.

Further studies ought to be directed at the development of improved pretreatment methods, metabolic and genetic engineering of fungal strains, and the establishment of biohydrogen production with other bioprocesses in a circular bioeconomy system. Process optimization and scalability can also be improved by the use of machine learning and process modeling. Fungal systems will primarily function as biological pretreatment and saccharification platforms within future biohydrogen biorefineries, enabling efficient bacterial hydrogen fermentation from low-value lignocellulosic and waste substrates.

References

Ahmed SF, Mofijur M, Nahrin M, Chowdhury SN, Nuzhat S, Alherek M, Rafa N, Ong HC, Nghiem L, Mahlia T (2021) Biohydrogen production from wastewater-based microalgae: progress and challenges. Int J Hydrog Energy 47(88):37321–37342. https://doi.org/10.1016/j.ijhydene.2021.09.178

Ali SS, Al-Tohamy R, Elsamahy T, Sun J (2024) Harnessing recalcitrant lignocellulosic biomass for enhanced biohydrogen production: recent advances, challenges, and future perspective. Biotechnol Adv 72:108344. https://doi.org/10.1016/j.biotechadv.2024.108344

Basak B, Saha S, Chatterjee P, Ganguly A, Chang W, Jeon B (2019) Pretreatment of polysaccharidic wastes with cellulolytic Aspergillus fumigatus for enhanced production of biohythane in a dual-stage process. Bioresour Technol 299:122592. https://doi.org/10.1016/j.biortech.2019.122592

Brar KK, Cortez AA, Pellegrini VO, Amulya K, Polikarpov I, Magdouli S, Kumar M, Yang Y, Bhatia SK, Brar SK (2022) An overview on progress, advances, and future outlook for biohydrogen production technology. Int J Hydrog Energy 47(88):37264–37281. https://doi.org/10.1016/j.ijhydene.2022.01.156

Bundhoo M, Mohee R, Hassan A (2015) Effects of pre-treatment technologies on dark fermentative biohydrogen production: a review. J Environ Manag 157:20–48. https://doi.org/10.1016/j.jenvman.2015.04.006

Fan Y, Zhang F, He K, Yu D, Chen H, Tian D, Shi Y, Li Z, Wang X (2024) Functional microorganisms in hydrogen production: mechanisms and applications. Bioresour Technol 419:132007. https://doi.org/10.1016/j.biortech.2024.132007

Hashemi S, Solli L, Aasen R, Lamb J, Horn S, Lien K (2022) Stimulating biogas production from steam-exploded birch wood using Fenton reaction and fungal pretreatment. Bioresour Technol 366:128190. https://doi.org/10.2139/ssrn.4204599

Kovács E, Szűcs C, Farkas A, Szuhaj M, Maróti G, Bagi Z, Rákhely G, Kovács KL (2022) Pretreatment of lignocellulosic biogas substrates by filamentous fungi. J Biotechnol 360:160–170. https://doi.org/10.1016/j.jbiotec.2022.10.013

Manuel C, Carlos Q, Carmen P, Marisela VL, Iván M (2022) Fungal solid-state fermentation of food waste for biohydrogen production by dark fermentation. Int J Hydrog Energy 47(70):30062–30073. https://doi.org/10.1016/j.ijhydene.2022.06.313

Pachaiappan R, Cornejo-Ponce L, Sagade AA, Mani M, Aroulmoji V, Rajan VF, Manavalan K (2024) A concise review of recent biohydrogen production technologies. Sustain Energy Technol Assess 62:103606. https://doi.org/10.1016/j.seta.2024.103606

Panahi HKS, Dehhaghi M, Guillemin GJ, Gupta VK, Lam SS, Aghbashlo M, Tabatabaei M (2022) A comprehensive review on anaerobic fungi applications in biofuels production. Sci Total Environ 829:154521. https://doi.org/10.1016/j.scitotenv.2022.154521

Pandey A, Park J, Ko J, Joo H, Raj T, Singh L, Singh N, Kim S (2022) Machine learning in fermentative biohydrogen production: advantages, challenges, and applications. Bioresour Technol 370:128502. https://doi.org/10.1016/j.biortech.2022.128502

Rady H, Ali S, El-Sheekh M (2024) Strategies to enhance biohydrogen production from microalgae: a comprehensive review. J Environ Manag 356:120611. https://doi.org/10.1016/j.jenvman.2024.120611

Ramprakash B, Lindblad P, Eaton-Rye JJ, Incharoensakdi A (2022) Current strategies and future perspectives in biological hydrogen production: a review. Renew Sust Energ Rev 168:112773. https://doi.org/10.1016/j.rser.2022.112773

Rittmann S, Herwig C (2012) A comprehensive and quantitative review of dark fermentative biohydrogen production. Microb Cell Factories 11:115. https://doi.org/10.1186/1475-2859-11-115

Saha R, Bhattacharya D, Mukhopadhyay M (2021) Enhanced production of biohydrogen from lignocellulosic feedstocks using microorganisms: a comprehensive review. Energy Convers Manage X 13:100153. https://doi.org/10.1016/j.ecmx.2021.100153

Sheng T, Zhao L, Gao L, Liu W, Wu G, Wu J, Wang A, Sheng T, Zhao L, Gao L, Liu W, Wu G, Wu J, Wang A (2018) Enhanced biohydrogen production from nutrient-free anaerobic fermentation medium with edible fungal pretreated rice straw. RSC Adv 8(41):22924–22930. https://doi.org/10.1039/c8ra03361g

Show K, Lee D, Tay J, Lin C, Chang J, Show K, Lee D, Tay J, Lin C, Chang J (2012) Biohydrogen production: current perspectives and the way forward. Int J Hydrog Energy 37(20):15616–15631. https://doi.org/10.1016/j.ijhydene.2012.04.109

Soares JF, Confortin TC, Todero I, Mayer FD, Mazutti MA, Soares JF, Confortin TC, Todero I, Mayer FD, Mazutti MA (2019) Dark fermentative biohydrogen production from lignocellulosic biomass: technological challenges and future prospects. Renew Sust Energ Rev 117:109484. https://doi.org/10.1016/j.rser.2019.109484

Srivastava N, Alhazmi A, Mohammad A, Haque S, Srivastava M, Pal DB, Singh R, Mishra P, Vo DVN, Yoon T, Gupta VK (2021) Biohydrogen production via integrated sequential fermentation using magnetite nanoparticles treated crude enzyme to hydrolyze sugarcane bagasse. Int J Hydrog Energy 47(72):30861–30871. https://doi.org/10.1016/j.ijhydene.2021.08.198

Zhao L, Cao G, Wang A, Ren H, Zhang K, Ren N (2014) Consolidated bioprocessing performance of Thermoanaerobacterium thermosaccharolyticum M18 on fungal pretreated cornstalk for enhanced hydrogen production. Biotechnol Biofuels 7(1):178. https://doi.org/10.1186/s13068-014-0178-7

Zhi Z, Wang H (2014) White-rot fungal pretreatment of wheat straw with Phanerochaete chrysosporium for biohydrogen production: simultaneous saccharification and fermentation. Bioprocess Biosyst Eng 37(7):1447–1458. https://doi.org/10.1007/s00449-013-1117-x

Zhou L, Li L (2016) Novel fungal consortium pretreatment of waste oat straw to enhance economic and efficient biohydrogen production. Ecocycles 2(2). https://doi.org/10.19040/ecocycles.v2i2.61

Chapter 10
Pigment Production Using Fungal Biotechnology: Recent Advances, Applications, and Challenges

Vivek Raj, Santosh Kumar Karn, Wajiha Saadi, Ajay Kumar Chauhan, Arun Lal Srivastav, and Manikant Tripathi

Abstract The increasing demand for renewable and toxic-free substitutes to artificial colorants have placed fungal pigments at the forefront of fungal biotechnology research. Natural colorants can be used as dyes for textiles, leather, and paper, as well as in food additives and cosmetics. Modern industry uses colorants or pigments sourced from flora as well as microorganisms. Non-conventional sources, namely filamentous fungal species (notably basidiomycetes and ascomycetes) as well as lichens, generate a variety of hues, encompassing chemical categories of pigments like carotenoids, azaphilones, polyketides, and melanins. Fungal biotechnology allows scalable pigment production that is economical and eco-friendly. This chapter highlights the potential of fungal pigments as a sustainable option over current sources. Including greener extraction methods for intracellular and extracellular pigments and explains how environmental concerns can drive interest in fungi. It also emphasizes multidisciplinary and biotechnological methods for overcoming the challenges of the pigment industry.

Keywords Fungi · Fungal biotechnology · Natural pigment production

V. Raj (✉) · W. Saadi
Department of Biotechnology, Patna Women's College, (Autonomous), Patna University, Patna, Bihar, India

S. K. Karn
Department of Biochemistry & Biotechnology, Sardar Bhagwan Singh University, Balawala, Dehradun, Uttarakhand, India

A. K. Chauhan
Department of Biotechnology, Sharda University, Knowledge Park II, Greater Noida, UP, India

A. L. Srivastav
Chitkara University School of Engineering and Technology, Chitkara University, Himachal Pradesh, India

M. Tripathi
Biotechnology Program, Dr. Rammanohar Lohia Avadh University, Ayodhya, Uttar Pradesh, India

M. Tripathi et al. (eds.), *Recent Advancements in Fungal Biotechnology Volume 1*, Fungal Biology, https://doi.org/10.1007/978-981-92-0339-0_10

10.1 Introduction

The contemporary food manufacturing sector mostly emphasizes on natural food colorants (Faustino et al. 2019). Nature has always been understood as a storehouse of biological objects, with the ability to create pigment. The natural organic pigments include betalains, anthocyanins, curcumin, chlorophylls, and carotenoids (Corrêa et al. 2019). A few challenges faced by the natural pigment industry are the availability of raw materials, stability, and sensitivity of the pigments (Galaffu et al. 2015). For deeper insights into the opportunity of pigment extraction from natural sources, existing knowledge on well-studied plant and animal sources are being extended to microorganisms, the potential source for bio-pigment generation (Sen et al. 2019). Microbial pigments provide numerous benefits compared to pigments derived from animal and plant . Among numerous microbial options, fungi and microalgae offer an array of hydrophilic bio-pigments, with ecological functions (Kalra et al. 2020); however, low yields of the algal cultures remain a drawback in fully exploiting their potential. In attempts to employ fungi in bio-pigment production, Basidiomycetes, which ancient cultures utilized for dyeing wool or silk, have been examined. Nonetheless, large manufacturing by such fungi are unattainable. For this reason, the industrial sector places greater emphasis on cultivating filamentous fungi, as these organisms are capable of being efficiently propagated under laboratory conditions to enable large-scale mass production (Kalra et al. 2020). Currently, filamentous fungi-derived pigments, including canthaxanthin and ankaflavin of *Monascus* sp. and Natural Red™ (earlier known as Arpink Red™) produced by *Penicillium oxalicum* var. *armeniaca*, are accessible (Afroz Toma et al. 2023). In this study, the role of filamentous fungi as pigment-producing organisms are examined, along with the key business challenges that restrict their industrial use.

10.2 Fungi as a Source of Pigments

Chemical stability, high light stability, and good output of fungi's chromogenic chemicals have received attention (Kalra et al. 2020; Chen et al. 2017). Fungal pigmentation comprises of metabolites from various classes, including oxopolyenes, naphthoquinones, quinones, carotenoids, melanins, hydroxyanthraquinones, azaphilones, and anthraquinones (Fig. 10.1) (Afroz Toma et al. 2023). Natural Red™ is the first commercial red hue derived from *Penicillium oxalicum* var. *armeniaca* CCM 8242, extracted from a soil-derived fungal strain (Afroz Toma et al. 2023). Most such colorants are biosynthetically common in many ascomycetes and are derivatives of polyketides. Alongside polyketide derivatives, filamentous fungi also contain other pigmented metabolites, including carotenoids, polyphenols, and terpenoids. Fungal pigments confer a variety of pharmacological functions to fungi (Kalra et al. 2020). Nonetheless, how they work physiologically and the conditions

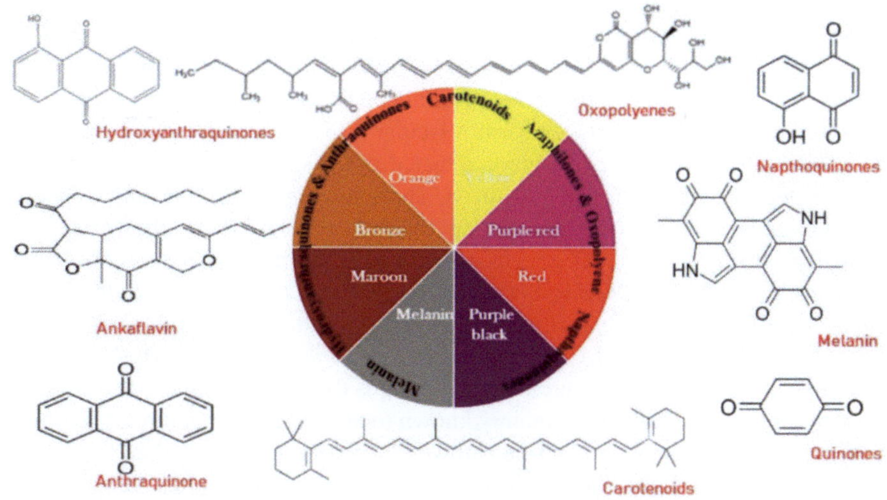

Fig. 10.1 Hues of colors produced by different classes of compounds

governing their generation remain unidentified (Sen et al. 2019). The pigment production in fungi can be increased by heterologous expression of color imparting gene, which could facilitate the bulk manufacturing. Moreover, the co-culturing or modifying the culture conditions can also enhance gene expression and increase the Titre, Rate, Yield (TRY) of certain pigments.

10.3 Pigmented Fungi and Biosynthesis of Pigments

Most of the information regarding fungal pigments were reported from four genera: *Penicillium, Aspergillus, Monascus,* and *Paecilomyces* (Lin and Xu 2020). Which can be classified as polyketides, azaphilones (derivatives of polyketides), melanins, and carotenoids (Lin and Xu 2020).

10.3.1 Carotenoids

"Carotene" comes from carota in Latin. Wackenroder coined the term after extracting and purifying the pigment of carrots (*Daucus carota*). Carotenoid names came from their root word "carotene," with Greek prefixes (like α, β, ε) added to indicate their two end groups. Fungal carotenoids are known to have distinctive colors as orange, red, and yellow hues. The molecular framework of carotenoids consist of eight linked isoprene residues arranged in an extended polyene structure, a configuration that significantly influences their physical and chemical behavior. This chain also contains conjugated double bonds that absorb light (Lin and Xu 2020). In humans and other mammals, certain carotenoids with β-ring structures, like β-carotene, zeaxanthin, and β-cryptoxanthin, plays a critical role as metabolic precursors that enable the body to produce vitamin A. Diets rich in these pigments have been observed to lower the risk of cardiovascular disease, some forms of cancer due to antioxidant nature, and even age-associated eye diseases such as cataracts and macular degeneration (Lin and Xu 2020). There are now over 200 fungal species known to generate carotenes (Lin and Xu 2020). Carotenoid biosynthesis has been reported in a range of fungal genera, notably species belonging to *Rhodotorula, Phycomyces, Blakeslea, Rhodosporium, Mucor, Fusarium, Xanthophyllomyces,* and *Neurospora.* Representative examples include *Rhodotorula mucilaginosa, Blakeslea trispora,* and *Neurospora crassa,* among others (Lin and Xu 2022). *Blakeslea trispora* is known to be a significant industrial fungus that produces β-carotene, a food pigment. The filamentous fungi mainly known to utilize the cyclization of lycopene to form β-carotene. In the production pathway the first terpenoid synthesis occurs as intermediate, isopentenyl pyrophosphate (IPP, C5), is converted to lycopene (reddish-colored intermediates of β-carotene biosynthesis in *B. trispora*) (Mussagy et al. 2019). The molecule acetyl-CoA can serve as a starting base in the mevalonate (MVA) pathway to form the fungal isopentenyl pyrophosphate (IPP) (Lin and Xu 2020; Mussagy et al. 2019). At the beginning of β-carotene biosynthesis, the compound is assembled through a stepwise process in which isopentenyl pyrophosphate (IPP, C5) molecules are successively linked. This sequential condensation produces a chain of intermediate products, farnesyl pyrophosphate (FPP, C15), geranyl pyrophosphate (GPP, C10), and eventually geranylgeranyl pyrophosphate (GGP, C20), that serve as key precursors in the pathway (Mussagy et al. 2019). A colorless molecule known as phytoene is the initial result of the reaction of two GGPP units via condensation, which has an aliphatic carotenoid-like structure. This catalyzed reaction is a phytoene synthase condensation reaction (Fig. 10.1). Phytoene is then changed into lycopene by the action of dehydrogenase. Further cyclization stages mediated by lycopene cyclase, and lycopene synthesizes β-carotene and γ-carotene (Fig. 10.1) (Hernández-Almanza et al. 2016).

10.3.2 Melanin

Fungal melanin is derived from polyketides and free radicals through a polymerization process, with the final color of compounds either black, brown, or dark green in hue. They are sensitive to acidic decomposition but not to oxidation. These are hydrophobic and negatively charged (Pombeiro-Sponchiado et al. 2017). They are found in the extracellular polymers that surround the cells as well as in the fungal cell walls. In addition to preserving metal ion balance and cell wall integrity, melanin synthesis and secretion are linked to fungal adaptation strategies (Ramachandra et al. 2023).

However, melanin may confer fungal virulence in certain mammalian pathogenic fungi, including *Aspergillus fumigatus, Candida glabrata, Candida albicans, Candida parapsilosis,* and *Cryptococcus neoformans*, suggesting a defense mechanism against the host. Through antigen masking, melanin helps harmful fungal species survive phagocytosis (Boral et al. 2018). Work carried out with mouse models infected by the human pathogenic fungus *Wangiella dermatitidis* has indicated that when melanin synthesis is reduced, the invasive hyphal form can be more easily removed. This reduction also makes the organism more prone to destruction by neutrophils, which in turn lowers its ability to cause disease. Fungal development is not dependent on melanin, i.e., main metabolism, since the same fungus may grow in lab conditions in both albino and melanized strains. Instead, they serve as a fungal defense mechanism, i.e., secondary metabolism. In fungi, melanin synthesis can occur in three distinct ways, giving rise to different structural forms. These include pigment embedded within the cell wall, melanin produced via the dihydroxynaphthalene (DHN) pathway, and a water-soluble extracellular pigment commonly referred to as pyomelanin—dihydroxyphenylalanine (DOPA) in melanin. *A. fumigatus* is known to synthesize the former two, whereas *C. neoformans* is known to synthesize the latter one (Perez-Cuesta et al. 2020). The breakdown of L-tyrosine/L--phenylalanine produces the dark brown pigment known as pyomelanin. L-tyrosine/L-phenylalanine degrades into an intermediate, homogentisate, which polymerizes into pyomelanin. This reaction is coupled with conidial germination (Perez-Cuesta et al. 2020). Fungi, including *A. fumigatus* and common mold, can synthesize melanin by the DHN pathway. A group of genes encodes this process on the second chromosome. Six genes make up the gene cluster: pksP/alb1, arp1, arp2, ayg1, abr1, and abr2 (Verde-Yáñez et al. 2023). The enzyme is transcribed by the first gene, pksP/alb1, polyketide synthase (PKS). In this biosynthetic pathway, the enzyme responsible for YWA1 formation produces a heptaketide naphthopyrone by condensing acetyl-CoA with successive units of malonyl-CoA through a "β-ketoacyl condensation sequence." Following this reaction, the enzyme encoded by the ayg1 gene facilitates a hydrolysis step that generates 1,3,6,8-tetrahydroxynaphthalene (1,3,6,8-THN). Thereafter, the pathway proceeds through several reduction steps, dehydration and aromatization, and oxidative polymerization. In the biosynthetic pathway, the intermediate compound 1,3,6,8-tetrahydroxynaphthalene (1,3,6,8-THN) undergoes enzymatic degradation mediated by Arp2p, leading to the formation of vermelone. This molecule is subsequently oxidized by the copper-containing oxidase Abr1p, producing 1,8-DHN (1,8-dihydroxynaphthalene), which serves as a direct precursor to melanin. Finally, laccase enzyme Abr2p catalyzes the polymerization of this product, completing the synthesis of DHN-melanin (Law et al. 2025).

A sequence of reduction-oxidation processes follows the beginning of DOPA-melanin by *C. neoformans*. Tyrosine, also known as L-3,4-dihydroxyphenylalanine (L-DOPA), is the first. An enzyme like laccase or tyrosinase oxidizes it to create DOPA-quinone. The reactions that follow are fast to produce the intermediate dihydroxyindole, which is then converted to melanin by polymerization (Perez-Cuesta et al. 2020). Recent discoveries indicate that, as is the case with the model fungus *C. neoformans*, the mushroom *Auricularia auricula* has melanin attached to the cell wall, which is DOPA-melanin. (Li et al. 2021). The physicochemical characteristics and functional groupings of fungal melanin, no matter how formed, are similar (Wu et al. 2018). *A. auricula* in the proper conditions is capable of synthesizing up to 11% of the total fungal dry mass of melanin (Liu et al. 2019). The study of *A. auricula* melanin is promising, as the use of mushrooms is beneficial compared to plant- and animal-based resources.

10.3.3 Polyketides

Two kinds of polyketide-based fungal pigments are anthraquinones and naphthoquinones (Lin and Xu 2020). In fungi, the production of polyketides takes place through a biosynthetic pathway that relies on the enzyme polyketide synthase (PKS). This enzyme utilizes malonyl-CoA as the extender molecule and acetyl-CoA as the starter unit to construct the complex polyketide structure (Venil et al. 2020). Anthraquinones are chemical compounds made up of three benzene rings that are connected to form a polycyclic aromatic structure. When functional groups such as $-CH_2OH$, $-OCH_3$, $-CH_3$, and $-OH$ are introduced, they give rise to hydroxyanthraquinones. Additionally, adjustments to the carbonyl groups and the double bonds within the aromatic rings lead to reduced derivatives. This structural modification is critical in influencing the biological activity and chemical reactivity of these compounds. They possess a wide range of colors as a result of their relatively short conjugated chromophores. Their substituted forms produce red and blue hues, whereas their unsubstituted forms provide a pale-yellow hue. Fungi are typically the source of anthraquinones, particularly HAQNs (hydroxyanthraquinones), like *Microsporum, Mycosphaerella rubella, Curvularia lunata, Emericella purpurea, Penicillium* spp., *Drechslera* spp., *Fusarium* spp., *Eurotium* spp., and *Aspergillus* spp. (Lin and Xu 2020; Venil et al. 2020). An anthraquinone-based natural colorant, Arpink Red™, is worth mentioning (Lin and Xu 2020). *Penicillium oxalicum* var. *armeniaca* is the fungus that produces it.

Various fungi also produce naphthoquinones. Their color varies between yellow, orange, and brown. Numerous secondary metabolites, including naphthoquinone, are induced by both abiotic and biotic stress (Lin and Xu 2022). *Fusarium fujikuroi* is a typical species of the genus *Fusarium*, which contains bikaverin, a vivid red pigment of the naphthoquinone class. Research on *F. fujikuroi* has revealed the existence of a gene cluster made up of six bik genes. For the production of bikaverin, this gene cluster produces a multipurpose polyketide synthase (Lin and Xu 2022). The bikaverin biosynthetic cluster, positioned on the fifth chromosome, comprises several functionally distinct genes. Among them, *Bik1, Bik2,* and *Bik3* are directly involved in the biosynthesis of bikaverin, whereas *Bik4* and *Bik5* act as regulatory elements controlling the pathway. In addition, *Bik6* produces a transporter protein responsible for facilitating bikaverin export. Owing to its notable antimicrobial and anti-proliferative characteristics, bikaverin is regarded as a promising compound for potential use in pharmaceutical formulations and various industrial processes (Dos Santos and Bicas 2021).

10.3.4 Azaphilones

Azaphilones are a morphologically varied class of fungal pigments that are polyketide derivatives, characterized by pyrone-quinone-based structures with an asymmetric center and a highly oxygenated bicyclic core (Antipova et al. 2023). In nature, azaphilone-synthesizing fungi are common. These include microscopic mold (fungi), like those found in the genera Talaromyces, Chaetomium, Monascus, and Penicillium, as well as fungi that do not look like mold, like those found in the genera Bulgaria (Bulgariaceae) and Creosphaeria (members of the Xylariaceae family), *Daldinia,* and *Hypoxylon. Monascus* is the best-known genus among these. Azaphilone pigments make fruiting bodies and/or mycelia appear green, red, or yellow. The genus *Monascus* contains three species: *M. purpureus, M. pilosus*, and *M. ruber* (Lin and Xu 2022). The six main azaphilones that make up *Monascus* colorants are monascin and ankaflavin (yellow), monascorubramine and rubropunctamine (violet), and rubropunctatin and monascorubrin (orange). Nevertheless, a small number of *Monascus* spp. produces the mycotoxin citrinin (Egea et al. 2023). For these reasons, there were disagreements about their food safety. Other species that produce azaphilones are *Pleosporales* sp., a marine-derived fungus that produces *Cladosporium perangustum FS62* and *pleosporalone A*, a marine sediment-derived fungus that produces *perangustols A* and *B* (Fan et al. 2016), and an Austdiol-producing endophytic strain of *Dothideomycete* sp. *Coniella fragariae*, a fungus that produces coniellins H and I from goose excrement; *Xylariales* sp. PSUES163, a fungus obtained from seagrass that produces *xylariphilone* (Fan et al. 2016), *Fusarium* sp., an endophytic fungus that produces fusarone; *Aspergillus fumigatus* 14–27, an *Aspergillus* produced from gorgons that produces pinophilin B (Zhang et al. 2021), *Phomopsis* sp., an endophytic bacterium that produces phomopsones A-C, and *Aspergillus niger* ATCC 1015, a type strain of *Aspergillus* that produces Azanigerones BC (Yang et al. 2020), and Pseudohalonectrin AB is produced by the aquatic fungus *Pseudohalonectria adversaria.* Fatty acid synthase (FAS) and polyketide synthase (PKS) are essential components of the polyketide pathway that produces Monascus azaphilone pigmentation (MonAzps) (Egea et al. 2023). Additionally, this route is acknowledged as a model system for researching fungal azaphilone production. MpPKS5, a non-ribosomal PKS enzyme with six domains, is where the pathway starts: acyl carrier protein transacylase (SAT), ketoacyl synthase (KS), acyltransferase (AT), acyl carrier protein (ACP), product template (PT), and reductive release (R) domain. The product of the mppD gene, i.e., serine hydrolase, facilitates the multifunctional MpPKS5 to form hexaketide benzaldehyde, which is a highly reactive intermediate. The next step is the production of FK17-P2a, the homologous alcohol and the first stable product in the biosynthetic pathway. This reaction is catalyzed by ketoreductase (product of the mppA gene).

FAD-dependent monooxygenase, a byproduct of the mppF gene, oxidizes FK17-P2a to create azanigerone E. The latter creates a pyranoquinone skeleton, which serves as the polyketide chromophore for the production of azaphilones. Simultaneously, the MpFas2 enzyme, which functions as the main fatty acid synthase in the pathway, is indispensable for pigment formation. It is responsible for producing short-chain 3-oxo-fatty acyl thioesters that act as building blocks for the side chains of azaphilone molecules. Within the biosynthetic cluster of Monascus purpureus, a specific O-acyl transferase encoded by the mppB gene is believed to facilitate the attachment of these fatty acyl groups, originating from MpFas2, to the C-4 hydroxyl position of the intermediate compound. This reaction gives rise to the orange pigments rubropunctatin and monascorubrin, along with the yellow counterparts monascin and ankaflavin as subsequent products. However, the structure of rubropunctatin and monascorubrin contains chemically unstable pyran rings. Because of this, they tend to undergo a spontaneous O-to-N substitution with naturally occurring amines present in the system. This transformation results in the generation of the red pigments rubropunctamine and monascorubramine, both of which exhibit distinctive vinylogous pyridine frameworks (Chen et al. 2019).

According to comparative transcriptome studies on *M. purpureus* YY-1, pigment production may be enhanced by carbon deficiency stress (Yang et al. 2020). Two carbon sources, i.e., glycerol and soluble starch, were tested in M. purpureus FAFU618 to contribute to the synthesis of MonAzPs (Huang et al. 2018). It was found that the intracellular and extracellular pigment formation was different in the glycerol group (GCG) and the soluble starch group (SSG). In the intracellular pigment profile analyzed using UPLC-QTOF-MS/MS, the compositional variations in the two groups were different. The ankaflavin and monascin levels in the GCG samples were increased significantly, but the SSG cultures were characterized by a higher concentration of rubropunctatin and monascorubrin. The expression of genes that regulate MonAzP biosynthesis was observed to show opposite transcriptional reactions through complementary real-time quantitative PCR analysis. In particular, there was a decreased expression of mppR2, mppR1, mppC, and mppA and an increased expression of mppE, mppB, mppD, MpFasA2, MpPKS5, and MpFasB2 transcripts, suggesting that changes in regulation in these pathways were among the reasons for pigment differences. Meanwhile, 27 independent levels of mycelial protein expression in GCG and SSG were found during this research; 18 of them were large and were linked to the process of translation, proteolysis, the glycolytic pathway, and energy production. All the other nine proteins, such as ribosomal proteins and the heat shock proteins (HSPs), turned out to be down-regulated in GCG (Huang et al. 2018).

10.4 Biotechnological Advances in Pigment Production

10.4.1 Metabolomics

The development of biotechnology has enabled screening procedures that allow the selection of strains and the exploitation of previously disregarded fungal strains. The screening of potential fungal strains through metabolomics gives a chance to classify the strains based on the unique metabolites. The identification of metabolites with the help of modern data analysis and chemoinformatics can assist in a systematic study of those molecules in the target species. They assist in the dereplication of the known molecules and in the targeting of new pigments that have a similar chromophore to a known pigment (Lin and Xu 2022). It was revealed that the technique based on the use of the X-hitting algorithm is useful when applied to the UV-vis spectra of metabolites (cross hits). It is a tool employed in the rapid screening of filamentous ascomycetous fungi of the genus *Penicillium* subgenus *Biverticillium* that do not turn out to synthesize citrinin or any other toxin (Ashok et al. 2022).

10.4.2 Metabolic Engineering

The molecular biology tools will be in a position to sequence the fungal genomes and assist in the identification of genes that are involved in the pigment production (Sankari et al. 2018). Among the most helpful techniques in finding new pigments is genome mining, which makes it possible to explore not only groups of genes that are not active in standard laboratory conditions but also the full potential of fungal strains (Nielsen and Nielsen 2017). These gene clusters can then be manipulated to produce more of a specific pigment, or one of these clusters can be expressed in a heterologous host to make an individual pigment in enormous amounts (Xu et al. 2022). Gene cloning and the expression of the necessary pigment synthesis are believed to be reliable and cost-efficient techniques with the use of microbial vectors, such as *Pichia pastoris, Corynebacterium glutamicum, Pseudomonas putida, E. coli*, and *Bacillus subtilis*. In *Rhodotorula mucilaginosa* KC8, metabolic engineering and mutagenesis have enhanced the carotenoid production, that is, of torularhodin and β-carotene (Wang et al. 2017). Another microbial host that can be heterologous is *Saccharomyces cerevisiae*. *S. cerevisiae* can also be engineered to synthesize a more recent pigment, betalain, by means of glucose as the starting material and supplementation of the culture medium with different amines (Grewal et al. 2018). Knocking out the mrPigF gene resulted in the production of ankaflavin and monascin pigments in *Monascus ruber* M7. This guided the active and selective synthesis of the required elements of MonAzPs (Chen et al. 2017). Similarly, the TDNA random mutagenesis was applied to the biosynthetic gene cluster of azaphilone pigment in *M. purpureus* (Pimenta et al. 2021). It was observed that the transcription of the transcriptional regulator gene (mppR1) and polyketide synthase gene (MpPKS5) was significantly repressed in the W13 albino mutant. Another albino mutant was also created by mutation of the MpPKS5 gene, showing that the mppR1 and MpPKS5 genes are necessary for the production of azaphilone. The MpPKS5 gene cluster has a number of domains, that is, organizations of MT, R, SAT, KS, PT, AT, and ACP. Other azaphilone polyketide synthases conserve these domains. The six azaphilone molecules that *Aspergillus niger* ATCC 1015 produced (azanigerones A–F) were also linked to the aza gene cluster. Through transcriptional analysis, deletion of polyketide synthase (PKS) genes and evaluation of complete fungal genome expression, in *F. verticillioides*, three new clusters of gene co-expression were identified (Pedersen et al. 2020). PKS genes associated with such clusters were functionally examined; perithecia (sexual fruiting bodies of fungi) appeared violet in color, and fusaric acid and mycotoxins were detected (Pedersen et al. 2020). A separate study focused on identifying the gene involved in producing 1,8-dihydroxynaphthalene (DHN) melanin in *Ascochyta rabiei*. To achieve this, researchers employed a degenerate PCR primer to isolate the arPKS1 gene, which encodes a polyketide synthase-related polypeptide. Similar enzymes in other ascomycetes are known to facilitate DHN-melanin formation. To verify that ArPKS1 directly contributes to melanin synthesis in *A. rabiei*, site-directed mutagenesis was used, resulting in pycnidial mutants that were deficient in melanin production. These findings highlight the essential role of this gene in fungal pigmentation (Kalra et al. 2020). The genetic modification of two strains of *Streptomyces galilaeus* (ATCC 31133 and ATCC 31671) that were originally known to produce anthracyclines, including aclacinomycin A and 2-hydroxyaklavinone, involved the use of DNA of *Streptomyces coelicolor*. These strains have the four genetic loci (actIII, actIV, actVII, and actI), which then developed the capability of producing anthraquinone compounds, such as aloesaponarin II (Bartel et al. 1990). The discovery of clusters of genes and the explanation of their promoters in regulation assists in the study of the molecular mechanism of pigment formation.

10.4.3 Culture Optimization

The One Strain Many Compounds (OSMAC) model has underlined that secondary metabolism of fungi is different based on the growing conditions (Romano et al. 2018). Therefore, culture enhancement is essential for improving the yield. This involves altering the media composition and controlling the operational factors. To determine how medium composition affects pigment yield, glutamic acid's impacts on monacolin K pigment formation and biosynthetic gene cluster expression were examined. It was noted that after the addition of glutamic acid to the medium, synthesis of monacolin K increased to 215.4 mg L^{-1}, i.e., a 3.5-fold increase compared to control. This was due to glutamic acid-induced mokB-mokI expression on the eighth day (Zhang et al. 2021).

10.5 Processing and Extraction of Natural Pigments

The nature and whereabouts of the desired metabolites are of great significance during the extraction processes. The downstream process of extracellular pigments is less complicated than that of the intracellular ones because the latter can only be recovered by employing special methods (Morales-Oyervides et al. 2017).

10.5.1 Extraction of Extracellular Pigments

Downstream processing of water-soluble and extracellular pigments of the submerged culture is simple. Extraction using water-soluble pigments are cost effective and considered safe compared to extraction in organic solvent. Nonetheless, insoluble compounds in water and intracellular compounds require solvents that are organic (Geow et al. 2021). The two main factors that regulate the extraction of extracellular pigments from submerged cultures are fermentation processing and fermentation optimization.

10.5.1.1 Fermentation Optimization

Fermentation optimization can be done by regulating the growth parameters of submerged culture by One Factor at Time (OFAT) and Response Surface Methodology (RSM). In this case, process parameters such as light exposure, the ratio of carbon to nitrogen, aeration, pH, and concentration in media, water-soluble and extracellular pigment production can be fine tune (Fig. 10.2). For instance, the concentration of extracellular pigments secreted by *Monascus* sp. is altered by varying the pH of the culture medium (Lin and Xu 2020). Likewise, a researcher highlighted that medium composition is essential for choosing extracellular pigments. They investigated how carbon in both complex and basic forms affected the biomass of mycelium and concluded that simple carbon sources like DMD (defined minimal dextrose broth) favored biomass production yet lowered synthesis of pigment (Lebeau et al. 2020). The biosynthesis of extracellular pigments is preferred over biomass formation in "complex carbon and nitrogen sources" like potato dextrose broth (PDB), which contains necessary cofactors like copper, manganese, zinc, phosphorus, iron, calcium, and magnesium.

Two submerged fermentation techniques are used commercially to find metabolites of interest and increase their production. The first approach is the fed-batch method. It enhances the circumstances necessary for better yellow pigment creation from several *Monascus sp.* (Lin and Xu 2020). The second strategy is the continuous mode, which regularly harvests the microbial cells at a predetermined rate and time while delaying the exponential phase by giving the cells new media (Rouf et al. 2017). Together, with suitable optimization techniques, they can generate

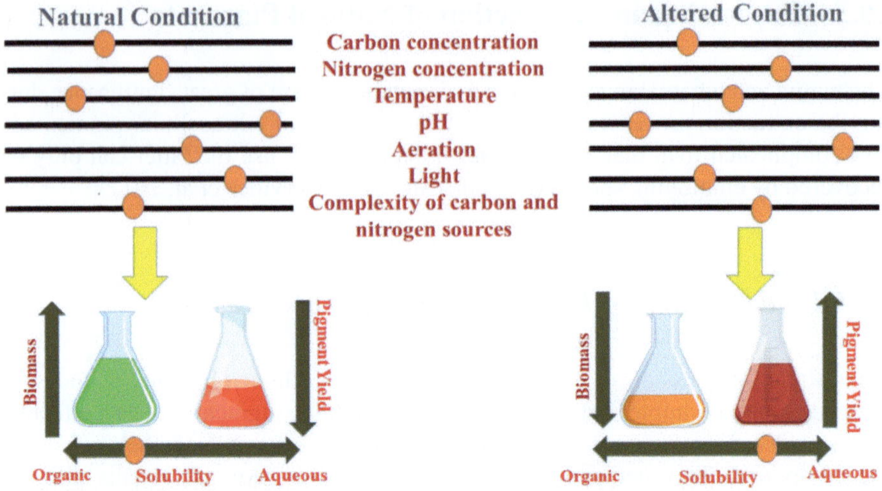

Fig. 10.2 Various factors affecting the growth of biomass and pigment yield

substantial quantities of extracellular colorants that can be readily recovered or processed without the requirement for solvent extraction.

10.5.1.2 Downstream Processing of Pigment

The aqueous two-phase system (ATPS) is the most commonly used technique for eliminating color compounds from fermented broth. It employs liquid-liquid fractionation and green ionic liquids. ATPS maintains the structure of molecules and creates a favorable environment for them and bioactivity, unlike the two phases of organic solvents used in traditional extraction techniques; ATPS are made of water (McQueen and Lai 2019). Using ATPS based on quaternary ammonium and imidazolium systems, red pigment was recovered from the fermented broth of *Penicillium purpurogenum* DPUA 1275. In an ideal extraction method, proteins (EETotal = 60.7 ± 2.8%) are extracted from the broth used for fermentation, and the red color (Kred = 24.4 ± 2.3) has strong partition coefficients thanks to the tetraethylammonium bromide ionic liquid (Xu et al. 2024).

10.5.2 *Extraction of Intracellular Pigments*

For the intracellular compounds fungal cell wall was broken by the action of chitinase enzyme as "Green extraction techniques" used for extracting hydrophobic compounds and intracellular compounds since they either need very little or no organic solvent. Many of these also function at low temperatures, which, notwithstanding their deterioration, help extract thermolabile pigments. These include methods like pressurized liquid extraction, ionic liquid-assisted extraction, supercritical CO_2 extraction, pulsed electric field-assisted extraction, ultrasound-assisted extraction, and microwave-assisted extraction (Lin and Xu 2020).

10.5.2.1 Ultrasound-Assisted Extraction (UAE)

Ultrasound-assisted extraction is recognized as a sustainable and highly effective method for isolating bioactive compounds. Most naturally occurring metabolites are labile at high temperatures, and this allows them to undergo damage due to heat generated by the ultrasonication. The process efficacy is improved by operating process at lower temperatures, preserving the integrity of such compounds. The method utilizes ultrasonic energy to enhance the rate at which solid samples and liquid solvents come into contact with each other, increasing the rate at which the solvents take up desired molecules. Sound waves cause microscopic ruptures in the cell structures due to the rapid changes in pressure, which result in the internal constituents of the cell diffusing into the solvent media with relative ease. It primarily uses sonic energy, which travels across the medium without being absorbed by the molecule. The medium allows ultrasound waves to pass through via vibrational motion of the molecules, which is induced by the pressure waves (Tiwari 2015).

10.5.2.2 Pressurized Liquid Extraction (PLE)

Accelerated solvent extraction is another name for pressurized liquid extraction. High pressure and temperature are used in this automated process to ensure complete extraction from solid matrices. Higher temperature increases the diffusion rate and solubility of the metabolites by disrupting the interaction of analytes with the matrix. High pressure facilitates the solvent's penetration into the sample. Together, these factors reduce solvent usage while removing metabolites from the matrix. Both the amount of extraction solvent used and the overall extraction time are decreased by PLE. By applying PLE may keep the molecule's closest possible vicinity, it has been acknowledged as the best technique for removing bioactive compounds from plants and fungi. The species of *Talaromyces* had the greatest yield of the red pigments, and it was demonstrated that the extraction process of PLE is far faster and more environmentally friendly when compared to the conventional extraction (Lebeau et al. 2020).

10.5.2.3 Microwave-Assisted Extraction (MAE)

The frequency range of microwave energy falls in the range of 300 megahertz to 300 gigahertz. This type of radiation is applied in a process called microwave-assisted extraction (MAE), during which the radiation enters the solvent-sample mixture and makes molecules inside it vibrate intermittently and create heat internally (Xiong et al. 2016). When using this approach instead of more conventional ones, the extraction time and solvent amount are decreased. Researchers often utilize MAE to extract biologically active molecules of plant or microbial materials in extracts of greater purity and productivity. Particle size, solid-liquid ratio, the substrate material, temperature, pressure, and solvents are some of the factors that contribute greatly to the effectiveness and specificity of MAE (Microwave-Assisted Extraction). The importance as well as the possibilities of water as a sole solvent or solvent-free methods of extracting bioactive metabolites have been reviewed by many. A green solvent, water in combination with MAE, helps to extract various metabolites of fungi and plants. It has been shown that MAE is effective in extracting pigments in botanical sources, and the findings can be applied in this field (Kalra et al. 2020).

10.5.2.4 Supercritical Extraction (SFE)

The last ten years have witnessed an increased use of supercritical extraction (with the use of CO_2) as an environmentally friendly extraction method. Liquid CO_2 is used in SFE to extract bioactive compounds from solid matrices (Khaw et al. 2017). It works on the principle that some gases that hit their critical temperature has a better solvation ability and develop into supercritical fluids, which possess both gaseous and liquid characteristics. The pressure and temperature used during the extraction method determine solubility of supercritical fluids. One of the major advantage of SFE is that it can be used at lower temperature range and thus enables the extraction of thermolabile chemicals. It is often referred to as a sustainable, green method of selectively isolating molecules. Non-polar chemicals are primarily extracted using supercritical CO_2; however, by adding additional co-solvents, such as ethanol, its polarity can be changed, allowing supercritical CO_2 to extract carotenoids from typically polar compounds like ethylene. The application of SFE (supercritical fluid extraction) to the large-scale and laboratory isolation of carotenoids has been extensively documented (Uwineza and Waśkiewicz 2020). The selection of parameters like co-solvent, flow rate, pressure, time, and temperature cannot be ignored and is crucial to the successful and accurate extraction of the target molecule.

10.5.2.5 Ionic Liquid–Assisted Extraction

Ionic liquids (ILs) are evolving into specialized extraction solvents for purifying bioactive compounds produced naturally. ILs are considered to be tunable designer solvents due to their variety of combinations of salts and are designed with a range of components, which overcome the narrow selectivity of organic solvents (Lebeau et al. 2020). Ionic liquids (ILs), owing to their charged molecular composition, are capable of interacting with electromagnetic fields. Due to this characteristic, they can be incorporated into the latest extraction methods, such as ultrasound-assisted extraction (UAE), IL-integrated UAE systems, and microwave-assisted extraction (MAE), also known as IL-integrated MAE. Moreover, protic ionic liquids (PILs) have been shown to be useful for disrupting cell walls and enhancing intracellular carotenoid production and extraction by *Rhodotorula glutinis* yeast (Mussagy et al. 2019).

10.5.2.6 Pulsed Electric Field (PEF) Assisted Extraction

This method of extraction, which falls under the category of non-thermal techniques, is predicated on what is known as electroporation, or electro-permeabilization. This approach involves the cell being exposed to a series of short high-voltage bursts that temporarily disrupt the integrity of the cell membrane, causing it to become more permeable. The degree of this change, whether short-term or long-term, depends on a number of operating parameters like pulse strength, pulse duration, frequency, and total energy contribution. When the membrane structure becomes more permeable, it is easier to perform diffusion of solvent molecules into the intracellular space, and as a result, the metabolites and other bioactive compounds are released faster. PEF is applied to selectively isolate metabolites through differentiation in the degree of treatment. The future of the application of PEF in pigment collection of algae and other matrices is brought into focus depending on the research (Fatima et al. 2024). Hydrophobic carotenoids and other pigments could be effectively recovered by the integration of solvent extractions using a dual-phase combination of organic solvents at various pH values. The 2-step extraction process results in high proportions of stable proteins during the first stage of PEF extraction and increased pigments during the second stage. This is a 2-stage PEF-aided procedure to enable the extraction to be carried out efficiently using minimal organic solvents (Fatima et al. 2024).

10.6 Conclusion

Finding a sustainable, profitable source of pigments that are less detrimental to the environment and human health is the primary objective of the pigment industry, particularly for food-grade pigments. The use of natural colorants have gained popularity again due to the current social trend of using natural ingredients and consumer worries about synthetic colors' detrimental impact on human health and the environment. The last few decades have recorded a gradual increase in the application of various biotechnological methods to provide tasty, nutritional, wholesome, and high sensory quality products, making this process cost competitive and applicable in large quantities. Although nature may offer a worthy source of safe pigments, serious limitations such as the availability of raw materials and inconsistency in pigment profiles due to plant-derived colors are driving the color industry into the prospective potential of pigments provided by microbial, especially fungal, sources.

Based on the merits of fungal diversity, it can be said that fungi are treated as cellular factories to produce pigments, and as a result of this, researchers are manipulating their functionality. Fungi obtained from different sources are known to produce pigments of a broad scope, usually associated with diversified biological functions and a wide variety of colors. The traditional technologies have produced pigments like monascin (derived from a fungus). Although these technologies have been developed, intensive research is being conducted on the discovery of alternative sources and innovative ways of producing these pigments through biotechnological methods that would be profitable in terms of production. It is thus imperative that more research should be done to enhance pigment characteristics of yield and composition through the most effective growth conditions, metabolic engineering methods, cost-effective organic media, various elicitors to trigger pigment synthesis, stabilization conditions to improve pigment application, and cost-effective ways to extract pigments on a large scale.

References

Afroz Toma M, Rahman MH, Rahman MS, Arif M, Nazir KNH, Dufossé L (2023) Fungal pigments: carotenoids, riboflavin, and polyketides with diverse applications. Journal of Fungi 9(4):454

Antipova TV, Zhelifonova VP, Zaitsev KV, Vainshtein MB (2023) Fungal azaphilone pigments as promising natural colorants. Microbiology 92(1):1–10

Ashok G, Mohan U, Boominathan M, Ravichandiran V, Viswanathan C, Senthilkumar V (2022) Natural pigments from filamentous fungi: production and applications. In: Industrially important fungi for sustainable development: volume 2: bioprospecting for biomolecules. Springer International Publishing, Cham, pp 651–678

Bartel PL, Zhu CB, Lampel JS, Dosch DC, Connors NC, Strohl WR, Beale Jr. JM, Floss HG (1990) Biosynthesis of anthraquinones by interspecies cloning of actinorhodin biosynthesis genes in streptomycetes: clarification of actinorhodin gene functions. J Bacteriol 172(9):4816–4826

Boral H, Metin B, Döğen A, Seyedmousavi S, Ilkit M (2018) Overview of selected virulence attributes in Aspergillus fumigatus, Candida albicans, Cryptococcus neoformans, Trichophyton rubrum, and Exophiala dermatitidis. Fungal Genet Biol 111:92–107

Chen W, Chen R, Liu Q, He Y, He K, Ding X, Kang L, Guo X, Xie N, Zhou Y, Lu Y (2017) Orange, red, yellow: biosynthesis of azaphilone pigments in Monascus fungi. Chem Sci 8(7):4917–4925

Chen W, Feng Y, Molnár I, Chen F (2019) Nature and nurture: confluence of pathway determinism with metabolic and chemical serendipity diversifies Monascus azaphilone pigments. Nat Prod Rep 36(4):561–572

Corrêa RCG, Garcia JAA, Correa VG, Vieira TF, Bracht A, Peralta RM (2019) Pigments and vitamins from plants as functional ingredients: current trends and perspectives. Adv Food Nutr Res 90:259–303

Dos Santos MC, Bicas JL (2021) Natural blue pigments and bikaverin. Microbiol Res 244:126653

Egea MB, Dantas LA, Sousa TLD, Lima AG, Lemes AC (2023) The potential, strategies, and challenges of Monascus pigment for food application. Front Sustain Food Syst 7:1141644

Fan Z, Sun ZH, Liu HX, Chen YC, Li HH, Zhang WM (2016) Perangustols A and B, a pair of new azaphilone epimers from a marine sediment-derived fungus Cladosporium perangustm FS62. J Asian Nat Prod Res 18(11):1024–1029

Fatima I, Munir M, Qureshi R, Hanif U, Gulzar N, Sheikh AA (2024) Advanced methods of algal pigments extraction: a review. Crit Rev Food Sci Nutr 64(27):9771–9788

Faustino M, Veiga M, Sousa P, Costa EM, Silva S, Pintado M (2019) Agro-food byproducts as a new source of natural food additives. Molecules 24(6):1056

Galaffu N, Bortlik K, Michel M (2015) An industry perspective on natural food colour stability. In: Colour additives for foods and beverages. Woodhead Publishing, pp 91–130

Geow CH, Tan MC, Yeap SP, Chin NL (2021) A review on extraction techniques and its future applications in industry. Eur J Lipid Sci Technol 123(4):2000302

Grewal PS, Modavi C, Russ ZN, Harris NC, Dueber JE (2018) Bioproduction of a betalain color palette in Saccharomyces cerevisiae. Metab Eng 45:180–188

Hernández-Almanza A, Montanez J, Martínez G, Aguilar-Jimenez A, Contreras-Esquivel JC, Aguilar CN (2016) Lycopene: progress in microbial production. Trends Food Sci Technol 56:142–148

Huang ZR, Zhou WB, Yang XL, Tong AJ, Hong JL, Guo WL, Li TT, Jia RB, Pan YY, Lin J, Lv XC (2018) The regulation mechanisms of soluble starch and glycerol for production of azaphilone pigments in Monascus purpureus FAFU618 as revealed by comparative proteomic and transcriptional analyses. Food Res Int 106:626–635

Kalra R, Conlan XA, Goel M (2020) Fungi as a potential source of pigments: harnessing filamentous fungi. Front Chem 8:369

Khaw KY, Parat MO, Shaw PN, Falconer JR (2017) Solvent supercritical fluid technologies to extract bioactive compounds from natural sources: a review. Molecules 22(7):1186

Law CW, Constad WH, Francis JH (2025) Phaeohyphomycosis, a pigment-producing fungus, mimicked melanoma on external examinations. Ophthalmology 132(8):e152

Lebeau J, Petit T, Fouillaud M, Dufossé L, Caro Y (2020) Aqueous two-phase system extraction of polyketide-based fungal pigments using ammonium-or imidazolium-based ionic liquids for detection purpose: a case study. J Fungi 6(4):375

Li J, Li Z, Zhao T, Yan X, Pang Q (2021) Proteomic analysis of Auricularia auricula-judae under freezing treatment revealed proteins and pathways associated with melanin reduction. Front Microbiol 11:610173

Lin L, Xu J (2020) Fungal pigments and their roles associated with human health. J Fungi 6(4):280

Lin L, Xu J (2022) Production of fungal pigments: molecular processes and their applications. J Fungi 9(1):44

Liu X, Hou R, Wang D, Mai M, Wu X, Zheng M, Fu J (2019) Comprehensive utilization of edible mushroom Auricularia auricula waste residue—extraction, physicochemical properties of melanin and its antioxidant activity. Food Sci Nutr 7(11):3774–3783

McQueen L, Lai D (2019) Ionic liquid aqueous two-phase systems from a pharmaceutical perspective. Front Chem 7:135

Morales-Oyervides L, Oliveira J, Sousa-Gallagher M, Méndez-Zavala A, Montañez JC (2017) Perstraction of intracellular pigments through submerged fermentation of Talaromyces spp. in a surfactant rich media: a novel approach for enhanced pigment recovery. J Fungi 3(3):33

Mussagy CU, Santos-Ebinuma VC, Gonzalez-Miquel M, Coutinho JA, Pereira JF (2019) Protic ionic liquids as cell-disrupting agents for the recovery of intracellular carotenoids from yeast Rhodotorula glutinis CCT-2186. ACS Sustain Chem Eng 7(19):16765–16776

Nielsen JC, Nielsen J (2017) Development of fungal cell factories for the production of secondary metabolites: linking genomics and metabolism. Synth Syst Biotechnol 2(1):5–12

Pedersen TB, Nielsen MR, Kristensen SB, Spedtsberg EML, Yasmine W, Matthiesen R, Kaniki SEK, Sørensen T, Petersen C, Muff J, Sondergaard TE (2020) Heterologous expression of the core genes in the complex fusarubin gene cluster of Fusarium Solani. Int J Mol Sci 21(20):7601

Perez-Cuesta U, Aparicio-Fernandez L, Guruceaga X, Martin-Souto L, Abad-Diaz-de-Cerio A, Antoran A, Buldain I, Hernando FL, Ramirez-Garcia A, Rementeria A (2020) Melanin and pyomelanin in aspergillus fumigatus: from its genetics to host interaction. Int Microbiol 23(1):55–63

Pimenta LP, Gomes DC, Cardoso PG, Takahashi JA (2021) Recent findings in azaphilone pigments. J Fungi 7(7):541

Pombeiro-Sponchiado SR, Sousa GS, Andrade JC, Lisboa HF, Gonçalves RC (2017) Production of melanin pigment by fungi and its biotechnological applications. Melanin 1(4):47–75

Ramachandra SK, Kaushik S, Austin H, Kumar R, Parthiban M (2023) The role of melanin in plant pathogenic fungi: insights into structure, biosynthesis, and function. Plant Biol Crop Res 6:1090

Romano S, Jackson SA, Patry S, Dobson AD (2018) Extending the "one strain many compounds" (OSMAC) principle to marine microorganisms. Mar Drugs 16(7):244

Rouf A, Kanojia V, Naik HR, Naseer B, Qadri T (2017) An overview of microbial cell culture. J Pharmacogn Phytochem 6(6):1923–1928

Sankari M, Rao PR, Hemachandran H, Pullela PK, Tayubi IA, Subramanian B, Gothandam KM, Singh P, Ramamoorthy S (2018) Prospects and progress in the production of valuable carotenoids: insights from metabolic engineering, synthetic biology, and computational approaches. J Biotechnol 266:89–101

Sen T, Barrow CJ, Deshmukh SK (2019) Microbial pigments in the food industry—challenges and the way forward. Front Nutr 6:7

Tiwari BK (2015) Ultrasound: a clean, green extraction technology. TrAC Trends Anal Chem 71:100–109

Uwineza PA, Waśkiewicz A (2020) Recent advances in supercritical fluid extraction of natural bioactive compounds from natural plant materials. Molecules 25(17):3847

Venil CK, Velmurugan P, Dufossé L, Renuka Devi P, Veera Ravi A (2020) Fungal pigments: potential coloring compounds for wide ranging applications in textile dyeing. J Fungi 6(2):68

Verde-Yáñez L, Vall-Llaura N, Usall J, Teixidó N, Torreblanca-Bravo È, Torres R (2023) Identification and biosynthesis of DHN-melanin related pigments in the pathogenic fungi Monilinia laxa, M fructicola, and M fructigena. J Fungi 9(2):138

Wang Q, Liu D, Yang Q, Wang P (2017) Enhancing carotenoid production in Rhodotorula mucilaginosa KC8 by combining mutation and metabolic engineering. Ann Microbiol 67(6):425–431

Wu Z, Zhang M, Yang H, Zhou H, Yang H (2018) Production, physicochemical characterization, and antioxidant activity of natural melanin from submerged cultures of the mushroom Auricularia auricula. Food Biosci 26:49–56

Xiong W, Chen X, Lv G, Hu D, Zhao J, Li S (2016) Optimization of microwave-assisted extraction of bioactive alkaloids from lotus plumule using response surface methodology. J Pharm Anal 6(6):382–388

Xu Y, Du X, Yu X, Jiang Q, Zheng K, Xu J, Wang P (2022) Recent advances in the heterologous expression of biosynthetic gene clusters for marine natural products. Mar Drugs 20(6):341

Xu Y, Xu J, Fan Z, Zhang S, Wu Y, Han R, Yu N, Tong X (2024) Effective separation of protein from Polygonatum cyrtonema crude polysaccharide utilizing ionic liquid tetrabutylammonium bromide. Front Chem 11:1287571

Yang ZJ, Zhang YF, Wu K, Xu YX, Meng XG, Jiang ZT, Ge M, Shao L (2020) New azaphilones, phomopsones AC with biological activities from an endophytic fungus, Phomopsis sp. CGMCC No. 5416. Fitoterapia *145*:104573

Zhang YH, Peng XY, Feng LX, Zhu HJ, Cao F, Wang CY (2021) A new epimer of azaphilone derivative pinophilin B from the gorgonian-derived fungus Aspergillus fumigatus 14–27. Nat Prod Res 35(13):2232–2238

Chapter 11
Recent Advancements in Fungal Biotechnology for Developing Biopolymers from Waste Biomass: Challenges and Perspectives

Sushma, Gaurav Mudgal, Jaspreet Kaur, Sanjoy K. Dhar, Abhijeet Singh, Sonia Kumari Shishodia, and Gholamreza Abdi

Abstract The production of plastics globally exceeded 400 million metric tons in 2022, and more than 60% of plastics come from fossil-origin polymers, taking centuries to degrade in the Earth's ecosystems. On the other hand, agro-industrial activities produce more than 5 billion tons of lignocellulosic and food sector waste every year, but most of it is not tapped properly and ends up being burned, increasing the carbon footprint. However, fungal biotechnology can provide a revolutionary approach in tapping the versatile biocatalytic abilities of fungi, including yeasts, for converting biomass waste into useful bioplastics like polyhydroxyalkanoates (PHAs), polylactic acid (PLA), fungal chitosan, and β-glucans. Bioplastics possess flexible properties, biocompatibility, and biodegradability, making them useful for

Sushma and Gaurav Mudgal contributed equally with all other contributors.

Sushma · G. Mudgal (✉) · S. K. Shishodia (✉)
University Institute of Biotechnology, Chandigarh University, Mohali, Punjab, India

J. Kaur
Department of Biotechnology, Kanya Maha Vidyalaya College, Jalandhar, Punjab, India

S. K. Dhar
Banophool Adibashi Green Heart College, Mirpur, Dhaka, Bangladesh

Bangladesh University of Health Sciences, Mirpur, Dhaka, Bangladesh

A. Singh
School of Bioscience and Technology (SBST), Vellore Institute of Technology, Vellore, Tamil Nadu, India

G. Abdi
Department of Biotechnology, Persian Gulf Research Institute, Persian Gulf University, Bushehr, Iran
e-mail: abdi@pgu.ac.ir

M. Tripathi et al. (eds.), *Recent Advancements in Fungal Biotechnology Volume 1*, Fungal Biology, https://doi.org/10.1007/978-981-92-0339-0_11

packaging, biomedical applications, and biocompatible textiles such as bio-cotton. Recent breakthroughs in systems biology studies and genome editing through CRISPR-Cas9 systems, as well as adaptive laboratory evolution (ALE) efforts, have made it possible to increase polymer production yields, substrate tolerance, and robustness in fungal host systems such as *Aspergillus niger*, *Rhizopus oryzae*, and *Yarrowia lipolytica.* This chapter reviews the current trends in biosynthesis using biopolymers from fungal sources and metabolically engineered approaches, particularly from waste biomass. This chapter identifies important challenges ranging from substrate diversity to enzyme inhibition and scale-up, while formulating holistic approaches that integrate strain design derived from omics analysis tools, bioreactor development, and closed-loop bioeconomy concepts. Insights into regulatory frameworks, life cycle analysis, and techno-economic analysis will be investigated in the context of the strategic importance of fungal biotechnology in moving the entire material science field toward sustainability.

Keywords Biofabrication · Filamentous fungi · Microbial valorization · Enzymatic hydrolysis · CRISPR-Cas9 · Solid-state fermentation · Bioreactor design · Omics integration · Life cycle assessment · Biodegradable composites

11.1 Introduction

Plastics are found all around us and are an essential material due to the following properties of plastic, such as being lightweight, durable, and useful for electrical insulation, and mostly because of its shape and size, many other properties can be altered as per need. Plastics have origins tracing back thousands of years. They are known as some of the most advanced materials in years, with the use of natural polymers such as rubber by the Mesoamericans in 1600 BC and shellac in India around 1000 BC. The modern period of plastics began in 1839 when Eduard Simon discovered polystyrene, although this was not commercialized until much later. In 1862, Alexander Parkes launched Parkesine, the first-ever man-made plastic, which came from cellulose, during the Great Exhibition of London. The International Exhibition held in London marked the transition point for material science. This was followed by the invention of celluloid by John Wesley Hyatt in 1868, which was used to replace ivory in billiard balls and also emerged as a crucial element of the film industry. The dawn of completely synthetic plastics occurred in 1907 when Leo Baekeland invented Bakelite, which is heat-resistant and chemically stable. He has thus contributed to the invention of electrical insulators and household goods.

This chapter is positioned in the context of the plastics problem, where fossil-based plastics are mainly produced and have lifetimes of hundreds of years in the environment, and it delves into the necessity of sustainable biopolymer technology, specifically in terms of the conversion of more than five billion tons of agricultural and food processing wastes per year through fungal and yeast-like microbes. This includes biopolymer production through PHAs, PLA, chitosan, and β-glucans, and it is characterized according to desirable mechanical properties, biocompatibility, and biodegradability for certain applications. Based on recent advancements in bioinformatics and genetic and evolutionary mutations in microbes, this chapter describes how to improve the productivity of biopolymers in hosts such as *Aspergillus niger*, *Rhizopus oryzae*, and *Yarrowia lipolytica*. In addition to basic research and innovation in biopolymer technologies, this chapter includes issues such as metabolism of heterogeneous substrates and fungal biocatalyst inhibition; in fact, it also includes techno-economic analysis and life cycle costing in terms of positioning biotechnologies in biopolymer research and development, influencing the circular bio-economy.

11.1.1 Global Plastic Crisis and Environmental Impacts

The global plastic crisis is one of the most dire environmental issues of the twenty-first century, with probable implications for ecosystems, biodiversity, and human health. Since the 1950s, global plastic production has risen from 1.5 million metric tons to over 460 million metric tons annually in 2019, a trajectory that accentuates both the material's utility and its surging environmental burden (Fulke et al. 2026). This exponential growth has resulted in an estimated 10 million metric tons of plastic entering aquatic environments annually, with projections indicating that, without intervention, cumulative ocean plastic pollution may reach between 75 and 200 million metric tons by 2050 (Williams et al. 2023). The persistence of conventional plastics, which take hundreds of years to degrade, results in widespread contamination from the Mariana Trench to Mount Everest, and it is so pervasive that scientists have proposed plastic as a geological marker of the Anthropocene epoch (González-Fernández and Medellín-Castillo 2024).

A key aspect of this problem concerns the fragmentation of macroplastics into microplastics (<5 mm) and nanoplastics (<100 nm) that pervades all ecosystems on the planet (Nayanathara Thathsarani Pilapitiya and Ratnayake 2024). These microplastics and nanoplastics are either produced by the primary input of microbeads in cosmetics and the subsequent formation of microplastics from the shedding of synthetic fibers during the laundry cycle, or through the secondary breakdown of larger plastic debris by UV light action, mechanical stress, and biological activity (Kumar and Krishan 2023). Microplastics and nanoplastics, once in the environment, find their way into the food web as organisms ingest these pollutants, which has already been observed in plankton and top predators (Nguyen et al. 2025). Microplastics and nanoplastics can cause several biological effects that include entanglement and obstruction, reduction in feeding efficiencies, reproductive decline, and bioaccumulation of toxic additives such as phthalates and Bisphenol A (Wang et al. 2024a). Also, microplastics/microbeads are vectors of persistent organic pollutants (POPs) and pathogenic microbes that pose the same level of threat by trophic transfer as the original contaminants themselves (Boccia et al. 2024).

Marine ecosystems are the most susceptible, as plastic debris may lead to habitat degradation, coral reef smothering, and interference with the benthic communities (Wang et al. 2024a). About 260,000 tons of plastic float on the ocean surface, while additional amounts settle in sediments or are consumed by marine organisms (Fulke et al. 2026). Riverine transport has an especially critical function in transporting land-based waste to oceans, although country-specific contributions are radically different because of population density, waste management infrastructure, and proximity to major waterways (Mai et al. 2023). At the same time, current recycling efforts cannot completely cover the desperate need; thus, only about 9% of all produced plastic is recycled, while 12% is incinerated, and an overwhelming 79% goes to landfills or to the natural environment (Walker and Fequet 2023).

The question of international waste plastic trade makes the treatment of the issue even more complicated, with countries with higher GDPs exporting large amounts of plastic waste, 46% of recyclable polyethylene waste in Europe alone, to countries with weaker regulations, which end up leaking into water systems despite regulations for recycling and proper disposal (Bishop et al. 2020). There is a major concern for environmental justice, especially for countries that contribute very little to plastic manufacturing but experience the major effects due to the unsustainable waste disposal practices imposed on the world by countries such as the US and China, which have weak regulations for the disposal of plastic waste; China has completely banned the importation of plastic waste from all countries since 2017, while the European Green Deal has made plastic packaging recyclable and reusable by 2030 for all European Union (EU) member states (Boccia et al. 2024; Stoett et al. 2024).

Scientific research is shifting its focus toward the development of systematic solutions based on the principles of a circular economy. These solutions include the improvement of chemical and feedstock recycling processes, the development of truly biodegradable substitute materials from renewable biomass, and the adoption of Extended Producer Responsibility schemes to promote sustainable product design (Vinod et al. 2020). Thermochemical conversion processes such as pyrolysis and gasification may have some potential for energy recovery from non-recyclable plastics, but these need careful planning and execution to reduce both greenhouse gas (GHG) emissions and hazardous byproducts (Rai et al. 2023). Simultaneously, there is a need for behavioral and system changes, which include the reduction of single-use plastics, upgrading of collection and sorting infrastructure, and promotion of public awareness, which need to be made the basic components of a holistic response strategy. The current modeling forecast reveals that satisfying international commitments, such as the Osaka Blue Ocean Vision of preventing further marine plastic leaks by 2050, would require collective reductions in plastic production, improved waste management, and advances in recycling rates globally (Higuchi and Isobe 2024). Unless a holistic response strategy is pursued collectively, the rising costs of plastic pollution are likely to prevent us from making collective progress on various United Nations Sustainable Development Goals, ranging from access to clean water, climate, and life below water, among others.

The increasingly global addiction to synthetic plastics has precipitated an environmental crisis that now requires urgent science-driven solutions; sustainable biopolymers represent a critical alternative to mitigate the pervasive contamination of ecosystems. Conventional petroleum-based plastics, which take hundreds of years to degrade, have accumulated in natural environments in a manner such that microplastic particles are even found in the most remote marine trenches, agricultural soils, and human tissues (Bishop et al. 2020; Priya et al. 2023). Complex limitations to mechanical recycling, being seen as one of the key mitigation strategies, are increasingly apparent, with the processing itself causing secondary pollution through the release of microplastics at shredding and washing stages, with recent estimates suggesting substantial discharges from recycling facilities globally (Suzuki et al. 2024). In addition, the export of plastic waste from high-income countries to regions lacking adequate waste management infrastructure exacerbates environmental injustices and increases leakage to aquatic systems, compromising global efforts to contain plastic pollution (Bishop et al. 2020). These various systemic failures reveal the need for a transition to materials that are intrinsically sustainable and meet circular economy principles.

11.1.2 Need for Sustainable Biopolymer Alternatives

Biobased and biodegradable polymers are part of a paradigm shift in material design, offering routes to reduce both fossil fuel dependency and persistent waste accumulation {Das, 2025 #176}. Recent advances have underlined the opportunity offered by biofibers, biopolymers, and biocomposites from renewable biomass sources such as cellulose, lignin, and microbial polyhydroxyalkanoates, which present comparable mechanical properties to conventional plastics while enabling end-of-life options through composting or anaerobic digestion (Vinod et al. 2020; Rao et al. 2025) {Singh, 2025 #174}. Conversely, plant-derived polymers have been developed, such as polylactic acid and polybutylene succinate, produced via fermentation from corn starch or sugarcane, allowing their carbon cycle to be closed when produced under an integrated biorefinery context {Garcia Simão, 2024 #20; Jayaprakash, 2025 #175}. These materials not only mitigate greenhouse gas emissions during their life cycle but also avoid the toxic leachates released upon the decomposition of conventional plastics, hence causing minimal ecotoxicological risk.

Beyond material substitution, next-generation strategies incorporate biological systems for value-added degradation and resource recovery. Microorganism-mediated plastic degradation, aided by enzymatically active strains like *Cunninghamella echinulata*, presents a promising route for cleaving recalcitrant polymer chains under ambient conditions, allowing in situ bioremediation of contaminated sites (Garcia Simão et al. 2024). Thermochemical conversion technologies, namely pyrolysis and gasification, can convert non-recyclable plastic waste, which also includes pandemic-related disposables such as personal protective equipment, into high-value energy vectors, including syngas, liquid hydrocarbons, and aromatic compounds, thus contributing to energy resilience and reducing landfill burden (Rai et al. 2023). Such hybrid approaches address specific United Nations SDGs, namely SDG 7 (Affordable and Clean Energy), SDG 12 (Responsible Consumption and Production), and SDG 13 (Climate Action), by connecting waste valorization to low-carbon industrial processes.

Despite these gains, production scale-up, actual biodegradability under real conditions, and economic viability of collection and processing infrastructures remain key challenges. Accordingly, policy interventions, especially extended producer responsibility (EPR), bans on single-use plastics, and financial incentives for green innovation, are necessary to speed up adoption (Garcia Simão et al. 2024). Public awareness campaigns should also be put in place, paralleled by global harmonization of regulations, so that behavioral change may be achieved with balanced implementation across countries. If materials science, microbiology, engineering, and policy fail to mount a coordinated response, the trajectory of plastic pollution will continue to threaten ecological integrity and human health well into the future.

11.1.3 Role of Fungal Biotechnology in Circular Bioeconomy

Fungal biotechnology has found its place at the core of the circular bioeconomy, as it provides data-driven solutions for the efficient bioconversion of biomass waste into valuable compounds. Fungi are unique in terms of their metabolism and are highly adept at degrading complex organic materials such as lignocellulose, industrial by-products, and agricultural and food industry residues through the production of a diverse array of surface-binding enzymes, which include cellulases, hemicellulases, and lignin-modifying oxidoreductases (Meyer et al. 2020; Khatami et al. 2026). The biochemical phenotype of fungi makes them essential for bioconversion in biorefinery processes, which are optimized using filamentous fungi such as *Aspergillus*, *Trichoderma*, and *Phanerochaete*. The biorefinery is optimized to focus on the bioconversion of inadequately used biomass and biomass residues to provide material solutions for decreasing pollution in the environment, which provides newer sources of renewable energy and materials (Jatwani et al. 2025). For example, white-rot fungi, such as *Phanerochaete chrysosporium*, are able to decompose complex lignin-containing materials in biomass, which further provides bioavailable sugar for biofuel conversion and production (Jatwani et al. 2025). These biochemical conversions and processes act as a prime example of the concept of the circular bioeconomy {Singh, 2025 #177}, which aims to realign the bioeconomy of the future to focus on the conversion of "take, make, and dispose" in biorefineries for the sustained use and reuse of biological resources.

The use of fungal fermentation methods serves as a precursor to synthesis following degradation processes. Filamentous fungi are considered nature's cell factories for the biosynthesis of various valuable compounds, ranging from the production of organic acids like citric acid and itaconic acid, enzymes, pigments, lipids, and secondary metabolites, particularly for pharmaceutical applications like antibiotics, antifungals, and antiviral compounds, among others, used in industrial applications (Meyer et al. 2020; Zou et al. 2023). Moreover, oleaginous yeasts like *Papiliotrema laurentii* can store their total lipids within their cellular compartments under nitrogen limitation. Therefore, these yeasts could be used for the production of microbial oils, particularly for the production of biodiesel components for industrial applications (Moreira Vieira et al. 2020). Moreover, advances in genetic engineering and synthetic biology have improved the yield, utilization, and metabolic ability to process non-food sources like glycerol, molasses, and even syngas, thereby increasing their potential for industrial production, since the primary products do not compete with the food chain for raw materials (Garg 2025). However, the advancement in artificial intelligence approaches is set to revolutionize the process for strain improvement (Garg 2025).

Apart from the use of fungi in the production of materials and chemicals, fungal biotech has considerable applications in sustainable agriculture practices and animal nutrition {Thakur, 2025 #172}. For example, solid-state fermentation technology using white-rot fungi has been demonstrated to enhance the digestibility of lignocellulosic materials employed in animal feed formulated for ruminants by decaying lignin-based inhibitors hindering enzymatic hydrolysis of cellulose and hemicellulose in these materials (Yan et al. 2025). Not only does this augment the efficacy of animal feeding, thereby enhancing animal nutrition and sustainable farming practices, but it also helps in reducing methane gas emissions from the enteric fermentation processes of ruminants, a major factor in climate change. Furthermore, the mycelium of these fungi has emerged as a useful biomaterial with diverse applications in the fabrication of biodegradable packaging materials, leather replicas, building materials, and textile materials (Meyer et al. 2020; Füting et al. 2021) {Mudgal, 2025 #173}. Mycelium materials do not require the use of materials science-based energy inputs in their production processes. Additionally, they do not generate substantial waste materials. Despite such benefits, some challenges exist for the scaling-up of fungal biotechnology processes for commercial applications. The rigidity and hardness of fungal cell walls create particular challenges for the recovery of products, requiring the adoption of appropriate and species- and form-dependent cell disruption methodologies (Mantri et al. 2025). Additionally, process cost efficiency, maintenance of sterilization during large-scale fermentation, and the heterogeneity of waste feedstocks continue to warrant inventive bioreactor and monitoring systems, as well as dynamic microbial consortia (Vieira et al. 2025). Environmental and social acceptance, as well as the regulatory environment, also affect the rate and applicability, especially for genetically modified organisms (GMOs), for use in the environment as a whole. The integration and alignment of technological innovation, policy agendas on bio-based industry, and emerging demand for the sustainability paradigm, however, emphasize the imperative role and value of fungal systems as a key enabler for the achievement of a low-carbon and resource-efficient future consistent with the United Nations Sustainable Development Goals (Meyer et al. 2020; Carmona Marques et al. 2025).

11.2 Waste Biomass: Composition, Availability, and Valorization Potential

11.2.1 *Types of Agro-Industrial Waste (Lignocellulosic, Food, Municipal)*

The utilization of waste streams into valuable bioproducts offers significant solutions for the aforementioned environmental issues linked to resource depletion and pollution {Mudgal, 2025 #178}. Moreover, the filamentous fungal group displays an extraordinary metabolic flexibility that helps them break down complex organic matter such as lignocellulosic biomass, which consists of agricultural crops, including wheat straw, rice husk, corn stover, sugarcane bagasse, and forestry products, by the production of a wide range of cellulases, hemicellulases, and lignin-oxidative enzymes (Khatami et al. 2026). Furthermore, these plant compounds are reinforced by lignin, which makes them resistant to biodegradation and enzymatic degradation by microorganisms; hence, pretreatment approaches are required for biological valorization (Jatwani et al. 2025). White-rot fungi, particularly *Phanerochaete chrysosporium*, display unique strategies for lignin degradation by the production of ligninases, peroxidases, and laccases, which are crucial in bioconversion reactions for the release of fermentable sugars for biopolymer production (Jatwani et al. 2025; Li et al. 2025).

Fermentation by fungi can be applied to the degradation of organic material but also for the biosynthesis of valuable biopolymers like polyhydroxyalkanoates (PHAs), microbial cellulose, chitin, or exopolysaccharides, where assimilated carbon is channeled into these products. Additionally, fungi have the ability to produce intracellular lipids or proteins, which can be utilized for the manufacture of bioplastics, packaging materials, or even food-grade films, in accordance with the rules of green chemistry and the concept of cradle-to-cradle (Garg 2025). For example, the oleaginous yeast *Papiliotrema laurentii*, which grows rich in intracellular lipids upon nutrient limitation, can serve as a sustainable source for the production of commercial biodiesel without displacing food production globally (Garg 2025). Furthermore, with recent breakthroughs in the field of synthetic biology, precision fungal engineering for the maximization of substrate conversion, enhanced compound productivity, or the enhancement of the pool of convertible biomass materials, such as industrial byproducts or municipal organic waste, has now become feasible for the industry (Garg 2025; Liu et al. 2025). Table 11.1 outlines various categories of waste and how fungi can be utilized for their valorization into value-added products.

Beyond their applications in material production, fungi also hold great importance as agents for waste minimization and resource recovery as part of the integrated biorefinery process lifecycle approach. Food industry wastes (such as the peels and trim and spent grains from the brewing) and dairy industry wastes (such as whey), rich in simple sugars, amino acids, and organic acids, respectively, make them uniquely suited for bioconversion by fungi without requiring extensive

Table 11.1 Agro-industrial waste types and fungal valorization potential

Waste type	Common examples	Key components (C:N ratio, sugars, lignin)	Fungal species/ enzymes used	Potential biopolymer products	Key references
Lignocellulosic	Wheat straw, rice husk, bagasse	Cellulose, hemicellulose, lignin	White-rot fungi (*Phanerochaete chrysosporium*), laccases, and peroxidases	PHAs, PLA, microbial cellulose	Pilafidis et al. (2022); Suryadi et al. (2022)
Food industry waste	Fruit peels, whey, spent grains	Simple sugars, amino acids, and organic acids	*Aspergillus Niger*, *Rhizopus oryzae*	Chitosan, exopolysaccharides	Pilafidis et al. (2022); Minervini et al. (2022)
Municipal organic waste	Kitchen scraps, yard trim	Mixed composition, high variability	Mixed fungal consortia, cocultures	Biodegradable films, fungal mycelium composites	Alemu et al. (2022); Barta et al. (2024)

PHAs polyhydroxyalkanoates, *PLA* polylactic acid

pretreatment processes (Ezeorba et al. 2024). Similarly, municipal solid organic waste streams, such as domestic kitchen and yard trim, now considered underutilized and readily accessible continuous resource streams, can also find new productive uses as biomaterials by aerobic fungal bioconversion rather than being dumped into the landfill or incinerated (Ezeorba et al. 2024). Through such processes, the use of fungi helps to abate greenhouse gas emissions, reliance on fossil fuel-based plastics, and microplastic pollution across the entire consumer products supply chain lifecycle (Carmona Marques et al. 2025).

Additionally, novel techniques like fungal cocultures, which refer to the controlled co-culture of two or more fungal strains, have been shown to be synergistic in improving the efficiency of lignocellulosic biomass degradation (Vieira et al. 2025). These systems capitalize on fungal enzyme and metabolism traits in a way that achieves a degradation efficiency even better than monoculture systems (Vieira et al. 2025). These systems are modeled upon natural biodegradation environments and are useful in advancing effective self-feeding bioreactors that could be scaled up.

Despite the above benefits, there remain obstacles in the scale-up of fungal bioprocessing, for instance, the complex cell wall structure in fungi, which is a hindrance to the extraction process of the final product; the variability in the nature of the wastes; and the cost associated with the maintenance of sterility and energy inputs (Liu et al. 2025; Ost et al. 2025). However, the integration of artificial intelligence, omics, and bioprocessing is propelling the advancement of strains and their fermentation, solidifying the relevance of the use of fungi in the pursuit of sustainability targets at the global level, including the United Nations Sustainable Development Goals, specifically goals associated with responsible consumption, action on climate change, and life on land (Garg 2025; Carmona Marques et al. 2025).

11.2.2 Physicochemical Properties Relevant to Fungal Fermentation

Physicochemical characteristics of organic waste biomass are important prerequisites for successful fermentation processes by fungi (Wang et al. 2023). These key factors include chemical composition, water content, size, and nutrient ratios, especially the C:N ratio (Li et al. 2016; Zadražil and Brunnert 1981; Wankhede et al. 2024). Lignocellulose, like agricultural and forestry by-products, is abundant, although its intrinsic resistance, offered by biogenic materials like lignin, cellulose, and hemicelluloses, might demand preprocessing to increase susceptibility to enzymes (Taherzadeh and Karimi 2008; Reginatto and Antônio 2015; Dhiman et al. 2024; Dessie et al. 2023). Agricultural and food wastes, however, have high water and sugar concentrations, making them easily fermentable, although they are also prone to contamination and decomposition (Escaramboni et al. 2021; Adhikari et al. 2012; Fisgativa et al. 2017). Municipal wastes are highly variable in their physicochemical composition, depending upon the pattern and extent of consumption in different regions (Kojima et al. 2021; Ping et al. 2012; Moretto et al. 2019).

The chemical composition of the waste biomass is of primary importance, as it influences the availability of carbon sources and other necessary nutrients to support fungal growth and metabolism (Wang et al. 2023; Wankhede et al. 2024). Lignocellulosic biomass, containing cellulose, hemicellulose, and lignin, offers a challenging resource to fungal biodegradation (Taherzadeh and Karimi 2008; Parvathy Eswari et al. 2023; Ramachandraiah et al. 2021). Lignin serves as a physical barrier, suppressing enzyme access to cellulose and hemicellulose (Parvathy Eswari et al. 2023). White-rot fungi show a strong capacity to degrade lignin by employing oxidative enzymes such as ligninases, manganese peroxidase, and lignin peroxidase, whereas brown-rot fungi employ an enzyme-less approach using hydroxyl radicals to alter lignin and make polysaccharides available (Parvathy Eswari et al. 2023).

Pretreatment approaches are generally used, and these enhance the lignocellulosic structure, thereby making it more fermentable (Taherzadeh and Karimi 2008; Reginatto and Antônio 2015; Basak et al. 2020). Biological approaches, which may use *Aspergillus fumigatus*, are cheap and eco-friendly, which makes them ideal for improving the use of polysaccharides (Basak et al. 2020). Using the fungal approach, it was demonstrated that the production of soluble sugars improves, which can easily undergo fermentation (Basak et al. 2020).

Moisture content is another major consideration in solid-state fermentation (SSF), since it influences the cultivation of fungi, sporulation, and the production of various end products, along with the transfer of heat and mass in the substrates (Zadražil and Brunnert 1981; Vauris et al. 2022). While the required level of moisture during fermentation depends on the chosen fungal species and the substrates being used, a lower level of moisture is considered insufficient for enhanced metabolic processes, while a higher level of moisture causes a lower porosity level, reduced aeration capacity, and the possibility of bacterial contamination (Wang et al. 2023). For example, the presence of *Rhizopus azygosporus* UICC 539 has proved the success of solid-state fermentation of sterile slurry/palm kernel cake substrates for animal feed preparation (Arsy et al. 2020).

Biomass particle size will affect the surface area, facilitating fungal growth and enzyme production (Vauris et al. 2022; Zhang et al. 2024; Hotz et al. 2023). A smaller particle size results in a higher surface area, which potentially increases nutrient uptake and enzyme production, though particles that are too fine cause compaction and thereby affect aeration (Vauris et al. 2022). The morphology of fungal hyphae, which form the filamentous structures comprising the fungal mycelia, is further influenced by the nature of biomass supports and their manipulation for optimal fermentation (Hotz et al. 2023; Novy et al. 2021).

Nutrient balancing, especially C:N ratios, is essential for proper fungal metabolism (Li et al. 2016; Melanouri et al. 2024). Fungi require a balanced ratio of carbon sources, necessary for energy production and growth, and nitrogen sources, necessary for protein and nucleic acid biosynthesis (Wang et al. 2023; Li et al. 2016). When C:N ratios become imbalanced, it can restrict fungal growth and enzyme production (Li et al. 2016). This was observed when endophytic fungal strains were optimized through fermentation media composition using single-factor optimization and orthogonal experimental design (Li et al. 2016). Endophytes from resilient plants possess assets like enzymes and bioactives (e.g., exopolysaccharides), as well as propensities to survive in harsh climates and possibly in similar adversities in industrial processes (Kaur et al. 2022; Kaur et al. 2023; Dhar et al. 2024).

Food waste, a potential resource used as a substrate for fungal fermentation and biosynthesis of enzymes, typically contains carbohydrates and proteins in a complex form; therefore, proper nutrient addition is necessary for its utilization by fermentation (Escaramboni et al. 2021). For example, the growth of *Rhizopus oligosporus* with a combination of food waste, sugarcane bagasse, wheat bran, and corn steep liquor in solid-state fermentation results in high levels of enzyme production (Escaramboni et al. 2021).

The presence of inhibitory substances like phenols, heavy metals, and detergents can cause significant inhibition of fungal growth and enzymatic activity (Sala et al. 2021). These substances are of great significance and are commonly found in municipal and industrial wastes (Kover et al. 2021; Saba et al. 2023). The effect of operating pH and hydraulic retention times on fungal growth and chemical oxygen demand removal was studied on a *Rhizopus microsporus* mold culture of corn-ethanol wastewater under different pH and hydraulic retention times (Jasti et al. 2009). On the other hand, wastewater from bulgur food industries has been treated with fungal species, including *Aspergillus parasiticus,* aimed at reducing chemical oxygen demand, proteins, and carbohydrates (Arikan et al. 2020). Agro-industrial wastewater, including rice husk, apple pomace, and orange skin, represents a potential substrate suitable for fungal biopesticide production, despite possible inhibitory substances (Sala et al. 2021).

These physicochemical properties can be optimized, usually by the blending of different waste types or by the appropriate pretreatment, to improve the accessibility and efficiency of fermentation processes (Wang et al. 2023; Wankhede et al. 2024; Taherzadeh and Karimi 2008; Kover et al. 2021). The co-fermentation of the mushroom agricultural waste substrate and the food waste in a 1:1 ratio, for example, has been found to optimize the production of anaerobic fermentation gases (Gu et al. 2022). The blending techniques allow the modulation of the C:N ratio and the compensation for the inhibitory influences, whereas the pretreatment achieves the physical or chemical alteration of the biomass to make it more accessible to fungi (Taherzadeh and Karimi 2008; Teo et al. 2024). The solid-state fermentation process for the production of cellulases, for instance, may require the modulation of factors such as the initial pH values, the initial moisture, and the inoculum size (Teo et al. 2024; Yoon et al. 2014). The conversion and utilization of the waste into value-added products by fungal fermentation, therefore, holds potential within the framework of a bioeconomy because it deals with the problems concerning the waste and resource recovery (Pilafidis et al. 2022; Kover et al. 2021; Anne et al. 2024; Ekloh and Yafetto 2024).

11.2.3 Regional Biomass Distribution and Utilization Trends

Biomass distribution and use are highly impacted by regional geographic and socio-economic patterns, which vary across continents (Calvert and Mabee 2014). Continental regions, for example, are dominated by agricultural residues, including rice husks and sugarcane bagasse (Shukla et al. 2023; Anu et al. 2023). Forestry residues and paper and municipal wastes, on the other hand, are produced predominantly in the European regions, while forestry and food processing wastes, including corn stover and food residues, are produced at high levels in the regions located in the continent of North America (Calvert and Mabee 2014). Regions that are considered developing suffer many constraints regarding the use of biomass, which can lead to open-burning practices or natural deterioration, thereby raising the levels of greenhouse gas emissions (Anu et al. 2023). On the other hand, developed countries are adopting a resource life cycle approach, with competing demands between bioenergy and the production of biopolymers favorably addressed based on sound policies (Ji et al. 2020; Wang and Tester 2023). Knowledge regarding regional biomass distribution is vital for establishing value chains, especially with the rising component involving the production of fungal biopolymers (Calvert and Mabee 2014).

Effective valorization of the waste biomass demands strong pretreatment methods to cope with the inherent recalcitrant nature of the biomass, which consists of cellulose, hemicellulose, and lignin (Lorenci Woiciechowski et al. 2020; Wu et al. 2022). Biomass with a high lignin content has a complex and cross-linked structure.

This structure also leads to a strong hindrance in the accessibility of the fermentable sugars (Wu et al. 2022; Ramos et al. 2020).

11.2.4 Pre-treatment Strategies for Enhanced Substrate Accessibility

As it is, the accessibility of the cellulose to the enzymes is reduced by the presence of lignin and hemicellulose. The purpose of pretreatment is to overcome the protective effect of biomass, thereby making cellulose more available to the enzymes and the biofuel production more favorable. There exist physical, chemical, or biological methods for the pretreatment, with recent approaches combining these methods for better productivity (Raina et al. 2024; Hassan et al. 2018; Li et al. 2022). These methods are briefly explained below. Physical pretreatment processes involve milling, extrusion, and steam explosion. The principal emphasis of these processes has been on the reduction of particle size and the augmentation of the specific surface area of the biomass. For example, hydrothermal pretreatment processes like the use of diluted acid steaming have been seen to raise the hydrogen production from *Chlorella pyrenoidosa* biomass substantially (Tan et al. 2021; Zhou et al. 2023; Xia et al. 2013). Autoclaving has also been considered a physical pretreatment method that has been investigated for the bio-hydrogen production capability of food waste (Bhurat et al. 2023).

Chemical pretreatment requires the action of chemicals, either in the form of dilute acid, alkaline, or organosolv, to dissolve hemicellulose and lignin (Lorenci Woiciechowski et al. 2020; Wan Azelee et al. 2023; Li et al. 2023a). Acid pretreatment, specifically with sulfuric acid, has been proven to show significant synergy when combined with hydrothermal pretreatment for biogas production from sugarcane bagasse substrate (Ahmad et al. 2021). Alkaline peroxide and organosolv approaches are able to remove lignin efficiently, thereby improving the susceptibility of biomass to enzymatic hydrolysis (Li et al. 2023a; Sharma and Sharma 2024). Organosolv processes, for example, showed the highest lignin removal of about 30% for the agricultural mix and 40% for the forestry mix while maintaining high holocellulose content (Sharma and Sharma 2024).

Biological pretreatments are based on the use of ligninolytic fungal cultures or enzyme blends, thereby providing eco-friendly alternatives that require lower energy and do not produce inhibitors in large amounts (Anu et al. 2023; Wan and Li 2012; Ummalyma et al. 2019). White-rot fungi, such as *Trametes versicolor* and *Phanerochaete chrysosporium*, are widely reported for lignin degradation, thereby increasing the accessibility of cellulose for hydrolytic enzyme-mediated saccharification, followed by fermentation steps (Wan and Li 2012; Ding et al. 2019; Sindhu et al. 2016). The lignin degradation mechanism is based on the fungal enzymatic system that depolymerizes lignin, which contains phenylpropanoid units (Javaid et al. 2019). This approach, which is eco-friendly and economically more favorable, usually takes a longer period than the latter methods (Anu et al. 2023; Li et al. 2022). Fungus-mediated pretreatment using *Aspergillus fumigatus* has been developed and optimized for raising the digestibility of polysaccharidic residues for biohythane production purposes (Basak et al. 2020). Moreover, biological pretreatment using lipid-producing fungi with mild acid treatment has also been successfully utilized on palm empty fruit bunches for the production of sugars, fungal oil, or bioethanol (Cheirsilp et al. 2023). Additionally, anaerobic fungi like *Orpinomyces* sp. strain C1A are also highly effective lignocellulosic biomass degraders, and the effects could be increased by hydrothermolysis pretreatment (Liggenstoffer et al. 2014). Mushroom farming, together with integrated pretreatment systems for ethanol production, could well set out the biorefinery pilot plants where the biotechnologies are fully eco-friendly and reliable (Ramamoorthy et al. 2023).

The aim of such pretreatment methods ultimately revolves around, besides improving accessibility of the substrate through modification of biomass and decreasing cellulose crystallinity, suppressing inhibitory compound formation, which may hinder fungal fermentation and subsequent production of biopolymers (Abraham et al. 2020; Silva Rabelo et al. 2023; Ji et al. 2022). Biocoordinated physicochemical methods of pretreatment aim at enhancing the enzymatic hydrolysis of lignocellulosic materials (Gu et al. 2022). The choice of suitable pretreatment techniques is largely influenced by the biomass being processed, the product of choice, and economic feasibility (Hassan et al. 2018; Abraham et al. 2020; K and Selvasembian 2024). Consolidated techniques, which are based on the combination of strengths possessed by other methods, are being used in modern biorefineries to strike optimal balances based on efficiency, economy, and sustainability considerations (Raina et al. 2024; Hassan et al. 2018; Ji et al. 2022). The trend is clear, as the current global focus is being steered toward sustainable use of biomass resources for the production of decentralized power and mitigation of global warming in agricultural nations where agricultural biomass is abundantly available, such as in India, thereby emphasizing the importance of prior knowledge of the distribution and competing demand for biomass resources (Anu et al. 2023). The treatment processing routes can be further classified post-pretreatment biomass conversion processes into thermochemical conversions (combustion, gasification, and pyrolysis) and biochemical conversions (anaerobic digestion and fermentation). This also highlights how there are numerous routes by which biomass can be converted, thereby justifying the need for effective pretreatment, depending upon the mode of conversion.

11.3 Fungal Systems for Biopolymer Production

Fungal systems have emerged with better adaptation and capabilities in terms of biopolymer production, providing sustainable options for petrochemical plastics (Luft et al. 2020; Shafiq et al. 2025; Nicolescu et al. 2023). Filamentous fungi and yeast-like fungi have their distinct characteristics and abilities in terms of metabolic pathways for biosynthetic production of various biopolymers, such as PHAs, PLA, chitosan, β-glucans, and melanins (Shafiq et al. 2025; Paul et al. 2020; Nandhini et al. 2022).

11.3.1 Filamentous Fungi and Yeast-Like Organisms: Classification and Characteristics

Fungi are a wide range of eukaryotes with microbial characteristics and unique morphology. Filamentous fungi, known for hyphal growth, form an important component in the bioeconomy owing to their importance in fermentation processes and the production of several enzymes, organic acids, and bioactive compounds (Ost et al. 2025). Their cell walls are made up of a hard inner component of chitin and beta-1,3–/beta-1,6-glucans, with a softer gel-like substance made up of polysaccharides and glycoproteins in the outer portion (Ost et al. 2025). This sophisticated structure is fundamental to their growth and development processes, also playing a pivotal part in biotechnological uses (Ost et al. 2025). For example, in *Aspergillus nidulans*, the existence of two genes, *agsA* and *agsB*, encodes α-1,3-glucan synthase, with *agsB* as the predominant α-1,3-glucan synthase enzyme supporting vegetative hypha development before aggregation occurs in mycelia (Miyazawa et al. 2018). Mycelial growth morphology in filamentous fungi can also be manipulated to increase the production levels of the metabolite through submerged fermentation processes with chitin biosynthesis disruption, altered cell growth, and mycelia development (Gong et al. 2023).

The yeast-like fungus *Aureobasidium pullulans* is a well-documented producer of biopolymers like pullulan, a key natural polymer that possesses non-toxic, non-immunogenic, non-carcinogenic, and non-mutagenic activities (Singh et al. 2017). The yeast-like fungus *Papiliotrema laurentii* is an oleaginous yeast, requiring the ability to accumulate high lipid concentrations, and has potential applications as a feedstock for the production of biofuels based on fatty acids (Moreira Vieira et al. 2020). Generally, oleaginous yeasts begin lipid accumulation during the late exponential growth phase, with the maximum lipid concentration achieved after 48 hours of growth (Moreira Vieira et al. 2020).

11.3.2 Biologically Important Bioplastics: PHAs, PLA, Chitosan

Fungal systems are highly productive in the synthesis of many valuable biopolymers, a few of which have been described here:

Polyhydroxyalkanoates (PHAs) PHAs are a group of biodegradable polyesters, produced through the fermentation process by several types of microorganisms, acting as a carbon and energy storage compound within the cells (Paul et al. 2020;

Kamravamanesh et al. 2018). PHAs offer a good alternative for general-use plastics because they possess the same material properties, compatibility, and complete biodegradability (Kamravamanesh et al. 2018; Kumari et al. 2022). Although biosynthesis is a common process among non-pathogenic bacteria, emerging technologies in synthetic biology target the enhancement of microbial production for the so-called "green" plastics, like PHAs (Anderson et al. 2018). The costly production of carbon sources for bacterial fermentation is still a challenge for PHA production (Kamravamanesh et al. 2018).

Polylactic Acid (PLA) PLA belongs to a group of bioplastics that are thermoplastic polyesters and are manufactured from renewable sources such as lactic acid building blocks; these can be fermented from sugars found in starch or sugar-containing plants (Nandhini et al. 2022; Phiri et al. 2023). Polylactic acid has applications in packaging materials, 3D printer filament, and healthcare because of its biodegradable nature and mechanical properties (Phiri et al. 2023). The production level of bioplastics such as polylactic acid was estimated to be 2.43 million tons in 2024 (Nandhini et al. 2022). The life cycle of PLA becomes even more relevant in the present scenario of biocycling. In the biocycling process, raw materials such as harvested materials are processed in a way that extracts starch from these materials. The extracted starch is further refined into a chemical composition so that bioplastics are manufactured. Once the bioplastic materials are used, they are further composted and given back in the form of mulch (Tran et al. 2024).

Chitosan It is a biopolymer that is present in the cell walls of fungal organisms, as well as the exoskeletons of insects and the shells of crustaceans (Abo Elsoud and El Kady 2019). Chitosan from fungal sources has been seen to be more preferable compared to sources obtained from crustaceans, owing to its non-seasonal availability, lack of allergens, and better control over the production process (Abo Elsoud and El Kady 2019). Chitosan has a wide range of applications, from biomedical to food and agriculture, owing to properties like biocompatibility, biodegradability, high antibacterial properties, and being non-toxic in nature (Abo Elsoud and El Kady 2019; Mousavi 2021; Christina et al. 2024). A range of species belonging to the category of Basidiomycota have been studied to conduct mycelial biomass harvest as well as extraction of chitosan from fungal sources by means of submerged and solid-state fermentation processes, respectively, and some of them include *Heterobasidion annosum*, *Phanerochaete chrysosporium*, and *Pleurotus ostreatus* (Irbe et al. 2023).

β-Glucans These are crucial polysaccharides in the cell walls of filamentous fungi (Ost et al. 2025; Anchundia et al. 2025). Fungal β-glucans are known for their medicinal uses and bioactivities, such as antioxidant, antimicrobial, anti-inflammatory, antiviral, antidiabetic, and anticoagulatory activities, which make them useful in drug development (Anchundia et al. 2025; Stoica et al. 2023). Specifically, the β-glucan of the fungus *Rhizopus oryzae* M10A1 has been investigated for optimal production in liquid potato starch waste (Anchundia et al. 2025).

Another example is the exobiopolymer (1 → 3)-β-D-glucan of *Ophiocordyceps dipterigena* BCC 2073, which is potentially used as an inducer of interleukin-8 synthesis (Kocharin et al. 2010). Another significant β-glucan is schizophyllan, whose production has been characterized using sago starch as a carbon source (Gunaji et al. 2020).

Melanin One of the important secondary metabolites produced by fungi is melanin, which is involved significantly in the adaptation of fungi to extreme stimuli from their environment, including conditions of extreme radiation (Lim et al. 2021; Ashok et al. 2021). Melanin is normally produced through a polyketide synthesis (PKS) gene cluster in certain filamentous fungi (Lim et al. 2021). This particular melanin-producing fungus, *Paecilomyces variotii,* isolated from Chornobyl, has been examined for its role in making them resistant to UV (Lim et al. 2021). *Auricularia auricula* is another significant fungus in which melanin is produced, where its production occurs due to the oxidation and polymerization of phenolic compounds (Wang et al. 2024a).

11.3.3 Biosynthetic Pathways

Fungi display highly developed biosynthetic pathways, which are pivotal to the fungal production of biopolymers. Enzymes are integral to these pathways, and the concepts of enzyme promiscuity, enzyme engineering, enzyme immobilization, or cell-free systems are essential for the development of bioprocesses (Katsimpouras and Stephanopoulos 2021). Synthetic biology is the key to controlling such metabolic pathways, which can be altered for microbial biopolymer production (Khan et al. 2024; Mattern et al. 2015; Vatsa et al. 2020). Engineered cells, like yeast and bacteria, are being developed with the ability to detect biomolecules and create biotherapeutics, demonstrating the controllability of the internal processes of these cells (Zhang and Fussenegger 2024). The biopolymers produced from fungi, like chitin and β-glucans, require a multitude of steps with the use of enzymatic components like chitin synthases and glucan synthases (Ost et al. 2025; Gong et al. 2023). Melanin production in fungi involves a polyketide pathway where the starting materials acetyl-CoA and malonyl-CoA are converted into phenolic compounds that are further processed into a polymeric structure (Wang et al. 2024b).

11.3.4 Native vs. Engineered Fungal Strains

Both native and biologically developed fungal strains are involved in biopolymer synthesis with various benefits. In native fungi, the main aim may be the ability to secrete a certain biopolymer. For instance, Basidiomycota species have a native capability to secrete chitosan (Irbe et al. 2023), and *Rhizopus oryzae* M10A1 secretes β-glucans (Anchundia et al. 2025). *Paecilomyces variotii* has the native capability to secrete melanin to protect against UV rays (Lim et al. 2021), while *Auricularia auricula* has the capability to secrete melanin (Wang et al. 2024b). However, the native capability to secrete biopolymers may not be sufficient at an industrial scale. Engineered fungal strains possess improved production abilities through metabolic engineering and synthetic biology (Anderson et al. 2018; Khan et al. 2024; Mattern et al. 2015; Zhang and Fussenegger 2024). Synthetic biology allows regulation of various regulatory components, biosynthetic pathways, and multi-domain biosynthetic enzymes such as polyketide synthases (Mattern et al. 2015; Abdi et al. 2024a). It provides a means of enhancing existing biosynthetic pathways and creating new ones, which contribute to improving the yield, purity, and specific characteristics of biopolymers (Anderson et al. 2018; Khan et al. 2024). Designing genetically modified fungal strains may result in improved production of biopolymers and their designs through control of cell factories (Khan et al. 2024). The process mainly involves strict control of metabolic pathways to generate biopolymers with desired properties such as strength, malleability, and biodegradability (Khan et al. 2024). Although synthetic biology has great potential, little information exists in the scientific field about applying these pathways to generate selectively designed biopolymers, especially for optimal food contact materials, mainly because of deployment restrictions related to polymers produced by potentially pathogenic strains (Khan et al. 2024). The need for engineered fungal strains in advancing microbial production of biopolymers as a safe substitute for plastics has great field potential in the coming years (Luft et al. 2020; Shafiq et al. 2025).

11.4 Biotechnological Advancements

Fungal biotechnology is an area of rapidly progressing development with great potential for the production of biopolymers, supported by advanced biotechnological approaches and methods (Luft et al. 2020; Subramanian et al. 2025; Roth et al. 2023; Villena et al. 2020). The combined use of metabolic engineering, synthetic biology, advanced gene editing methodologies such as CRISPR-Cas9 and RNA interference, multi-omics analysis, and bioreactor/fermentation technology will play pivotal roles in improving the productivity and biotechnological sustainability of biopolymer production with fungi (Garg 2025; Wang et al. 2023; Wu et al. 2025; Tiwari and Dufossé 2023; Medeleanu et al. 2025) (Table 11.2).

11.4.1 Metabolic Engineering and Synthetic Biology Tools

Metabolic engineering entails the directed manipulation of metabolic pathways in fungal hosts with the aim of enhancing the production, purity, and functionality of target biopolymers (Medeleanu et al. 2025). Metabolic engineering has harnessed

Table 11.2 Biotechnological tools for enhancing fungal biopolymer production

Technique/ tool	Mechanism/ approach	Example fungal host	Outcome/ advantage	Key references
CRISPR-Cas9 editing	Targeted gene knock-in/ knock-out	*Aspergillus niger, Pseudomonas putida*	Higher yield of PHAs, improved substrate tolerance	Arab et al. (2025); Zhu (2026); Hu et al. (2026); Jin et al. (2022)
RNAi	Silencing competing metabolic pathways	*Rhizopus oryzae*	Improved substrate utilization, reduced byproducts, and other purposes	Morales-Barron et al. (2025); Meussen et al. (2012); Shirazi et al. (2015)
Adaptive laboratory evolution (ALE)	Long-term selection under stress	*Yarrowia lipolytica*	Enhanced tolerance to inhibitors, robustness in fermentation	Xia et al. (2024); Sun et al. (2025); Sha et al. (2023); Chen et al. (2024)
Omics integration	Genomics, transcriptomics, proteomics	Multiple fungal strains	Systems-level optimization, pathway discovery	Wijayawardene et al. (2023); Li et al. (2023b)

the natural metabolic diversity and versatility of fungi, which display the capability to utilize diverse materials such as organic biomass and waste materials and to produce various useful products (Subramanian et al. 2025; Villena et al. 2020). For example, filamentous fungi display a remarkable capacity to produce various biopolymers, including enzymes, organic acids, lipids, and others (Subramanian et al. 2025; Giavasis et al. 2019; Adrio and Demain 2003). Metabolic engineering entails optimization of existing metabolic pathways and the construction of new ones to direct metabolic flux for the production of specific biopolymers (Medeleanu et al. 2025). Metabolic engineering entails understanding metabolic limitations and selection of new metabolic routes independent of existing routes, which include optimization of key genes of important enzymes through overexpression (Katsimpouras and Stephanopoulos 2021; Park et al. 2021). For example, flavone aglycone production enhancement in *Scutellaria baicalensis* hairy root cultures was realized through metabolic engineering, which entailed upregulation of gene expression of flavonoid biosynthetic pathways using maize transcription factor Lc and *Arabidopsis* PAP1 (Park et al. 2021).

Synthetic biologists offer a valuable ally in this effort, offered by metabolic engineers through the intentional design and construction of biological systems or by redesigning existing systems for novel uses (Khan et al. 2024; Wu et al. 2025; Tiwari and Dufossé 2023). From a biopolymer perspective, biologists engage in activities that offer gene expression control, regulation of pathways in biopolymer-producing organisms or microbes, or implementation of a synthetic gene circuit in biopolymer producers (Khan et al. 2024; Tiwari and Dufossé 2023). For example, biologists play a critical role in improving the productivity and biopolymer quality related to proteins in yeast and filamentous fungi biomass because this microbial protein has far-reaching applications (Wu et al. 2025). Through biopolymer manipulation related to microbes in biodegradable biopolymer synthesis or modification, biologists can specifically tailor their biopolymer properties in a predetermined fashion regarding properties like biodegradability or functional groups (Khan et al. 2024; Medeleanu et al. 2025).

11.4.2 CRISPR-Cas9, RNAi, and Adaptive Laboratory Evolution

The CRISPR-Cas9 gene editing system has transformed the field of fungal biotechnology because it has given a precise way of gene modification (Pal et al. 2024; Pant et al. 2022; Mondal et al. 2021). It has the ability to carry out gene knockouts, insertions, and precise base modifications. This has paved the way for precise gene modification of fungi (Morio et al. 2020; Woodcraft et al. 2023; El-Sayed et al. 2017). In biopolymer production, the use of CRISPR-Cas9 gene editing has the ability to carry out modifications such as improving production, modifying composition, improving stress tolerance, and reducing the production of undesirable biopolymer compounds (Pal et al. 2024; Woodcraft et al. 2023). For example, the use of the CRISPR-Cas9 gene editing tool has the ability to introduce precision in improving the cellulolytic ability and biopolymer production of cellulase in fungi, specifically *Trichoderma reesei* cells, using gene modification approaches targeting genes responsible for the regulation of cellulase activity (Pal et al. 2024; Pant et al. 2022; Mondal et al. 2021; Singh et al. 2024). The use of CRISPR-Cas9 multiplexing has the ability to modify multiple genes or alleles. It has the ability to introduce precision into pathway engineering (Borah et al. 2024). RNA interference (RNAi) is yet another method that has great potential in the regulation of genes, using the specific silencing of gene expression (Stavridou et al. 2024). This would specifically be helpful in the silencing of genes that are involved in competitive metabolic pathways or genes that code enzymes that degrade the target biopolymer (Stavridou et al. 2024). Unlike the use of CRISPR/Cas9, which targets the DNA level, RNAi is an alternative method that can specifically regulate genes post-transcriptionally, with the ability to temporarily or dose-dependently suppress genes.

Adaptive laboratory evolution (ALE) is a non-GMO technique that consists of the growth of microorganisms under selective pressure to evolve new strains that display the desired phenotype. Such a technique is able to provide drastic increases in biopolymer production, stress resistance, and substrate uptake without the direct implementation of genetic engineering (Stavridou et al. 2024). ALE is often combined with genetic engineering to optimize existing strains and identify new genes that control favorable traits. To illustrate the integration of fungal enzymatic degradation, synthetic biology tools, and circular bioeconomy principles, Fig. 11.1 presents a schematic of biopolymer production pathways from agro-industrial waste.

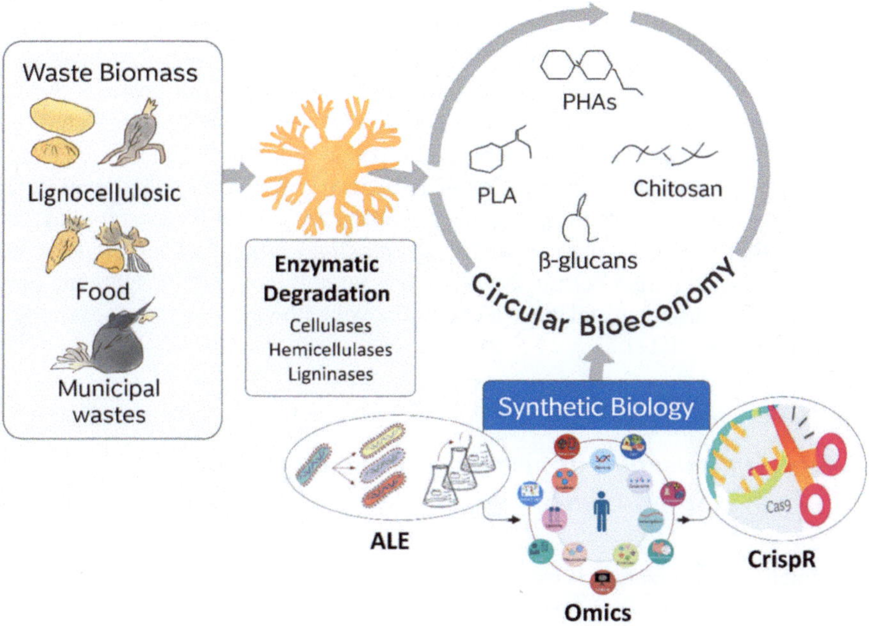

Fig. 11.1 Schematic representation of fungal biopolymer production pathways from agro-industrial waste. Waste biomass sources—including lignocellulosic, food, and municipal waste—are enzymatically degraded by fungi using cellulases, hemicellulases, and ligninases. Resulting metabolites are converted into biopolymers such as polyhydroxyalkanoates (PHAs), polylactic acid (PLA), chitosan, and β-glucans. Synthetic biology tools, including CRISPR, adaptive laboratory evolution (ALE), and omics integration, enhance yield and substrate tolerance. The circular bioeconomy framework emphasizes sustainable material recovery and closed-loop production

11.4.3 Omics Integration: Genomics, Transcriptomics, Proteomics

The integration of multi-omics technology, such as genomics, transcriptomics, and proteomics, helps create a holistic understanding of *mycota* and their metabolism (Sanches et al. 2024; Xie et al. 2025a; Abdi et al. 2024b). Such holistic knowledge is the key to designing strains and searching for biopolymer optimization targets. Edible fungi genomics aids in the development of superior fungal species by identifying genes responsible for growth, development, or the production of premium compounds (Xie et al. 2025a, b). Transcriptomics is the study of the entire set of transcripts in a cell to reveal gene expression profiles (Zhang et al. 2021). By comparing the transcriptome of the cell under various conditions (such as those for biopolymer production), one can identify genes that are upregulated or downregulated, thereby suggesting their role in the biopolymer synthesis pathway. For example, transcriptomics has been applied to evaluate possible pathways for salt resistance in pecans based on NaCl concentrations (Zhang et al. 2021). Proteomics involves the study of the total proteins produced in the organism (Sanches et al. 2024; Xie et al. 2025a, b). Proteins are considered the workhorses of cells, and proteomics could provide information about enzyme and structural proteins that are directly involved in the synthesis of biopolymers. Protein expression and modification could provide mechanisms to optimize enzyme function.

Integrating these omics datasets helps in understanding the system-level fungal metabolism, thereby permitting the identification of crucial genes, proteins, and pathways that could be engineered for the maximal production of biopolymers. These analyses can also permit the predictive modeling of the biosynthetic pathways of natural products and the selection of novel pathways or genes for modification (Borah et al. 2024).

11.4.4 Optimization of Fermentation: Comparison of SSF and SmF

Fermentation optimization is essential for the economic-scale production of fungal biopolymers. Two main processes are commonly practiced, namely, Solid-State Fermentation (SSF) and Submerged Fermentation (SmF) (Mascarin et al. 2024; Bustos-Vázquez et al. 2025). The growth of microorganisms on solid supports with minimal to no free water present is called SSF (Bustos-Vázquez et al. 2025; Eras-Muñoz et al. 2024). SSF is attractive due to higher volumetric productivity rates, lower energy demands, reduced water use, and the capacity to ferment agro-industrial residues as substrate, aiming to implement the circular economy mindset (Chai et al. 2022; Gómez-Ramos et al. 2024). SSF is more relevant to fungal species because of their filamentous growth pattern that easily spans and colonizes substrates without agglomeration (Subramanian et al. 2025; Eras-Muñoz et al. 2024). Recently reported developments in SSF have been practical and attractive for biosurfactant production to surpass problems linked to the widespread use of antifoam agents, an obstacle often present in SmF fermentation (Bustos-Vázquez et al. 2025; Eras-Muñoz et al. 2024). SSF reactor designs have gained substantial importance to help promote this technique by designing packed-bed reactors to increase efficiency and resolve all existing concerns and challenges (Gómez-Ramos et al. 2024; Barrios-Nolasco et al. 2024). Biomass yields from SSF have been reported to be substantial when employing specially designed custom reactors. Specifically, *Yarrowia lipolytica* 2.2ab has proved to be promising when grown within a so-designed packed-tray bioreactor. Submerged Fermentation (SmF): This is the process of culturing microorganisms in liquid media. SmF is generally preferred for the commercial production of various fungal compounds, such as enzymes, polysaccharides, and organic acids, because the submerged fermentation process can be easily controlled with respect to various process conditions, such as pH,

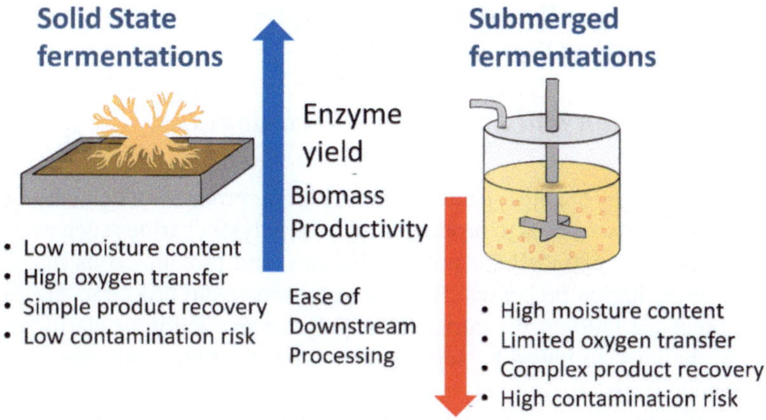

Fig. 11.2 Comparative schematic of solid-state fermentation (SSF) and submerged fermentation (SmF) systems used in fungal biopolymer production. SSF involves fungal growth on solid substrates under low moisture conditions, offering high oxygen transfer and reduced contamination risk. SmF utilizes liquid media in stirred bioreactors, enabling higher biomass productivity but requiring complex downstream processing and stringent sterility. Key differences in enzyme yield, scalability, and operational parameters are highlighted

temperature, aeration, and agitation rate, to result in large-scale production with higher commercial feasibility and applicability. Figure 11.2 compares solid-state fermentation (SSF) and submerged fermentation (SmF), highlighting their operational differences and implications for fungal biopolymer production.

Nonetheless, submerged fermentation can be energy-intensive and requires the use of an antifoaming agent, thus increasing the downstream processing cost. Recent improvements in submerged fermentation have been realized in the mass production of entomopathogenic fungi. It can be very effective as far as the commercialization and use of the fungi are concerned. Bioreactor design is highly significant for both SSF and SmF. It is used for optimal conditions for the growth of fungi and for the production of biopolymers (Mascarin et al. 2024; Gómez-Ramos et al. 2024). Optimum design of a bioreactor considers issues like the transfer of matter (oxygen, substrate), heat transfer, agitation, and monitoring. Bioreactors for SSF include design aspects for a homogeneous distribution of temperature and water, preventing the problem of channeling, and aeration for the optimal growth of fungi. Packed-bed and packed-tray bioreactors, characterized by a shorter packing height and wall-cooling jackets, are under development for SSF applications (Gómez-Ramos et al. 2024; Barrios-Nolasco et al. 2024). In the case of SmF, bioreactors include design parameters for effective agitation and aeration for a homogeneous mixture, which is a primary necessity for the optimal growth of a high density of fungi (Mascarin et al. 2024).

11.5 Downstream Processing Techniques for Polymer Characterization

11.5.1 Extraction and Purification Techniques

Fungal biotechnology enables sustainable biopolymer production that uses filamentous fungi and yeasts as tools to produce structural polysaccharides such as chitin or chitosan, β-glucans, extracellular polysaccharides, and proteinaceous materials in controlled fermentation processes. Biopolymer downstream processes are initiated with cell breakage tailored for intracellular polymer extraction or clarification for exopolysaccharides, followed by separation and purification conducted in an optimized manner suited for biopolymer yield, molecular structure, and scalability. Sequences of acid and alkali treatment are commonly employed for the extraction of chitin and chitosan from fungal mycelia, demineralization, deproteinization, and deacetylation, whereas ethanol or acetone precipitation is readily used for concentrating exopolysaccharides from fungal broth with minimal temperature impact. MF/UF membrane processes decrease solvent usage along with molecular weight fractionation, whereas ion-exchange/gel filtration provides polishing that meets biomedical-purity requirements, eliminating pigments, proteins, and endotoxins. These processes rely on the underlying chemistry of biopolymers such as acetylation, branching, or charge moieties, their final use, and regulatory constraints governing residues, which, in turn, are conditioned during fermentation control in relation to pH, oxygen transfer, and feeding practice to control composition, stability, and batch-to-batch consistency.

11.5.2 Physicochemical Characterization: FTIR, SEM, TGA, DSC

Physicochemical analysis is warranted for the confirmation of the identity, purity, and processing ability of fungal biopolymer materials. Fourier Transform InfraRed (FTIR) spectroscopy, Scanning Electron Microcrospy (SEM), Thermogravimetric Analysis (TGA), and Diffrential Scanning Calorimetry (DSC) analyses have come together to provide a set of complementary techniques. FTIR spectroscopy allows the mapping of functional groups and polymer backbone absorption for chitosan—the amide I/II absorption band and the broad O-H/N-H absorption band together with the shifts related to the degree of deacetylation. For β-glucans, the absorption is related to the presence of β-(1–3)/(1–6) bonds between sugars—for the biopolymer structure and function correlations. The surface topography of the biopolymer films is provided by the SEM analysis. The biopolymer thermal transitions and decomposition patterns are provided by TGA analysis. The glass transition temperature of the biopolymer films can be obtained from the DSC analysis. The integration of these methods with the biopolymer film rheology and gel permeation chromatography would make the biopolymer films more informative concerning weight average distributions of the biopolymer films, which play a critical role in the biopolymer film.

11.5.3 Mechanical and Biodegradation Properties

Biodegradation properties and mechanical strengths act as critical drivers in the usage of the polymer in packaging, medical, agricultural, and filtration materials. The tensile strength, stiffness, and resistance to elongation and fracture vary as a function of the level of acetylation, molecular weight, crystallinity, plasticizer content, and processing method; the strengths of fungal chitosan films have the same values as plant polysaccharide films when optimized in crystallinity and humidity, while the viscoelastic properties of β-glucans are well-suited to wound dressings and scaffolding in tissue engineering. Biodegradation is carried out via enzymatic and hydrolysis reactions, the rates of which vary as a function of the operating conditions (temperature and humidity level, presence of inoculum), microstructure, and crystallinity and have been standardized in composting, soil, and aquatic environments to measure mass loss, rates of CO_2 release, and mechanical properties at specific time points (composting, soil, aquatic test conditions, and so on). Biomechanical aspects, such as the lysozyme-accelerated hydrolysis of deacetylated chitin and the depolymerization of β-glucans by glucanases present in fungi, enable the customization of the material properties to have appropriate rates of degradation. Polymer mechanical properties must have an optimal rate of degradation to prevent premature failure and persistence in the use and end-of-life scenarios.

11.5.4 Standards and Quality Control for Industrial Applications

There is a need to maintain quality in industrial applications based on universally established criteria. The mechanical properties of thin films and sheets may be measured using ASTM D882 guidelines, whereas DMA studies can be done in addition to thermal profiling; decomposition in composting is measured using ASTM D6400 and ISO 14855 guidelines, whereas aquatic disintegration is covered by ISO 18830/19679. Biocompatibility studies of medical devices and healthcare applications may refer to ISO10993 guidelines (cytotoxicity, sensitization, irritation, and systemic toxicity), where sterilizable compatibility and residual analyses may be included in early stages of development (Sharma et al. 2021). In-process control of batch quality may be done through control of biosynthesis (concerning pH, DO levels, and substrate feeding of biosynthesis); purification processes (extraction and purification); and release testing of molecular weight, degree of acetylation (DA), residual protein/metallic, and pigment levels, in addition to endotoxin and microbial contamination levels. In biomedical applications, US Food and Drug Administration (FDA) and European Medicines Agency (EMA) regulations need thorough risk management strategies and may involve testing requirements; thus, storage stability may also be covered based on FDA development guidelines. In non-biomedical applications (packaging and agriculture), it is important to focus on FDA regulations regarding contact with foods, in addition to environmental regulations, whereas detailed specification sheets can be used in addition to ISO9001/13485 guidelines. These technologies provide a strong foundation for fungal biopolymers to serve as substitutes and complements to petroleum-based polymers on the basis of strong purification and analytical procedures, deliberate mechanical design, and quality systems in place. The blending of cross-linking material sciences and bioprocessing and knowledge of regulations facilitates the translation of laboratory-scale research to plant-scale use to achieve specific performance and end-of-life properties.

11.6 Challenges and Limitations

Fungal biotechnology presents some attractive solutions for the production of biopolymers through green technologies, but various challenges and constraints affect the successful use of the technique on a larger scale. The constraints range from substrate diversity to scale-up problems, through which solutions all need to be sought.

11.6.1 Substrate Variability and Inhibitory Compounds

Substrate choice affects fungal biopolymer yield and characteristics. Substrate sources such as agricultural residues, cellulosic material, and by-products from the food sector are desirable, but their heterogeneous nature brings variability to nutrient availability and inhibitory concentrations. Phenolic and furfural derivatives of lignin can be inhibitory to fungal metabolism, thereby affecting biopolymer formation. Substrate differences with respect to season and regions add to the problem of process standardization. Other process steps, like substrate treatment (alkali hydrolysis, enzymatic saccharification), can reduce inhibitors but are more complicated. Therefore, the trade-off between the affordability and consistency of the substrate is still a major bottleneck.

11.6.2 Scale-Up Bottlenecks and Economic Viability

Scalability from laboratory to pilot and then industrial scales entails technology and cost issues. Process requirements in fermentation engineering (oxygen transfer rates, shear rates, viscosity) increase in complexity from laboratory to industrial scales. Downstream processing (extraction, purification) in biotechnology is energy-intensive and therefore not very competitive with plastics derived from petroleum. Capital costs of specialized equipment and facilities remain significant, whereas selling prices of biopolymers are still above those of conventional plastics.

11.6.3 Regulatory and Policy Barriers

Industrial use of biopolymers from fungi involves adherence to very rigid regulatory requirements. Biomedical uses require compliance with ISO 10993 standards and FDA/EMA regulatory requirements that involve very extensive biocompatibility and safety testing. Food contact requirements involve compliance with migration standards and allergenicity evaluation. Regulatory ambiguity with regard to bioplastic classification (as biodegradable, compostable, or recyclable) can pose an obstacle to market positioning. Such regulatory elements create delays in commercialization and require active interaction with policymakers and standards-developing institutions.

11.6.4 Environmental Trade-Offs and Lifecycle Concerns

Even with the biodegradability of fungal-based biopolymers, it is important to carefully weigh environmental trade-offs. Fossil fuel-based downstream processing can counterbalance any sustainability advantages (Seetharaman et al. 2025). Moreover, waste-to-energy strategies in downstream processing could counter sustainability advantages gained through fungal-based biopolymers. Life cycle assessments (LCAs) show that fungal biopolymers decrease plastic waste in the environment but may increase GHG emissions when produced through fossil fuel-based energy. End-of-life processes (composting, anaerobic digestion, and soil burial) are not equal in efficiency and can be incompletely degraded to form microplastic-like materials. Land-use change associated with growing substrate can clash with food production. Hence, rigorous life cycle assessment is critical to guarantee real increases in environmental value.

The environmental benefits of fungal biopolymer products hinge upon the ability of the biopolymer compounds and products derived from fungi to have positive environmental benefits. However, the production of fungal biopolymer compounds and products is faced with various challenges. For the products and compounds to achieve environmental benefits, the challenges must be approached from different dimensions. Different strategies must be employed to address these challenges.

11.7 Future Perspectives

Fungal biotechnology for biopolymer production is on the threshold of a revolutionary period, whereby future outlooks emphasize integrative biorefinery paradigms, digital biotechnology, governance, and innovative applications of biopolymer research in material science for sustainable applications. A critical area of future outlook for anti-biopolymer resistance is, of course, integrative biorefinery, whereby fungal biopolymer production is incorporated into multi-product biorefinery platforms aimed at valorizing agricultural residues, lignocellulosic biomass, and food processing residues. Through the combined production of biopolymers, biofuels, organic acids, and enzymes, biorefineries have greatly aided in maximizing resource efficiency and economic returns, thereby diminishing waste generation. In addition to biorefinery applications, biotechnology for fungal biopolymer production is expected to benefit from artificial intelligence (AI) and machine learning (ML) for designing new strains and process optimization. Predictive software applications will be useful for designing novel fungal biopolymer-producing strains for maximizing biopolymer production, optimizing biopolymer pathways, and optimizing biopolymer production processes. Recommendations in policy and global collaborative efforts will be just as vital. There is a need for standardized global norms in terms of biodegradability, compostability, and biocompatibility to overcome regulatory diversities that slow down the pace of commercialization. Subsidization, carbon credits, and green purchases by governments will act as ways to encourage the use of fungal biopolymers in various nations, while global collaboration efforts by academia, industries, and governments will help in building equitable access to sustainable technologies. The vision of fungal biopolymers in sustainable material science is very broad and extends toward finding suitable, environmentally friendly alternatives to replace petrochemical plastics in packaging, textiles, biomedical devices, agriculture, and other areas because of their biodegradability and functional tunability, in accordance with principles of the circular economy. Further development in nanostructuring, composite design, and functionalization will extend their application in high-performance materials, connecting sustainability with innovation.

11.8 Conclusion

The importance of fungal biopolymer studies has highlighted crucial findings: downscaling and processing, characterization and profiling, and quality control are very important basics to work on. The strategic approach to research and development focuses on standardization, process intensification, artificial intelligence, and harmonization to ensure credibility in the environmental realm. Conclusion considerations recognize the crucial place of fungal biotechnology for sustainability: a renewable, biodegradable, and versatile material to transform industries and solve ecological issues. The intersection of scientific discovery and artificial intelligence will allow fungal biopolymers to translate from promising scientific entities in a lab to mainstream industrial applications, initiating a future where material sciences and sustainability meet.

References

Abdi G, Firdous S, Vasantharekha R, Singh GB, Seetharaman B, Mudgal G, Jain M, Tarighat MA (2024a) Progress on synthetic genomics. In: Singh V (ed) Advances in genomics: methods and applications. Springer Nature Singapore, Singapore, pp 181–197

Abdi G, Patil N, Mishra P, Tariq M, Dhariwal R, Parveen S, Dave K, Jain M, Tarighat MA, Mudgal G (2024b) Genomic precision: unveiling the transformative role of genome editing in advancing genomics research and applications. In: Singh V (ed) Advances in genomics: methods and applications. Springer Nature Singapore, Singapore, pp 265–306

Abo Elsoud MM, El Kady EM (2019) Current trends in fungal biosynthesis of chitin and chitosan. Bull Natl Res Cent 43. https://doi.org/10.1186/s42269-019-0105-y

Abraham A, Mathew AK, Park H, Choi O, Sindhu R, Parameswaran B, Pandey A, Park JH, Sang B-I (2020) RETRACTED: pretreatment strategies for enhanced biogas production from lignocellulosic biomass. Bioresour Technol 301:122725. https://doi.org/10.1016/j.biortech.2019.122725

Adhikari BK, Trémier A, Barrington S, Martinez J (2012) Biodegradability of municipal organic waste: a respirometric test. Waste Biomass Valorizat 4:331–340. https://doi.org/10.1007/s12649-012-9135-5

Adrio JL, Demain AL (2003) Fungal biotechnology. Int Microbiol 6:191–199. https://doi.org/10.1007/s10123-003-0133-0

Ahmad F, Silva V, Sakamoto IK, Sun J, Silva EL, Varesche MBA (2021) Bioprospecting sulfuric acid assisted hydrothermal pretreatment of sugarcane bagasse and microbial community structure for methane production. Bioenergy Res 15:631–649. https://doi.org/10.1007/s12155-021-10268-2

Alemu D, Tafesse M, Mondal AK (2022) Mycelium-based composite: the future sustainable biomaterial. Int J Biomater 2022:8401528. https://doi.org/10.1155/2022/8401528

Anchundia M, León-Revelo G, Santacruz S, Torres F (2025) Production of β-Glucans from Rhizopus oryzae M10A1 by optimizing culture conditions using liquid potato starch waste. Polymers 17:1283. https://doi.org/10.3390/polym17091283

Anderson LA, Islam MA, Prather KLJ (2018) Synthetic biology strategies for improving microbial synthesis of "green" biopolymers. J Biol Chem 293:5053–5061. https://doi.org/10.1074/jbc.tm117.000368

Anne T, Julia B, Rémy B, Sylvaine B, Mathilde B, Manon VB, Isabelle Z, Etienne P (2024) Bio-waste: a unique and valuable resource with significant recovery, prevention and management challenges. Waste Biomass Valorizat 16:2985–2998. https://doi.org/10.1007/s12649-024-02841-5

Anu, Alokika, Kuhad RC, Rapoport A, Kumar V, Singh D, Kumar V, Tiwari SK, Ahlawat S, Singh B (2023) Biological pretreatment of lignocellulosic biomass: an environment-benign and sustainable approach for conversion of solid waste into value-added products. Crit Rev Environ Sci Technol 54:771–796. https://doi.org/10.1080/10643389.2023.2277670

Arab B, Chen J, Khusnutdinova AN, Chou CP, Liu Y (2025) Advancing bio-recycling of nylon monomers through CRISPR-assisted engineering. Environ Technol Innovation 39:104267. https://doi.org/10.1016/j.eti.2025.104267

Arikan EB, Bouchareb EM, Dizge N (2020) Investigation of fungal treatment potential for bulgur cooking process wastewater. Bioresour Technol Rep 11:100468. https://doi.org/10.1016/j.biteb.2020.100468

Arsy SL, Oetari A, Sjamsuridzal W (2020) Solid-state fermentation of sterile slurry and palm kernel cake (PKC) mixture using Rhizopus azygosporus UICC 539. IOP Conf Ser: Earth Environ Sci 483:12029. https://doi.org/10.1088/1755-1315/483/1/012029

Ashok G, Mohan U, Boominathan M, Ravichandiran V, Viswanathan C, Senthilkumar V (2021) Natural pigments from filamentous fungi: production and applications. Fungal Biol:651–678. https://doi.org/10.1007/978-3-030-85603-8_19

Barrios-Nolasco A, Castillo-Araiza CO, Huerta-Ochoa S, Reyes-Arreozola MI, Buenrostro-Figueroa JJ, Prado-Barragán LA (2024) Evaluating the performance of Yarrowia lipolytica 2.2ab in solid-state fermentation under bench-scale conditions in a packed-tray bioreactor. Fermentation 10:344. https://doi.org/10.3390/fermentation10070344

Barta D-G, Simion I, Tiuc A-E, Vasile O (2024) Mycelium-based composites as a sustainable solution for waste management and circular economy. Materials (Basel) 17:404

Basak B, Saha S, Chatterjee PK, Ganguly A, Woong Chang S, Jeon B-H (2020) Pretreatment of polysaccharidic wastes with cellulolytic Aspergillus fumigatus for enhanced production of biohythane in a dual-stage process. Bioresour Technol 299:122592. https://doi.org/10.1016/j.biortech.2019.122592

Bhurat KS, Banerjee T, Bobde PV, Sureshchandra Bhurat S (2023) Effect of chemical, physical, and biological pre-treatment of food wastes on bio-hydrogen production by dark anaerobic fermentation under mesophilic conditions. Energy Sour A Recov Utilizat Environ Effects 45:1017–1029. https://doi.org/10.1080/15567036.2023.2174615

Bishop G, Styles D, Lens PNL (2020) Recycling of European plastic is a pathway for plastic debris in the ocean. Environ Int 142:105893. https://doi.org/10.1016/j.envint.2020.105893

Boccia P, Mondellini S, Mauro S, Zanellato M, Parolini M, Sturchio E (2024) Potential effects of environmental and occupational exposure to microplastics: an overview of air contamination. Toxics 12:320. https://doi.org/10.3390/toxics12050320

Borah A, Singh S, Chattopadhyay R, Kaur J, Bari VK (2024) Integration of CRISPR/Cas9 with multi-omics technologies to engineer secondary metabolite productions in medicinal plant: challenges and prospects. Funct Integrat Genomics 24. https://doi.org/10.1007/s10142-024-01486-w

Bustos-Vázquez MG, Rodríguez-Durán LV, Pichardo-Sánchez MA, Rodríguez-Durán NR, Rodríguez-Durán NA, Trujillo-Ramírez D, Torres-de los Santos R (2025) Recent advances in biosurfactant production in solid-state fermentation. Fermentation 11:592. https://doi.org/10.3390/fermentation11100592

Calvert K, Mabee W (2014) Spatial analysis of biomass resources within a socio-ecologically heterogeneous region: identifying opportunities for a mixed feedstock stream. ISPRS Int J Geo Inf 3:209–232. https://doi.org/10.3390/ijgi3010209

Carmona Marques P, Fernandes PCB, Sampaio P, Silva J (2025) Advances in biotechnology in the circular economy: a path to the sustainable use of resources. Sustainability 17:6391. https://doi.org/10.3390/su17146391

Chai Y, Bai M, Chen A, Peng L, Shao J, Luo S, Deng Y, Yan B, Peng C (2022) Valorization of waste biomass through fungal technology: advances, challenges, and prospects. Ind Crop Prod 188:115608. https://doi.org/10.1016/j.indcrop.2022.115608

Cheirsilp B, Billateh A, Intasit R, Upaichit A, Boonsawang P, Louhasakul Y (2023) Fungal pretreatment and acid post-treatment for fractionation and biovalorization of palm biomass wastes into fungal oil, bioethanol, and lactic acid. Ind Crop Prod 196:116503. https://doi.org/10.1016/j.indcrop.2023.116503

Chen C, Li Y-W, Chen X-Y, Wang Y-T, Ye C, Shi T-Q (2024) Application of adaptive laboratory evolution for Yarrowia lipolytica: a comprehensive review. Bioresour Technol 391:129893. https://doi.org/10.1016/j.biortech.2023.129893

Christina K, Subbiah K, Arulraj P, Krishnan SK, Sathishkumar P (2024) A sustainable and eco-friendly approach for environmental and energy management using biopolymers chitosan, lignin and cellulose — a review. Int J Biol Macromol 257:128550. https://doi.org/10.1016/j.ijbiomac.2023.128550

Dessie W, Luo X, He F, Liao Y, Duns GJ, Qin Z (2023) Lignin valorization: a crucial step towards full utilization of biomass, zero waste and circular bioeconomy. Biocatal Agric Biotechnol 51:102777. https://doi.org/10.1016/j.bcab.2023.102777

Dhar SK, Kaur J, Singh GB, Chauhan A, Tamang J, Lakhara N, Asyakina L, Atuchin V, Mudgal G, Abdi G (2024) Novel Bacillus and Prestia isolates from Dwarf century plant enhance crop yield and salinity tolerance. Sci Rep 14:14645. https://doi.org/10.1038/s41598-024-65632-x

Dhiman S, Khanna K, Kour J, Singh AD, Bhardwaj T, Devi K, Sharma N, Kumar V, Bhardwaj R (2024) Landfill bacteriology: role in waste bioprocessing elevated landfill gaseselimination and heat management. J Environ Manag 354:120364. https://doi.org/10.1016/j.jenvman.2024.120364

Ding C, Wang X, Li M (2019) Evaluation of six white-rot fungal pretreatments on corn stover for the production of cellulolytic and ligninolytic enzymes, reducing sugars, and ethanol. Appl Microbiol Biotechnol 103:5641–5652. https://doi.org/10.1007/s00253-019-09884-y

Ekloh E, Yafetto L (2024) Fermentation and valorization of watermelon (Citrullus lanatus) rind wastes into livestock feed using Aspergillus niger and Mucor sp. Sci Afr 23. https://doi.org/10.1016/j.sciaf.2023.e02035

El-Sayed ASA, Abdel-Ghany SE, Ali GS (2017) Genome editing approaches: manipulating of lovastatin and taxol synthesis of filamentous fungi by CRISPR/Cas9 system. Appl Microbiol Biotechnol 101:3953–3976. https://doi.org/10.1007/s00253-017-8263-z

Eras-Muñoz E, Gea T, Font X (2024) Carbon and nitrogen optimization in solid-state fermentation for sustainable sophorolipid production using industrial waste. Front Bioeng Biotechnol 11. https://doi.org/10.3389/fbioe.2023.1252733

Escaramboni B, Garnica BC, Abe MM, Palmieri DA, Fernández Núñez EG, de Oliva Neto P (2021) Food waste as a feedstock for fungal biosynthesis of amylases and proteases. Waste Biomass Valorizat 13:213–226. https://doi.org/10.1007/s12649-021-01511-0

Ezeorba TPC, Okeke ES, Mayel MH, Nwuche CO, Ezike TC (2024) Recent advances in biotechnological valorization of agro-food wastes (AFW): optimizing integrated approaches for sustainable biorefinery and circular bioeconomy. Bioresour Technol Rep 26:101823. https://doi.org/10.1016/j.biteb.2024.101823

Fisgativa H, Tremier A, Le Roux S, Bureau C, Dabert P (2017) Understanding the anaerobic biodegradability of food waste: relationship between the typological, biochemical and microbial characteristics. J Environ Manag 188:95–107. https://doi.org/10.1016/j.jenvman.2016.11.058

Fulke AB, Bhanushali S, Jadhav HS (2026) Global marine plastic pollution: sources, distribution, implications on human health and mitigation strategies. Cont Shelf Res 296:105578. https://doi.org/10.1016/j.csr.2025.105578

Füting P, Barthel L, Cairns TC, Briesen H, Schmideder S (2021) Filamentous fungal applications in biotechnology: a combined bibliometric and patentometric assessment. Fungal Biol Biotechnol 8. https://doi.org/10.1186/s40694-021-00131-6

Garcia Simão RdC, Rocha PMC, Martins JTK, Turkiewicz M, Plewka J, da-Conceição Silva JL, Maller A, Kadowaki MK, Costa-Júnior ÁPS (2024) Exploring biodegradable alternatives: microorganism-mediated plastic degradation and environmental policies for sustainable plastic management. Arch Microbiol 206. https://doi.org/10.1007/s00203-024-04170-6

Garg S (2025) The importance of fungal biotechnology for sustainable applications. Trends Biotechnol. https://doi.org/10.1016/j.tibtech.2025.06.010

Giavasis I, Seviour RJ, Hudman P, McNeil B (2019) Fungal bioproducts for use in food: polysaccharides, organic acids, and Mycoprotein. In: Advances in food bioproducts and bioprocessing technologies, pp 511–548. https://doi.org/10.1201/9780429331817-25

Gómez-Ramos GA, Prado-Barragán LA, Rivera BA, Huerta-Ochoa S, Rivera VM, Couder-García M, García-Hernández RdlÁ, Castillo-Araiza CO (2024) Bioreactor engineering for circular economy: bioactive compound production in solid-state fermentation—review. Chem Eng Technol 48. https://doi.org/10.1002/ceat.202400289

Gong Z, Zhang S, Liu J (2023) Recent advances in chitin biosynthesis associated with the morphology and secondary metabolite synthesis of filamentous fungi in submerged fermentation. J Fungi 9:205. https://doi.org/10.3390/jof9020205

González-Fernández LA, Medellín-Castillo NA (2024) Plastic waste and its eco-friendly management. In: Soni R, Debbarma P, Suyal DC, Goel R (eds) Advanced strategies for biodegradation of plastic polymers. Springer Nature Switzerland, Cham, pp 225–259

Gu H, Feng L, Zhen X (2022) Study on the stability of anaerobic digestion of food waste and the waste mushroom substrate based on SBR reactor and Netlogo simulation. J Mater Cycles Waste Manag 25:793–809. https://doi.org/10.1007/s10163-022-01559-7

Gunaji RG, Junin R, Bandyopadhyay S (2020) Production and characterization of biopolymer schizophyllan using sago starch as a carbon source. J Phys Conf Ser 1529:52063. https://doi.org/10.1088/1742-6596/1529/5/052063

Hassan SS, Williams GA, Jaiswal AK (2018) Emerging technologies for the pretreatment of lignocellulosic biomass. Bioresour Technol 262:310–318. https://doi.org/10.1016/j.biortech.2018.04.099

Higuchi C, Isobe A (2024) Reduction scenarios of plastic waste emission guided by the probability distribution model to avoid additional ocean plastic pollution by the 2050s. Mar Pollut Bull 207:116791. https://doi.org/10.1016/j.marpolbul.2024.116791

Hotz EC, Bradshaw AJ, Elliott C, Carlson K, Dentinger BTM, Naleway SE (2023) Effect of agar concentration on structure and physiology of fungal hyphal systems. J Mater Res Technol 24:7614–7623. https://doi.org/10.1016/j.jmrt.2023.05.013

Hu D, Zhao R, Lin Y, Jiang C (2026) CRISPR-Cas9 gene editing in aspergillus: from pathogenesis to metabolic engineering. Biology (Basel) 15:53

Irbe I, Andze L, Blumfelde M, Filipova I, Verovkins A, Zoldners J (2023) Harvesting mycelial biomass of selected basidiomycetes for chitosan biopolymer extraction. Polymers 15:3548. https://doi.org/10.3390/polym15173548

Jasti N, Rasmussen ML, Khanal SK, Pometto AL, van Leeuwen JH (2009) Influence of selected operating parameters on fungal biomass production in corn-ethanol wastewater. J Environ Eng 135:1106–1114. https://doi.org/10.1061/(asce)ee.1943-7870.0000099

Jatwani S, Aggarwal F, Sarma SJ, Khare D (2025) From waste to wealth: innovative biotechnological approaches in Phanerochaete chrysosporium for a circular bioeconomy. Total Environ Eng 5:100044. https://doi.org/10.1016/j.teengi.2025.100044

Javaid R, Sabir A, Sheikh N, Ferhan M (2019) Recent advances in applications of acidophilic fungi to produce chemicals. Molecules 24:786. https://doi.org/10.3390/molecules24040786

Ji X, Liu Y, Meng J, Wu X (2020) Global supply chain of biomass use and the shift of environmental welfare from primary exploiters to final consumers. Appl Energy 276:115484. https://doi.org/10.1016/j.apenergy.2020.115484

Ji W, Peng Q, Fang X, Li Z, Li Y, Xu C, Zhao S, Li J, Chen R, Mo G et al (2022) Structures of a deltacoronavirus spike protein bound to porcine and human receptors. Nat Commun 13:1467. https://doi.org/10.1038/s41467-022-29062-5

Jin FJ, Wang BT, Wang ZD, Jin L, Han P (2022) CRISPR/Cas9-based genome editing and its application in Aspergillus species. J Fungi (Basel, Switzerland) *8*. https://doi.org/10.3390/jof8050467

K A, Selvasembian R (2024) Potential and opportunities of waste biomass valorization toward sustainable biomethane production. ChemBioEng Rev 11. https://doi.org/10.1002/cben.202400004

Kamravamanesh D, Lackner M, Herwig C (2018) Bioprocess engineering aspects of sustainable Polyhydroxyalkanoate production in cyanobacteria. Bioengineering 5:111. https://doi.org/10.3390/bioengineering5040111

Katsimpouras C, Stephanopoulos G (2021) Enzymes in biotechnology: critical platform technologies for bioprocess development. Curr Opin Biotechnol 69:91–102. https://doi.org/10.1016/j.copbio.2020.12.003

Kaur J, Mudgal G, Chand K, Singh GB, Perveen K, Bukhari NA, Debnath S, Mohan TC, Charukesi R, Singh G (2022) An exopolysaccharide-producing novel agrobacterium pusense strain JAS1 isolated from snake plant enhances plant growth and soil water retention. Sci Rep 12:21330. https://doi.org/10.1038/s41598-022-25225-y

Kaur J, Mudgal G, Negi A, Tamang J, Singh S, Singh GB, Bose K. JC, Debnath S, Wadaan MA, Farooq Khan M et al (2023) Reactive black-5, congo red and methyl orange: chemical degradation of azo-dyes by agrobacterium. 15:1664

Khan MR, Torrieri E, Allais F, Fadlallah S (2024) Can synthetic biology really empower microbial biopolymers as efficient food contact materials? Trends Food Sci Technol 143:104250. https://doi.org/10.1016/j.tifs.2023.104250

Khatami K, Qazanfarzadeh Z, Jiménez-Quero A (2026) Fungal fermentation: the blueprint for transforming industrial side streams and residues. Bioresour Technol 440:133426. https://doi.org/10.1016/j.biortech.2025.133426

Kocharin K, Rachathewee P, Sanglier J-J, Prathumpai W (2010) Exobiopolymer production of Ophiocordyceps dipterigena BCC 2073: optimization, production in bioreactor and characterization. BMC Biotechnol 10:51. https://doi.org/10.1186/1472-6750-10-51

Kojima Y, Nakakubo R, Iwabuchi K (2021) Thermophilic dry methane fermentation of municipal solid waste with various compositions. J Mater Cycles Waste Manag 24:311–320. https://doi.org/10.1007/s10163-021-01321-5

Kover A, Kraljić D, Marinaro R, Rene ER (2021) Processes for the valorization of food and agricultural wastes to value-added products: recent practices and perspectives. Syst Microbiol Biomanuf 2:50–66. https://doi.org/10.1007/s43393-021-00042-y

Kumar A, Krishan G (2023) Microplastic pollutants in aquatic ecosystems: present and future challenges. Water 16:102. https://doi.org/10.3390/w16010102

Kumari SVG, Pakshirajan K, Pugazhenthi G (2022) Recent advances and future prospects of cellulose, starch, chitosan, polylactic acid and polyhydroxyalkanoates for sustainable food packaging applications. Int J Biol Macromol 221:163–182. https://doi.org/10.1016/j.ijbiomac.2022.08.203

Li X-J, Feng X-X, Jin H-Q, Xu Q-Q, Lin F-C, Zhang C-L (2016) Optimization of solid-state fermentation media for growth of two endophytic fungal strains. Microbiology

Li X, Shi Y, Kong W, Wei J, Song W, Wang S (2022) Improving enzymatic hydrolysis of lignocellulosic biomass by bio-coordinated physicochemical pretreatment—a review. Energy Rep 8:696–709. https://doi.org/10.1016/j.egyr.2021.12.015

Li R, Zheng Y, Zhao X, Yong Q, Meng X, Ragauskas A, Huang C (2023a) Recent advances in biomass pretreatment using biphasic solvent systems. Green Chem 25:2505–2523. https://doi.org/10.1039/d3gc00271c

Li J, Wiebenga A, Lipzen A, Ng V, Tejomurthula S, Zhang Y, Grigoriev IV, Peng M, de Vries RP (2023b) Comparative genomics and transcriptomics analyses reveal divergent plant biomass-degrading strategies in fungi. J Fungi (Basel) 9:860

Li J, Pi C, Zhang J, Jiang F, Bao T, Gao L, Wu X (2025) Fungal bioconversion of lignin-derived aromatics: pathways, enzymes, and biotechnological potential. Biotechnol Adv 83:108624. https://doi.org/10.1016/j.biotechadv.2025.108624

Liggenstoffer AS, Youssef NH, Wilkins MR, Elshahed MS (2014) Evaluating the utility of hydrothermolysis pretreatment approaches in enhancing lignocellulosic biomass degradation by the anaerobic fungus Orpinomyces sp. strain C1A. J Microbiol Methods 104:43–48. https://doi.org/10.1016/j.mimet.2014.06.010

Lim S, Bijlani S, Blachowicz A, Chiang Y-M, Lee M-S, Torok T, Venkateswaran K, Wang CCC (2021) Identification of the pigment and its role in UV resistance in Paecilomyces variotii, a Chernobyl isolate, using genetic manipulation strategies. Fungal Genet Biol 152:103567. https://doi.org/10.1016/j.fgb.2021.103567

Liu L, Wang F, Wang L, Jiang X, Zhang H (2025) Addressing challenges in microbial manufacturing: systematic microbial biotechnology. Int J Hydrogen Energ 6:100871. https://doi.org/10.1016/j.xinn.2025.100871

Lorenci Woiciechowski A, Dalmas Neto CJ, de Souza P, Vandenberghe L, de Carvalho Neto DP, Novak Sydney AC, Letti LAJ, Karp SG, Zevallos Torres LA, Soccol CR (2020) Lignocellulosic biomass: acid and alkaline pretreatments and their effects on biomass recalcitrance—conventional processing and recent advances. Bioresour Technol 304:122848. https://doi.org/10.1016/j.biortech.2020.122848

Luft L, Confortin TC, Todero I, Zabot GL, Mazutti MA (2020) An overview of fungal biopolymers: bioemulsifiers and biosurfactants compounds production. Crit Rev Biotechnol 40:1059–1080. https://doi.org/10.1080/07388551.2020.1805405

Mai L, Sun X, Zeng EY (2023) Country-specific riverine contributions to marine plastic pollution. Sci Total Environ 874:162552. https://doi.org/10.1016/j.scitotenv.2023.162552

Mantri BU, Vahidinasab M, Berensmeier S (2025) Engineering strategies for fungal cell disruption in biotechnological applications. Eng Life Sci 25. https://doi.org/10.1002/elsc.70061

Mascarin GM, Golo PS, de Souza Ribeiro-Silva C, Muniz ER, de Oliveira Franco A, Kobori NN, Fernandes ÉKK (2024) Advances in submerged liquid fermentation and formulation of entomopathogenic fungi. Appl Microbiol Biotechnol 108. https://doi.org/10.1007/s00253-024-13287-z

Mattern DJ, Valiante V, Unkles SE, Brakhage AA (2015) Synthetic biology of fungal natural products. Front Microbiol 6. https://doi.org/10.3389/fmicb.2015.00775

Medeleanu ML, Călinoiu L-F, Martău G-A, Vodnar D-C (2025) Metabolic engineering of microorganisms for tailor-made biopolymer production: a review. Int J Biol Macromol 330:147922. https://doi.org/10.1016/j.ijbiomac.2025.147922

Melanouri EM, Papanikolaou S, Diamantopoulou P (2024) Mortierella Ramanniana lipid fermentation wastewater as an innovative maceration liquid medium for sustainable solid-state cultivation of higher fungi. Waste Biomass Valorizat 15:6903–6925. https://doi.org/10.1007/s12649-024-02575-4

Meussen BJ, de Graaff LH, Sanders JPM, Weusthuis RA (2012) Metabolic engineering of Rhizopus oryzae for the production of platform chemicals. Appl Microbiol Biotechnol 94:875–886. https://doi.org/10.1007/s00253-012-4033-0

Meyer V, Basenko EY, Benz JP, Braus GH, Caddick MX, Csukai M, de Vries RP, Endy D, Frisvad JC, Gunde-Cimerman N et al (2020) Growing a circular economy with fungal biotechnology: a white paper. Fungal Biol Biotechnol 7. https://doi.org/10.1186/s40694-020-00095-z

Minervini F, Comitini F, De Boni A, Fiorino GM, Rodrigues F, Tlais AZA, Carafa I, De Angelis M (2022) Sustainable and Health-Protecting Food Ingredients from Bioprocessed Food by-Products and Wastes. 14:15283

Miyazawa K, Yoshimi A, Kasahara S, Sugahara A, Koizumi A, Yano S, Kimura S, Iwata T, Sano M, Abe K (2018) Molecular mass and localization of α-1,3-Glucan in cell wall control the degree of hyphal aggregation in liquid culture of Aspergillus nidulans. Front Microbiol 9. https://doi.org/10.3389/fmicb.2018.02623

Mondal S, Halder SK, Mondal KC (2021) Tailoring in fungi for next generation cellulase production with special reference to CRISPR/CAS system. Syst Microbiol Biomanuf 2:113–129. https://doi.org/10.1007/s43393-021-00045-9

Morales-Barron BM, Larios-Serrato V, Morales-García YE, Quintero-Hernández V, Estrada-de los Santos P, Muñoz-Rojas J (2025) Effect of Pseudomonas protegens EMM-1 Against Rhizopus oryzae in interactions with Mexican Autochthonous Red Maize. Life (Basel) 15:554

Moreira Vieira N, Zandonade Ventorim R, de Moura Ferreira MA, Barcelos de Souza G, Menezes de Almeida EL, Pereira Vidigal PM, Nunes Nesi A, Gomes Fietto L, Batista da Silveira W

(2020) Insights into oleaginous phenotype of the yeast Papiliotrema laurentii. Fungal Genet Biol 144:103456. https://doi.org/10.1016/j.fgb.2020.103456

Moretto G, Valentino F, Pavan P, Majone M, Bolzonella D (2019) Optimization of urban waste fermentation for volatile fatty acids production. Waste Manag 92:21–29. https://doi.org/10.1016/j.wasman.2019.05.010

Morio F, Lombardi L, Butler G (2020) The CRISPR toolbox in medical mycology: state of the art and perspectives. PLoS Pathog 16. https://doi.org/10.1371/journal.ppat.1008201

Mousavi H (2021) A comprehensive survey upon diverse and prolific applications of chitosan-based catalytic systems in one-pot multi-component synthesis of heterocyclic rings. Int J Biol Macromol 186:1003–1166. https://doi.org/10.1016/j.ijbiomac.2021.06.123

Nandhini R, Sivaprakash B, Rajamohan N, Vo D-VN (2022) Lignin and polylactic acid for the production of bioplastics and valuable chemicals. Environ Chem Lett 21:403–427. https://doi.org/10.1007/s10311-022-01505-x

Nayanathara Thathsarani Pilapitiya PGC, Ratnayake AS (2024) The world of plastic waste: a review. Cleaner Mater *11*:100220. https://doi.org/10.1016/j.clema.2024.100220

Nguyen D, Nguyen M-K, Pham M-T, La DD, Chang SW, Kim SS, Nguyen DD (2025) Plastic debris in bird nests: sources, ecological impacts, and bioindicator potential for monitoring microplastic pollution. Environ Geochem Health 47. https://doi.org/10.1007/s10653-025-02688-2

Nicolescu CM, Bumbac M, Buruleanu CL, Popescu EC, Stanescu SG, Georgescu AA, Toma SM (2023) Biopolymers produced by lactic acid bacteria: characterization and food application. Polymers 15:1539. https://doi.org/10.3390/polym15061539

Novy V, Nielsen F, Cullen D, Sabat G, Houtman CJ, Hunt CG (2021) The characteristics of insoluble softwood substrates affect fungal morphology, secretome composition, and hydrolytic efficiency of enzymes produced by Trichoderma reesei. Biotechnol Biofuels 14. https://doi.org/10.1186/s13068-021-01955-5

Ost KJ, Student M, Cord-Landwehr S, Moerschbacher BM, Ram AFJ, Dirks-Hofmeister ME (2025) Cell walls of filamentous fungi—challenges and opportunities for biotechnology. Appl Microbiol Biotechnol 109. https://doi.org/10.1007/s00253-025-13512-3

Pal V, Sharma D, Vishwakarma P, Samanta D, Singh KA, Nagar J, Sani RK, Rai R (2024) CRISPR/Cas9: a cutting-edge tool for cellulase enhancement in fungi. Fungal Biol Rev 50:100394. https://doi.org/10.1016/j.fbr.2024.100394

Pant S, Ritika u, Nag P, Ghati A, Chakraborty D, Maximiano MR, Franco OL, Mandal AK, Kuila A (2022) Employment of the CRISPR/Cas9 system to improve cellulase production in Trichoderma reesei. Biotechnol Adv *60*:108022. https://doi.org/10.1016/j.biotechadv.2022.108022

Park CH, Xu H, Yeo HJ, Park YE, Hwang G-S, Park NI, Park SU (2021) Enhancement of the flavone contents of Scutellaria baicalensis hairy roots via metabolic engineering using maize Lc and Arabidopsis PAP1 transcription factors. Metab Eng 64:64–73. https://doi.org/10.1016/j.ymben.2021.01.003

Parvathy Eswari A, Ravi YK, Kavitha S, Rajesh Banu J (2023) Recent insight into anaerobic digestion of lignocellulosic biomass for cost effective bioenergy generation. e-Prime Adv Electr Eng Electron Energy 3:100119. https://doi.org/10.1016/j.prime.2023.100119

Paul E, Bessière Y, Dumas C, Girbal-Neuhauser E (2020) Biopolymers production from wastes and wastewaters by mixed microbial cultures: strategies for microbial selection. Waste Biomass Valorizat 12:4213–4237. https://doi.org/10.1007/s12649-020-01252-6

Phiri R, Mavinkere Rangappa S, Siengchin S, Oladijo OP, Dhakal HN (2023) Development of sustainable biopolymer-based composites for lightweight applications from agricultural waste biomass: a review. Adv Industr Eng Polymer Res 6:436–450. https://doi.org/10.1016/j.aiepr.2023.04.004

Pilafidis S, Diamantopoulou P, Gkatzionis K, Sarris D (2022) Valorization of agro-industrial wastes and residues through the production of bioactive compounds by macrofungi in liquid state cultures: growing circular economy. Appl Sci 12:11426. https://doi.org/10.3390/app122211426

Ping S, Fangran L, Yongjun H, Yanbo L, Jun T (2012) Municipal domestic waste fermentation for dehydrating before incineration. Environ Sanitat Eng

Priya AK, Muruganandam M, Imran M, Gill R, Vasudeva Reddy MR, Shkir M, Sayed MA, AlAbdulaal TH, Algarni H, Arif M et al (2023) A study on managing plastic waste to tackle the worldwide plastic contamination and environmental remediation. Chemosphere 341:139979. https://doi.org/10.1016/j.chemosphere.2023.139979

Rai PK, Sonne C, Song H, Kim K-H (2023) Plastic wastes in the time of COVID-19: their environmental hazards and implications for sustainable energy resilience and circular bio-economies. Sci Total Environ 858:159880. https://doi.org/10.1016/j.scitotenv.2022.159880

Raina N, Chuetor S, Elalami D, Tayibi S, Barakat A (2024) Biomass valorization for bioenergy production: current techniques, challenges, and pathways to solutions for sustainable bioeconomy. Bioenergy Res 17:1999–2028. https://doi.org/10.1007/s12155-024-10792-x

Ramachandraiah A, Kaushik J, Gowda L, Chigadannavar PS, Krishnappa R, Venkatesh S, Sivadas S, Judith S, Ramesh UM, Quadri Z et al (2021) Prospect of biofuel production by fungus. In: Fungi bio-prospects in sustainable agriculture, environment and nano-technology, pp 285–313. https://doi.org/10.1016/b978-0-12-821394-0.00011-1

Ramamoorthy NK, Vengadesan V, Pallam RB, Sadras SR, Sahadevan R, Sarma VV (2023) A pilot-scale sustainable biorefinery, integrating mushroom cultivation and in-situ pretreatment-cum-saccharification for ethanol production. Preparative Biochem Biotechnol 53:954–967. https://doi.org/10.1080/10826068.2022.2162922

Ramos LP, Suota MJ, Fockink DH, Pavaneli G, da Silva TA, Łukasik RM (2020) Enzymes and biomass pretreatment. In: Recent advances in bioconversion of lignocellulose to biofuels and value-added chemicals within the biorefinery concept, pp 61–100. https://doi.org/10.1016/b978-0-12-818223-9.00004-7

Rao PM, Singh AO, Sriniketha Narayanan K, Shankar R, Mudgal G, Radha P (2025) Cellulose-based packaging: an emerging sustainable packaging option for future use. In: Kesari KK, Prakash C, Khalid M, Negi A (eds) Functionalized cellulose materials: sustainable manufacturing and applications. Springer Nature Switzerland, Cham, pp 89–115

Reginatto V, Antônio RV (2015) Fermentative hydrogen production from agroindustrial lignocellulosic substrates. Braz J Microbiol 46:323–335. https://doi.org/10.1590/s1517-838246220140111

Roth MG, Westrick NM, Baldwin TT (2023) Fungal biotechnology: from yesterday to tomorrow. Front Fungal Biol 4. https://doi.org/10.3389/ffunb.2023.1135263

Saba B, Bharathidasan AK, Ezeji TC, Cornish K (2023) Characterization and potential valorization of industrial food processing wastes. Sci Total Environ 868:161550. https://doi.org/10.1016/j.scitotenv.2023.161550

Sala A, Vittone S, Barrena R, Sánchez A, Artola A (2021) Scanning agro-industrial wastes as substrates for fungal biopesticide production: use of Beauveria bassiana and Trichoderma harzianum in solid-state fermentation. J Environ Manag 295:113113. https://doi.org/10.1016/j.jenvman.2021.113113

Sanches PHG, de Melo NC, Porcari AM, de Carvalho LM (2024) Integrating molecular perspectives: strategies for comprehensive multi-omics integrative data analysis and machine learning applications in Transcriptomics, proteomics, and metabolomics. Biology 13:848. https://doi.org/10.3390/biology13110848

Seetharaman B, Balu UR, Mudgal G, Firdaus S, Dash S, Mishra RK, Ramasamy V (2025) The piteous price of progress: the environmental and health costs of modern society. In: Prakash C, Kesari KK, Negi A (eds) Sustainable development goals towards environmental toxicity and green chemistry: environment and sustainability. Springer Nature Switzerland, Cham, pp 291–313

Sha Y, Zhou L, Wang Z, Ding Y, Lu M, Xu Z, Zhai R, Jin M (2023) Adaptive laboratory evolution boosts Yarrowia lipolytica tolerance to vanillic acid. J Biotechnol 367:42–52. https://doi.org/10.1016/j.jbiotec.2023.03.006

Shafiq M, Zeb L, Xue B, Wan Y, Wen Z, Wang Y (2025) Fungal biopolymers in 3D bioprinting: current status, challenges and future advancements. Chem Eng J 523:168797. https://doi.org/10.1016/j.cej.2025.168797

Sharma P, Sharma N (2024) RSM approach to pre-treatment of lignocellulosic waste and a statistical methodology for optimizing bioethanol production. Waste Manag Bull 2:49–66. https://doi.org/10.1016/j.wmb.2023.12.004
Sharma A, Chand K, Singh GB, Mudgal G (2021) Toxicity with waste-generated ionizing radiations: blunders behind the scenes. In: Kesari KK, Jha NK (eds) Free radical biology and environmental toxicity. Springer International Publishing, Cham, pp 305–325
Shirazi F, Kontoyiannis DP, Ibrahim AS (2015) Iron starvation induces apoptosis in rhizopus oryzae in vitro. Virulence 6:121–126. https://doi.org/10.1080/21505594.2015.1009732
Shukla A, Kumar D, Girdhar M, Kumar A, Goyal A, Malik T, Mohan A (2023) Strategies of pretreatment of feedstocks for optimized bioethanol production: distinct and integrated approaches. Biotechnol Biofuels Bioprod 16. https://doi.org/10.1186/s13068-023-02295-2
Silva Rabelo CAB, Vidal AM, Casanova-González R, Romero-García JM, Varesche MBA, Romero I, Castro E (2023) Vine shoots pre-treatment strategies for improved hydrogen production and metabolites redistribution in Clostridium butyricum. Ind Crop Prod 204:117317. https://doi.org/10.1016/j.indcrop.2023.117317
Sindhu R, Binod P, Pandey A (2016) RETRACTED: biological pretreatment of lignocellulosic biomass—an overview. Bioresour Technol 199:76–82. https://doi.org/10.1016/j.biortech.2015.08.030
Singh RS, Kaur N, Rana V, Kennedy JF (2017) Pullulan: a novel molecule for biomedical applications. Carbohydr Polym 171:102–121. https://doi.org/10.1016/j.carbpol.2017.04.089
Singh H, Janiyani K, Gangawane A, Pandya S, Jasani S (2024) Engineering cellulolytic fungi for efficient lignocellulosic biomass hydrolysis: advances in mutagenesis, gene editing, and nanotechnology with CRISPR-Cas innovations. *Discover* Appl Sci 6. https://doi.org/10.1007/s42452-024-06405-z
Stavridou E, Karapetsi L, Nteve GM, Tsintzou G, Chatzikonstantinou M, Tsaousi M, Martinez A, Flores P, Merino M, Dobrovic L et al (2024) Landscape of microalgae omics and metabolic engineering research for strain improvement: an overview. Aquaculture 587:740803. https://doi.org/10.1016/j.aquaculture.2024.740803
Stoett P, Scrich VM, Elliff CI, Andrade MM, de M. Grilli N, Turra A (2024) Global plastic pollution, sustainable development, and plastic justice. World Dev 184:106756. https://doi.org/10.1016/j.worlddev.2024.106756
Stoica RM, Moscovici M, Lakatos ES, Cioca LI (2023) Exopolysaccharides of fungal origin: properties and pharmaceutical applications. PRO 11:335. https://doi.org/10.3390/pr11020335
Subramanian V, Adler MJ, Benyamin M, Pullen RM, Servinsky MD, Kozlowski MT, Decker SR (2025) From the bench to the reactor: engineered filamentous fungi for biochemical and biomaterial production. Biotechnol Biofuels Bioprod 18. https://doi.org/10.1186/s13068-025-02712-8
Sun M-L, Fang D, Zeng D, Wang Z-Z, Huang C-C, Zhang L, Fan Y-C, Liao S, Cao Y, Li K et al (2025) Leveraging stress tolerance property to enhance succinic acid production and cell metabolic activity of Yarrowia lipolytica. ACS Sustain Chem Eng 13:5210–5219. https://doi.org/10.1021/acssuschemeng.4c10404
Suryadi H, Judono JJ, Putri MR, Eclessia AD, Ulhaq JM, Agustina DN, Sumiati T (2022) Biodelignification of lignocellulose using ligninolytic enzymes from white-rot fungi. Heliyon 8:e08865. https://doi.org/10.1016/j.heliyon.2022.e08865
Suzuki G, Uchida N, Tanaka K, Higashi O, Takahashi Y, Kuramochi H, Yamaguchi N, Osako M (2024) Global discharge of microplastics from mechanical recycling of plastic waste. Environ Pollut 348:123855. https://doi.org/10.1016/j.envpol.2024.123855
Taherzadeh MJ, Karimi K (2008) Pretreatment of Lignocellulosic wastes to improve ethanol and biogas production: a review. Int J Mol Sci 9:1621–1651. https://doi.org/10.3390/ijms9091621
Tan J, Li Y, Tan X, Wu H, Li H, Yang S (2021) Advances in pretreatment of straw biomass for sugar production. Front Chem 9. https://doi.org/10.3389/fchem.2021.696030
Teo HL, Wahab RA, Zainal-Abidin MH, Mark-Lee WF, Huyop F, Susanti E, Mahat NA, Azman AR (2024) Statistically assisted optimisation for the simultaneous production of Trichoderma

harzianum and aspergillus fumigatus cellulolytic enzymes. Biomass Convers Biorefinery 15:3827–3843. https://doi.org/10.1007/s13399-023-05222-5

Tiwari P, Dufossé L (2023) Focus and insights into the synthetic biology-mediated chassis of economically important fungi for the production of high-value metabolites. Microorganisms 11:1141. https://doi.org/10.3390/microorganisms11051141

Tran H, Gurnani R, Kim C, Pilania G, Kwon H-K, Lively RP, Ramprasad R (2024) Design of functional and sustainable polymers assisted by artificial intelligence. Nat Rev Mater 9:866–886. https://doi.org/10.1038/s41578-024-00708-8

Ummalyma SB, Supriya RD, Sindhu R, Binod P, Nair RB, Pandey A, Gnansounou E (2019) Biological pretreatment of lignocellulosic biomass—current trends and future perspectives. Second Third Generat Feedstocks:197–212. https://doi.org/10.1016/b978-0-12-815162-4.00007-0

Vatsa PP, Shri M, Bhargava P, Joshi CG, Joshi M (2020) Recent developments in synthetic biology toolbox. In: Engineering of microbial biosynthetic pathways, pp 111–132. https://doi.org/10.1007/978-981-15-2604-6_7

Vauris A, Valcauda S, Husson F, Coninck JD (2022) A novel method to assess heat transfer and impact of relevant physicochemical parameters for the scaling up of solid state fermentation systems. Biotechnol Rep 36. https://doi.org/10.1016/j.btre.2022.e00764

Vieira RIM, Peixoto AdS, Monclaro AV, Ricart CAO, Filho EXF, Miller RNG, Gomes TG (2025) Fungal Coculture: unlocking the potential for efficient bioconversion of Lignocellulosic biomass. J Fungi *11*:458. https://doi.org/10.3390/jof11060458

Villena GK, Kitazono AA, Hernández-Macedo ML (2020) Bioengineering fungi and yeast for the production of enzymes, metabolites, and value-added compounds. Fungal Biol:209–237. https://doi.org/10.1007/978-3-030-41870-0_9

Vinod A, Sanjay MR, Suchart S, Jyotishkumar P (2020) Renewable and sustainable biobased materials: an assessment on biofibers, biofilms, biopolymers and biocomposites. J Clean Prod 258:120978. https://doi.org/10.1016/j.jclepro.2020.120978

Walker TR, Fequet L (2023) Current trends of unsustainable plastic production and micro(nano) plastic pollution. TrAC Trends Anal Chem 160:116984. https://doi.org/10.1016/j.trac.2023.116984

Wan Azelee NI, Mahdi HI, Cheng Y-S, Nordin N, Illias RM, Rahman RA, Shaarani SM, Bhatt P, Yadav S, Chang SW et al (2023) Biomass degradation: challenges and strategies in extraction and fractionation of hemicellulose. Fuel 339:126982. https://doi.org/10.1016/j.fuel.2022.126982

Wan C, Li Y (2012) Fungal pretreatment of lignocellulosic biomass. Biotechnol Adv 30:1447–1457. https://doi.org/10.1016/j.biotechadv.2012.03.003

Wang K, Tester JW (2023) Sustainable management of unavoidable biomass wastes. Green Energy Resour 1:100005. https://doi.org/10.1016/j.gerr.2023.100005

Wang J, Huang Z, Jiang Q, Roubík H, Xu Q, Gharsallaoui A, Cai M, Yang K, Sun P (2023) Fungal solid-state fermentation of crops and their by-products to obtain protein resources: the next frontier of food industry. Trends Food Sci Technol 138:628–644. https://doi.org/10.1016/j.tifs.2023.06.020

Wang Y, van Putten R-J, Tietema A, Parsons JR, Gruter G-JM (2024a) Polyester biodegradability: importance and potential for optimisation. Green Chem 26:3698–3716. https://doi.org/10.1039/d3gc04489k

Wang J, Ma Z, Wang C, Chen W (2024b) Melanin in Auricularia auricula: biosynthesis, production, physicochemical characterization, biological functions, and applications. Food Sci Biotechnol 33:1751–1758. https://doi.org/10.1007/s10068-024-01542-y

Wankhede L, Bhardwaj G, Pereira GVdM, Soccol CR, Brar SK (2024) Raw material selection for sustainable fermentation-derived alternative protein production: a review. Syst Microbiol Biomanuf 5:1–14. https://doi.org/10.1007/s43393-024-00294-4

Wijayawardene NN, Boonyuen N, Ranaweera CB, de Zoysa HKS, Padmathilake RE, Nifla F, Dai DQ, Liu Y, Suwannarach N, Kumla J et al (**2023**) OMICS and other advanced technologies in mycological applications. J Fungi (Basel, Switzerland) *9*. https://doi.org/10.3390/jof9060688

Williams JM, Nitzsche MP, Bromberg L, Qu Z, Moment AJ, Hatton TA, Park A-HA (2023) Hybrid thermo-electrochemical conversion of plastic wastes commingled with marine biomass to value-added products using renewable energy. Energy Environ Sci 16:5805–5821. https://doi.org/10.1039/D3EE02461J

Woodcraft C, Chooi Y-H, Roux I (2023) The expanding CRISPR toolbox for natural product discovery and engineering in filamentous fungi. Nat Prod Rep 40:158–173. https://doi.org/10.1039/d2np00055e

Wu Z, Peng K, Zhang Y, Wang M, Yong C, Chen L, Qu P, Huang H, Sun E, Pan M (2022) Lignocellulose dissociation with biological pretreatment towards the biochemical platform: a review. Mater Today Bio 16:100445. https://doi.org/10.1016/j.mtbio.2022.100445

Wu X, Wu Y, Lv X, Li J, Liu L, Du G, Chen J, Liu Y (2025) Synthetic biology and protein processing engineering-driven microbial protein supply by yeast and filamentous fungi. Trends Food Sci Technol 166:105408. https://doi.org/10.1016/j.tifs.2025.105408

Xia A, Cheng J, Ding L, Lin R, Huang R, Zhou J, Cen K (2013) Improvement of the energy conversion efficiency of Chlorella pyrenoidosa biomass by a three-stage process comprising dark fermentation, photofermentation, and methanogenesis. Bioresour Technol 146:436–443. https://doi.org/10.1016/j.biortech.2013.07.077

Xia K, Chen Y, Liu F, Zhao X, Sha R, Huang J (2024) Adaptive responses of erythritol-producing Yarrowia lipolytica to thermal stress after evolution. Appl Microbiol Biotechnol 108:263. https://doi.org/10.1007/s00253-024-13103-8

Xie M, Wang J, Wang F, Wang J, Yan Y, Feng K, Chen B (2025a) A review of genomic, transcriptomic, and proteomic applications in edible fungi biology: current status and future directions. J Fungi 11:422. https://doi.org/10.3390/jof11060422

Xie M, Wang F, Zhou X, Zeng J, Zhou Y, Wei T, Chen B-X (2025b) Multi-omics approaches for advancing edible fungi: from genomics-enabled strain development to metabolomics-guided postharvest preservation. Food Biosci 71:107161. https://doi.org/10.1016/j.fbio.2025.107161

Yan Q, Datsomor O, Zhao W, Chen W, Wei C, Wei D, Gao X, Qin C, Gu Q, Zou C et al (2025) Enhancing agricultural sustainability by improving the efficiency of Lignocellulosic biomass utilization in the ruminant diet via solid-state fermentation with white-rot fungi: a review. Microorganisms 13:1708. https://doi.org/10.3390/microorganisms13071708

Yoon LW, Ang TN, Ngoh GC, Chua ASM (2014) Fungal solid-state fermentation and various methods of enhancement in cellulase production. Biomass Bioenergy 67:319–338. https://doi.org/10.1016/j.biombioe.2014.05.013

Zadražil F, Brunnert H (1981) Investigation of physical parameters important for the solid state fermentation of straw by white rot fungi. Eur J Appl Microbiol Biotechnol 11:183–188. https://doi.org/10.1007/bf00511259

Zhang X, Fussenegger M (2024) Structural materials meet synthetic biology in biomedical applications. Mater Today 72:163–182. https://doi.org/10.1016/j.mattod.2023.12.008

Zhang J, Jiao Y, Sharma A, Shen D, Wei B, Hong C, Zheng B, Pan C (2021) Transcriptomic analysis reveals potential pathways associated with salt resistance in pecan (Carya illinoensis K. Koch). J Biotechnol 330:17–26. https://doi.org/10.1016/j.jbiotec.2021.02.001

Zhang Y, Liu M, Wu F, Li J, Azi F, Yu K, He Z, Wang P, Wu C, Lu B et al (2024) Fungal fermentation improves the nutritional quality, flavor characteristic and physicochemical property of highland barley bran. J Cereal Sci 120:104055. https://doi.org/10.1016/j.jcs.2024.104055

Zhou H, Zhao Q, Jiang J, Wang Z, Li L, Gao Q, Wang K (2023) Enhancing of pretreatment on high solids enzymatic hydrolysis of food waste: sugar yield, trimming of substrate structure. Bioresour Technol 379:128989. https://doi.org/10.1016/j.biortech.2023.128989

Zhu Y (2026) The potential and innovative applications of CRISPR gene editing technology in enzyme gene development. Enzym Microb Technol 195:110799. https://doi.org/10.1016/j.enzmictec.2025.110799

Zou G, Wei Y, Ji X-J (2023) Fungal green biotechnology and biomanufacturing. Front Microbiol 14:1334993

Chapter 12
Meta-omics Approaches in Fungal Biotechnology for Energy Production: Challenges and Perspectives

Roshini Singh, Kulsoom Fatima, and Surya Mishra

Abstract The growing global demand for sustainable and renewable energy has heightened research into microbial systems capable of efficient biomass conversion. Among them, fungi have been recognized as a highly promising and versatile group of organisms due to their exceptional enzymatic systems, diverse metabolic pathways, and ability to degrade complex organic materials. Recent advancements in meta-omics, including metagenomics, metatranscriptomics, metaproteomics, and metabolomics, have transformed the understanding of fungal communities and their functional roles in bioenergy production. These integrative approaches enable a comprehensive characterization of fungal diversity, metabolic regulation, enzyme engineering, and metabolic flow alterations across diverse environmental conditions. This chapter provides an in-depth overview of the applications of meta-omics tools in fungal biotechnology to increase bioenergy production. It discusses how metagenomic analyses facilitate the identification of novel genes and enzymes with industrial relevance, how metatranscriptomics explains gene expression patterns during biomass degradation, how metaproteomics identifies active protein networks involved in metabolic processes, and how metabolomics reveals key intermediates and end-products in energy metabolism. By combining recent findings and technological advancements, the chapter highlights integrative strategies using multi-omics data to construct systems-level models that optimize fungal metabolic networks for enhanced energy production.

Keywords Fungal biotechnology · Meta-omics · Bioenergy production · Metabolic networks · Systems biology

R. Singh
Department of Biotechnology, SR Institute of Management and Technology, Lucknow, Uttar Pradesh, India

K. Fatima
Department of Biomedical Engineering, Wayne State University, Detroit, MI, USA

S. Mishra (✉)
Department of Biotechnology, Goel Institute of Technology & Management, Lucknow, Uttar Pradesh, India

M. Tripathi et al. (eds.), *Recent Advancements in Fungal Biotechnology Volume 1*, Fungal Biology, https://doi.org/10.1007/978-981-92-0339-0_12

12.1 Introduction

The rapid growth of the global population, industrialization, and urbanization has led to an unprecedented increase in energy demand. Traditionally, this demand has been met primarily through fossil fuels such as coal, petroleum, and natural gas. While these resources have supported economic development for decades, their continued use poses serious environmental, financial, and social challenges. Fossil fuel combustion is a major contributor to greenhouse gas emissions, air pollution, climate change, and ecosystem degradation (IPCC 2022). In addition, fossil fuels are finite resources, raising concerns about long-term energy security and price volatility.

Sustainable and renewable energy sources are therefore essential to meet present energy needs without compromising the ability of future generations to meet theirs. Renewable energy systems rely on naturally replenished resources such as solar, wind, hydropower, and biomass. Among these, biomass-based energy is unique because it directly converts biological material into fuels and chemicals that can substitute liquid fossil fuels. Bioenergy has the added advantage of utilizing agricultural residues, forestry waste, and industrial byproducts, thereby promoting waste valorization and circular economy principles (Hall and Scrase 1998).

Fungi play a central role in natural carbon cycling and biomass decomposition. In terrestrial ecosystems, filamentous fungi are among the primary decomposers of plant material, breaking down complex polymers such as cellulose, hemicellulose, and lignin. This natural capability makes fungi particularly valuable for bioenergy applications, where efficient biomass conversion is a critical requirement (Kirk and Farrell 1987). One of the most important advantages of fungi is their ability to secrete large quantities of extracellular enzymes. Unlike many bacteria, fungi can penetrate solid substrates using their hyphal networks, enabling them to access and effectively degrade insoluble plant materials. White-rot and brown-rot fungi are recognized for their lignin-degrading capabilities, which are crucial for releasing fermentable sugars from lignocellulosic biomass (Hatakka 2001).

Fungi are also metabolically versatile. Certain yeast species, such as *Saccharomyces cerevisiae*, are widely used for bioethanol production due to their efficient fermentation pathways and industrial robustness. Other fungal groups, including oleaginous fungi, can accumulate large amounts of intracellular lipids, making them promising candidates for biodiesel production (Ratledge and Wynn 2002). Additionally, fungi can participate in anaerobic digestion systems, contribute to biogas production, and assist in the generation of bio-oil and bio-hydrogen when integrated with thermochemical processes.

Beyond individual species, fungi often function within complex microbial communities, where synergistic interactions enhance biomass degradation and energy yields. Their ecological adaptability, enzyme diversity, and scalability make fungal systems a cornerstone of next-generation bioenergy technologies. Traditional microbiological techniques have relied heavily on culturing organisms in laboratory conditions. However, it is now well recognized that a significant fraction of microbial diversity cannot be readily cultured using standard methods. This limitation has restricted our understanding of complex microbial systems involved in natural and engineered bioenergy processes. The emergence of meta-omics approaches has transformed microbial biotechnology by enabling culture-independent analysis of entire microbial communities (Handelsman et al. 1998).

In the context of fungal biotechnology, meta-omics has enabled researchers to uncover previously unknown fungal species, novel enzymes, and metabolic pathways involved in biomass conversion. For example, metagenomic studies of forest soils, compost systems, and anaerobic digesters have revealed extensive fungal diversity and functional redundancy that would otherwise remain hidden (Hess et al. 2011). Metatranscriptomics and metaproteomics analyses further allow identification of actively expressed genes and enzymes under specific environmental conditions, offering insights into real-time system performance.

Integrative omics allows researchers to link genetic potential (metagenomics) with functional expression (metatranscriptomics and metaproteomics) and actual biochemical outcomes (metabolomics). For example, the presence of lignocellulolytic genes alone does not guarantee efficient biomass degradation; integrative analysis can reveal whether these genes are actively expressed, are translated into functional enzymes, and contribute to metabolite production (VerBerkmoes et al. 2009).

Furthermore, integrative omics is essential for understanding fungal interactions within mixed microbial consortia. By revealing how fungi cooperate or compete with bacteria and archaea, researchers can design more efficient and resilient bioenergy systems. As bioenergy technologies move toward industrial-scale implementation, integrative omics will remain a foundational tool for bridging basic fungal biology with applied energy solutions.

12.2 Fungal Systems in Bioenergy Production

12.2.1 Industrially Relevant Fungal Species

Fungi represent a diverse and industrially significant group of microorganisms that have long been exploited for biotechnology applications. In the context of bioenergy production, certain fungal species have gained prominence due to their robustness, metabolic efficiency, and ability to produce valuable enzymes or biofuels at scale. These fungi can be categorized into filamentous fungi and yeasts, each contributing differently to bioenergy systems.

Among yeasts, *S. cerevisiae* remains the most extensively used organism for bioethanol production. Its advantages include high ethanol tolerance, rapid fermentation rates, genetic tractability, and a long history of safe industrial use. Although *S. cerevisiae* cannot naturally degrade lignocellulose, it plays a critical role in downstream fermentation once fermentable sugars are released (Walker and Stewart

2016). Other non-conventional yeasts, such as *Scheffersomyces stipitis* and *Kluyveromyces marxianus*, can ferment pentose sugars like xylose, making them valuable for lignocellulosic bioethanol processes.

Filamentous fungi are particularly important in the upstream stages of bioenergy production. Species such as *Trichoderma reesei, Aspergillus niger,* and *Penicillium* spp. are widely used for the industrial production of cellulases and hemicellulases. *T. reesei*, in particular, is considered a benchmark organism for cellulase production due to its high secretion capacity and well-characterized enzyme systems (Peterson and Nevalainen 2012). White-rot fungi such as *Phanerochaete chrysosporium* and *Pleurotus ostreatus* are notable for their lignin-degrading abilities, which are essential for pretreatment-free biomass conversion strategies.

Oleaginous fungi, including *Mortierella, Mucor, and Yarrowia lipolytica*, have attracted attention for biodiesel applications due to their ability to accumulate lipids ranging from 20% to 50% of their dry biomass. These species can utilize a wide range of carbon sources, including industrial and agricultural wastes, making them suitable for sustainable biofuel production (Ageitos et al. 2011).

12.2.2 Lignocellulose Degradation Potential

Lignocellulosic biomass is the most abundant renewable organic resource on Earth and consists primarily of cellulose, hemicellulose, and lignin. While cellulose and hemicellulose are polysaccharides that can be converted into fermentable sugars, lignin is a complex aromatic polymer that provides structural rigidity and resistance to microbial attack. Efficient degradation of lignocellulose is therefore one of the major challenges in bioenergy production.

Fungi are uniquely adapted to overcome this challenge. In natural ecosystems, filamentous fungi act as primary decomposers of plant material, especially in forest soils and decaying wood. Their hyphal growth allows them to physically penetrate plant cell walls, while their extracellular enzyme systems chemically break down complex polymers. White-rot fungi are the only known organisms capable of complete lignin mineralization, making them particularly valuable for biomass processing (Baldrian 2008).

Brown-rot fungi, although unable to fully degrade lignin, rapidly depolymerize cellulose through a combination of enzymatic action and non-enzymatic oxidative mechanisms. Soft-rot fungi, often found in moist environments, degrade cellulose and hemicellulose while modifying lignin structure. Together, these fungal strategies represent a powerful biological toolkit for lignocellulose conversion.

In bioenergy applications, fungal pretreatment of biomass can significantly reduce the need for harsh chemical or thermal pretreatment methods. Biological pretreatment using fungi is energy-efficient, environmentally friendly, and capable of selectively removing lignin while preserving carbohydrates. However, the slow growth rate of fungi and long processing times remain challenges that ongoing research seeks to address through strain improvement and process optimization.

12.2.3 Fungal Enzyme Systems: Ligninases, Cellulases, Hemicellulases

The ability of fungi to convert lignocellulosic biomass into simpler molecules is primarily mediated by their extracellular enzyme systems. These enzymes are broadly classified into lignin-degrading enzymes (ligninases), cellulose-degrading enzymes (cellulases), and hemicellulose-degrading enzymes (hemicellulases). Together, they act synergistically to dismantle the plant cell wall matrix.

Ligninases include lignin peroxidase (LiP), manganese peroxidase (MnP), versatile peroxidase (VP), and laccase. These oxidative enzymes generate free radicals that break the complex aromatic structure of lignin, making cellulose and hemicellulose more accessible. White-rot fungi are the primary producers of these enzymes, and their non-specific oxidative mechanisms allow them to attack a wide range of lignin structures (Leonowicz et al. 2001).

Efficient cellulose degradation requires the coordinated action of all three enzyme types, a feature well-developed in fungi such as *T. reesei*. The diversity of hemicellulases reflects the structural complexity of hemicellulose and highlights the metabolic versatility of fungi. The industrial relevance of fungal enzymes lies in their high efficiency, stability, and adaptability to different substrates. Advances in protein engineering and systems biology are further enhancing enzyme performance, enabling more cost-effective bioenergy production.

12.2.4 Fungal Consortia and Synergistic Interactions

In natural environments, fungi rarely function in isolation. Instead, they exist as part of complex microbial communities where interactions with bacteria, archaea, and other fungi shape ecosystem function. These interactions often result in synergistic effects that enhance biomass degradation and energy production beyond what individual species can achieve alone.

Fungal consortia can improve lignocellulose breakdown by dividing metabolic labor among different organisms. For example, one fungal species may specialize in lignin degradation, while another focuses on cellulose hydrolysis. This division of labor reduces metabolic burden and increases overall system efficiency. Additionally, fungi can create favorable conditions for bacterial partners by modifying pH, releasing simple sugars, or increasing substrate accessibility (de Boer et al. 2005).

However, managing consortia presents challenges related to stability, competition, and process control. Understanding interspecies communication, nutrient exchange, and stress responses is essential for designing reliable consortium-based systems. This has driven growing interest in meta-omics approaches to decode the functional dynamics of fungal communities in bioenergy applications.

12.3 Meta-Omics Approaches in Fungal Biotechnology

12.3.1 Metagenomics

High-throughput sequencing and metagenomic approaches have enabled the resolution of previously uncharacterized "dark taxa" and uncultured fungal lineages, facilitating both functional annotation and taxonomic identification of complex environmental communities. Deep metagenomic analyses allow the recovery of fungal metagenome-assembled genomes (MAGs) that exhibit extensive repertoires of carbohydrate-active enzymes (CAZymes) and secreted effector proteins (Ettinger et al. 2025). Such genomic features support the prediction of fungal lifestyles, including symbiotic associations and degradative capacities, based on genome content. In the context of solid-state fermentations, metagenomic studies have successfully linked dominant fungal taxa, such as *Aspergillus* species, to specific enzyme complements involved in plant cell wall degradation, thereby directly associating microbial community structure with bioprocess performance and functional outcomes (Wijayawardene et al. 2023).

12.3.2 Metatranscriptomics

Soil metatranscriptomics analyses reveal guild-specific expression patterns of plant cell wall–degrading and fungal cell wall–degrading enzymes, alongside pronounced ectomycorrhizal expression of nitrogen transporters (Auer et al. 2024). These findings elucidate competitive dynamics for carbon and nitrogen during decomposition processes. Integrative transcriptome studies in white-rot fungi demonstrate coordinated expression of carbohydrate-active enzymes (CAZymes), including lytic polysaccharide monooxygenases (LPMOs) and lignin peroxidases, which are closely associated with early-stage lignocellulose degradation and concomitant detoxification responses (Miyauchi et al. 2017).

Furthermore, combined transcriptome–metabolome analyses in cultivated mushrooms and gut-associated fungi reveal substrate-dependent induction of carbohydrate metabolism pathways and oxidoreductase enzymes, thereby mapping active metabolic routes that underpin energy generation and biomass yield (Li et al. 2025).

12.3.3 Metaproteomics

Metaproteomics analyses have identified more than 300 community-derived proteins, predominantly originating from *Aspergillus* species. These proteins include a wide range of plant cell wall–degrading enzymes and oxidative enzymes associated with polyphenol modification, thereby highlighting the functional contributions of dominant fungi to fermentation outcomes. Moreover, integrated metagenomic and metaproteomic approaches enable the assignment of specific enzymes to their corresponding microbial hosts. This linkage provides a mechanistic basis for the rational design of tailored enzyme consortia and for optimizing microbial community composition to enhance efficiency and performance in industrial biomass processing applications (Zhao et al. 2015).

12.3.4 Metabolomics

Fungal metabolomics and integrative multi-omics analyses have revealed lignin-associated aromatic compound catabolism as well as extracellular fatty acid and phenylpropanoid biosynthetic pathways in white-rot fungi, underscoring their direct relevance to bioproduct and biofuel development (Adejor et al. 2025). The combined application of metabolomics with genomics and transcriptomics has consistently enabled the identification of biosynthetic gene clusters and corresponding secondary metabolites with significant potential for industrial and pharmaceutical exploitation.

Furthermore, stable-isotope ^{13}C metabolic flux analysis provides a quantitative framework for mapping carbon partitioning between energy production and biomass formation. This approach has revealed both active and bypassed steps within central metabolic pathways in pathogenic yeasts and is readily transferable to the metabolic engineering and optimization of biofuel-producing fungal strains (Heckart et al. 2025).

12.4 Integrative Multi-Omics for Systems-Level Understanding

Integrating genomic, transcriptomic, proteomic, and metabolomic data provides a comprehensive systems-level perspective of cells and microbial consortia, enabling the development of predictive metabolic models and supporting rational strain and community engineering. Common multi-omics integration strategies include element- or feature-based, pathway-based, and model- or network-based approaches. These strategies are frequently applied in a stepwise manner, progressing from the genome through the epigenome, transcriptome, proteome, metabolome, and ultimately to the phenotype (Subramanian et al. 2020). Integration methodologies encompass correlation and co-expression analyses, joint dimensionality reduction techniques, network inference, and machine-learning- or artificial intelligence-based data fusion approaches. Metabolomics often serves as a central anchoring layer, as it directly links upstream regulatory processes to pathway activity and observable phenotypic outcomes (Pinu et al. 2019).

Genome-scale metabolic models (GEMs) encode gene–protein–reaction (GPR) relationships and enable constraint-based simulations, such as flux balance analysis (FBA), to predict metabolic behavior. Omics-constrained GEMs further refine these predictions: transcriptomic data facilitate the construction of condition-specific models using FBA or parsimonious FBA (pFBA); proteomic data enable enzyme-constrained modeling frameworks, such as the Generic Enzyme-Constraint method using Kinetic Objectives (GECKO); and metabolomic or fluxomic data provide flux bounds and experimental validation (Dahal et al. 2020). GEM platforms are now available for a wide range of organisms, including microbes, plants, and animals, and are extensively applied in metabolic engineering, bioenergy research, and disease modeling (Wörheide et al. 2021).

Projecting multi-omics data onto metabolic or interaction networks facilitates the identification of pathway bottlenecks, alternative metabolic routes, and key regulatory or control nodes influencing carbon flux. Constraint-based analyses performed on GEMs, when integrated with multi-omics data, reveal optimal strategies for redirecting carbon toward desired products or away from toxic intermediates (Yan et al. 2017).

The integration of multi-omics datasets with GEMs enables in silico strain design, rational target selection, and reliable prediction of engineering outcomes, such as enhanced biofuel or value-added chemical production. Artificial intelligence and machine-learning approaches further enhance model reconstruction, parameter estimation, and predictive accuracy from large-scale multi-omics datasets, thereby accelerating design–build–test–learn cycles (Helmy et al. 2020). Multi-omics integration, anchored in genome-scale metabolic models and systems biology frameworks, enables network-level analysis of carbon flux and supports predictive, AI-augmented metabolic engineering. This integrated approach has become central to the rational design and optimization of microbial strains and consortia for advanced biotechnological applications.

12.5 Technological Platforms and Computational Tools

High-throughput omics technologies, together with advanced artificial intelligence (AI) tools, now enable end-to-end acquisition, processing, analysis, and interpretation of large-scale fungal and multi-omics datasets. These integrated workflows span sequencing and mass spectrometry–based data generation, cloud-enabled bioinformatic pipelines, and machine learning–driven data integration and interpretation (Baldrian et al. 2021).

12.5.1 High-Throughput Sequencing Technologies

Next-generation sequencing (NGS) and RNA sequencing (RNA-Seq) constitute core platforms for genomic and transcriptomic analyses, while long-read sequencing technologies, such as Oxford Nanopore, enable untargeted shotgun metagenomics and improved resolution of fungal genomes and taxonomic identification. High-throughput sequencing (HTS) further underpins metabarcoding, metagenomics, and phylogenomics, facilitating the discovery of previously uncharacterized "dark taxa" and providing insights into global fungal diversity across diverse environmental samples (Aragona et al. 2022).

12.5.2 Mass Spectrometry for Proteomics and Metabolomics

Liquid chromatography–mass spectrometry (LC–MS)-based proteomics, metabolomics, and lipidomics enable global and targeted quantification of thousands of proteins and metabolites with high sensitivity and broad dynamic range. Recent methodological advances include single-cell and spatially resolved mass spectrometry approaches, such as mass spectrometry imaging (MSI) and laser ablation electrospray ionization mass spectrometry (LAESI-MS), which allow subcellular-level proteomic and metabolomic analyses and dynamic in situ measurements (Birhanu 2023). Mass spectrometry–based proteomics is increasingly being translated into clinical and diagnostic workflows, supported by robust quantification strategies and advanced data analysis software platforms.

12.5.3 Bioinformatics Pipelines and Databases

Large-scale repositories, including the Sequence Read Archive (SRA), along with specialized multi-omics databases, serve as long-term archives for high-throughput sequencing and mass spectrometry datasets, enabling data reuse and meta-analyses. Cloud-based and web-enabled platforms, such as Majorbio Cloud, Analyst Suite, and BiomiX, provide standardized and preconfigured pipelines for the analysis of genomic, metagenomic, transcriptomic, proteomic, and metabolomic data. These platforms offer interactive reports, visualization tools, and user-friendly interfaces that lower the barrier to comprehensive multi-omics analysis (Ren et al. 2022).

12.5.4 Machine Learning and AI-Driven Omics Interpretation

Machine learning and deep learning methods play an increasingly central role in omics data interpretation, supporting tasks such as feature selection, classification, clustering, biomarker discovery, and predictive modeling across transcriptomic, proteomic, metabolomic, and integrated multi-omics datasets. Applications include retention time and tandem mass spectrometry (MS/MS) spectrum prediction, peptide and protein identification, pathway inference, patient stratification, and multi-scale modeling of genotype–environment–phenotype relationships (Reel et al. 2021).

To ensure biological interpretability alongside predictive performance, there is growing emphasis on interpretable machine learning frameworks and explainable AI approaches. These methods enhance transparency and facilitate the extraction of mechanistic insights, thereby supporting hypothesis generation and informed decision-making in fungal biology and biotechnology.

12.6 Applications of Meta-Omics in Bioenergy Generation

12.6.1 Discovery of High-Efficiency Biomass-Degrading Enzymes

One of the most impactful contributions of meta-omics to bioenergy research is the discovery of novel, high-efficiency enzymes involved in biomass degradation. Traditional enzyme discovery relied on culturing individual organisms and screening their enzymatic activity under laboratory conditions. While successful to a degree, this approach overlooks the vast majority of microorganisms that are unculturable or exhibit different behaviors in natural environments. Metagenomics has fundamentally changed this paradigm by allowing direct access to the collective genetic material of microbial communities.

Through metagenomic sequencing of environments rich in lignocellulosic degradation, such as forest soils, compost heaps, termite guts, and anaerobic digesters, researchers have identified thousands of previously unknown genes encoding cellulases, hemicellulases, and auxiliary oxidative enzymes. Many of these enzymes exhibit superior properties, including higher thermal stability, broader pH tolerance, and increased catalytic efficiency compared to conventionally known enzymes (Berlemont and Martiny 2016). For fungal systems, metagenomics has revealed extensive enzyme diversity within the phyla Basidiomycota and Ascomycota communities that would otherwise remain hidden.

Beyond gene discovery, functional metagenomics enables direct expression and screening of environmental DNA libraries for desired activities. This approach has led to the identification of multifunctional enzymes capable of acting on multiple substrates, reducing the need for complex enzyme cocktails. When combined with metatranscriptomics and metaproteomics, researchers can further identify which enzymes are actively expressed during biomass degradation, ensuring that discovered enzymes are not only present but also functionally relevant under industrial conditions.

The integration of these meta-omics approaches accelerates enzyme discovery pipelines and supports the development of tailored enzyme mixtures for specific biomass feedstocks, ultimately reducing processing costs and improving bioenergy yields.

12.6.2 Optimizing Fungal Metabolism for Bioethanol Production

Bioethanol production from lignocellulosic biomass requires efficient conversion of complex plant polymers into fermentable sugars, followed by their transformation into ethanol. While yeasts such as *S. cerevisiae* dominate industrial fermentation, filamentous fungi play a critical role in upstream biomass processing. Meta-omics approaches have become essential tools for optimizing fungal metabolism across both stages.

Metatranscriptomics analyses allow researchers to monitor gene expression patterns in fungi during different phases of biomass degradation and fermentation. These studies have revealed dynamic regulation of carbohydrate-active enzymes, sugar transporters, and central carbon metabolism pathways in response to substrate availability and environmental stress (Cherry and Fidantsef 2003). By identifying bottlenecks in metabolic flux, researchers can target specific genes for overexpression or deletion to enhance sugar release and ethanol productivity.

Metabolomics further complements these findings by providing direct measurements of intracellular and extracellular metabolites. Accumulation of inhibitory compounds such as organic acids, furans, and phenolics can be detected and linked to reduced fermentation efficiency. Integrative analysis enables the redesign of fungal strains with improved tolerance to these inhibitors, either through metabolic engineering or adaptive laboratory evolution.

In co-culture systems involving fungi and yeasts, meta-omics reveals cross-feeding mechanisms and regulatory interactions that influence ethanol yield. Such insights support the development of consolidated bioprocessing strategies, where enzyme production, biomass hydrolysis, and fermentation occur in a single integrated system. This represents a major step toward economically viable second-generation bioethanol production.

12.6.3 Lipid-Rich Fungi for Biodiesel Generation

Oleaginous fungi have emerged as promising biological platforms for biodiesel production due to their ability to accumulate large quantities of lipids, primarily in the form of triacylglycerols. These lipids can be transesterified into fatty acid methyl esters, which are chemically similar to conventional diesel fuel. Meta-omics approaches have played a central role in understanding and enhancing lipid accumulation in fungal systems (Sartaj et al. 2023).

Metagenomic and genomic studies have identified key genes involved in fatty acid synthesis, elongation, and storage in oleaginous fungi such as *Y. lipolytica* and *Mortierella alpina*. Metatranscriptomics reveals how nutrient limitation, particularly nitrogen deprivation, triggers a metabolic shift from biomass growth to lipid storage (Ratledge 2014). This shift is accompanied by increased expression of acetyl-CoA-generating enzymes and lipid biosynthesis pathways.

Metabolomic profiling provides insight into carbon flux redistribution during lipid accumulation, highlighting the balance between energy production, membrane synthesis, and storage lipid formation. These insights have enabled rational metabolic engineering strategies to increase lipid yield, alter fatty acid composition, and improve substrate utilization (Sun et al. 2025).

Importantly, meta-omics has demonstrated that lipid-producing fungi can utilize low-cost feedstocks, including agricultural residues, industrial effluents, and organic wastes. This aligns biodiesel production with sustainability goals by reducing competition with food resources and minimizing waste. As a result, oleaginous fungi represent a viable and scalable alternative to plant-based biodiesel systems.

12.6.4 Engineering Fungal Consortia for Waste-to-Energy Conversion

Waste-to-energy systems rely on the conversion of organic waste streams into usable energy carriers, such as biogas, bioethanol, or bio-oil. Fungal consortia are particularly well-suited for these systems due to their ability to degrade heterogeneous and recalcitrant waste materials. Meta-omics has become indispensable for designing, monitoring, and optimizing such consortia (Wang and Tester 2023).

Metagenomic analyses provide detailed information on the taxonomic composition and functional potential of fungal communities within waste-processing systems. This information helps identify key degraders, keystone species, and redundant functions that contribute to system resilience. Metatranscriptomics and metaproteomics further reveal how community activity shifts in response to changes in waste composition, temperature, or loading rate (Zhang et al. 2020).

By integrating these data, researchers can rationally assemble synthetic consortia with complementary metabolic capabilities. For example, lignin-degrading fungi can be paired with cellulolytic fungi and fermentative microbes to create efficient, multi-stage conversion pathways. Meta-omics also aids in early detection of instability or functional decline, enabling corrective interventions before system failure occurs (Zhang et al. 2025b).

Such engineered consortia offer significant advantages over single-organism systems, including higher conversion efficiency, greater robustness, and improved tolerance to fluctuating environmental conditions. These properties are critical for real-world waste-to-energy applications, where feedstock composition is often unpredictable.

12.6.5 Biogas, Hydrogen, and Bio-Oil Production

Beyond liquid biofuels, fungal meta-omics has contributed to the optimization of gaseous and thermochemical bioenergy pathways. In anaerobic digestion systems, fungi act as primary hydrolyzers, breaking down complex polymers into simpler compounds that can be further converted into methane by archaeal communities. Meta-omics has clarified the specific roles fungi play in these systems and how their activity influences overall biogas yield (Vanwonterghem et al. 2014).

Metatranscriptomics studies have shown that fungal enzyme expression is highly responsive to operational parameters such as temperature, pH, and retention time. This knowledge allows fine-tuning of reactor conditions to maximize hydrolysis rates and methane production. Similarly, in bio-hydrogen systems, fungal pretreatment enhances substrate availability for hydrogen-producing bacteria, indirectly improving hydrogen yields.

In thermochemical processes such as pyrolysis, fungal pretreatment alters biomass composition by reducing lignin content and modifying cellulose crystallinity. Meta-omics analysis helps link these biochemical changes to improved bio-oil quality and reduced formation of inhibitory byproducts. Thus, fungal systems contribute both biologically and indirectly to advanced bioenergy pathways (Kumar et al. 2021).

Overall, meta-omics provides a powerful framework for understanding and enhancing the contribution of fungi across diverse bioenergy platforms. By integrating molecular-level insights with process-level optimization, these approaches support the development of efficient, sustainable, and scalable bioenergy technologies.

12.7 Challenges and Limitations

Meta-omics approaches have significantly advanced fungal biotechnology; however, numerous limitations still delay their repetitive implementation and industrial transformation. Major challenges include the inherent complexity of fungal genomes and communities, incomplete reference databases, challenges in multi-omics data integration and interpretation, sophisticated techniques used during sample preparation and sequencing, and high costs associated with large-scale production.

12.7.1 Complexity of Fungal Genomes and Communities

Fungi exhibit extensive phenotypic diversity, genome plasticity, and a high proportion of unculturable species, complicating comprehensive genomic and functional characterization. In species-rich environments, sequencing is often impossible for individual taxa, leading to reduced sensitivity in taxonomic and functional assessment, particularly when bacterial populations dominate fungal communities. Furthermore, intragenomic sequence similarity and the presence of cryptic species hinder accurate gene–species linkage in both metagenomic and metatranscriptomics analyses.

12.7.2 Data Integration and Interpretation

The generation of large, multi-layered datasets from genomics, transcriptomics, proteomics, and metabolomics presents substantial challenges for integration and interpretation. Robust bioinformatic pipelines and standardized analytical frameworks for fungal multi-omics remain limited. In addition, sophisticated techniques, low power, and inappropriate analytical methods can decrease the production rate and hinder the biological interpretation.

12.7.3 Limited Reference Databases

A considerable proportion of fungal diversity, often referred to as "dark taxa" and many fungal secondary metabolites remain absent from curated reference databases, limiting accurate taxonomic classification and functional annotation. Some widely used resources, such as UNITE, MycoCosm, and FungiDB, represent only a limited amount of known fungal diversity data and suffer from poor interoperability, which limits comparative study and meta-analyses.

12.7.4 Technical Constraints in Sample Preparation and Sequencing

RNA instability poses a major challenge in fungal meta-omics, as RNA degradation and preservation strategies can introduce significant community-level biases. Moreover, methodological choices—including the selection of internal transcribed spacer (ITS) subregions (ITS1 vs. ITS2), amplicon sequencing versus shotgun sequencing, and short-read versus long-read platforms—substantially affect the taxonomic classification, quantitative accuracy, and downstream interpretation.

12.7.5 Cost, Scalability, and Industrial Translation

The high costs associated with multi-omics sequencing, data storage, computational approaches, and specialized bioinformatics expertise limit the scalability of these approaches in industrial pipelines. Additionally, the lack of standardized, validated workflows and comprehensive techno-economic assessments hampers the translation of discovery-driven fungal meta-omics into robust and scalable bioprocesses.

12.8 Future Perspectives

Future advances in fungal meta-omics are expected to transform current technical and conceptual limitations into powerful opportunities for biological discovery and industrial innovation. Large-scale sequencing techniques targeting thousands of high-quality fungal genomes and environmental metagenomes will substantially expand and improve the taxonomic and phylogenetic classification, thereby strengthening the foundation for community-level and functional annotation. In parallel, long-read and hybrid assembly strategies optimized for complex and mixed systems such as lichens and other symbiotic consortia will enable more accurate genome reconstruction from heterogeneous fungal communities (Arya et al. 2025).

The integration of genomics, transcriptomics, proteomics, and metabolomics using advanced bioinformatics, artificial intelligence, and machine-learning approaches is expected to yield predictive models for secondary metabolite biosynthesis, pathogenicity, and stress adaptation. Comparative multi-omics across fungal samples, including *Sordariomycetes*, will further support evolutionary, functional, and trait-based inferences directly relevant to industrial and environmental biotechnology.

The development of curated reference databases that integrate genomic sequences, genomic markers, metabolite structures, enzyme functions, and ecological traits will progressively reduce the prevalence of "dark taxa" and improve annotation accuracy in fungal meta-omics datasets. Community-driven databases featuring interactive visualization tools and machine-learning-based functional and interaction predictions are also anticipated to play a central role in future fungal biotechnology research (Hyde et al. 2024).

Advances in protocols for nucleic acid and metabolite stabilization, host or background DNA depletion, and standardized metabarcoding pipelines such as optimized ITS workflows and long-read barcode sequencing will reduce technical bias and enhance cross-study comparability. Moreover, continued reductions in the cost of long-read sequencing and the emergence of portable sequencing platforms are expected to enable real-time monitoring of industrial fermentation processes, environmental bioreactors, and fungal production systems.

12.9 Conclusion

Meta-omics has transformed fungal biotechnology by enabling comprehensive, systems-level insight into fungal communities and their roles in biomass conversion and bioenergy production. By integrating metagenomics, metatranscriptomics, metaproteomics, and metabolomics, researchers can identify novel genes and enzymes, map active metabolic pathways, and construct predictive systems biology–based and genome-scale metabolic models to optimize carbon flux and biofuel yields.

Despite these advances, challenges such as complex fungal genomes, limited reference databases, technical constraints in sample preparation and sequencing, and difficulties in data integration still limit full industrial translation. Continued development of high-throughput technologies, robust bioinformatic pipelines, and AI-driven analytical tools, alongside cost-effective workflows, will be crucial. Overall, meta-omics offers a powerful framework to accelerate rational strain engineering and process optimization, paving the way for more efficient and sustainable fungal-based bioenergy and bioproducts.

References

Adejor J, Tumukunde E, Li G, Shehu TA, Wu L, Jiang Z, Wang S (2025) Stepping out of the dark: how metabolomics shed light on fungal biology. FEMS Microbiol Rev 49:fuaf028. https://doi.org/10.1093/femsre/fuaf028

Ageitos JM, Vallejo JA, Veiga-Crespo P, Villa TG (2011) Oily yeasts as oleaginous cell factories. Appl Microbiol Biotechnol 90(4):1219–1227. https://doi.org/10.1007/s00253-011-3200-z

Aragona M, Haegi A, Valente MT, Riccioni L, Orzali L, Vitale S, Luongo L, Infantino A (2022) New-generation sequencing technology in diagnosis of fungal plant pathogens: a dream comes true? J Fungi 8(7):737. https://doi.org/10.3390/jof8070737

Arya P, Sekhon G, Kaur M, Chosyang S, Kushwaha N, Singh B (2025) Multi-omics research in fungus: current progress and future prospects. In: Elsevier eBooks, pp 161–211. https://doi.org/10.1016/b978-0-443-13263-6.00007-9

Auer L, Buée M, Fauchery L, Lombard V, Barry KW, Clum A, Copeland A, Daum C, Foster B, LaButti K, Singan V, Yoshinaga Y, Martineau C, Alfaro M, Castillo FJ, Imbert JB, Ramírez L, Castanera R, Pisabarro AG, Finlay R, Lindahl B, Olson Å, Séguin A, Kohler A, Henrissat B, Grigoriev IV, Martin FM (2024) Metatranscriptomics sheds light on the links between the functional traits of fungal guilds and ecological processes in forest soil ecosystems. New Phytol 242(4):1676–1690. https://doi.org/10.1111/nph.19471

Baldrian P (2008) Enzymes of saprotrophic basidiomycetes. Br Mycolog Soc Symp Ser 28:19–41. https://doi.org/10.1016/S0275-0287(08)80004-5

Baldrian P, Větrovský T, Lepinay C, Kohout P (2021) High-throughput sequencing view on the magnitude of global fungal diversity. Fungal Divers 114(1):539–547. https://doi.org/10.1007/s13225-021-00472-y

Berlemont R, Martiny AC (2016) Genomic potential for polysaccharide deconstruction in bacteria. Appl Environ Microbiol 82(19):5873–5882. https://doi.org/10.1128/AEM.03718-14

Birhanu AG (2023) Mass spectrometry-based proteomics as an emerging tool in clinical laboratories. Clin Proteomics 20(1):32. https://doi.org/10.1186/s12014-023-09424-x

Cherry JR, Fidantsef AL (2003) Directed evolution of industrial enzymes: an update. Curr Opin Biotechnol 14(4):438–443. https://doi.org/10.1016/S0958-1669(03)00099-5

Dahal S, Yurkovich JT, Xu H, Palsson BO, Yang L (2020) Synthesizing systems biology knowledge from omics using genome-scale models. Proteomics 20(17–18):e1900282. https://doi.org/10.1002/pmic.201900282

de Boer W, Folman LB, Summerbell RC, Boddy L (2005) Living in a fungal world: impact of fungi on soil bacterial niche development. FEMS Microbiol Rev 29(4):795–811. https://doi.org/10.1016/j.femsre.2004.11.005

Ettinger CL, Eisen JA, Stajich JE (2025) Eukaryotic MAGs recovered from deep metagenomic sequencing of the seagrass, Zostera marina, include a novel chytrid in the order Lobulomycetales. bioRxiv (Cold Spring Harbor Laboratory). https://doi.org/10.1101/2025.02.11.637735

Hall DO, Scrase JI (1998) Will biomass be the environmentally friendly fuel of the future? Biomass Bioenergy 15(4–5):357–367. https://doi.org/10.1016/S0961-9534(98)00030-0

Handelsman J, Rondon MR, Brady SF, Clardy J, Goodman RM (1998) Molecular biological access to the chemistry of unknown soil microbes: a new frontier for natural products. Chem Biol 5(10):R245–R249. https://doi.org/10.1016/S1074-5521(98)90108-9

Hatakka A (2001) Biodegradation of lignin. In: Steinbüchel A (ed) Biopolymers, vol 1. Wiley-VCH, pp 129–180. https://doi.org/10.1002/3527600035.bpol1005

Heckart A, Cocuron J, Ray SC, Matheny GF, Rappleye CA, Alonso AP (2025) Mapping carbon utilization pathways in Histoplasma capsulatum through 13 C-metabolic flux analysis. mSystems 10(10):e0056925. https://doi.org/10.1128/msystems.00569-25

Helmy M, Smith D, Selvarajoo K (2020) Systems biology approaches integrated with artificial intelligence for optimized metabolic engineering. Metab Eng Commun 11:e00149. https://doi.org/10.1016/j.mec.2020.e00149

Hess M, Sczyrba A, Egan R, Kim TW, Chokhawala H, Schroth G, Luo S, Clark DS, Chen F, Zhang T, Mackie RI, Pennacchio LA, Tringe SG, Visel A, Woyke T, Wang Z, Rubin EM (2011) Metagenomic discovery of biomass-degrading genes and genomes from cow rumen. Science 331(6016):463–467. https://doi.org/10.1126/science.1200387

Hyde KD, Baldrian P, Chen Y, Chethana KWT, De Hoog S, Doilom M, De Farias ARG, Gonçalves MFM, Gonkhom D, Gui H, Hilário S, Hu Y, Jayawardena RS, Khyaju S, Kirk PM, Kohout P, Luangharn T, Maharachchikumbura SSN, Manawasinghe IS et al (2024) Current trends, limitations and future research in the fungi? Fungal Divers 125(1):1–71. https://doi.org/10.1007/s13225-023-00532-5

IPCC (2022) Climate change 2022: mitigation of climate change. Cambridge University Press. https://doi.org/10.1017/9781009157926

Kirk TK, Farrell RL (1987) Enzymatic "combustion": the microbial degradation of lignin. Ann Rev Microbiol 41:465–505. https://doi.org/10.1146/annurev.mi.41.100187.002341

Kumar R, Kim TH, Basak B, Patil SM, Kim HH, Ahn Y, Yadav KK, Cabral-Pinto MM, Jeon B (2021) Emerging approaches in lignocellulosic biomass pretreatment and anaerobic bioprocesses for sustainable biofuels production. J Clean Prod 333:130180. https://doi.org/10.1016/j.jclepro.2021.130180

Leonowicz A, Matuszewska A, Luterek J, Ziegenhagen D, Wojtaś-Wasilewska M, Cho NS, Hofrichter M, Rogalski J (2001) Biodegradation of lignin by white rot fungi. Fungal Genet Biol 27(2–3):175–185. https://doi.org/10.1006/fgbi.1999.1150

Li J, Luo Y, Yu B, He J, Wang H, Wang Q, Chen D (2025) Multi-omics analysis provides insights into mechanisms of intestinal fungi adaptation to dietary carbohydrates. Curr Res Microb Sci 9:100451. https://doi.org/10.1016/j.crmicr.2025.100451

Miyauchi S, Navarro D, Grisel S, Chevret D, Berrin J, Rosso M (2017) The integrative omics of white-rot fungus Pycnoporus coccineus reveals co-regulated CAZymes for orchestrated lignocellulose breakdown. PLoS One 12(4):e0175528. https://doi.org/10.1371/journal.pone.0175528

Peterson R, Nevalainen H (2012) Trichoderma reesei RUT-C30—thirty years of strain improvement. Microbiology 158(1):58–68. https://doi.org/10.1099/mic.0.054031-0

Pinu FR, Beale DJ, Paten AM, Kouremenos K, Swarup S, Schirra HJ, Wishart DS (2019) Systems biology and multi-omics integration: viewpoints from the metabolomics research community. Metabolites 9(4):76. https://doi.org/10.3390/metabo9040076

Ratledge C (2014) The role of malic enzyme as the provider of NADPH in oleaginous microorganisms. FEMS Microbiol Rev 38(4):564–579. https://doi.org/10.1111/1574-6976.12057

Ratledge C, Wynn JP (2002) The biochemistry and molecular biology of lipid accumulation in oleaginous microorganisms. Adv Appl Microbiol 51:1–51. https://doi.org/10.1016/S0065-2164(02)51000-5

Reel PS, Reel S, Pearson E, Trucco E, Jefferson E (2021) Using machine learning approaches for multi-omics data analysis: a review. Biotechnol Adv 49:107739. https://doi.org/10.1016/j.biotechadv.2021.107739

Ren Y, Yu G, Shi C, Liu L, Guo Q, Han C, Zhang D, Zhang L, Liu B, Gao H, Zeng J, Zhou Y, Qiu Y, Wei J, Luo Y, Zhu F, Li X, Wu Q, Li B et al (2022) Majorbio cloud: a one-stop, comprehensive bioinformatic platform for multiomics analyses. iMeta 1(2):e12. https://doi.org/10.1002/imt2.12

Sartaj K, Prasad R, Matsakas L, Patel A (2023) Transforming recalcitrant wastes into biodiesel by oleaginous yeast: an insight into the metabolic pathways and multi-omics landscape. Chem Eng J 474:145625. https://doi.org/10.1016/j.cej.2023.145625

Subramanian I, Verma S, Kumar S, Jere A, Anamika K (2020) Multi-omics data integration, interpretation, and its application. Bioinformatics and Biology Insights 14:1177932219899051. https://doi.org/10.1177/1177932219899051

Sun Y, Yan K, Zhang Y, Han Y, Hao L, Gao Y, Wang H, Chu H, Ye J, Liu Y, Yang Y (2025) Advanced multifunctional nano-delivery platform focusing on treating diseases related to lipid metabolism via targeted intervention in various lipid metabolic processes. Mil Med Res 12(1):87. https://doi.org/10.1186/s40779-025-00672-6

Vanwonterghem I, Jensen PD, Ho DP, Batstone DJ, Tyson GW (2014) Linking microbial community structure, interactions, and function in anaerobic digesters using metagenomics. ISME J 8(5):1074–1085. https://doi.org/10.1038/ismej.2014.50

VerBerkmoes NC, Denef VJ, Hettich RL, Banfield JF (2009) Systems biology: functional analysis of natural microbial consortia using community proteomics. Nat Rev Microbiol 7(3):196–205. https://doi.org/10.1038/nrmicro2080

Walker GM, Stewart GG (2016) Saccharomyces cerevisiae in the production of fermented beverages. Beverages 2(4):30. https://doi.org/10.3390/beverages2040030

Wang K, Tester JW (2023) Sustainable management of unavoidable biomass wastes. Green Energy Resour 1(1):100005. https://doi.org/10.1016/j.gerr.2023.100005

Wijayawardene NN, Boonyuen N, Ranaweera CB, de Zoysa HKS, Padmathilake RE, Nifla F, Dai D-Q, Liu Y, Suwannarach N, Kumla J, Bamunuarachchige TC, Chen H-H (2023) OMICS

and other advanced technologies in mycological applications. J Fungi 9(6):688. https://doi.org/10.3390/jof9060688
Wörheide MA, Krumsiek J, Kastenmüller G, Arnold M (2021) Multi-omics integration in biomedical research — a metabolomics-centric review. Anal Chim Acta 1141:144–162. https://doi.org/10.1016/j.aca.2020.10.038
Yan J, Risacher SL, Shen L, Saykin AJ (2017) Network approaches to systems biology analysis of complex disease: integrative methods for multi-omics data. Brief Bioinform 19(6):1370–1381. https://doi.org/10.1093/bib/bbx066
Zhang Y, Yu X, Li Y, Chen L (2020) Metatranscriptomic insights into fungal roles in organic waste degradation. Appl Environ Microbiol 86(17):e00991–e00920. https://doi.org/10.1128/AEM.00991-20
Zhang W, Wen Y, Wang Z, Diao C, Liu Z (2025b) The Fungi;Bacteria interaction mechanism of microbial consortium during efficient lignin degradation based on metabolomics analysis. Molecules 30(3):508. https://doi.org/10.3390/molecules30030508
Zhao M, Zhang D, Su X, Duan S, Wan J, Yuan W, Liu B, Ma Y, Pan Y (2015) An integrated metagenomics/metaproteomics investigation of the microbial communities and enzymes in solid-state fermentation of Pu-erh tea. Sci Rep 5(1):10117. https://doi.org/10.1038/srep10117

Chapter 13
CRISPR–Cas9 Applications in Fungi for Developing Energy and Biomaterials: Challenges and Perspectives

Gaurav Mudgal, Jaspreet Kaur, Sanjoy K. Dhar, and Kübra Sağlam

Abstract The increasing demand for both sustainable energy and bio-based materials has given rise to an imperative need for precision tools in the engineering of microbial platforms. Among these, Clustered Regularly Interspaced Short Palindromic Repeats (CRISPR)–Cas9, where Cas9 refers to CRISPR-associated protein 9, has revolutionized fungal biotechnology by enabling targeted genome editing with an unprecedented level of accuracy, efficiency, and scalability. Fungi, particularly filamentous species and oleaginous yeasts, are potential candidates that offer very promising perspectives for biofuel, organic acid, and functional biomaterial synthesis from renewable feedstocks. However, large and complex genomes, variable transformation efficiencies, and limited genetic toolkits have historically impeded their industrial development. This chapter provides a detailed overview of the transformative role of CRISPR–Cas9 in overcoming such limitations, explaining its applications in metabolic rewiring, pathway optimization, and regulatory circuit modulation across fungal systems such as *Aspergillus nodulans*, *Trichoderma reesei*, and *Yarrowia lipolytica*. Attention has been given to recent advances in multiplex editing, base editing, and CRISPRi/a strategies for increased lipid accumulation, cellulase production, and biopolymer biosynthesis. The chapter also addresses key challenges, including off-target effects, delivery systems, and intellectual property barriers, while proposing integrative frameworks that combine CRISPR with omics data, machine learning, and bioprocess engineering. By synthesizing current

G. Mudgal (✉)
University Institute of Biotechnology, Chandigarh University, Mohali, Punjab, India

J. Kaur
Department of Biotechnology, Kanya Maha Vidyalaya College, Jalandhar, Punjab, India

S. K. Dhar
Banophool Adibashi Green Heart College, Mirpur, Dhaka, Bangladesh

Bangladesh University of Health Sciences, Mirpur, Dhaka, Bangladesh

K. Sağlam (✉)
Department of Food Engineering, Faculty of Agriculture, Tekirdağ Namik Kemal University, Tekirdağ, Turkey

M. Tripathi et al. (eds.), *Recent Advancements in Fungal Biotechnology Volume 1*, Fungal Biology, https://doi.org/10.1007/978-981-92-0339-0_13

advances and future directions, this chapter positions CRISPR-enabled fungal platforms as pivotal agents in the development of next-generation bioenergy and biomaterials, aligned with circular economy and climate resilience goals.

Keywords Genome editing · Oleaginous yeasts · Metabolic rewiring · Biofuel synthesis · Cellulase enhancement · Synthetic biology · sgRNA design · Bioprocess integration · Regulatory circuits · Prime editing

13.1 Introduction

The current state of the world is beset by a crisis in both energy and materials, and this has far-reaching implications for economic, environmental, and human well-being. Global energy demand increased at an above-average level in 2024, and this has been largely driven by an acceleration in electricity demand, thanks to industrialization, transport electrification, and an increased demand from digital and AI applications (IEA 2024). The increased dependency on fossil fuel energy has continued to contribute to an escalation in carbon emissions. Energy-related CO_2 emissions reached record levels in 2024, according to findings from the International Energy Agency, and this has further strained calls for increased focus on transitioning to renewable energy (IEA 2024).

Geopolitics and issues with the supply chain have also meant that the fossil fuel market is becoming more volatile. The materials crisis is also a significant problem. The world is producing more than 400 million tons of plastic every year, with less than 10% being recycled. A huge amount is being dumped into landfills and the ocean. Plastic, a product of fossil fuels, continues breaking down in the environment for thousands of years, causing the crisis of microplastics, which find their way into the food chain and the human health system (Sharma et al. 2021; Seetharaman et al. 2025; Singh et al. 2025a; Mudgal et al. 2025a).

The rewards of tackling these crises are enormous. The shift to renewable energy alternatives has the potential to cut global greenhouse gas emissions by over 70% by the middle of this century, whereas the shift from petroleum-based plastics to biodegradable materials has the potential to cut plastic waste by several hundreds of millions of tons every year (IRENA 2025). In economic terms, the renewable energy industry already provides jobs to over 13 million people globally. The growth of bio-based materials will create tens of millions of jobs in agriculture, biotechnology, and industry. Socially, the reduction of overdependence on fossil fuels and plastics will help mitigate health-related issues caused by air pollution, climate change-related natural disasters, and exposure to toxic materials. To accomplish these, innovation and discovery of biological solutions capable of providing sustainable energy and materials will be needed.

Fungi are among the most fascinating biological models to provide such a solution. Both filamentous fungi and oil-producing yeasts have native enzymatic capabilities to break down lignocellulosic materials, a cheap, renewable, and plentiful substrate that comes from agricultural residues and wood waste, converting them into fermentable sugars. These can be easily processed into bioproducts such as bioethanol, biodiesel, and biogas, which provide a clean, safe, and more sustainable fuel upgrade to fossil fuels. Certain oil-producing yeasts, *Papiliotrema laurentii*, are capable of accumulating lipids that constitute more than 60% of their cell mass, directly applicable to biodiesel manufacture. Moreover, mycelium, or vegetative filaments, from fungal cells can be used in microbial fuel cells, where electrons from external sources are drawn to extracellular microorganisms, reducing organic waste while generating electricity. Finally, in materials sciences, mycelium-based biomaterials are made, derived from their fungal roots, which are very light, biodegradable, resistant to flames, and thus applicable to packaging, insulation, construction materials, and other products that use former packaging materials, though this application is still under development to suit commercial standards.

Fungal-based foams can even replace Styrofoam materials, and mycelium-based colorant compounds from *Monascus ruber* M1 are non-toxic bioproducts useful to replace existing, toxic colors currently in use. The global payoff of deploying fungal systems is multifaceted. Environmentally, fungal biofuels could displace millions of barrels of oil annually, reducing emissions and reliance on geopolitically unstable fossil fuel markets. Biodegradable fungal materials could replace a significant fraction of petroleum-derived plastics, cutting plastic pollution and easing waste management burdens. Economically, the use of fungal biotechnology is cost-effective, as it uses low-cost feed sources like agricultural waste. Socially, the use of fungi is geared toward applying the principles of the circular economy, which focuses on the local use of resources. An added aspect of the use of fungi is the fact that they can be produced on a large scale through cultivation either within a controlled bioreactor or through the use of waste materials.

Therefore, the fuel and biomaterial crisis has been driven by the unsustainable use of fossil fuels, the escalating demand for energy, and the problems associated with the use of synthetic plastic materials. The problems associated with the crisis are catastrophic, but the opportunities associated with the use of sustainable materials are equally revolutionary. Fungi, with their diverse metabolism, offer a brilliant natural solution to the problems associated with the fuel and biomaterial crisis. Through the metabolism offered by fungi, the problems associated with the crisis, problems of energy security and pollution, in addition to the problems of climate change and sustainable development, would be addressed.

The CRISPR–Cas9 genome editing tool has revolutionized fungal biotech, offering precisely what is needed to accelerate the development of fungi as biofuel feedstocks and biomaterials. To start with, fungi are arguably the most versatile type of microbial cell factories that occur naturally and possess inherent biochemical potential for the degradation of complicated lignocellulosic biomass and the production of diverse value-added compounds like lipids, organic acids, enzymes, and secondary metabolites (Riaz et al. 2025; Villena et al. 2020). Such potential makes filamentous fungi and yeasts leaders in industrial biotech, especially with the advent of circular economies that mainly focus on the usage of renewable instead of fossil resources. However, natural fungal strains always have inherent sub-optimal rates of productivity, limited capabilities for substrate utilization, and inappropriate patterns of byproduct production, for which certain genetic modifications must be made to enhance their potential.

However, the main challenge with traditional methods is addressed by CRISPR–Cas9, which provides a very efficient, scalable, and programmable platform for strain engineering. In contrast to traditional mutagenesis methods, which rely on random approaches that are often labor-intensive, CRISPR–Cas9 technology enables site-specific modifications in the form of gene knockouts, knock-ins, point mutations, and transcriptional regulation at an unprecedented level of precision and throughput (Song et al. 2019; Ullah et al. 2020). This possibility has been exploited to improve cellulase and hemicellulase production in important industrial species such as *Trichoderma reesei* and *Rasamsonia emersonii*, which have, in turn, enhanced enzymatic hydrolysis of plant biomass into fermentable sugars for subsequent biofuel synthesis (Pant et al. 2022; Singh et al. 2024). In oleaginous yeasts like *Papiliotrema laurentii*, lipid biosynthesis pathways have been modified with CRISPR, enabling increased accumulation of triacylglycerides responsible for the efficient production of biodiesel (Moreira Vieira et al. 2020).

Apart from biofuel production, CRISPR–Cas9 genome-edited fungi are being used as a non-hazardous, environmentally benign source of biomaterials. Mycelium composites, obtained from the hyphae of fungi cultivated on agricultural waste substrates, are biodegradable and produce no hazardous byproducts, making them ideal "green" foams and plastics. Optimization of genome functions using CRISPR technology could improve the rate of mycelium growth, mechanical properties, and ability to survive harsh environmental conditions, allowing engineered fungi to be used as specific materials in industry. CRISPR technology has also improved the overproduction of natural pigments in pigment-producing fungi like *Monascus ruber*, which can serve as food-safe, non-toxic dyes to replace hazardous chemical dyes (Yoon et al. 2014).

Despite these advances, several challenges still serve as barriers to the wider application of CRISPR–Cas9 in fungal systems. The delivery of CRISPR components remains one of the major bottlenecks due to difficulties with species having strong cell walls or less efficient transformation protocols. Recent innovations with nanoparticle-mediated delivery platforms, including lipid-based vectors, gold nanoparticles, and DNA nanoclews, have started to overcome this barrier by increasing the efficiency of transfection and reducing cytotoxicity. Off-target effects are also a concern, although strategies using high-fidelity Cas9 variants, RNP complexes, and transient expression systems have made significant improvements in editing specificity. Perhaps more concerning is the continued regulatory uncertainty around genetically edited organisms, particularly those that involve transgenic elements, which is impacting commercial timelines and public acceptability.

In the future, the integration of CRISPR–Cas9 with multi-omics technologies will allow systems-level understanding of fungal metabolism and regulatory networks that will guide the rational design of superior industrial strains. Inducible promoters, synthetic circuits, and dynamic control mechanisms have great potential for further optimization of metabolic fluxes and the alleviation of the fitness cost problem during the production of bioproducts. As research is underway for non-homologous end-joining-deficient fungal hosts and marker-free editing systems, the field is getting closer to transgene-free, self-cloning strains that possibly face fewer regulatory hurdles.

This chapter will look in depth into the current industry, challenges, and perspectives in the use of CRISPR–Cas9 technologies in fungi for the production of fuel and biomaterials. It will assess the various species of fungi, the technologies of gene delivery, the efficiency of gene editing, and the safety and regulation, as well as the imminent futures capable of forming the next frontier of the fungal biomanufacturing industry.

13.2 Fungal Systems for Bioenergy and Biomaterials

13.2.1 Role of Fungi in Biofuel Production

Fungal systems are gradually being identified for their immense potential for the enhancement of bioenergies and biomaterials, mainly due to their versatility and ability to metabolize a wide range of materials (Villena et al. 2020; Hinneburg et al. 2025; Chai et al. 2022; Agrawal et al. 2021; Agrawal and Verma 2021). Fungal systems are a very important area of material and geochemical degradation and have long been harnessed for preparing various material components of foods and beverages, up to now for more than a thousand years (Villena et al. 2020; Kapahi et al. 2021). However, the current biotechnology and implementation of the newly discovered CRISPR–Cas9 system for the first time as a powerful gene-editing method have brought a revolutionary change into the utilization and manipulation of the fungal system for the purpose of sustainable energy and material application processes due to its high precision and efficacy and the predominant genomic variability of the organisms involved in the target application processes themselves (Song et al. 2019; Ullah et al. 2020; Singh et al. 2024; Pal et al. 2024; Subramanian et al. 2025; Nødvig et al. 2015; Yang et al. 2024). A detailed description of the various fungal species utilized for biofuel and material production processes is provided in Table 13.1.

Table 13.1 Comprehensive overview of fungal species utilized in biofuel and biomaterial production systems

Fungal species and composites	Category	Application	Mechanism	Efficiency/yield	Advantages	Limitations	Key findings	References
Aspergillus niger	Filamentous fungi	Bioethanol	Cellulase and hemicellulase production for lignocellulosic biomass hydrolysis	High enzyme yields (10–15 FPU/mL)	GRAS status, robust growth, and high enzyme secretion	Requires pretreatment of feedstock, sensitive to inhibitors	Efficient conversion of agricultural waste to fermentable sugars with 75–85% hydrolysis efficiency	Kumar et al. (2016)
Trichoderma reesei	Filamentous fungi	Bioethanol	Synergistic cellulase system (CBH I, CBH II, endoglucanases)	90% cellulose conversion efficiency	Industry standard for cellulase production, well-characterized genetics	Low β-glucosidase activity requires supplementation	Genetic engineering increased enzyme production by 40%, reducing costs by 30%	Peterson and Nevalainen (2012)
Cryptococcus curvatus	Oleaginous yeast	Biodiesel	Lipid accumulation (60–70% DCW) through nitrogen limitation	Lipid productivity: 4.5 g/L/day	High lipid content, uses diverse carbon sources, and has rapid growth	Downstream processing costs require optimization	Biodiesel met ASTM D6751 standards with excellent cold flow properties	Meng et al. (2014)
Rhodosporidium toruloides	Oleaginous yeast	Biodiesel	De novo lipogenesis from lignocellulosic hydrolysates	Lipid yield: 0.22 g/g glucose	Tolerates inhibitors, grows on waste streams, high lipid accumulation	Lower growth rate than conventional yeasts	Produced biodiesel from corn stover hydrolysate with 68% lipid content	Park et al. (2018)

Ganoderma lucidum	White-rot fungi	Mycelium biomaterials	Mycelial binding of agricultural residues through hyphal networks	Complete colonization in 7–14 days	Strong mechanical properties, fire-resistant, and biodegradable	Growth time requires sterile conditions and moisture sensitivity	Tensile strength of 0.5–1.5 MPa, comparable to expanded polystyrene	Holt et al. (2012)
Pleurotus ostreatus	White-rot fungi	Mycelium biomaterials	Hyphal aggregation and chitin-glucan matrix formation	Rapid colonization (5–10 days)	Fast growth, uses diverse substrates, and has good insulation properties	Variable mechanical properties require post-processing	Thermal conductivity 0.05–0.07 W/mK, suitable for building insulation	Jones et al. (2020)
Trametes versicolor	White-rot fungi	Lignin degradation	Laccase and peroxidase secretion for lignin depolymerization	70–80% lignin removal in 21 days	Non-selective lignin degradation, low nutrient requirements	Slow process: cellulose co-degradation	Pretreatment improved bioethanol yields by 45% from wheat straw	Bourbonnais et al. (1995)
Fusarium oxysporum	Filamentous fungi	Biobutanol	ABE fermentation pathway modification	Butanol titer: 12–15 g/L	Higher alcohol tolerance, genetic tractability	Lower yields than Clostridium; pathogenic strains exist	Engineered strain produced 18 g/L of butanol from sugarcane bagasse	Liu et al. (2017)
Mucor circinelloides	Oleaginous fungi	Biodiesel & omega-3	PUFA biosynthesis via desaturase/elongase pathways	GLA content: 15–20% of total lipids	Produces valuable PUFAs, grows rapidly, and uses low-cost substrates	Requires controlled fermentation and extraction complexity	Co-production of biodiesel and high-value gamma--linolenic acid	Eroshin et al. (2000)

(continued)

Table 13.1 (continued)

Fungal species and composites	Category	Application	Mechanism	Efficiency/ yield	Advantages	Limitations	Key findings	References
Neurospora crassa	Filamentous fungi	Mycoprotein	High protein biomass production (45–55% protein content)	Biomass productivity: 2.5 g/L/h	Model organism, rapid growth, complete amino acid profile	RNA content requires reduction, and processing is needed	Protein quality comparable to animal protein with superior fiber content	Souza Filho et al. (2017)
Schizophyllum commune	White-rot fungi	Biopolymers (schizophyllan)	Extracellular β-glucan production with a triple helix structure	Yield: 8–12 g/L in submerged culture	Biocompatible, immunomodulatory, biodegradable	High viscosity complicates processing and is expensive to produce	Enhanced drug delivery efficiency by 300% in cancer treatment studies	Kumari et al. (2008)
Mortierella isabellina	Oleaginous fungi	Biodiesel & nutraceuticals	Arachidonic acid (ARA) biosynthesis through Δ6-desaturase	Lipid content: 50–60% DCW; ARA: 8–10% of lipids	Produces omega-6 fatty acids, grows on waste glycerol	Oxygen--demanding, foam formation in bioreactors	Converted crude glycerol to ARA-rich oils with 55% lipid accumulation	Sagia et al. (2020)
Fomes fomentarius	White-rot fungi	Leather alternatives	Dense mycelial mat formation with hyphal fusion	Material thickness: 2–10 mm in 14–21 days	Leather-like texture, dyeable, sustainable alternative	Moisture sensitivity requires chemical treatment for durability	Mechanical properties similar to natural leather after the tanning process	Sivaprasad et al. (2021)

Yarrowia lipolytica	Oleaginous yeast	Biodiesel & omega-3	Genetic engineering for EPA/DHA production via metabolic pathway insertion	Lipid productivity: 5.8 g/L/day; EPA: 15% of lipids	GRAS organism, genetic tools available, secretes lipases	Requires complex media, oxygen--intensive	Engineered strain produced marine omega-3s at an industrial scale	Xue et al. (2013)
Pycnoporus cinnabarinus	White-rot fungi	Bioethanol pretreatment	High laccase production for lignin oxidation without peroxidases	95% lignin modification in 15 days	No mediator required for thermostable laccases	Slower than chemical pretreatment	Biological pretreatment increased glucose release by 52% from poplar wood	Eggert et al. (1996)
Aureobasidium pullulans	Yeast-like fungi	Biopolymers (pullulan)	Exopolysaccharide production from starch and sucrose	Pullulan yield: 60–80 g/L	Biodegradable, oxygen barrier properties, film-forming	Melanin co-production, viscosity issues	Edible films showed a 95% oxygen barrier, extending food shelf life by 40%	Singh et al. (2009)
Cunninghamella elegans	Oleaginous fungi	Biodiesel	Lipid biosynthesis with high C18:1 and C18:2 content	Lipid content: 40–45% DCW	Optimal fatty acid profile for biodiesel, fast growth	Lower lipid content than some competitors	Biodiesel met EN 14214 standards with excellent oxidative stability	Bellou et al. (2016)

(continued)

Table 13.1 (continued)

Fungal species and composites	Category	Application	Mechanism	Efficiency/ yield	Advantages	Limitations	Key findings	References
Penicillium chrysogenum	Filamentous fungi	Enzyme production	Cellulase, lipase, and protease secretion for biomass processing	Multi-enzyme production in a single fermentation	Industrial workhorse, well-studied, GRAS status	Lower cellulase activity than specialized fungi	The enzyme cocktail achieved 68% saccharification of pretreated rice straw	Quevedo--Amador et al. (2024)
Lentinula edodes	White-rot fungi	Biomaterials & bioethanol	Lignin peroxidase and manganese peroxidase production	60–70% delignification in 28 days	Edible species, selective lignin degradation, produce lentinan	Slow-growing, temperature sensitive	Dual application: pretreatment increased ethanol yield by 38%, and mycelium formed packaging material	Lubbers (2025)
Rhizopus oryzae	Filamentous fungi	Bioethanol	Direct conversion of starch to ethanol via amylolytic fermentation	Ethanol yield: 0.45–0.48 g/g starch	No separate hydrolysis step; uses starchy materials directly	Lower ethanol tolerance produces organic acids	One-step fermentation of cassava starch achieved 95% theoretical yield	Derman et al. (2022)
Phanerochaete chrysosporium	White-rot fungi	Lignin valorization	Lignin peroxidases and H_2O_2--generating oxidases	Complete mineralization of lignin model compounds	Most studied white-rot fungi produce aromatic compounds	Requires secondary metabolism induction, a slow process	Converted lignin to vanillin and syringaldehyde at 12% yield	Bilal et al. (2017)

Zymomonas mobilis	Bacteria (comparison)	Bioethanol	Entner-Doudoroff pathway with high ethanol tolerance	Ethanol productivity: 5–10 g/L/h	Higher yields than fungi, faster fermentation	Narrow substrate range compared to fungi	Reference organism: Fungi show advantages in enzyme production and waste utilization	Rogers et al. (2007)
Mycelium composites	Multi-species systems	Construction materials	Fungal binding of agricultural waste creates structural materials	Compressive strength: 0.5–2.0 MPa	Carbon-negative, lightweight, moldable into complex shapes	Variable properties require growth time, moisture control	MycoBoard panels showed fire resistance Class B and were 30% lighter than conventional materials	Attias et al. (2019)
Aspergillus oryzae	Filamentous fungi	Enzyme production	Amylase and protease secretion for starch and protein processing	Enzyme activity: 1000+ U/mL	GRAS status, high protein secretion capacity, and koji fermentation	Requires sterile conditions, sensitive to contamination	Industrial enzymes reduced bioethanol processing costs by 25%	Machida et al. (2005)

The table summarizes key fungal organisms across multiple taxonomic groups, including filamentous fungi, oleaginous yeasts, and white-rot basidiomycetes, detailing their specific applications in bioethanol, biodiesel, biobutanol production, and novel biomaterial development. Mechanistic processes, efficiency of conversion, feasibility, existing drawbacks, and landmark research findings are discussed along with the corresponding literature references

13.2.2 Fungal Networks in Microbial Fuel Cells

The application of fungi is primarily related to their application in the production of various biofuels, including bioethanol, biodiesel, and biogas (Chai et al. 2022; Kapahi et al. 2021). Fungi have adequate enzyme systems, such as cellulases and ligninases, which are required for the application of lignocellulosic biomass (LCB), a non-digestible and abundant material that is renewable (Singh et al. 2024; Pal et al. 2024; Wang et al. 2020). For instance, the oleaginous yeast strain *Papiliotrema laurentii* UFV-1 possesses the ability to produce a considerable quantity of lipids that are utilized to produce biofuels derived from fatty acids directly (Moreira Vieira et al. 2020). The primary emphasis of optimizing the oleaginous trait is to enhance the lipid production of *P. laurentii* UFV-1 and reach a maximum lipid production within 48 hours of the incubation period (Moreira Vieira et al. 2020). Furthermore, the production of bioelectricity using fungal organisms is possible using microbial fuel cells (MFC), which occurs due to the high mycelial density that performs well during the process of extracellular electron transfer (Umar et al. 2024). This approach is quite useful and has the ability to provide a platform for the production of bioelectricity and biodegradation processes synchronously.

13.2.3 Mycelium-Based Biomaterials

When it comes to biomaterials, fungi are a revolutionary step ahead in terms of producing natural materials as well as novel material classes themselves (Hinneburg et al. 2025). For instance, bioplastics and mycelium-based composites derived from fungal mycelium can replace traditional materials (Elsacker et al. 2025). These ELMs take advantage of the self-assembling and adaptive nature of the fungi, which allows for new biofabrication methods (Elsacker et al. 2025). Besides mycelium, fungi produce a broad range of biomass-soluble primary and secondary biomolecules, which have plenty of applications in medicine, agriculture, etc. For instance, one has industrial enzymes, pigments, vitamins, organic acids, lipids, and antibiotics (Villena et al. 2020; Agrawal et al. 2021; Agrawal and Verma 2021). Furthermore, the constant finding of cryptic/uncharacterized secondary metabolite biosynthetic gene clusters shows that fungi still have immense potential concerning new compound production (Jiang et al. 2021; Leal et al. 2024).

13.2.4 CRISPR–Cas9 for Fungal Optimization

Looking ahead, the CRISPR–Cas9 system is poised to remain at the forefront of genetic engineering, offering transformative possibilities for advancing fungal platforms in industrial biotchnology (Song et al. 2019; Ullah et al. 2020; Pal et al. 2024; Subramanian et al. 2025; Yang et al. 2024; Kumar et al. 2021; Wang and

Coleman 2019; Chao-min et al. 2017; Shi et al. 2017; Liu et al. 2015; Li and Liu 2017; Liao et al. 2021). The capacity for precise and efficient genome editing is expected to accelerate further improvements in production efficiency, stress resilience, and substrate metabolism of fungi. Prospectively, this technology will continue to refine gene function and regulatory networks, particularly in fungi, where conventional genetic approaches often fall short in speed and effectiveness (Pant et al. 2022; Liu et al. 2015). Such advancements may open new avenues for sustainable industrial processes, novel therapeutic applications, and enhanced biotechnological innovation.

The CRISPR–Cas9 system has come to the forefront as a state-of-the-art genetic engineering technique, which has greatly influenced the improvement of fungal systems for industrial use (Song et al. 2019; Subramanian et al. 2025; Nødvig et al. 2015; Kumar et al. 2021; Chao-min et al. 2017; Shi et al. 2017; Krappmann 2016; Zhang et al. 2020). Presently, this technology enables the efficient and precise editing of the genome and helps to significantly promote various advantages, including the improvement of the production efficiency and stress resistance, and the enhanced substrate metabolism of fungi. For instance, it has been successfully used in the improvement of gene function and regulation, particularly in filamentous fungi, in which conventional genetic methods can sometimes be ineffective and time-consuming (Song et al. 2019; Shi et al. 2017; Deng et al. 2017).

13.2.5 Challenges in Delivery and Off-Target Effects

Despite the vast potential, some hurdles and issues also exist, which are to be considered to fully benefit from the potential offered by the CRISPR–Cas9 system. One of the most important hurdles associated with the delivery of the CRISPR–Cas9 system to the fungal cells lies in their complex cell walls. Some nano-platforms, such as lipid nanoparticles, gold nanoparticles, DNA nanoclews, and polymer nanoparticles, have been studied because of their potential to act as effective carriers to deliver the gene editors to the fungal cells. As stated by Singh et al. in 2024, gold nanoparticles and other nanoparticles are effective in ensuring the delivery of the Cas9 gene editors to the fungal cells (Singh et al. 2024). The nanoparticles contain the CRISPR–Cas9 system, which easily enters the fungal cells to release the system into the nucleus to edit genes (Singh et al. 2024; Eoh and Gu 2019). Some biomaterials have also been recognized as an effective alternative to deliver Cas9 due to their versatility (Eoh and Gu 2019).

A second key challenge relates to off-target effects, where the guide RNA target sequences are complementary to the target site, causing mutations (Garrigues et al. 2023). Although the CRISPR–Cas9 system has high specificity, the assessment of the off-target mutations could be critical for the safety and specificity of the genome modifications. Studies on non-homologous end-joining (NHEJ) deficient fungal strains are being carried out to tackle the issue and reduce the effect of off-target mutational errors (Garrigues et al. 2023). In addition to the abovementioned challenges, the complexity of fungal genomes and the relative efficacies of fungal genetic transformability between species also constitute hurdles (Ullah et al. 2020; Wang and Coleman 2019).

13.2.6 *Multi-Omics and Advanced Editing Integrations*

The future outlook for the application of fungal systems using CRISPR–Cas9 includes the integration of multi-omics (genomics, transcriptomics, proteomics, and metabolomics) to achieve an understanding of the metabolism and regulation of the fungal systems (Grigaitis et al. 2021). The application of CRISPR–Cas9 for the optimization of fungal systems will also be improved based on the advanced understanding and strategy being applied. The application of advanced CRISPR approaches and systems, which include base editing and prime editing, will also improve the application of fungal systems (Liao et al. 2021). Through CRISPR–Cas9, the potential for combating pathogenic fungi that cause diseases in humans will also be realized through the application of the technology to produce novel approaches and treatments for the diseases (Román et al. 2019). The application of advanced CRISPR–Cas9 will play a critical role in realizing the potential of fungal systems for the sustainability of a bright future for the production of bioenergy and biomaterials.

13.3 CRISPR–Cas9 Technology: Fundamentals

13.3.1 *Origins of CRISPR-Cas9 Systems*

The CRISPR–Cas9 system has been a novel genome-editing technology, which has profoundly impacted genetic studies and their applications in different biological models, as reported (Song et al. 2019; Ullah et al. 2020; Pal et al. 2024). Originally identified in bacteria and archaea as an adaptive immune system, CRISPR (Clustered Regularly Interspaced Short Palindromic Repeats) has been recognized to protect bacteria against bacteriophages and plasmids (Chao-min et al. 2017). When repurposed as an approach to genome editing, it utilizes an RNA-directed DNA endonuclease from bacteria, Cas9 (CRISPR-associated protein 9), which targets specific DNA loci (Singh et al. 2024; Chao-min et al. 2017).

13.3.2 *Mechanism of Cas9 and Guide RNAs*

The basic mechanism by which CRISPR–Cas9 works is based on the Cas9 enzyme and a guide RNA (gRNA) molecule. The gRNA is a synthesized RNA molecule composed of two separate components: a CRISPR RNA (crRNA) that binds to the complementary DNA sequence by base-pair complementarity and a trans-activating CRISPR RNA (tracrRNA) that binds to the Cas9 enzyme. The gRNA and Cas9 enzyme can also be made as a single synthesized molecule known as a single guide RNA (sgRNA) molecule. The gRNA molecule searches for a specific site on the genome, usually a 20-base region that is complementary to the crRNA, and this is usually followed by a Protospacer Adjacent Motif (PAM) region (NGG in *Streptococcus pyogenes* Cas9) (Singh et al. 2024). The PAM is an integral component needed for Cas9 binding and cleavage and protects the host genome against targeting by the immune system.

13.3.3 DNA Repair Pathways: NHEJ Versus HDR

Upon identifying the target DNA sequence by the gRNA, the Cas9 protein makes a double-strand break (DSB) at this specific location. This DSB is then repaired by the cell's own DNA repair machinery, which is mostly non-homologous end joining (NHEJ) and homology-directed repair (HDR). NHEJ is an error-prone repair model that often introduces small insertions and deletions (indels) at this locus, causing frameshift mutations and early termination codons, thereby knocking out the gene function. On the other hand, HDR is a more precise repair model that uses a homologous DNA template to mend this DSB (Singh et al. 2024). By using an external DNA template that carries desired modifications, scientists can use this HDR process to introduce these specific modifications with high precision (Singh et al. 2024). The flexibility offered by CRISPR–Cas9 arises from its ease of gRNA design and modification, making it easy to target any location of choice in the genome by changing the 20-nucleotide crRNA sequence of gRNA.

13.3.4 Nanoparticle-Based Delivery Systems

The efficient delivery of CRISPR–Cas9 components into fungal cells is considered one of the major hurdles in the use of CRISPR–Cas9 in fungal systems, especially in filamentous fungi (Yang et al. 2024; Wang and Coleman 2019). The fungal cell is encased in a hard cell wall. The hard cell wall of fungal cells creates an obstruction against the uptake of nucleic acids or proteins (Yang et al. 2024). The classic methods of genetic transformation of fungi can be time-consuming (Song et al. 2019; Kumar et al. 2021). However, progress is being achieved in the field of delivery systems. Nanoparticle delivery systems, for example, are currently being pursued actively to increase the uptake of Cas9 protein/Cas9 mRNA/Cas9 plasmid as well as guide RNA into fungal cells (Singh et al. 2024; Yang et al. 2024). The novel approach of this thought-provoking scheme of using nanoparticle-mediated CRISPR–Cas9 delivery in fungal systems is represented in Fig. 13.1.

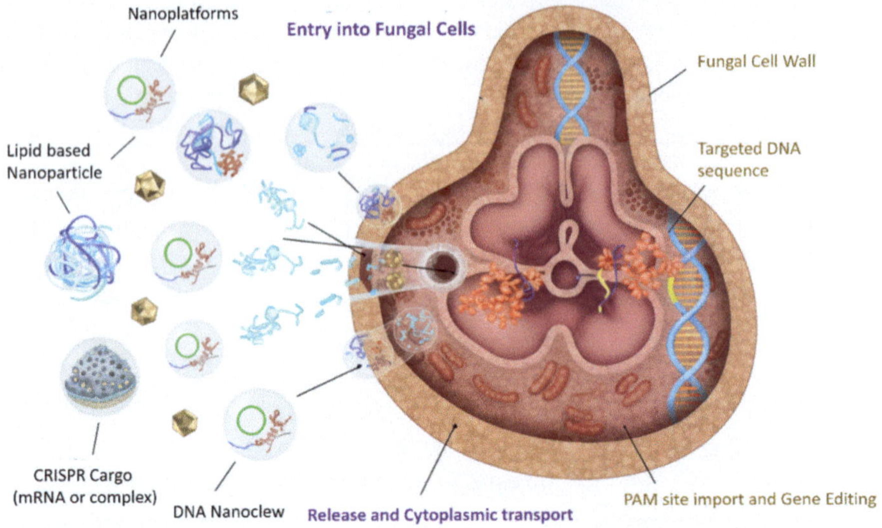

Fig. 13.1 Schematic illustration of nanoparticle-mediated CRISPR–Cas9 genome editing in fungi. A number of nano-formulations, such as lipid nanoparticles (LNPs), gold nanoparticles (AuNPs), DNA nanoclews, and polymeric nanoparticles, have been developed to encapsulate various CRISPR constructs such as plasmids encoding Cas9 and gRNA, mRNA-gRNA transcripts, and ribonucleoprotein complexes (RNPs). The nanoparticles can easily bypass the cell wall barrier and enter fungal cells through the process of endocytosis and other direct entry methods. Once inside the cell, the CRISPR components enter the nucleus, where the Cas9 nuclease locates the PAM locus for performing double-stranded breaks for gene editing

13.3.5 Managing Off-Target Effects

As specific as the CRISPR–Cas9 system generally is, it can occasionally yield off-target effects where the gRNA binds to sequences similar but not identical to the targeted sequence, hence causing unintended genomic modifications (Yang et al. 2024). Such off-target effects are of prime importance to be mitigated with respect to ensuring specificity and safety in genome editing, particularly in industrial or therapeutic applications. The generation of high-fidelity Cas9 variants and gRNA design algorithms stands at the heart of present research, focusing on improving the aforementioned features for the purpose of reducing such off-target activity. Furthermore, the added complexity of fungal genomes, including variations in genetic transformation efficiency across different fungal species, is yet another challenge that needs ongoing optimization strategies, depending on each particular species (Wang et al. 2020; Wang and Coleman 2019). Sophisticated CRISPR technologies, including base editing and prime editing, further expand precision in fungal genome engineering, promoting subtle and diverse genetic alterations (Yang et al. 2024).

13.4 Applications in Fungal Bioengineering

13.4.1 Lipid Pathway Enhancement for Biofuel Production

CRISPR–Cas9 facilitates the targeted reengineering of lipid pathways in oil-producing yeasts to boost triacylglyceride (TAG) storage to provide tailored fatty acids for the production of biodiesel and oleochemicals (Kapoor et al. 2025; Utomo et al. 2021). These enhancements can be achieved mainly through the deletion or repression of β-oxidation/competing carbon sink pathways; the overproduction of acetyl-CoA carboxylase (ACC), ATP citrate lyase (ACL), or diacylglyceride acyl-transferase (DGAT); or the introduction of exogenous desaturases or elongases to alter the levels of double bonds (Kapoor et al. 2025; Skrekas 2022). These modifications can be complemented by process techniques like nitrogen limitations or fed-batch media feeding to direct the flux of carbon into storage lipids without sacrificing growth rates (Skrekas 2022). In practice, there are considerations in optimizing growth in conjunction with the storage of lipids in redox and carbon flux.

13.4.2 Enzyme Overexpression for Biomass Conversion (Cellulases and Hemicellulases)

In filamentous fungi, CRISPR-Cas9-based editing enables the rapid pursuit of high-performance enzyme-producing strains for lignocellulosic saccharification (Singh et al. 2024; Shen et al. 2024). Approaches span the use of promoter replacement for enhanced or inducible gene expression, multi-copy integration of prioritized cellulase and hemicellulase gene expression, repression of transcriptional repressors, and redesign of secretion signals and chaperones (Shen et al. 2024; Ma et al. 2025). Impacts include increased levels and/or increased activity with lower enzyme doses for subsequent processing (Singh et al. 2024). Critical constraints on performance—secretion and enzyme stability—can be overcome by co-expressing foldases, protease inhibitors, and/or components for the secretion machinery or by properly aligning enzyme mixes with substrate composition (Ma et al. 2025).

13.4.3 Biopolymer Biosynthesis (PHAs, Polyesters, Melanin, and Other Functional Polymers)

The host fungus can be engineered for the production of biopolymers, either in the form of intracellular or extracellular bioplastics, using heterologous metabolic routes and redirecting metabolic precursor flows. Such examples would involve the use of genetic engineering involving PHA synthases from bacteria for PHA bioplastics (Todorov et al. 2025), redirecting glycogen or acetyl-CoA metabolism toward the generation of precursors for the bioplastic poly(lactic-co-glycolic acid) (PLGA) (Todorov et al. 2025), or manipulating cell wall formation for increased yields of bioplastic chitin or β-glucan-based bioplastics (Tupe et al. 2022; Zannat et al. 2025). All these aspire toward manipulating precursor availability, enzymes for biopolymer formation, and export or separation strategies for biopolymers exhibiting desired properties (Das et al. 2025; Mudgal et al. 2025b; Rao et al. 2025; Singh et al. 2025b; Thakur et al. 2025; Jayaprakash et al. 2025), while taking note of the impact of biopolymer formation on the physiology of the cells (Todorov et al. 2025).

13.4.4 Regulatory Modulation: CRISPRi/a, Base Editing, and Multiplex Strategies

More advanced CRISPR systems offer a high degree of control over gene expression or sequence modification beyond double-strand breaks. CRISPR interference (CRISPRi) or CRISPR activation (CRISPRa) systems are used for the reversible repression or activation of target genes (Teng et al. 2024). Base editing tools can be used to introduce single-base changes to the genome in order to fine-tune the kinetic constants or regulatory elements (Teng et al. 2024). Multiplex editing can be used to simultaneously rewire entire biochemical pathways to improve their flux distributions (Teng et al. 2024). These systems offer dynamic control in various stages of the fermentation process; however, their design must be accompanied by effective delivery tools and the assessment of off-target profiles. A harmonized strategy in host organisms, compatibility with feedstocks, and process design are essential for seamless exploitation of CRISPR-based engineering tools. Host organisms can be selected according to the type of molecule to be produced: oil-producing yeast and filamentous organisms for most compounds; engineered uptake and detoxification strategies for irregular waste streams and typical inhibitors; and design modifications in line with fermentation type (SSF vs. SmF), nutrient sources, and process recovery strategies. The produced organisms can then be tested for lab metrics such as titer, rate, and yield and then up-scaled. Lastly, add aspects that include biosafety and marker-free editing.

CRISPR-enabled interventions transform fungi into versatile cell factories for lipids, enzymes, and biopolymers by enabling targeted pathway rewiring, expression control, and multiplexed edits (Teng et al. 2024). To realize industrial impact, there is a need for integrating strain engineering with process optimization, feedstock realism, and recovery steps, together with extensive validation and considerations on regulatory strategy (Teng et al. 2024).

13.5 Advanced Strategies and Integrative Approaches

Advanced CRISPR strategies, implemented beyond single gene interventions, are now combined with systems biology and evolutionary approaches to construct hardy fungal platforms (Ma et al. 2025). The subsections that follow describe how these integrative methods strengthen biofuel and biomaterial production.

13.5.1 Multiplex Genome Editing and Pathway Refactoring

Multiplex CRISPR technology allows various loci changes to be made simultaneously to rewire whole metabolic routes, thus reducing iterative cycles of design-build-test and allowing coordinated control of precursor supply, cofactor balance, and byproduct sinks (Ma et al. 2025). Advantages include faster strain construction, reduced regulatory burden per edit, and the ability to test combinatorial designs in a single campaign. Practical implementation requires robust multiplex guide design, efficient delivery systems, and strategies to minimize off-target effects and chromosomal instability (Kaul et al. 2025). Figure 13.2 shows a compelling view of the merits of this nuanced approach.

13.5.2 Adaptive Laboratory Evolution and Evolutionary Engineering

Adaptive Laboratory Evolution (ALE) can be used in a complementary manner to rational editing for selecting desirable, but hard to predict, properties: inhibitor tolerance, substrate range, or productivity/growth trade-offs. By implementing ALE cycles along with precise CRISPR modifications, it will be feasible to fix modified phenotypes and identify helpful mutations for further rationalization. The ideal strategy will be based on alternating cycles of directed editing and ALE, followed by whole-genome sequencing for the purpose of incorporating adaptive mutations for the next-generation strains (Kaul et al. 2025).

13.5.3 Omics Integration and Machine Learning Guided Design

The integration of genomics, transcriptomics, proteomics, and metabolomics helps to get the systems-level topology of bottlenecks and regulation points, omics data inputs, constraint-based models, and machine learning tools to rank edits, model

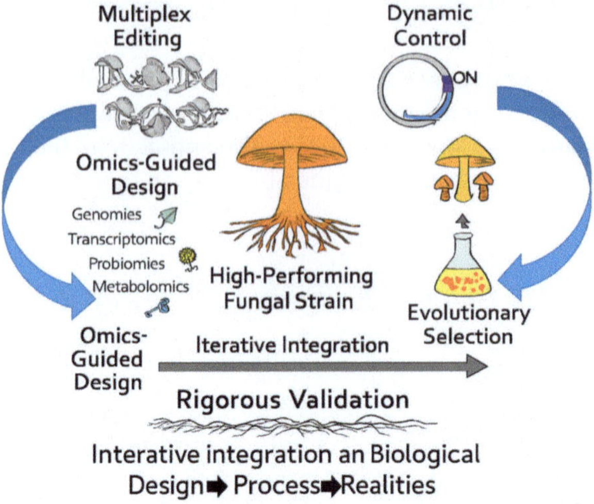

Fig. 13.2 Holistic framework for designing resilient and high-performing fungal strains through innovative CRISPR technology approaches. This figure describes four different approaches: Multiplex Editing (upper left), which utilizes a set of guide RNAs to redirect fungal metabolism; Dynamic Control (upper right), which makes use of CRISPRi/a switches, biosensors, and conditional promoters to control phases; Omics-guided Designs (lower left), which apply genomic, transcriptomic, proteomic, and metabolomic analyses to understand fungal weaknesses and selectively target them; and Evolutionary Selection (lower right), which applies adaptive evolution to make fungal strains more resilient. Each quadrant revolves around the central notion of a high--performing fungal strain, represented with branched hyphae, which represent industrial robustness. The curved arrows represent how each quadrant revolves around a perpetual loop related to the realities of biological design, while the bottom quadrant emphasizes the need to validate all approaches through reactors and recovery

enzyme targets, and predict the change in flux in an industrial setting (Abdi et al. 2024a, b). This closes the data-driven loop of design, build, test, and learn to develop things faster and with greater predictability as one transitions from the lab to the pilot scale.

13.5.4 Synthetic Biology Circuits and Dynamic Control

Dynamic control platforms (inducible promoters, CRISPRi/a switches, metabolite sensors) make possible phase-specific control, such as growth vs. production phase control, in which cell mass production and product formation are uncoupled (Ma et al. 2025). The CRISPRi/a system offers reversible and adjustable control over gene expression without genome modification for permanent disruption of gene function. This is particularly important in industrial applications in which flexibility and regulatory requirements are considerations (Kaul et al. 2025).

Table 13.2 Comparative attributes of advanced approaches

Strategy	Primary goal	Typical outcome	Implementation note	References
Multiplex CRISPR	Rewire pathways	Coordinated flux changes	Requires robust gRNA arrays	Arab et al. (2025), Zhu (2026), Hu et al. (2026), and Jin et al. (2022)
ALE	Improve robustness	Tolerance to inhibitors	Best combined with sequencing	Xia et al. (2024), Sun et al. (2025), Sha et al. (2023), and Chen et al. (2024)
Omics + ML	Target prioritization	Predictive edits	Needs high-quality datasets	Abdi et al. (2024a, b)
Dynamic CRISPRi/a	Phase control	Higher overall productivity	Use inducible systems	Wijayawardene et al. (2023) and Li et al. (2023)

13.5.5 Process-Aware Strain Engineering and Scale-up Considerations

Engineering needs to integrate fermentation methods (SSF and SmF), feedstock diversity, and recovery of product. Engineering strains tolerant to inhibitors found in natural feedstocks and effort put toward secretion efficiency for those producing extracellular compounds, and engineering focused on traits simplifying product separation are the need of the hour, and pilot work and techno-economics will define the changes to pursue (Derman et al. 2022; Ma et al. 2025). Various comparative attributes of the above advanced approaches have been outlined in Table 13.2.

13.6 Challenges and Ethical Considerations

Despite the revolutionary potential of CRISPR–Cas9 systems in fungal biotechnology, there are numerous technical and ethical obstacles that have to be overcome before the adoption of this technology can proceed at the industrial level.

13.6.1 Technical Barriers

One of the major hurdles is related to the efficiency and stability of the genetic transformation procedures in fungi. In the case of many filamentous fungi, the genome structure can be complex in nature, the organism can be highly heterogeneous, and the rate of homologous recombination can be less efficient. The delivery methods, like electroporation, Agrobacterium-mediated transformation, or protoplast fusion, may be less consistent in their results. Additionally, the challenge of off-target effects can also pose a concern in the design of guide RNA. Although the

CRISPR–Cas9 system holds great promise in fungal biotechnology, some technical as well as social issues must be addressed prior to the adoption of this technology on an industrial scale. These issues range from the molecular or technical front to the social aspects.

13.6.2 Regulatory and Intellectual Property Landscape

The regulatory regimes that cover the application of genome-edited organisms are also evolving very quickly. Although it is true that there are countries that treat CRISPR-modified strains as GMOs that are already covered by the regulations in place now, there are also countries that do not consider modifications that do not involve foreign DNA. Moreover, the IP landscape in the application of CRISPR technology is quite complicated, where there are overlapping patents that could create a problem for accessibility by academia and small businesses.

13.6.3 Biosafety and Public Perception

The use or dissemination of genetically engineered fungal strains could have biosafety issues, especially when considering the interactions and horizontal gene transfer. The strains used for industrial purposes need to be constructed using techniques of containment, including auxotrophy and marker-free editing. On the other hand, public acceptance could also play an important role, since a lack of faith in genetic engineering could make the adoption of CRISPR-based biofuels and biomaterials rather challenging.

13.6.4 Data Sharing and Open-Source Toolkits

The pace at which the field is advancing has created a lot of data in the areas of genome studies, transcriptome analysis, and proteome analysis in the field of CRISPR. The sharing of data has been an uneven process that has affected the reproducibility and collaboration that were anticipated in the field. The use of open-source platforms and the adoption of a universal method in the application and development of technologies in the field of CRISPR have been a challenge. The application and development of fungal biotechnology also need consideration in terms of equitable distribution in order to avert the dominance of the field by a few corporations and institutions in the world. The issues associated with the use of CRISPR–Cas9 in fungal biotechnology are not merely scientific in nature but have regulatory, ethical, and social dimensions too. It is necessary to address the issues of the efficiency of gene transfer, the possibility of off-target effects, and the genetic stability of the cell simultaneously with the disputes over IP rights, biosafety, and the development of public confidence.

13.7 Future Perspectives

The rapid development of CRISPR reagents continues to revolutionize the face of fungal biotechnology while opening new perspectives toward the synthesis of biofuels and biomaterials. Looking ahead, several emerging strategies and integrative frameworks are expected to define the next decade of research and industrial translation.

13.7.1 CRISPR 2.0: Prime Editing and Novel Cas Variants

Future development will likely shift away from the classical Cas9 editing to prime editing, or even to other nucleases such as Cas12 and Cas13. These systems offer higher precision, a decrease in off-target effects, and more editing possibilities, including RNA targeting and programmable base changes. In fungi, this will allow for fine metabolic rewiring without sacrificing genome stability and accelerate strain development for complex pathways.

13.7.2 Integration into Circular Bioeconomy Models

CRISPR-based and fungal biofuel production processes will find prominence in future circular bioeconomy models in which waste biomass is utilized as feedstock for producing biofuel, polymers, and valuable organic chemicals. In future CRISPR-Cas9 fungal models, attempts will continue to optimize fungi that would harbor inhibitor resistance and tolerance mechanisms. Such engineered fungi are anticipated to become central enablers of ring-shaped bioeconomy systems, reinforcing sustainability and restoring efficiency.

13.7.3 Policy Recommendations and Global Collaboration

The future of CRISPR fungal technology depends not only on the function of technical achievements but also on harmonization and collaboration on the legislative and international levels. It is necessary, for instance, to have guidelines and resource availability for more equitable means of innovation in the CRISPR fungal toolbox, and it is crucial for collaboration among academia, the private sector, and the legislative bodies for massive scaling in the area of fungal energy and material needs.

13.7.4 Vision for CRISPR-Enabled Fungal Platforms in Sustainable Development

In the long run, the capabilities offered by the CRISPR system in fungi can be imagined to integrate a range of applications in cell factories. For instance, the future applications of these cell factories in the development of biofuels and biomaterials will rely on the engineering of the fungus based on precision editing techniques. Moreover, the applications in the development of a greener Earth will involve the disposal of waste. Future outlook foresees the shift from the modification of individual genes to the overall engineering of the systems of the fungi at a higher level. Development of prime editors, the circular economy, and the regulatory structure will ensure the continuation of CRISPR-based fungal biotechnology at the forefront of sustainable energy and materials.

13.8 Conclusion

The advent of the CRISPR–Cas9 system has caused a paradigm shift in the field of fungal biotechnology and significantly improved the efficiency of genome modification. This technology has opened up novel avenues for biofuel production, organic acid production, enzymes, and biomaterials with the efficiency of targeted modifications in the fungal host system. The design of fungal metabolism to improve enzyme secretion and fine-tune the regulatory circuits makes them excellent cell factories to meet the demands of global challenges related to energy and materials. At the technical level, the CRISPR system has overcome many limitations in traditional fungal genetics, such as low transformation efficiencies and poor genetic tool sets. Multiplex editing approaches in combination with base editing and CRISPRi/a have enabled scientists to simultaneously optimize multiple pathways with decreased off-target effects. The ability to dynamically control gene expression has been made possible with these approaches in fungi. Together with omics and machine learning strategies in synthetic biology, scientists are able to develop high-performing strains for a given application through a system-level approach. The rapid redesign strategies that exist within the system are based on accurate modification strategies available in CRISPR.

It is, however, important to recognize the implications that go beyond the biological aspects and into the broader area of biotechnology that CRISPR makes possible in fungi. The potential of these fungi, which can be derived from agro-industrial wastage streams for biofuel and biological products, becomes essential for implementation within the context of the bioeconomy paradigm for strategies on adaptation and mitigation for climate change. A great influence for this exists, as these organisms lead the conversion toward bio-friendly substitutes for fossil raw materials. It would be essential for this to happen in a way that is openly accessible and distributed for communal use and knowledge sharing and to avoid implications of CRISPR and its benefits within closed settings of secrecy and exclusivity.

Looking ahead, the future of fungal biotechnology will evolve at the intersection of innovation and responsible practices. Prime editing and new Cas mutants will provide even greater specificity, and adaptive laboratory evolution and process-aware engineering will improve strain robustness and scalability. On the other hand, international cooperation and engagement will drive the rapid and responsible adoption of these technologies. The hope is that CRISPR-modified fungi will provide a key enabling technology in sustainable development and will connect energy, materials science, and environmental approaches. In conclusion, CRISPR–Cas9 has brought about a paradigm shift for fungal biotech, thus enabling it to progress from marginal development to revolutionary outcomes. With convergent advances in cutting-edge genome modification, visionary thinking, and a sense of industrial realities, this area of study is all set to harness robust fungal systems that are more aligned to meet the increasing demand for biofuels and biomaterials but also collectively work toward creating a sustainable future. The future will depend on maintaining a balance between making revolutionary headway with CRISPR–Cas9 by creating real benefits for scientific, industrial, and global communities.

References

Abdi G, Firdous S, Vasantharekha R, Singh GB, Seetharaman B, Mudgal G, Jain M, Tarighat MA (2024a) Progress on synthetic genomics. In: Singh V (ed) Advances in genomics: methods and applications. Springer Nature Singapore, Singapore, pp 181–197

Abdi G, Patil N, Mishra P, Tariq M, Dhariwal R, Parveen S, Dave K, Jain M, Tarighat MA, Mudgal G (2024b) Genomic precision: unveiling the transformative role of genome editing in advancing genomics research and applications. In: Singh V (ed) Advances in genomics: methods and applications. Springer Nature Singapore, Singapore, pp 265–306

Agrawal K, Verma P (2021) Fungal metabolites: a recent trend and its potential biotechnological applications. In: New and future developments in microbial biotechnology and bioengineering, pp 1–14. https://doi.org/10.1016/b978-0-12-821005-5.00001-6

Agrawal K, Alam A, Verma P (2021) Bioprospecting and applications of fungi: a game changer in present scenario. Fungal Biol:1–25. https://doi.org/10.1007/978-3-030-68260-6_1

Arab B, Chen J, Khusnutdinova AN, Chou CP, Liu Y (2025) Advancing bio-recycling of nylon monomers through CRISPR-assisted engineering. Environ Technol Innovation 39:104267. https://doi.org/10.1016/j.eti.2025.104267

Attias N, Danai O, Tarazi E, Pereman I, Grobman YJ (2019) Implementing bio-design tools to develop mycelium-based products. Des J 22:1647–1657. https://doi.org/10.1080/14606925.2019.1594997

Bellou S, Triantaphyllidou I-E, Aggeli D, Elazzazy AM, Baeshen MN, Aggelis G (2016) Microbial oils as food additives: recent approaches for improving microbial oil production and its polyunsaturated fatty acid content. Curr Opin Biotechnol 37:24–35. https://doi.org/10.1016/j.copbio.2015.09.005

Bilal M, Asgher M, Iqbal HMN, Hu H, Zhang X (2017) Biotransformation of lignocellulosic materials into value-added products—a review. Int J Biol Macromol 98:447–458. https://doi.org/10.1016/j.ijbiomac.2017.01.133

Bourbonnais R, Paice MG, Reid ID, Lanthier P, Yaguchi M (1995) Lignin oxidation by laccase isozymes from Trametes versicolor and role of the mediator 2,2′-azinobis(3-ethylbenzthiazoline-6-sulfonate) in kraft lignin depolymerization. Appl Environ Microbiol 61:1876–1880. https://doi.org/10.1128/aem.61.5.1876-1880.1995

Chai Y, Bai M, Chen A, Peng L, Shao J, Luo S, Deng Y, Yan B, Peng C (2022) Valorization of waste biomass through fungal technology: advances, challenges, and prospects. Ind Crop Prod 188:115608. https://doi.org/10.1016/j.indcrop.2022.115608

Chao-min YIN, Xiu-zhi FAN, De-fang SHI, Hong GAO (2017) CRISPR/Cas genome editing technology and its application in fungi. Biotechnol Bull 33:58–65

Chen C, Li Y-W, Chen X-Y, Wang Y-T, Ye C, Shi T-Q (2024) Application of adaptive laboratory evolution for Yarrowia lipolytica: A comprehensive review. Bioresour Technol 391:129893. https://doi.org/10.1016/j.biortech.2023.129893

Das D, Singh A, Adari R, Mudgal G, Radha P (2025) Biocompatible cellulose derivatives: green chemistry and its Sustainable applications. In: Kesari KK, Prakash C, Khalid M, Negi A (eds) Functionalized cellulose materials: sustainable manufacturing and applications. Springer Nature Switzerland, Cham, pp 159–190

Deng H, Gao R, Liao X, Cai Y (2017) CRISPR system in filamentous fungi: current achievements and future directions. Gene 627:212–221. https://doi.org/10.1016/j.gene.2017.06.019

Derman E, Abdulla R, Marbawi H, Sabullah MK, Gansau JA, Ravindra P (2022) Simultaneous saccharification and fermentation of empty fruit bunches of palm for bioethanol production using a microbial consortium of S. cerevisiae and T. harzianum. Fermentation 8:295

Eggert C, Temp U, Eriksson KE (1996) The ligninolytic system of the white rot fungus Pycnoporus cinnabarinus: purification and characterization of the laccase. Appl Environ Microbiol 62:1151–1158. https://doi.org/10.1128/aem.62.4.1151-1158.1996

Elsacker E, Saluena Martin J, Sangosanya AO, Verstuyft A, Van Wylick A, Peeters E (2025) Gradients of aliveness and engineering: a taxonomy of fungal engineered living materials. Adv Mater. https://doi.org/10.1002/adma.202502728

Eoh J, Gu L (2019) Biomaterials as vectors for the delivery of CRISPR–Cas9. Biomater Sci 7:1240–1261. https://doi.org/10.1039/c8bm01310a

Eroshin VK, Satroutdinov AD, Dedyukhina EG, Chistyakova TI (2000) Arachidonic acid production by Mortierella alpina with growth-coupled lipid synthesis. Process Biochem 35:1171–1175. https://doi.org/10.1016/S0032-9592(00)00151-5

Garrigues S, Peng M, Kun RS, de Vries RP (2023) Non-homologous end-joining-deficient filamentous fungal strains mitigate the impact of off-target mutations during the application of CRISPR/Cas9. mBio. https://doi.org/10.1128/mbio.00668-23

Grigaitis P, Olivier BG, Fiedler T, Teusink B, Kummer U, Veith N (2021) Protein cost allocation explains metabolic strategies in Escherichia coli. J Biotechnol 327:54–63. https://doi.org/10.1016/j.jbiotec.2020.11.003

Hinneburg H, Gu S, Naseri G (2025) Fungal innovations—advancing sustainable materials, genetics, and applications for industry. J Fungi 11:721. https://doi.org/10.3390/jof11100721

Holt GA, Mcintyre G, Flagg D, Bayer E, Wanjura JD, Pelletier MG (2012) Fungal mycelium and cotton plant materials in the manufacture of biodegradable molded packaging material: evaluation study of select blends of cotton byproducts. J Biobased Mater Bioenergy 6:431–439. https://doi.org/10.1166/jbmb.2012.1241

Hu D, Zhao R, Lin Y, Jiang C (2026) CRISPR-Cas9 gene editing in Aspergillus: from pathogenesis to metabolic engineering. Biology 15:53

IEA (2024) Growth in global energy demand surged in 2024 to almost twice its recent average. https://www.iea.org/news/growth-in-global-energy-demand-surged-in-2024-to-almost-twice-its-recent-average

IRENA (2025) Tripling renewable power and doubling energy efficiency by 2030: crucial steps towards 1.5°C: key recommendations and introduction. https://www.irena.org/Digital-Report/Tripling-renewable-power-and-doubling-energy-efficiency-by-2030

Jayaprakash M, Basha SMJ, Jaiganesh S, Kumar SSKS, Jayaraman L, Mudgal G (2025) Collagen fusion with natural biopolymers in food packaging: an In-depth review from conceptualization to consumer application. In: Kesari KK, Prakash C, Khalid M, Negi A (eds) Functionalized cellulose materials: sustainable manufacturing and applications. Springer Nature Switzerland, Cham, pp 65–87

Jiang C, Lv G, Tu Y, Cheng X, Duan Y, Zeng B, He B (2021) Applications of CRISPR/Cas9 in the synthesis of secondary metabolites in filamentous fungi. Front Microbiol 12. https://doi.org/10.3389/fmicb.2021.638096

Jin FJ, Wang BT, Wang ZD, Jin L, Han P (2022) CRISPR/Cas9-based genome editing and its application in Aspergillus species. J Fungi (Basel) 8. https://doi.org/10.3390/jof8050467

Jones M, Mautner A, Luenco S, Bismarck A, John S (2020) Engineered mycelium composite construction materials from fungal biorefineries: a critical review. Mater Des 187:108397. https://doi.org/10.1016/j.matdes.2019.108397

Kapahi M, Rani R, Kohli K (2021) Fungal biorefineries for biofuel production for sustainable future energy systems. Fungal Biol:477–496. https://doi.org/10.1007/978-3-030-68260-6_17

Kapoor SA, Choudhary P, Kasana RC (2025) Exploring CRISPR/Cas9-mediated gene editing advances in conventional and non-conventional yeast species. Appl Biochem Biotechnol 197:7083–7122. https://doi.org/10.1007/s12010-025-05382-2

Kaul N, Ranga P, Yadav S, Choudhary P, Dahiya A, Sehrawat DR, Mann A (2025) CRISPR/Cas-based fungal genome engineering for secondary metabolite production: progress and challenges. In: Mann A, Kumar N, Kumar A, Chandra P, Sanwal SK, Sheoran P (eds) Cutting edge technologies for developing future crop plants. Springer Nature Singapore, Singapore, pp 339–348

Krappmann S (2016) CRISPR-Cas9, the new kid on the block of fungal molecular biology. Med Mycol 55:16–23. https://doi.org/10.1093/mmy/myw097

Kumar A, Gautam A, Dutt D (2016) Biotechnological transformation of lignocellulosic biomass in to industrial products: an overview. Adv Biosci Biotechnol Adv 7:149–168. https://doi.org/10.4236/abb.2016.73014

Kumar A, Chadha S, Rath D (2021) CRISPR-Cas9 system for functional genomics of filamentous fungi: applications and challenges. In: Fungi bio-prospects in sustainable agriculture, environment and nano-technology, pp 541–576. https://doi.org/10.1016/b978-0-12-821734-4.00016-2

Kumari M, Survase SA, Singhal RS (2008) Production of schizophyllan using Schizophyllum commune NRCM. Bioresour Technol 99:1036–1043. https://doi.org/10.1016/j.biortech.2007.02.029

Leal K, Rojas E, Madariaga D, Contreras MJ, Nuñez-Montero K, Barrientos L, Goméz-Espinoza O, Iturrieta-González I (2024) Unlocking fungal potential: the CRISPR-Cas system as a strategy for secondary metabolite discovery. J Fungi 10:748. https://doi.org/10.3390/jof10110748

Li H, Liu G (2017) The application of CRISPR/Cas9 in genome editing of filamentous fungi. Yi chuan = Hereditas 39:355

Li J, Wiebenga A, Lipzen A, Ng V, Tejomurthula S, Zhang Y, Grigoriev IV, Peng M, de Vries RP (2023) Comparative genomics and transcriptomics analyses reveal divergent plant biomass-degrading strategies in fungi. J Fungi 9:860

Liao B, Chen X, Zhou X, Zhou Y, Shi Y, Ye X, Liao M, Zhou Z, Cheng L, Ren B (2021) Applications of CRISPR/Cas gene-editing technology in yeast and fungi. Arch Microbiol 204. https://doi.org/10.1007/s00203-021-02723-7

Liu R, Chen L, Jiang Y, Zhou Z, Zou G (2015) Efficient genome editing in filamentous fungus Trichoderma reesei using the CRISPR/Cas9 system. Cell Discov 1. https://doi.org/10.1038/celldisc.2015.7

Liu S, Qureshi N, Hughes SR (2017) Progress and perspectives on improving butanol tolerance. World J Microbiol Biotechnol 33:51. https://doi.org/10.1007/s11274-017-2220-y

Lubbers RJM (2025) An updated perspective on the aromatic metabolic pathways of plant-derived homocyclic aromatic compounds in Aspergillus niger. Microorganisms 13:1718

Ma B, Li Y, Wang T, Li D, Jia S (2025) Advances in CRISPR/Cas9-based gene editing in filamentous fungi. J Fungi (Basel) 11:350

Machida M, Asai K, Sano M, Tanaka T, Kumagai T, Terai G, Kusumoto K-I, Arima T, Akita O, Kashiwagi Y et al (2005) Genome sequencing and analysis of Aspergillus oryzae. Nature 438:1157–1161. https://doi.org/10.1038/nature04300

Meng Y, Yao C, Xue S, Yang H (2014) Application of Fourier transform infrared (FT-IR) spectroscopy in determination of microalgal compositions. Bioresour Technol 151:347–354. https://doi.org/10.1016/j.biortech.2013.10.064

Moreira Vieira N, Zandonade Ventorim R, de Moura Ferreira MA, Barcelos de Souza G, Menezes de Almeida EL, Pereira Vidigal PM, Nunes Nesi A, Gomes Fietto L, Batista da Silveira W (2020) Insights into oleaginous phenotype of the yeast Papiliotrema laurentii. Fungal Genet Biol 144:103456. https://doi.org/10.1016/j.fgb.2020.103456

Mudgal G, Sivaji S, Kaur J, Dhar SK, Ramasamy V, Seetharaman B, Mishra RK, Govindappa V, Bose JC, Singh GB et al (2025a) Strategic conservation plan. In: Abdel Latef AAH, Zayed EM, Omar AA (eds) Sustainable remediation for pollution and climate resilience. Springer Nature Singapore, Singapore, pp 139–206

Mudgal G, Rashi, Kaur J, Kaur J, Singh S, Dhar SK, Vasanthrekha R, Singh A, Thakur H, Kaur S et al (2025b) Game changer: how cellulose-based bioleather is transforming the market. In: Kesari KK, Prakash C, Khalid M, Negi A (eds) Functionalized cellulose materials: sustainable manufacturing and applications. Springer Nature Switzerland, Cham, pp 277–295

Nødvig CS, Nielsen JB, Kogle ME, Mortensen UH (2015) A CRISPR-Cas9 system for genetic engineering of filamentous fungi. PLoS One 10. https://doi.org/10.1371/journal.pone.0133085

Pal V, Sharma D, Vishwakarma P, Samanta D, Singh KA, Nagar J, Sani RK, Rai R (2024) CRISPR/Cas9: a cutting-edge tool for cellulase enhancement in fungi. Fungal Biol Rev 50:100394. https://doi.org/10.1016/j.fbr.2024.100394

Pant S, Ritika u, Nag P, Ghati A, Chakraborty D, Maximiano MR, Franco OL, Mandal AK, Kuila A (2022) Employment of the CRISPR/Cas9 system to improve cellulase production in Trichoderma reesei. Biotechnol Adv 60:108022. https://doi.org/10.1016/j.biotechadv.2022.108022

Park Y-K, Nicaud J-M, Ledesma-Amaro R (2018) The engineering potential of Rhodosporidium toruloides as a workhorse for biotechnological applications. Trends Biotechnol 36:304–317. https://doi.org/10.1016/j.tibtech.2017.10.013

Peterson R, Nevalainen H (2012) Trichoderma reesei RUT-C30–thirty years of strain improvement. Microbiology (Reading, England) 158:58–68. https://doi.org/10.1099/mic.0.054031-0

Quevedo-Amador RA, Escalera-Velasco BP, Arias AMR, Reynel-Ávila HE, Moreno-Piraján JC, Giraldo L, Bonilla-Petriciolet A (2024) Application of waste biomass for the production of biofuels and catalysts: a review. Clean Techn Environ Policy 26:943–997. https://doi.org/10.1007/s10098-023-02728-4

Rao PM, Singh AO, Sriniketha Narayanan K, Shankar R, Mudgal G, Radha P (2025) Cellulose-based packaging: an emerging Sustainable packaging option for future use. In: Kesari KK, Prakash C, Khalid M, Negi A (eds) Functionalized cellulose materials: sustainable manufacturing and applications. Springer Nature Switzerland, Cham, pp 89–115

Riaz S, John AJ, Samuel MS, Ethiraj S (2025) Recent developments and emerging methodologies in the pre-treatment of lignocellulosic biomass. Sustain Chem Environ 11:100285. https://doi.org/10.1016/j.scenv.2025.100285

Rogers PL, Jeon YJ, Lee KJ, Lawford HG (2007) Zymomonas mobilis for fuel ethanol and higher-value products. In: Olsson L (ed) Biofuels. Springer Berlin Heidelberg, Berlin, Heidelberg, pp 263–288

Román E, Prieto D, Alonso-Monge R, Pla J (2019) New insights of CRISPR technology in human pathogenic fungi. Future Microbiol 14:1243–1255. https://doi.org/10.2217/fmb-2019-0183

Sagia S, Sharma A, Singh S, Chaturvedi S, Nain PKS, Nain L (2020) Single cell oil production by a novel yeast Trichosporon mycotoxinivorans for complete and ecofriendly valorization of paddy straw. Electron J Biotechnol 44:60–68. https://doi.org/10.1016/j.ejbt.2020.01.009

Seetharaman B, Balu UR, Mudgal G, Firdaus S, Dash S, Mishra RK, Ramasamy V (2025) The piteous price of progress: the environmental and health costs of modern society. In: Prakash

C, Kesari KK, Negi A (eds) Sustainable development goals towards environmental toxicity and green chemistry: environment and sustainability. Springer Nature Switzerland, Cham, pp 291–313

Sha Y, Zhou L, Wang Z, Ding Y, Lu M, Xu Z, Zhai R, Jin M (2023) Adaptive laboratory evolution boosts Yarrowia lipolytica tolerance to vanillic acid. J Biotechnol 367:42–52. https://doi.org/10.1016/j.jbiotec.2023.03.006

Sharma A, Chand K, Singh GB, Mudgal G (2021) Toxicity with waste-generated ionizing radiations: blunders behind the scenes. In: Kesari KK, Jha NK (eds) Free radical biology and environmental toxicity. Springer International Publishing, Cham, pp 305–325

Shen Q, Ruan H, Zhang H, Wu T, Zhu K, Han W, Dong R, Ming T, Qi H, Zhang Y (2024) Utilization of CRISPR-Cas genome editing technology in filamentous fungi: function and advancement potentiality. Front Microbiol 15:1375120. https://doi.org/10.3389/fmicb.2024.1375120

Shi T-Q, Liu G-N, Ji R-Y, Shi K, Song P, Ren L-J, Huang H, Ji X-J (2017) CRISPR/Cas9-based genome editing of the filamentous fungi: the state of the art. Appl Microbiol Biotechnol 101:7435–7443. https://doi.org/10.1007/s00253-017-8497-9

Singh RS, Saini GK, Kennedy JF (2009) Downstream processing and characterization of pullulan from a novel colour variant strain of Aureobasidium pullulans FB-1. Carbohydr Polym 78:89–94. https://doi.org/10.1016/j.carbpol.2009.03.040

Singh H, Janiyani K, Gangawane A, Pandya S, Jasani S (2024) Engineering cellulolytic fungi for efficient lignocellulosic biomass hydrolysis: advances in mutagenesis, gene editing, and nanotechnology with CRISPR-Cas innovations. Discover Appl Sci 6. https://doi.org/10.1007/s42452-024-06405-z

Singh GB, Dey S, Mudgal G (2025a) Microbial contributions to a circular economy. In: Kaur S, Dhiman S, Tripathi M (eds) Microbial metabolomics: recent developments, challenges and future opportunities. Springer Nature Singapore, Singapore, pp 419–439

Singh A, Kaur S, Thakur H, Rashi, Kashyap S, Lyudmila A, Mudgal G (2025b) Unveiling the transformative power of smart cellulosic nanomaterials: revisiting potential promises to sustainable future. In: Kesari KK, Prakash C, Khalid M, Negi A (eds) Functionalized cellulose materials: sustainable manufacturing and applications. Springer Nature Switzerland, Cham, pp 1–42

Sivaprasad S, Byju SK, Prajith C, Shaju J, Rejeesh CR (2021) Development of a novel mycelium bio-composite material to substitute for polystyrene in packaging applications. Mater Today Proc 47:5038–5044. https://doi.org/10.1016/j.matpr.2021.04.622

Skrekas C (2022) CRISPR-based technologies for high throughput metabolic engineering of yeast. Chalmers University of Technology

Song R, Zhai Q, Sun L, Huang E, Zhang Y, Zhu Y, Guo Q, Tian Y, Zhao B, Lu H (2019) CRISPR/Cas9 genome editing technology in filamentous fungi: progress and perspective. Appl Microbiol Biotechnol 103:6919–6932. https://doi.org/10.1007/s00253-019-10007-w

Souza Filho PF, Zamani A, Taherzadeh MJ (2017) Production of edible fungi from potato protein liquor (PPL) in airlift bioreactor. Fermentation 3:12

Subramanian V, Adler MJ, Benyamin M, Pullen RM, Servinsky MD, Kozlowski MT, Decker SR (2025) From the bench to the reactor: engineered filamentous fungi for biochemical and biomaterial production. Biotechnol Biofuels Bioprod 18. https://doi.org/10.1186/s13068-025-02712-8

Sun M-L, Fang D, Zeng D, Wang Z-Z, Huang C-C, Zhang L, Fan Y-C, Liao S, Cao Y, Li K et al (2025) Leveraging stress tolerance property to enhance succinic acid production and cell metabolic activity of Yarrowia lipolytica. ACS Sustain Chem Eng 13:5210–5219. https://doi.org/10.1021/acssuschemeng.4c10404

Teng Y, Jiang T, Yan Y (2024) The expanded CRISPR toolbox for constructing microbial cell factories. Trends Biotechnol 42:104–118. https://doi.org/10.1016/j.tibtech.2023.06.012

Thakur H, Kaur S, Singh A, Singh GB, Mudgal G (2025) Smart sustainable materials: a blueprint for a better tomorrow. In: Kesari KK, Prakash C, Khalid M, Negi A (eds) Functionalized cellulose materials: sustainable manufacturing and applications. Springer Nature Switzerland, Cham, pp 253–276

Todorov J, Pantić M, Kozarski M, Lazić V, Todorović N, Obradović M, Daković A, Krajišnik D, Milašinović N, Mirković M (2025) Chitosan–glucan biopolymer design: extraction from champignons with improved antioxidant and antimicrobial features. Processes 13:3937

Tupe SG, Deshmukh SK, Zambare RB, Tripathi AA, Deshpande MV (2022) Biopolymers from fungi and their applications. In: Deshmukh SK, Deshpande MV, Sridhar KR (eds) Fungal biopolymers and biocomposites: prospects and avenues. Springer Nature Singapore, Singapore, pp 3–14

Ullah M, Xia L, Xie S, Sun S (2020) CRISPR/Cas9-based genome engineering: A new breakthrough in the genetic manipulation of filamentous fungi. Biotechnol Appl Biochem 67:835–851. https://doi.org/10.1002/bab.2077

Umar A, Mubeen M, Ali I, Iftikhar Y, Sohail MA, Sajid A, Kumar A, Solanki MK, Kumar Divvela P, Zhou L (2024) Harnessing fungal bio-electricity: a promising path to a cleaner environment. Front Microbiol 14. https://doi.org/10.3389/fmicb.2023.1291904

Utomo JC, Hodgins CL, Ro D-K (2021) Multiplex genome editing in yeast by CRISPR/Cas9 – a potent and agile tool to reconstruct complex metabolic pathways. Front Plant Sci 12:719148. https://doi.org/10.3389/fpls.2021.719148

Villena GK, Kitazono AA, Hernández-Macedo ML (2020) Bioengineering fungi and yeast for the production of enzymes, metabolites, and value-added compounds. Fungal Biol:209–237. https://doi.org/10.1007/978-3-030-41870-0_9

Wang Q, Coleman JJ (2019) Progress and challenges: development and implementation of CRISPR/Cas9 technology in filamentous fungi. Comput Struct Biotechnol J 17:761–769. https://doi.org/10.1016/j.csbj.2019.06.007

Wang S, Zou C, Lou C, Yang H, Mei M, Jing H, Cheng S (2020) Effects of hemicellulose, cellulose and lignin on the ignition behaviors of biomass in a drop tube furnace. Bioresour Technol 310:123456. https://doi.org/10.1016/j.biortech.2020.123456

Wijayawardene NN, Boonyuen N, Ranaweera CB, de Zoysa HKS, Padmathilake RE, Nifla F, Dai DQ, Liu Y, Suwannarach N, Kumla J et al (2023) OMICS and other advanced technologies in mycological applications. J Fungi (Basel) 9. https://doi.org/10.3390/jof9060688

Xia K, Chen Y, Liu F, Zhao X, Sha R, Huang J (2024) Adaptive responses of erythritol-producing Yarrowia lipolytica to thermal stress after evolution. Appl Microbiol Biotechnol 108:263. https://doi.org/10.1007/s00253-024-13103-8

Xue Z, Sharpe PL, Hong S-P, Yadav NS, Xie D, Short DR, Damude HG, Rupert RA, Seip JE, Wang J et al (2013) Production of omega-3 eicosapentaenoic acid by metabolic engineering of Yarrowia lipolytica. Nat Biotechnol 31:734–740. https://doi.org/10.1038/nbt.2622

Yang P, Condrich A, Lu L, Scranton S, Hebner C, Sheykhhasan M, Ali MA (2024) Genetic engineering in bacteria, fungi, and oomycetes, taking advantage of CRISPR. DNA 4:427–454. https://doi.org/10.3390/dna4040030

Yoon LW, Ang TN, Ngoh GC, Chua ASM (2014) Fungal solid-state fermentation and various methods of enhancement in cellulase production. Biomass Bioenergy 67:319–338. https://doi.org/10.1016/j.biombioe.2014.05.013

Zannat A, Eason I, Wylie B, Rogers RD, Berton P, Shamshina JL (2025) Comparative analysis of chitin isolation techniques from mushrooms: toward sustainable production of high-purity biopolymer. Green Chem 27:3217–3233. https://doi.org/10.1039/D4GC06388K

Zhang S, Guo F, Yan W, Dai Z, Dong W, Zhou J, Zhang W, Xin F, Jiang M (2020) Recent advances of CRISPR/Cas9-based genetic engineering and transcriptional regulation in industrial biology. Front Bioeng Biotechnol 7. https://doi.org/10.3389/fbioe.2019.00459

Zhu Y (2026) The potential and innovative applications of CRISPR gene editing technology in enzyme gene development. Enzym Microb Technol 195:110799. https://doi.org/10.1016/j.enzmictec.2025.110799

Chapter 14
Artificial Intelligence Applications in Bioenergy and Biomaterials Development via Fungal Biotechnology: Challenges and Perspectives

Surabhi Done, Giridhar Babu Anam, and Burragoni Sravanthi Goud

Abstract The convergence of artificial intelligence (AI) and fungal biotechnology, referred to as biotechnological artificial intelligence (Bio-AI), is rapidly emerging as a powerful approach for advancing sustainable innovations in bioenergy and biomaterials. AI techniques such as machine learning, deep learning, and multivariate analysis are being integrated into fungal biotechnology pipelines to enhance strain selection, guide metabolic engineering, monitor fermentation processes, and predict bioactive compound production. AI is particularly effective in handling complex, high-dimensional datasets generated by omics technologies (genomics, proteomics, and metabolomics), accelerating the discovery of fungal secondary metabolites and high-value bio-based materials. In the bioenergy sector, AI applications are expanding to include feedstock characterization, prediction of conversion efficiency, and supply chain optimization. Despite these advancements, significant challenges remain, including data heterogeneity, limited standardized datasets, issues of model interpretability, and the need for stronger integration between computational models and wet-lab experimentation. Ethical considerations and interdisciplinary collaboration are also essential to ensure responsible and effective deployment. This chapter explores the transformative potential of AI at the intersection of fungal biotechnology, bioenergy, and biomaterials in driving the transition toward a sustainable bio-based economy.

Keywords Artificial intelligence · Bioenergy · Biomaterials · Fungal biotechnology · Machine learning · Metabolic engineering · Omics technologies · Smart biorefineries

S. Done · G. B. Anam (✉)
Department of Basic Sciences, SR University, Warangal, India
e-mail: a.giridharbabu@sru.edu.in

B. S. Goud
Department of Chemical Engineering, Yeungnam University, Gyeongsan, Gyeongbuk, South Korea

M. Tripathi et al. (eds.), *Recent Advancements in Fungal Biotechnology Volume 1*, Fungal Biology, https://doi.org/10.1007/978-981-92-0339-0_14

14.1 Introduction

Concurrent environmental challenges confront society, including climate change, biodiversity loss, and the unsustainability of linear fossil fuel-based manufacturing systems, exceeding the limits of which conventional manufacturing systems can operate. Mitigating these pressures demands a shift to circular, decarbonized biomanufacturing platforms (Dhiman et al. 2024; Hinneburg et al. 2025). Fungal biotechnology is particularly well-suited to the realization of this paradigm shift, as fungi possess a wide substrate tolerance and biosynthetic capability, converting low-value residues to valuable complex compounds or structured biomass, thus delivering circular economy approaches that effectively diminish material cycles and life cycle footprints (Madusanka et al. 2024; Dahiya et al. 2024). Filamentous fungi and yeasts have been in use as industrial workhorses for decades, with applications in food fermentation, enzymatic synthesis, and pharmaceuticals; these organisms have established cultivation methods and regulatory precedent (both of which mitigate translational barriers compared to novel microbial platforms) that are not present in potentially isolated microbes. Filamentous fungi, such as *Trichoderma*, *Aspergillus,* and *Talaromyces*, produce high-impact extracellular enzyme systems to attack lignocellulosic and other complex biopolymers and contribute to residue valorization toward a sustainable bioeconomy (Pan et al. 2025; Kovács 2025). These fungi can utilize agri-industrial wastes to produce biofuel, including biogas, bioethanol, and biodiesel, and other high-value-added products with minimal pretreatment that will promote circularity and a low carbon footprint in several bioprocessing operations (Khatami et al. 2025). Fungi also produce structurally varied polymers and secondary metabolites, such as chitin, β-glucans, and polyketides, that are proficient precursors for biomaterials, drugs, and specialty chemicals (Madusanka et al. 2024).

With the rise of "Bio-AI," an inclusive and unifying term referring to the combined usage of big data, machine learning (ML), AI, deep neural networks, and data automation processes (Wainaina and Taherzadeh 2023). AI approaches enable rapid mining of fungal genomes for novel enzymes and biosynthetic gene clusters, predictive optimization of feedstock pre-processing and fermentation control, and techno-economic trade-off analyses that guide scale-up decisions (Dhiman et al. 2024; Umashankar and Nygård 2024). With high-throughput wet-lab automation and digital twins of bioprocesses, both deploying AI, iterative design–build–test–learn cycles that were infeasible previously are now possible. The merging of computational intelligence into the biological capability of fungi is a key focus in this chapter, which provides potential guidance for researchers and industry stakeholders to develop carbon-efficient AI-based fungal bioprocesses that aid the circular bioeconomy (Shin et al. 2025; Pan et al. 2025).

14.2 Overview of Fungal Biotechnology in Bioenergy and Biomaterials

Fungal biotechnology is important for the advancement of sustainable bioenergy and biomaterials production by facilitating the conversion of lignocellulosic biomass into fuels, chemicals, and materials. Thanks to their well-developed enzymatic machineries, metabolic flexibility, and mycelial structures, fungi provide the backbone of integrated biorefining models that contribute to improved resource use efficiency, increased product variability, and a circular bioeconomy.

14.2.1 Industrial Importance of Fungi

Fungi have formed the basis of human industry for many decades, from conventional products such as bread, beer, and cheese to modern biotechnological applications, including industrial enzymes, organic acids, pharmaceuticals, and bioactive metabolites (Füting et al. 2021). Filamentous fungi and industrial yeasts remain key components of many biomanufacturing pipelines owing to their strong secretion capabilities, amenability to large-scale fermentations, and wide metabolic versatility (Vadivel et al. 2024; Yélamos et al. 2025). The main benefits include extracellular enzyme secretion for consolidated bioprocessing, secondary metabolism for high-value compounds, and fungal biomass or mycelium architectures with direct use as sustainable biomaterials. These characteristics altogether make fungi ideal platforms for the production of commodities and specialties in integrated biorefineries (Füting et al. 2021; Mandal and Krishnan 2021).

14.2.2 Fungal Feedstocks and Lignocellulosic Degradation

Lignocellulosic wastes (agricultural straw, corn stover, sawdust, and agro-industrial waste) are rich sources of carbon that can be obtained at low cost for sustainable bioprocessing. Fungi have evolved to break down plant cell walls via coordinated enzyme systems such as cellulases, hemicellulases, pectinases, and lignin-modifying oxidases (Su et al. 2025). *Trichoderma reesei* or *Aspergillus* spp. has been routinely used in the industry for producing lignocellulosic enzyme cocktails for biomass pretreatment and saccharification. The extracellular nature of fungal depolymerization allows them to act as colonizers as well as converters of solid substrates and minimizes the need for high-energy-cost hydrolysis steps (Vadivel et al. 2024).

14.2.3 *Fungal-Derived Bioenergy*

Fungi are important for bioenergy due to their ability to produce enzymes for the degradation of lignocellulosic biomass and transform renewable resources into energy such as bioethanol, biogas, and lipid-based biodiesel. Extracellular enzymes and metabolic versatility come with the ability to rapidly degrade biomass, and this is a desirable trait for sustainable and integrated bioenergy concepts.

14.2.3.1 Bioethanol Production

Fungi mainly produce cellulases and hemicellulases for pretreated lignocellulose for hydrolysis to fermentable sugars during bioethanol production. These sugars are then bioconverted to ethanol by yeast or engineered fungal strains. Optimizing enzyme cocktails tailored to feedstock composition is still an essential requirement, as the substrate heterogeneity strongly affects saccharification efficiency and, consequently, ethanol yield (Vadivel et al. 2024; Su et al. 2025).

14.2.3.2 Biogas Production via Anaerobic Digestion

Fungal pretreatment and co-culture strategies enhance anaerobic digestion by partially disrupting lignin, increasing substrate porosity, and releasing accessible carbon. These effects improve microbial accessibility and digestion efficiency, leading to higher methane yields and improved process stability (Karthikeyan et al. 2024).

14.2.3.3 Biodiesel and Oleochemical Production

Lipid-producing oleaginous fungi and yeasts, such as *Yarrowia lipolytica* and *Rhodosporidium toruloides,* store intracellular lipids in the form of single-cell oils. The lipids could be upgraded to biodiesel or high-value oleochemicals, especially when grown on low-cost waste sources, demonstrating scalability and economic potential (Wusnah et al. 2023).

14.2.4 *Biomaterials from Fungi*

Fungi enable sustainable production of biomaterials, including mycelium composites, chitin–β-glucan biopolymers, and food-enzyme-based biodegradable plastics from agri-wastes. Lightweight low-carbon mycelium composites for packaging, insulation, and construction offer comparable performance to traditional materials and full biodegradability (Madusanka et al. 2024; Patel and Sharma 2023; De et al.

2025). Fungal cell walls are renewable sources of chitin and β-glucans that can be used for films, coatings, or biomedical purposes and offer potential benefits over crustacean-based sources in terms of allergen safety and supply (Vadivel et al. 2024). Moreover, some fungal enzymes and metabolic pathways enable biosynthesis of biodegradable polymers, including polyhydroxyalkanoates that could serve as a greener alternative to plastic (Fadipe et al. 2025).

14.2.4.1 Integrated Bioenergy–Biomaterials Perspective

Fungal bioenergy and bio-based product pathways are synergistic in integrated biorefineries. Residues are valorized in the form of structural materials composed of solid waste and spent liquors, which can be used for fermentation and biogas or lipid biofuel production. Integrating fungal enzymology with computational optimization, sensor networks, and life cycle assessment also improves efficiency, thus leading to economically feasible and low-waste biorefinery systems (Füting et al. 2021; Vadivel et al. 2024; Mandal and Krishnan 2021).

14.3 Artificial Intelligence Technologies and Tools Relevant to Biotechnology

Artificial intelligence (AI) has emerged as an essential tool of contemporary biotechnology, providing analytical and predictive power, in many cases beyond classic statistical modeling or manual experimental workflows (Gomes et al. 2024; Helleckes et al. 2022). In fungal biotechnology, which encompasses genome sequences, transcriptomic and metabolomic profiles, sensor streams from bioprocesses, as well as product property measurements, AI tools unveil non-linear relationships that are not evident using other approaches (Angermueller et al. 2016). These approaches expedite hypothesis generation, lower experimental costs, and facilitate data-driven decision-making for strain development, process optimization, and product design (Fig. 14.1). Therefore, AI has been integrated into emerging bioenergy and biomaterials pipelines developed on fungal systems (Rathore et al. 2022).

14.3.1 Machine Learning Approaches

Machine learning (ML) is a method that enables computational systems to automatically learn models from data, using algorithms that attempt to transform the information into the prediction or classification of an object using certain input features. These approaches are suitable when the input data exhibits a structure, like

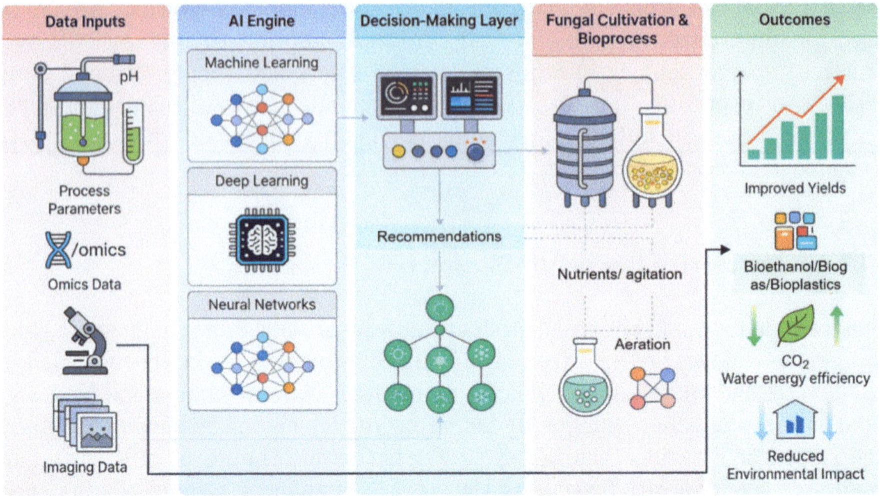

Fig. 14.1 Conceptual framework illustrating artificial intelligence-driven optimization of fungal biotechnology for bioenergy and biomaterials development. This figure integrates data inputs, AI engines, decision-making layers, fungal cultivation processes, and sustainability outcomes. Data inputs include real-time bioreactor process parameters (pH, temperature, aeration, agitation, and feed rates), multi-omics datasets capturing strain-level metabolic states, and imaging data describing fungal morphology and growth behavior. These inputs are processed by an AI engine integrating machine learning and deep learning models to learn relationships with target outputs such as yield, titer, and productivity. A decision-making layer translates model outputs into actionable recommendations, including process adjustments, strain or media selection, and, in advanced systems, autonomous control via digital twins. The fungal bioprocess module implements these decisions during cultivation and downstream processing. Resulting bioenergy and biomaterials outputs are optimized alongside sustainability outcomes, including improved resource efficiency, reduced emissions, and circular bioeconomy integration

tables of experimental observations, and have been applied in biotechnology to great effect because they are efficient, reliable, and easy to interpret (Gomes et al. 2024). The random forest (RF) model integrates many decision trees to predict accurately when the data are noisy. RF models are frequently used to make enzyme yield predictions from fermentation parameters, optimize culture media, classify fungal strains by phenotypic criteria, and determine important factors regulating lipid accumulation in oleaginous fungi (Helleckes et al. 2022). Support vector machines (SVMs) are useful for large-variable-small-sample sets, as commonly occur in omics data, and are used for fungal species identification, metabolite classification, and early detection of process disturbance. Furthermore, decision trees and boosting have recently been employed for characterizing feedstock quality, modeling biomass degradation, and predicting bioenergy process economics (Prakash et al. 2018).

14.3.2 Deep Learning Architectures

Deep learning (DL) is an advanced form of machine learning that leverages multi-layer artificial neural networks to learn specific important features automatically from raw data. In contrast to traditional ML techniques, deep learning has reduced dependence on handcrafted features and is especially well-suited for complex data such as images, time-series signals, and biological networks (Angermueller et al. 2016). Convolutional neural networks (CNNs) were developed to investigate images and spatial patterns and are popular in the analysis of fungal colony morphology, microscopic imaging data, and mycelial networks. Such models can be used to automatically identify fungi, measure growth, and predict material properties of mycelium composites (Yang et al. 2023). Recurrent neural networks (RNNs) such as the long short-term memory (LSTM) and gated recurrent unit (GRU) models are essential tools for these types of time-driven data obtained during fermentation, e.g., pH, oxygen, and substrate consumption. These models are used for estimating biomass growth, enzyme yields, and lipid accumulation, which helps smart process control (Rathore et al. 2022). More recently, graph neural networks (GNNs) were exploited to model how the biological entities interact with each other in predicting enzyme-substrate interactive activities as well as metabolic pathways and fungal biosynthetic gene clusters (Gaudelet et al. 2021).

14.3.3 Multivariate Statistics and Chemometrics

Multivariate statistical techniques and chemometrics remain imperative as key complementary tools to ML and DL, especially in the context of spectroscopy-based fungal analysis (Gomes et al. 2024). Methods like principal component analysis (PCA), partial least squares (PLS), orthogonal partial least squares-discriminant analysis (OPLS-DA), and soft independent modeling of class analogy (SIMCA) have been commonly used to quickly determine the composition of fungal biomass, such as lipid, chitin, and protein content. These tools provide access for quantifying lignocellulosic degradation using near-infrared (NIR) and Fourier-transform infrared (FTIR) spectroscopy, monitoring fermentation states with Raman spectroscopy, or evaluating quality attributes of mycelium-based biomaterials (Helleckes et al. 2022). Despite having less ability to flexibly learn from data than deep learning techniques, they are required in regulatory and quality control settings due to their interpretability, low sample requirements, and statistical soundness.

14.3.4 Digital Twins and Automation in Bioprocessing

Digital twins, a digital replica of a living or non-living physical bioprocess that incorporates real-time sensor data, are increasingly revolutionizing fungal fermentation and biorefinery processes (Rathore et al. 2022). Digital twins create and keep current predictions on process performance by integrating sensor measurements (e.g., temperature, pH, aeration rate, and optical density) with machine learning models and mechanistic equations. These capabilities allow the optimization, early detection of faults, predictive maintenance, and evaluation of scenarios affecting instantaneous performance, such as feedstock variability or changes in pretreatment incidents that may affect product yield. Automation platforms, such as robotic liquid handling complexes, automated strain libraries, and microfluidic bioreactors, are complementary to these AI-guided workflows by enabling the high-throughput experiments necessary for robust model training (Yang et al. 2023).

14.3.5 AI in Metabolic Engineering

AI-based approaches are revolutionizing metabolic engineering as they facilitate predictive designing by genetic modifications, pathway flux distributions, and gene essentiality patterns (Zampieri et al. 2019). Genome-scale metabolic models (GEMs) describe organism-spanning biochemical reaction networks and can be used to simulate growth and metabolite production for a range of environmental and genetic perturbations. When combined with machine learning methods like reinforcement learning or constraint-based optimization, GEMs enable the discovery of engineering targets for enhanced lipid biosynthesis, better secretion of lignocellulolytic enzymes, and redirecting carbon from pathways toward biopolymer precursors (Rathore et al. 2022). Next-generation computational strain design platforms such as OptKnock, AutoML-enabled design systems, and clustered regularly interspaced short palindromic repeats (CRISPR) guide RNA prediction tools, in combination with multi-omics datasets, have been developed to enable rapid precision fungal engineering for different feedstock sources and industrial bioprocess configurations (Zampieri et al. 2019).

14.4 AI-Driven Advancements in Fungal Bioenergy Production

Fungal biotechnology is a critical component of sustainable bioenergy generation owing to fungal enzymatic diversity, physiological tolerance, and the ability to valorize lignocellulosic biomass by fungi (Patil et al. 2025). However, traditional optimization methods are slow to develop and have issues related to feedstock variation as well as predictability. AI tackles these challenges by facilitating data-driven optimization of feedstock preprocessing, enzyme discovery, fermentation control, and process scale-up, which in turn reduces development timelines and enhances process robustness (Anbarasu et al. 2024; Zobel-Roos et al. 2020) (Fig. 14.1).

14.4.1 AI-Optimized Fungal-Mediated Feedstock Preprocessing

Fungal biotechnology is at the forefront of sustainable bioenergy, based on the enzymatic diversity of fungi, their physiological robustness, and potential to valorize the lignocellulosic biomass (Patil et al. 2025). Yet the traditional optimization methods have slow development time, variations in feedstock, and a lack of predictability. Artificial intelligence (AI) can help meet these challenges by facilitating data-driven optimization throughout the feedstock pre-processing stage, enzyme identification and production process, fermentation control, and subsequent scale-up, thus shortening development timescales while rendering processes more robust (Anbarasu et al. 2024; Zobel-Roos et al. 2020). Feedstock pre-processing is an essential upstream process in the production of fungal bioenergy and biomaterials, as it dictates substrate accessibility, efficiency of enzymatic hydrolysis, extent of fungal colonization, and overall process yield. Lignocellulosic by-products, agro-industrial residues, and organic wastes are highly heterogeneous in terms of composition, recalcitrance, and inhibitor profiles (Wan and Li 2012), where empirical optimization of mechanical, chemical, thermal, or biological pretreatment has been inefficient and energy-intensive. AI enables data-driven optimization of fungal feedstock preprocessing, considering compositional, physicochemical, and process-level information. Integrating the AI models with spectroscopic approaches, including near-infrared (NIR), FTIR, and Raman spectroscopy, makes it possible to noninvasively predict the recalcitrance of lignocellulosic biomass, as well as water-soluble inhibitors from biomass such as PCA (Prakash et al. 2018). Ensembles of machine learning approaches, along with deep learning architectures and Bayesian optimization frameworks, also identify optimum pretreatment parameters such as temperature, residence time, chemical dosage, and biological activity to lower energy input and increase efficiency of fungal enzymes and downstream saccharification. These advantages are evident in white-rot fungal pretreatment systems, where selective delignification is highly variable with respect to substrate composition and process severity (Wan and Li 2012; Janga et al. 2023) (Table 14.1).

Table 14.1 Artificial intelligence applications in fungal bioenergy and biomaterials: challenges and future perspectives

Application area	AI techniques used	Key achievements	Major challenges	Future perspectives	References
Fungal bioenergy production (ethanol, biodiesel, biohydrogen)	ML, deep learning, hybrid mechanistic–ML models, reinforcement learning	Optimization of enzyme production, fermentation yield, lipid accumulation, and bioethanol pathways	Small datasets, bioprocess variability, and limited model interpretability	Autonomous bioenergy fermentations, AI-driven digital twins, predictive scale-up	Wainaina and Taherzadeh (2023); Anbarasu et al. (2024)
Lignocellulosic biomass pretreatment & valorization	Ensemble ML, Bayesian optimization, predictive analytics	AI-guided fungal pretreatment, improved sugar release, waste valorization	Feedstock heterogeneity, sensor fouling, transferability across substrates	Adaptive pretreatment systems, real-time substrate routing	Khatami et al. (2025); Dhiman et al. (2024)
Strain engineering & metabolic pathway design	Deep learning, generative AI, genome-scale ML models	Accelerated enzyme discovery, pathway optimization, and improved strain productivity	Limited annotations, dual-use risks, and regulatory uncertainty	AI-driven fungal biofoundries, explainable genome engineering	Zampieri et al. (2019); Rives et al. (2021)
Mycelium-based biomaterials & composites	ML regression, deep learning, computational material design	Tailored mechanical, thermal, and structural properties of mycelium composites	Reproducibility, lack of material standards, and scale-up failures	Programmable living materials, AI-guided material certification	Madusanka et al. (2024)
Smart biorefineries & process automation	IoT-integrated AI, digital twins, reinforcement learning	Real-time monitoring, autonomous fermentation control, predictive maintenance	High capital cost, legacy infrastructure, and regulatory gaps	Fully autonomous fungal biorefineries, industry 4.0 integration	Gargalo et al. (2020); Zobel-Roos et al. (2020)
Environmental remediation & circular bioeconomy	ML prediction models, AI-assisted system design	Optimized fungal bioremediation, wastewater treatment, and biosorption	Environmental variability, limited field-scale validation	Carbon-negative fungal systems, AI-managed circular ecosystems	Janga et al. (2023); Patil et al. (2025)
Data integration & modeling frameworks	Multimodal AI, causal ML, explainable AI (XAI)	Integration of omics, imaging, and process data	Data heterogeneity, overfitting, black-box models	Open fungal databases, interpretable and trustworthy AI	Angermueller et al. (2016); Hellecks et al. (2022)

14.4.2 AI-Assisted Enzyme Discovery and Omics Integration

Fungal genome prospecting and multi-omics integration are becoming more and more important in the discovery of enzymes and the development of strains for fungal biotechnology (Table 14.1). These are now found to be excellent strains for genome mining of genes involved in hydrolysis and metabolism of biomass. Breakthroughs in AI have made this process easier than ever. Artificial intelligence (AI)-enabled genome mining by employing convolutional neural networks, transformer models, and graph neural networks facilitates the discovery and functional prediction of fungal carbohydrate-active enzymes (CAZymes) better than conventional homology-driven annotation methods through modeling complex sequence and spatial features (Rives et al. 2021). Multi-omics analysis, including transcriptomics, proteomics, metabolomics, and secretomics, allows for a comprehensive insight into fungal physiology. The regulatory hubs governing enzyme secretion and metabolic flux under different growth conditions can be predicted by network analysis and dimensionality-reduction methods (Wörheide et al. 2020). This knowledge base, in tandem with a rational development of new industrial fungal strains (e.g., *Trichoderma reesei*, *Aspergillus niger*), enables increased enzyme production and improved substrate specificity in adaptive secretomes under substrate conditions (Singh et al. 2025).

14.4.3 Intelligent Fermentation Control

In computerized fermentation monitoring, real-time and in-process sensors are used to track several critical physical (pH, temperature), chemical (dissolved oxygen), and biological (biomass, substrate utilization, and product release) factors. These sensors produce real-time data streams that capture the instantaneous physiological status of fungal cultures and provide the basis for a data-driven process optimization (Table 14.1). Integrated with artificial intelligence, the sensor-based monitoring enables predictive and adaptive operation of complex fermentation processes. Based on AI-based sensor analytics and recurrent neural networks, a real-time monitoring and prediction of fungal fermentation kinetics becomes feasible by capturing nonlinear time-dependent process information (Kadlec et al. 2009). In addition to monitoring, a reinforcement learning-driven controller adapts the core operations (aeration, feeding strategy, pH control, and induction time) in real-time with improved performance compared to a traditional PID-based controller for maintaining efficient metabolic states of the culture (Rajasekhar et al. 2024). Concurrently, AI-based contamination detection and anomaly recognition improve process reliability by spotting the anomalous sensor patterns that lead to batch failures by enhancing the process stability for overall productivity improvement in industrial fungal bioprocesses (Zobel-Roos et al. 2020).

14.4.4 AI-Based Technoeconomic Modeling, Scale-Up, and Key Recent Breakthroughs

The underpinning of scaling up fungal bioenergy and biomaterials from the laboratory to industry is not only biological performance but also economic feasibility. For even the most efficient of fungal strains or enzymes, it is still possible that their performance may be suboptimal when scaled up commercially due to feedstock variability, process uncertainty, and scale-based restrictions. Technoeconomic assessment (TEA) thus represents an imperative approach to access capital investment, operational costs, product pricing, and economic risks in fungal bioprocesses. State-of-the-art artificial intelligence algorithms have strongly improved traditional TEA frameworks in recent years. Machine-learning-assisted process technoeconomic models combine bioproduct productivity measures with cost variance, uncertainty in feedstock availability, and other scale-based limitations to provide prognostications on performance indicators like minimum fuel selling price, return-on-investment, and environmental effects (Fasahati et al. 2019; Rives et al. 2021). Risk modeling helps assist with decision-making in investing and deploying fungal bioenergy platforms such as cellulosic ethanol, fungal biodiesel, biogas, and biohydrogen (Table 14.1).

Meanwhile, significant achievements also involve AI-optimized enzyme cocktails, thermostable cellulase discovery with protein language models, AI-guided lipid pathway engineering in oleaginous fungi, predictive white-rot fungal pretreatment performance models, autonomous fermentation platforms, and machine-learning-assisted detoxification of lignocellulose-derived inhibitors. Together, this progress enhances scale-up readiness and facilitates industrial translation (Anbarasu et al. 2024; Wu et al. 2024).

14.5 AI-Enabled Biomaterials Innovation from Fungi

Fungi are living organisms with the potential to create a strong mycelial network and diverse natural polymers that may be used as environmentally friendly materials. Artificial intelligence (AI) combined with fungal biotechnology has accelerated the design of biomaterials with a range of properties, which can be exploited in various industries. Mycelium composites and hyphae-derived biopolymers have intrinsic benefits like complete biodegradability, low energy demand, and fitting within the principles of a circular bioeconomy (Attias et al. 2020). The development of a practical artificial bone implant, as an example of advanced biomaterial applications, has historically been pursued by labor-intensive trial and error with slow growth and difficult morphogenesis (Table 14.1). With the recent development of artificial intelligence (AI, e.g., machine/deep learning), it is now possible to quickly design materials and predict properties or identify bioactive metabolites in high-throughput. These data-informed strategies help expedite the transition from benchtop formulations to truly scalable, commercializable biomaterials, establishing fungal systems as potential platforms for a sustainable materials revolution (Alaneme et al. 2023; Wainaina and Taherzadeh 2023; Patil et al. 2025).

14.5.1 Material Property Prediction Through Machine Learning (ML)

Fungi with the ability to generate compact mycelial structures and develop different natural polymers, which may be utilized as eco-friendly materials. The integration of AI with fungal biotechnology is driving a rapid expansion of biomaterials that are programmable in their characteristics and technologically applicable (Table 14.1). Mycelium-based composites and myco-derived biopolymers have inherent properties, including low energy demand, compatible with the circular bioeconomy concept and decomposition (Attias et al. 2020). Historically, tuning of this class of material required laborious trial and error due to slow-growing, complex morphogenesis. Recent AI approaches, such as machine learning and deep learning, allow for efficient material design or property prediction and explicit high-throughput bioactive metabolite discovery. Such data-driven strategies have expedited the process of scaling laboratory to commercial biomaterials, demonstrating that fungal systems are attractive platforms for sustainable materials development (Alaneme et al. 2023; Wainaina and Taherzadeh 2023; Patil et al. 2025). Material properties play a significant role in the study of fungal biomaterials and their behavior under mechanical, thermal, and environmental stresses. The number of applications of machine learning models, such as Random Forest, Gradient Boosting, Support Vector Regression, and Artificial Neural Networks, to predict key properties of fungal biomaterials—compressive strength, elastic modulus, thermal conductivity, density, and porosity, as well as biodegradation rates—is growing. Such predictions are obtained from high-dimensional input parameters, which include substrate composition, fungal species or strain, inoculum density, and culture environmental growth conditions (Appels et al. 2019), leading to data-driven optimized material performance. The addition of structural descriptors, such as pore size distribution, network connectivity, and hyphal orientation, enhances prediction accuracy for the combination of ML models with advanced imaging modalities, including microCT and optical coherence tomography (Gautam et al. 2025).

For example, in mycelium-based packaging and insulation materials, ML models have been used to predict compressive strength and thermal conductivity from growth conditions and substrate composition, enabling the design of lightweight, biodegradable packaging foams and insulation panels with performance comparable to expanded polystyrene while significantly reducing environmental impact (Attias et al. 2020; Alaneme et al. 2023). Beyond prediction, generative ML approaches enable automated selection of substrate formulations and cultivation parameters, supporting scalable and application-specific biomaterial design (Scott-Fordsmand and Amorim 2022).

14.5.2 AI-Supported Design of Mycelium-Based Composites

Fungal composites (mycelium-based) are materials that have been developed through the growth of a fungal network throughout an organic feedstock, allowing for porous and biodegradable structures. Artificial intelligence (AI)-based design methods have become widely applied to improve their mechanical performance, hydrophobicity, and long-term durability by correlating the growth condition with the material behavior (Camilleri et al. 2025). Colony expansion, hyphal branching, and substrate colonization patterns are predicted using convolutional neural networks, facilitating uniform growth and early defect minimization in composite production (Alaneme et al. 2023). Mold geometry, aeration, and humidity control, as well as curing strategies, are additionally optimized through reinforcement learning and evolutionary algorithms to balance strength with water resistance in end-use materials (Alaneme et al. 2023). Similarly, AI models that combine metabolomic and structural data facilitate the prediction of hydrophobized hyphae, fire retardancy, and durability to expedite the development of high-performance mycelium composites for fashion, sustainable packaging, and automotive interiors (Jones et al. 2020) (Table 14.1).

14.5.3 Bioactive Secondary Metabolite Discovery

Fungi synthesize a large variety of small bioactive molecules (known as secondary metabolites) that can confer specific properties to biomaterials. While these metabolites confer remarkable material properties, numerous fungal biosynthetic gene clusters (BGCs) are cryptic and do not express under conventional cultivation conditions. Recent developments in artificial intelligence have completely transformed this field. Deep learning applied to LC–MS/MS data with tools like DeepBGC, antiSMASH 6.0, and PRISM allows prediction and functional annotation of such clusters (Rios-Martinez et al. 2023). The metabolites detected have been reported to aid tensile reinforcement, antimicrobial efficacy, sun protection factor, pigmentation, and self-healing characteristics. For instance, antibacterial secondary metabolites integrated in mycelium-based packaging can inhibit food-spoiling microbes and thereby extend shelf life without the use of synthetic additives. Other secondary metabolites stimulate micro-crack closure in biocomposites (Rios-Martinez et al. 2023). AI-driven unlocking of silent pathways, such as CRISPR activation methods, provides an even wider chemical design space for tailor-made fungal biomaterials (Tripathi et al. 2025) (Table 14.1).

14.5.4 Smart Biomaterials for Packaging, Construction, and Electronics

Fungal-derived materials, i.e., mycelium, can be engineered to respond to mechanical, environmental, or functional needs. Artificial intelligence (AI) is playing an ever-growing role in the advancement of different fields. In packaging, predictive models use machine learning to optimize foam density, resilience, and biodegradation rates for protecting fragile or temperature-sensitive goods with the least environmental impact (Gautam et al. 2025). In the area of construction and architecture, deep learning–aided finite element modeling is used to simulate the mechanical strength, thermal insulation, and acoustic performance of fungal bricks, panels, and load-bearing composites (Alaneme et al. 2023). Experimental initiatives, such as FUNGAR, are further proving mycelial structures to be responsive structures to the environment with self-healing properties (Sydor et al. 2021). For example, in the context of bioelectronics developments, AI can inform the design optimization of conductivity, porosity, and circuit geometry on a biodegradable fungal substrate for wearable devices as well as environmental sensors (Gautam et al. 2025).

14.5.5 Market and Commercialization Trends

For industrial applications of fungal biomaterials, scalability, batch-to-batch reproducibility, and regulatory compliance are strong factors in market success. In the field of fungal biomaterials production, AI addresses scalability, reproducibility, and regulatory issues (Wainaina and Taherzadeh 2023). Digital twins simulate full production pipelines to maximize yield, consistency, and efficiency, while computer vision and hyperspectral imaging provide real-time quality control and contaminant detection (Kadlec et al. 2009). AI-powered optimization integrates biodegradability, mechanical integrity, and safety standards for faster penetration into fashion, construction, electronics, interior decoration, and food packaging industries (Patil et al. 2025). From a regulatory point of view, AI-driven standardization and traceability tools help compliance in biodegradability standards, mechanical safety requirements, and chemical regulations (e.g., food-contact materials and building products). These features are stimulating the commercialization of fungal biomaterials in packaging, construction, interior design, electronics, and fashion (Patil et al. 2025).

14.6 Smart Biorefineries: Convergence of IoT, Automation, and AI

Artificial intelligence (AI), Internet of Things (IoT), automation, robotics, and digital control systems are being integrated into fungal biomanufacturing, leading to smart biorefineries driving real-time sensing, autonomous decision-making, and continuous optimization of operations (Scheper et al. 2020; Cheng et al. 2022). As opposed to classical biorefineries for which static control strategies are based on manual monitoring, Industry 4.0–capable fungal biorefineries apply dense IoT sensor networks that monitor a variety of signals related to dissolved oxygen, carbon dioxide evolution, viscosity development, heat evolution, substrate degradation, and contamination events during fermentation (Gargalo et al. 2020). AI-based models describe nonlinear relationships between fungal growth, morphology, and environmental factors, enabling predictive maintenance, dynamic nutrient supply, aeration control, and remote plant operation (Montague et al. 1989; Chen et al. 2025). These features help in the sustainable and high-yield production of fungal biofuels, enzymes, organic acids, and mycelium-based biomaterials, as well as improving the superior operational stability of processes and economic performance (Mondal et al. 2022).

AI is also key to real-time decision-making, supply chain control, and circular resource management at smart fungal biorefineries (Butean et al. 2025). Near infrared, Raman, and hyperspectral data in machine learning models can be used for predictive feedstock evaluation based on the prediction of fermentable sugar content, as well as for minimizing pretreatment, optimizing fungal strain selection, and increasing product yields. This results in near-zero waste routing of heterogeneous biomass streams (Prakash et al. 2018; Mahmoodi-Eshkaftaki et al. 2024). Logistics, storage, routing, and carbon footprint in the supply chain are optimized using reinforcement learning and a digital twin framework to secure feedstock availability for constant processing (Cheng et al. 2022). Actuators (agitation, ventilation, feeding) are manipulated according to online model predictive control strategies that exploit virtual bioprocess sensors. Fermentation platforms are controlled by AI algorithms based on recurrent neural networks. Computer vision and anomaly detection–based early-warning systems save batches by detecting contamination, abnormal morphology, gas evolution, thermal shift, or degradation in pH (Chen et al. 2025). Self-optimizing processes also incorporate digital twins, robotics, and autonomous downstream processing for filtration, separation, purification technology development toward drying, testing, and robotic handling of solid substrates and mycelium formation (Scheper et al. 2020). Such smart biorefineries are progressively combined with the integration of renewable energy, AI-managed heat recovery, circular design concepts valorizing fungal biomass by recycling CO_2 and effluents, and optimizing resource flows across the web of interconnected bioprocess ecosystems (Butean et al. 2025) (Table 14.1).

14.7 Challenges and Limitations

Table 14.1 provides an overview of challenges and limitations of the application of artificial intelligence in fungal bioenergy and biomaterial production. Despite the significant progress, the embedding of AI in fungal biotechnology is still confronted by scientific, operational, and organizational barriers that prevent actual deployment (Pet and Hu 2025). One of the important hindrances may be data complexity. Fungal biotechnological systems produce highly heterogeneous data ranging from genomics, multi-omics, imaging, and spectroscopy to fermentation time series and biomaterial performance metrics, which are frequently noisy, incomplete, and inadequately standardized (Wainaina and Taherzadeh 2023; Helleckes et al. 2022). The paucity of genomic annotation and functional labels, as well as uncharacterized biosynthetic gene clusters (BGCs), all contribute to reducing the predictive accuracy, whereas the small sample size (i.e., experimentally specific) increases overfitting risks and limits model generalization (Angermueller et al. 2016; Tripathi et al. 2025). While recent approaches like transfer learning, data augmentation, or hybrid modeling hold high potential, the unavailability of big, harmonized AI-ready datasets is a basic bottleneck (Mondal et al. 2022).

14.7.1 Model Interpretability and Interdisciplinary Integration

Most high-performing AI models operate as a "black box" mechanism, which restricts mechanistic understanding and adds to the complexity of regulatory approval, strain validation, safe scale-up, and biomaterial certification (Rathore et al. 2022; Cheng et al. 2022). Without transparent AI methodologies like SHAP analysis, attention mechanisms, or causal models, AI-powered decision-making in bioenergy fermenters and in bioprocess control may result in a lack of focus on metabolic behavior, resulting in less confident stakeholders. Other issues arise because of the lack of connectivity between fungal biology practitioners and AI researchers, in which the inconsistent data recording frequently results in non-standardized metadata. In addition, older bioreactor systems often produce generic results of limited usefulness, and poorly specified modeling objectives make it difficult to use AI-based knowledge effectively (Gargalo et al. 2020).

Regulatory, economic, and ethical issues also define the constraints of AI-driven fungal biofoundries and smart biorefineries. AI-based strain and pathway design raises dual-use issues requiring responsible governance, biosafety monitoring, and ethical deployment guidelines even for generally low-risk fungal systems (Butean et al. 2025). Greater automation will likely replace some classical bioprocess labor, necessitating workforce retraining and fair transition plans. High costs of capital for sensors, robotics, data platforms, and computational infrastructure limit proliferation, especially among startups and small operations (Scheper et al. 2020). Meanwhile, regulation pathways for AI-guided strain engineering/autonomous process control/digital twins are still piecemeal with little standardization on metadata, material testing, or certification. Experiences of past failures indicate that over-trusting predictions without investigations or a scarcity of mechanistic knowledge on which predictions are based can undermine AI projects. Thus, hybrid modeling methods, intensive benchmarking, and human monitoring on a regular basis are highly demanded (Zobel-Roos et al. 2020; Mondal et al. 2022).

14.8 Future Perspectives

14.8.1 AI-Driven Platforms, Biofoundries, and Predictive Systems

The convergence of fungal biotechnology and AI could revolutionize sustainable biomanufacturing, gaining autonomy over slow trial-and-error experiments toward programmable, autonomous reactors. From a technological perspective, AI-based fungal biofoundries are becoming combined systems of strain engineering, automated microfermentation, robotic measurements of phenotyping, and online monitoring (Table 14.1). These systems can be used in closed-loop design-build-test-learn cycles, which can develop strains of fungi more quickly to serve as the source of bioenergy, biomaterials, and biochemical products (Wainaina and Taherzadeh 2023; Helleckes et al. 2022). Machine-learning-assisted bioprocess optimization, monitoring, and control improve productivity and robustness with adaptive fermentation strategies, soft sensors, and predictive fault detection at various scales (Mondal et al. 2022; Butean et al. 2025). Concurrently, multiscale predictive modeling and hybrid mechanistic–AI digital twins connect molecular and cellular data to bioreactor and material scales, mitigating scale-up risks within solid-state and submerged fungal fermentations (Zobel-Roos et al. 2020; Cheng et al. 2022). Progress in deep learning for protein function prediction additionally strengthens AI-driven enzyme discovery and pathway engineering in fungi (Rives et al. 2021; Tripathi et al. 2025).

14.8.2 Circular Bioeconomy, Global Collaboration, and Responsible Innovation

AI-driven fungal platforms are likely to be pivotal players in a carbon-negative circular bioeconomy that underpins zero-waste biorefineries, industrial waste valorization, and sustainable production of energy (Table 14.1). Optimization of fungal fermentation processes by AI systems that can effectively convert industrial sidestreams and lignocellulosic wastes into fuels, chemicals, and functional biomaterials with better environmental compatibility and more economic viability has been predicted (Khatami et al. 2025; Anbarasu et al. 2024). AI-empowered fungal biorefineries for smart processing, real-time optimization, and the circular economy will further establish fungi's role toward sustainable industrial ecology (Patil et al. 2025; Butean et al. 2025). One way to accomplish the vision also lies in international cooperation for setting up open-access and standardized fungal databases in areas of genomic, enzymatic, metabolomic, process, and material information, covering the fragmentation issue that is still prevalent (Wainaina and Taherzadeh 2023; Helleckes et al. 2022). Similarly important are responsible innovation frameworks that encompass ethical governance, biosecurity, and workforce re-skilling in AI-enabled fungal biotechnology for the future bioeconomy (Scheper et al. 2020; Patil et al. 2025) to guarantee people will transparently regulate and deploy AI-powered fungal biotechnology sustainably.

14.9 Conclusion

The interface between AI and fungal biotechnology is fostering a new era of sustainable biomanufacturing. Due to the metabolic flexibility, the enzymatic potential, and the capability of feeding on low-value substrates, fungi can be attractive factories for bioenergy and biomaterial production. When implemented along with AI, these characteristics are further advanced, affording predictive bioprocesses that are energy efficient and autonomous. AI advances the entire range of biomanufacturing from genome mining and discovery of novel enzymes to fermentation control, prediction of material properties, and fully automated biorefineries. Advances in strain engineering using machine learning and deep learning that minimize the search for patterns in omics data. Additionally, reinforcement learning, real-time sensor networks, and cyber-physical automation enhance precision and scalability. There are still challenges ahead due to the lack of data, black-box algorithms, regulatory issues, and the digital–experiment knowledge gap. Novel tools are providing better accessibility, reliability, and safety through initiatives such as open-access databases, interpretable AI frameworks, and automated wet-lab platforms. In the future, AI-based fungal biofoundries, digital twins, and carbon-negative industrial ecosystems will become transformative innovations, with fungi at their core in the circular bioeconomy. Thus, the combination of AI and fungal systems is steering us toward antifragile, resource-efficient, and regenerative industries equipped to tackle pressing planetary challenges.

References

Alaneme KK, Anaele JU, Oke TM, Kareem SA, Adediran M, Ajibuwa OA, Anabaranze YO (2023) Mycelium based composites: a review of their bio-fabrication procedures, material properties and potential for green building and construction applications. Alex Eng J 83:234–250. https://doi.org/10.1016/j.aej.2023.10.012

Anbarasu K, Thanigaivel S, Sathishkumar K, Alam MM, Al-Sehemi AG, Devarajan Y (2024) Harnessing artificial intelligence for sustainable bioenergy: revolutionizing optimization, waste reduction, and environmental sustainability. Bioresour Technol 418:131893. https://doi.org/10.1016/j.biortech.2024.131893

Angermueller C, Pärnamaa T, Parts L, Stegle O (2016) Deep learning for computational biology. Mol Syst Biol 12(7):878. https://doi.org/10.15252/msb.20156651

Appels FVW et al (2019) Fabrication factors influencing mechanical properties of mycelium-based composites. Mater Des 161:64–71. https://doi.org/10.1016/j.matdes.2018.11.027

Attias N, Danai O, Abitbol T, Tarazi E, Ezov N, Pereman I, Grobman YJ (2020) Mycelium bio-composites in industrial design and architecture: comparative review and experimental analysis. J Clean Prod 246:119037. https://doi.org/10.1016/j.jclepro.2019.119037

Butean A, Cutean I, Barbero R, Enriquez J, Matei A (2025) A review of artificial intelligence applications for biorefineries and bioprocessing: from data-driven processes to optimization strategies and real-time control. PRO 13(8):2544. https://doi.org/10.3390/pr13082544

Camilleri E, Narayan S, Lingam D, Blundell R (2025) Mycelium-based composites: an updated comprehensive overview. Biotechnol Adv 79:108517. https://doi.org/10.1016/j.biotechadv.2025.108517

Chen Y, Yip YS, Tran HT, Shin S, Kim WS (2025) Artificial intelligence-driven robotic sensing system for noninvasive crop health monitoring and autonomous irrigation management. Adv Intell Syst 7(7). https://doi.org/10.1002/aisy.202500198

Cheng Y, Bi X, Xu Y, Liu Y, Li J, Du G, Lv X, Liu L (2022) Artificial intelligence technologies in bioprocess: opportunities and challenges. Bioresour Technol 369:128451. https://doi.org/10.1016/j.biortech.2022.128451

Dahiya S, Rapoport A, Singh B (2024) Biotechnological potential of lignocellulosic biomass as substrates for fungal xylanases and its bioconversion into useful products: a review. Fermentation 10(2):82. https://doi.org/10.3390/fermentation10020082

De G, Yang L, Lee J, Wu Y-H, Tian Z, Qin Z (2025) Mycelium–coir-based composites for sustainable building insulation. J Mater Chem A 13:9694–9707. https://doi.org/10.1039/D4TA07869A

Dhiman S, Kaur P, Narang J, Mukherjee G, Thakur B, Kaur S, Tripathi M (2024) Fungal bioprocessing for circular bioeconomy: exploring lignocellulosic waste valorization. Mycology 15(4):538–563. https://doi.org/10.1080/21501203.2024.2316824

Fadipe TY, Amobonye A, Pillai S (2025) Sustainable polyhydroxyalkanoates in the bioeconomy: a review of recent advances in production innovations, economic feasibility, and patents landscape. Int J Biol Macromol 337:149528. https://doi.org/10.1016/j.ijbiomac.2025.149528

Fasahati P, Liu JJ, Ohlrogge JB, Saffron CM (2019) Process design and economics for production of advanced biofuels from genetically modified lipid-producing sorghum. Appl Energy 239:1459–1470. https://doi.org/10.1016/j.apenergy.2019.01.143

Füting P, Barthel L, Cairns TC et al (2021) Filamentous fungal applications in biotechnology: a combined bibliometric and patentometric assessment. Fungal Biol Biotechnol. https://doi.org/10.1186/s40694-021-00131-6

Gargalo CL, Udugama I, Pontius K, Lopez PC, Nielsen RF, Hasanzadeh A, Mansouri SS, Bayer C, Junicke H, Gernaey KV (2020) Towards smart biomanufacturing: a perspective on recent developments in industrial measurement and monitoring technologies for bio-based production processes. J Ind Microbiol Biotechnol 47(11):947–964. https://doi.org/10.1007/s10295-020-02308-1

Gaudelet T, Day B, Jamasb AR, Soman J, Regep C, Liu G, Hayter JBR, Vickers R, Roberts C, Tang J, Roblin D, Blundell TL, Bronstein MM, Taylor-King JP (2021) Utilizing graph machine learning within drug discovery and development. Brief Bioinform 22(6). https://doi.org/10.1093/bib/bbab159

Gautam S, Verma M, Lakhanpal TS (2025) Machine-learning driven design of bio-based active food packaging films with improved mechanical properties. Sustain Food Technol 3(6):1705–1722. https://doi.org/10.1039/d5fb00198f

Gomes A, Gonçalves B, Inglês B, Silvério S, Pinto CA, Saraiva JA (2024) Potential impacts of Artificial intelligence (AI) in biotechnology. Appl Sci 14(24):11801. https://doi.org/10.3390/app142411801

Helleckes LM, Hemmerich J, Wiechert W, Von Lieres E, Grünberger A (2022) Machine learning in bioprocess development: from promise to practice. Trends Biotechnol 41(6):817–835. https://doi.org/10.1016/j.tibtech.2022.10.010

Hinneburg H, Gu S, Naseri G (2025) Fungal innovations – advancing sustainable materials, genetics, and applications for industry. J Fungi 11(10):721. https://doi.org/10.3390/jof11100721

Janga JK, Reddy KR, Raviteja K (2023) Integrating artificial intelligence, machine learning, and deep learning approaches into remediation of contaminated sites: a review. Chemosphere 345:140476. https://doi.org/10.1016/j.chemosphere.2023.140476

Jones M, Mautner A, Luenco S, Bismarck A, John S (2020) Engineered mycelium composite construction materials from fungal biorefineries: a critical review. Mater Des 187:108397. https://doi.org/10.1016/j.matdes.2019.108397

Kadlec P, Gabrys B, Strandt S (2009) Data-driven soft sensors in the process industry. Comput Chem Eng 33(4):795–814. https://doi.org/10.1016/j.compchemeng.2008.12.012

Karthikeyan PK, Bandulasena HCH, Radu T (2024) A comparative analysis of pre-treatment technologies for enhanced biogas production from anaerobic digestion of lignocellulosic waste. Ind Crop Prod 215:118591. https://doi.org/10.1016/j.indcrop.2024.118591

Khatami K, Qazanfarzadeh Z, Jiménez-Quero A (2025) Fungal fermentation: the blueprint for transforming industrial side streams and residues. Bioresour Technol 440:133426. https://doi.org/10.1016/j.biortech.2025.133426

Kovács E (2025) Anaerobic fungi: effective warriors in lignocellulosic biomass degradation and fermentation. FEMS Microbiol Ecol 101(11). https://doi.org/10.1093/femsec/fiaf108

Madusanka C, Udayanga D, Nilmini R et al (2024) A review of recent advances in fungal mycelium based composites. Discov Mater 4:13. https://doi.org/10.1007/s43939-024-00084-8

Mahmoodi-Eshkaftaki M, Mahbod M, Ghenaatian HR (2024) Non-destructive estimation of biomass characteristics: combining hyperspectral imaging data with neural networks. Renew Energy 224:120137. https://doi.org/10.1016/j.renene.2024.120137

Mandal S, Krishnan R (2021) Fungi: the budding source for biomaterials. Microb Biosyst 6(1):55–65. https://doi.org/10.21608/mb.2021.92305.1038

Mondal PP, Galodha A, Verma VK, Singh V, Show PL, Awasthi MK, Lall B, Anees S, Pollmann K, Jain R (2022) Review on machine learning-based bioprocess optimization, monitoring, and control systems. Bioresour Technol 370:128523. https://doi.org/10.1016/j.biortech.2022.128523

Montague G, Morris A, Ward A (1989) Fermentation monitoring and control: a perspective. Biotechnol Genet Eng Rev 7(1):147–188. https://doi.org/10.1080/02648725.1989.10647858

Pan R, Liu S, Chen Y et al (2025) Biosynthesis of biobased products from lignocellulose using filamentous fungi-integrated microbial consortia. Biotechnol Biofuels Bioprod 18:111. https://doi.org/10.1186/s13068-025-02715-5

Patel JY, Sharma J (2023) A review of mycelium-based bio-composites and their possible application in architecture. ShodhKosh J Visual Perform Arts 4(2SE):213–225. https://doi.org/10.29121/shodhkosh.v4.i2SE.2023.676

Patil SV, Mohite BV, Koli SH, Hasek T, Bachhav TS (2025) Fungal biotechnology and the circular economy. In: Fungal biology, pp 91–129. https://doi.org/10.1007/978-3-032-06478-3_5

Pet I, Hu P (2025) The molecular revolution in fungal diagnostics: bridging gaps across clinical, agricultural, and environmental mycology. Mycology Int J Fungal Biol:1–28. https://doi.org/10.1080/21501203.2025.2569931

Prakash M, Sarin J, Rieppo L, Afara I, Töyräs J (2018) Accounting for spatial dependency in multivariate spectroscopic data. Chemom Intell Lab Syst 182:166–171. https://doi.org/10.1016/j.chemolab.2018.09.010

Rajasekhar N, Radhakrishnan TK, Mohamed SN (2024) Reinforcement learning based temperature control of a fermentation bioreactor for ethanol production. Biotechnol Bioeng 121(10):3114–3127. https://doi.org/10.1002/bit.28784

Rathore AS, Nikita S, Thakur G, Mishra S (2022) Artificial intelligence and machine learning applications in biopharmaceutical manufacturing. Trends Biotechnol 41(4):497–510. https://doi.org/10.1016/j.tibtech.2022.08.007

Rios-Martinez C, Bhattacharya N, Amini AP, Crawford L, Yang KK (2023) Deep self-supervised learning for biosynthetic gene cluster detection and product classification. PLoS Comput Biol 19(5):e1011162. https://doi.org/10.1371/journal.pcbi.1011162

Rives A, Meier J, Sercu T, Goyal S, Lin Z, Liu J, Guo D, Ott M, Zitnick CL, Ma J, Fergus R (2021) Biological structure and function emerge from scaling unsupervised learning to 250 million protein sequences. Proc Natl Acad Sci 118(15). https://doi.org/10.1073/pnas.2016239118

Scheper T, Beutel S, McGuinness N, Heiden S, Oldiges M, Lammers F, Reardon KF (2020) Digitalization and bioprocessing: promises and challenges. Adv Biochem Eng Biotechnol 176:57–69. https://doi.org/10.1007/10_2020_139

Scott-Fordsmand JJ, Amorim MJ (2022) Using machine learning to make nanomaterials sustainable. Sci Total Environ 859(Pt 2):160303. https://doi.org/10.1016/j.scitotenv.2022.160303

Shin H-J, Ro H-S, Kawauchi M et al (2025) Review on mushroom mycelium-based products and their production process: from upstream to downstream. Bioresour Bioprocess 12:3. https://doi.org/10.1186/s40643-024-00836-7

Singh PP, Nagar P, Chakraborty S, Jaiswar D (2025) Fungal enzymes: latest developments in production and applications in industry. In: Fungal biology, pp 373–402. https://doi.org/10.1007/978-3-032-04519-5_15

Su S, Liu L, Chen J et al (2025) Biosynthesis of biobased products from lignocellulose using filamentous fungi-integrated microbial consortia. Biotechnol Biofuels Bioprod 18:111. https://doi.org/10.1186/s13068-025-02715-5

Sydor M, Bonenberg A, Doczekalska B, Cofta G (2021) Mycelium-based composites in art, architecture, and interior design: a review. Polymers 14(1):145. https://doi.org/10.3390/polym14010145

Tripathi N, Hérisson J, Faulon J (2025) Machine learning in predictive biocatalysis: a comparative review of methods and applications. Biotechnol Adv 84:108698. https://doi.org/10.1016/j.biotechadv.2025.108698

Umashankar P, Nygård Y (2024) Filamentous fungi as emerging cell factories for the production of aromatic compounds. Fungal Biol Biotechnol 11:19. https://doi.org/10.1186/s40694-024-00188-z

Vadivel D, Cartabia M, Scalet G, Buratti S, Di Landro L, Benedetti A, Auricchio F, Babbini S, Savino E, Dondi D (2024) Innovative chitin-glucan based material obtained from mycelium of wood decay fungal strains. Heliyon 10(7):e28709. https://doi.org/10.1016/j.heliyon.2024.e28709

Wainaina S, Taherzadeh MJ (2023) Automation and artificial intelligence in filamentous fungi-based bioprocesses: a review. Bioresour Technol 369:128421. https://doi.org/10.1016/j.biortech.2022.128421

Wan C, Li Y (2012) Fungal pretreatment of lignocellulosic biomass. Biotechnol Adv 30(6):1447–1457. https://doi.org/10.1016/j.biotechadv.2012.03.003

Wörheide MA, Krumsiek J, Kastenmüller G, Arnold M (2020) Multi-omics integration in biomedical research—a metabolomics-centric review. Anal Chim Acta 1141:144–162. https://doi.org/10.1016/j.aca.2020.10.038

Wu C, Wang Y, Tao L (2024) Machine learning-enabled techno-economic uncertainty analysis of sustainable aviation fuel production pathways. Chem Eng J Adv 20:100650. https://doi.org/10.1016/j.ceja.2024.100650

Wusnah W, Akbar MD, Supardan MD, Haryani S, Yunardi Y (2023) An overview of the potential utilisation of oleaginous yeast for biodiesel feedstock and wastewater treatment. IOP Conf Ser Earth Environ Sci 1182(1):012018. https://doi.org/10.1088/1755-1315/1182/1/012018

Yang J, Yue H, Pan L, Feng J, Zhao S, Suwannarangsee S, Champreda V, Liu C, Zhao X (2023) Fungal strain improvement for efficient cellulase production and lignocellulosic biorefinery: current status and future prospects. Bioresour Technol 385:129449. https://doi.org/10.1016/j.biortech.2023.129449

Yélamos AM, Marcos JF, Manzanares P, Garrigues S (2025) Harnessing filamentous fungi for enzyme cocktail production through Rice bran bioprocessing. J Fungi 11(2):106. https://doi.org/10.3390/jof11020106

Zampieri G, Vijayakumar S, Yaneske E, Angione C (2019) Machine and deep learning meet genome-scale metabolic modeling. PLoS Comput Biol 15(7):e1007084. https://doi.org/10.1371/journal.pcbi.1007084

Zobel-Roos S, Schmidt A, Uhlenbrock L, Ditz R, Köster D, Strube J (2020) Digital twins in biomanufacturing. In: Advances in biochemical engineering, biotechnology, vol 176, pp 181–262. https://doi.org/10.1007/10_2020_146

Chapter 15
Metabolic Engineering in Fungal Bioprocesses for Biofuel Production from Waste and Their Applications

Surya Mishra, Anuma Singh, and Roshini Singh

Abstract The increasing global demand for energy and the profound adverse environmental impacts of fossil fuels have accelerated the search for efficient, renewable, and sustainable energy alternatives. Among various biological systems, fungi have emerged as promising cell factories for biofuel production owing to their robust enzymatic machinery, metabolic flexibility, and ability to utilize diverse waste substrates. Metabolic engineering of fungi has become a powerful strategy to enhance biofuel yields by optimizing metabolic pathways, redirecting carbon flux, and improving substrate utilization efficiency. This chapter provides a comprehensive overview of recent advances in metabolic engineering approaches applied to fungal bioprocesses for the production of biofuels from waste materials. It highlights the use of genetic modification tools such as Clustered Regularly Interspaced Short Palindromic Repeats (CRISPR)–Cas systems (where Cas refers to CRISPR-associated proteins), gene overexpression, pathway reconstruction, and regulatory network optimization to enhance the biosynthesis of bioethanol, biodiesel, and other energy-rich compounds. Furthermore, the chapter discusses bioprocess design, optimization strategies, and the integration of metabolic engineering with systems biology and computational modeling for improved process efficiency. Current challenges, including strain stability, scalability, and economic feasibility, are addressed along with emerging opportunities through synthetic biology, artificial intelligence, and circular bioeconomy frameworks. Overall, the chapter highlights the essential role of metabolically engineered fungi in transforming waste

S. Mishra
Department of Biotechnology, Goel Institute of Technology & Management, Lucknow, Uttar Pradesh, India

A. Singh
Centre for Applied Bioscience Research (CABR), Munster Technological University (MTU), Tralee, Kerry, Ireland

R. Singh (✉)
Department of Biotechnology, SR Institute of Management and Technology, Bakshi Ka Talab, Lucknow, Uttar Pradesh, India

M. Tripathi et al. (eds.), *Recent Advancements in Fungal Biotechnology Volume 1*, Fungal Biology, https://doi.org/10.1007/978-981-92-0339-0_15

into sustainable energy resources, contributing to the development of next-generation biorefineries and a greener global energy landscape.

Keywords Metabolic engineering · Fungal biotechnology · Biofuel production · Waste valorization · Genetic modification · Carbon flux redirection · Sustainable biorefineries

15.1 Introduction

The growing global energy demand due to rapid industrialization, urban expansion, and continuous population growth has intensified the reliance on fossil fuels, resulting in escalating greenhouse gas emissions, accelerating climate change, and widespread ecological degradation. These challenges have highlighted the urgent need to transition from non-renewable, carbon-intensive energy systems toward sustainable alternatives that can ensure long-term environmental stability and energy security. Renewable energy technologies such as solar, wind, and bioenergy have emerged as critical components of this shift toward renewable sources, offering significant potential to reduce carbon emissions and mitigate the impacts of climate change. However, the large-scale implementation of renewable energy requires continual technological innovation, enhanced resource utilization strategies, and strong policy frameworks to overcome economic and logistical barriers (Holechek et al. 2022).

Among renewable energy sources, biofuels derived from lignocellulosic biomass, commonly referred to as second-generation biofuels, have gained significant attention due to their characteristics like high energy potential, sustainability, low cost, and eco-friendliness. Unlike first-generation biofuels, which are derived from edible crops, lignocellulosic biofuels depend on abundant waste resources, thereby reducing environmental burden and promoting circular resource use. Their integration into modern biorefineries supports the dual goals of waste reduction and clean energy generation, aligning with the urgencies of global sustainability.

Fungi, particularly filamentous fungi and anaerobic gut fungi, play an essential role in the bioconversion of lignocellulosic materials into fermentable sugars and downstream biofuels (Deshpande and Dhakephalkar 2025). They are capable of degrading cellulose, hemicellulose, lignin, and other complex plant polymers through enzymatic digestion, making them essential biocatalysts in sustainable biomass processing. Moreover, fungal cocultures and synthetic consortia have set a benchmark in enzyme secretion, synergistic biomass hydrolysis, and overall process efficiency, further strengthening their role within the circular bioeconomy. The metabolic versatility of fungi also enables them to utilize diverse low-cost feedstocks, including food waste, agricultural waste, and marine biomass, and to integrate effortlessly into waste treatment streams, enhancing the feasibility of eco-friendly, large-scale biofuel production. Marine-derived and obligate anaerobic fungi contribute additional advantages through their unique enzymatic profiles, which facilitate biomass degradation under extreme or hypoxic conditions, thereby broadening the substrate spectrum for industrial applications.

Genetic modifications have enabled researchers to optimize carbon utilization, enhance tolerance to process inhibitors, and ultimately boost biofuel yields (Pabbathi et al. 2025). The advent of CRISPR–Cas genome editing systems, comprehensive multi-omics platforms, and synthetic biology approaches has transformed fungal engineering into a highly predictable and efficient process. These advanced tools enable systematic rewiring of regulatory networks, reconstruction of metabolic pathways, and development of optimized fungal models with enhanced robustness and productivity. Integrative approaches that combine genome editing, systems biology, and computational modeling are rapidly accelerating the creation of high-performance fungal strains tailored for industrial-scale bioprocesses, reinforcing fungi's importance as next-generation cell factories in the emerging global bioeconomy (Adegboye et al. 2021; Garg et al. 2024).

15.2 Fungi as Efficient Cell Factories for Biofuel Production

Fungi have emerged as highly efficient and versatile cell factories for sustainable biofuel production, owing to their notable biodiversity, metabolic adaptability, and ability to degrade complex plant biomass. Industrially relevant fungal species encompass major taxonomic groups, including *Ascomycota*, *Basidiomycota*, *Zygomycota*, and anaerobic *Neocallimastigomycota*, each contributing unique physiological and biochemical characteristics that make them necessary for modern bioenergy systems. Filamentous fungi such as *Aspergillus niger*, *Trichoderma reesei*, and *Phanerochaete chrysosporium* are widely known for their excellent enzyme secretion and lignocellulose-degrading abilities. These features place them at the forefront of large-scale biofuel processes, particularly those relying on the conversion of plant biomass into fermentable substitutes for bioethanol, biodiesel, and other advanced biofuels (Saye et al. 2021; Panahi et al. 2022).

Anaerobic fungi belonging to *Neocallimastigomycota* are gaining significant research interest due to their natural ability to degrade lignocellulose under anaerobic conditions. These fungi generate cellulolytic and hemicellulolytic enzyme systems that operate effectively in anaerobic conditions, expanding the range of usable substrates and enabling biofuel production in environments where conventional aerobic fungi show limited performance. Their complementary metabolic abilities make them promising candidates for next-generation consolidated bioprocesses and anaerobic biorefineries (Ren et al. 2024; Singhania et al. 2025).

Anaerobic fungi also possess cellulosomes, a unique multi-enzyme complex structure that organizes cellulases and accessory enzymes into a highly efficient biomass-degrading mechanism. Unlike aerobic fungi, fungal cellulosomes anchor directly to plant fibers, leading to rapid degradation of lignocellulose. These specialized enzymatic complexes provide opportunities to engineer synthetic or hybrid cellulosomes for enhanced industrial biomass conversion.

A major advantage of fungi is their capacity to secrete highly efficient extracellular enzyme systems, including cellulases, hemicellulases, ligninases, pectinases, and a wide range of accessory proteins. These enzyme consortia act synergistically to break down the complex plant cell wall components into fermentable monosaccharides. Particularly, lytic polysaccharide monooxygenases (LPMOs) catalyze oxidative cleavage reactions that substantially enhance the degradation of recalcitrant cellulose. Industrially relevant species such as *T. reesei* and *A. niger* have been intensively engineered for high-yield enzyme production and large-scale lignocellulosic bioconversion processes (Contato and Conte-Junior 2025).

Beyond enzymatic efficiency, fungi exhibit exceptional metabolic versatility, enabling them to utilize a wide range of low-cost and renewable waste as substrates. Agricultural waste residues, such as spent mushroom substrate, food processing waste, and municipal solid waste (MSW), can be converted into fermentable sugars and further processed into biofuels. The ability of fungi to thrive on diverse lignocellulosic feedstocks aligns with the principles of the circular bioeconomy and supports waste-to-energy transitions (Dhiman et al. 2024).

Recent studies have demonstrated that fungal consortia can outperform monocultures in enzyme production, substrate utilization, and biofuel yields (Ramachandraiah et al. 2020). Integrating these biological systems with modern waste management infrastructures not only reduces environmental pollution but also improves the techno-economic feasibility of large-scale biofuel production (Leong et al. 2022; Vieira et al. 2025).

15.3 Metabolic Engineering Strategies in Fungi

Metabolic engineering of fungi has become a basis of modern industrial biotechnology, enabling the targeted improvement of native metabolic pathways and the construction of new biosynthetic routes. Recent advances in synthetic biology, systems biology, and high-throughput genome editing have significantly expanded the ability to redesign fungal cells to increase biofuel yields. This section outlines the major strategies employed to modulate fungal metabolism, improve strains, and tailor fungal hosts for specific industrial and environmental applications (Tamano and Yoshimi 2021).

15.3.1 Gene Overexpression and Knockout Approaches

Gene overexpression and targeted gene disruption remain essential tools in fungal metabolic engineering. By strategically increasing the expression of key pathway enzymes, researchers can increase the flux toward desired metabolic intermediates or end products. Overexpressing rate-limiting enzymes, transporters, or cofactor-regeneration systems frequently leads to substantial improvements in metabolite accumulation. Conversely, gene knockout or knockdown strategies help eliminate competing pathways that divert carbon away from the target product (Dagariya et al. 2025).

Traditional methods such as homologous recombination have been used for gene manipulation in fungi for a long period of time; however, their efficiency was often limited by low transformation frequencies and strain-specific constraints. RNA interference (RNAi) and advanced molecular tools have enabled post-transcriptional silencing of genes, allowing for partial or conditional reduction of gene expression. These approaches provide more control over metabolic pathways, making it possible to redirect metabolic flux with greater precision. Together, gene overexpression and knockout strategies remain essential components of any engineered fungal production system and strain improvement (Salazar-Cerezo et al. 2023).

15.3.2 CRISPR–Cas Systems for Precision Editing

The advent of CRISPR–Cas technologies has revolutionized fungal biotechnology by offering fast, accurate, and multiplexed genome engineering with unprecedented ease. CRISPR–Cas9 (CRISPR-associated protein 9) has been widely adopted in filamentous fungi for deleting genes, inserting heterologous sequences, and introducing point mutations to enhance enzyme activity or alter regulatory regions. The emergence of alternative Cas proteins such as Cas12a (Cpf1) has further broadened the toolkit, enabling staggered DNA cuts and improved editing efficiency in certain fungal hosts.

CRISPR interference (CRISPRi) and CRISPR activation (CRISPRa) systems provide powerful tools for fine-tuning gene expression without altering genomic sequences. These programmable regulatory platforms allow repression or activation of endogenous genes, permitting dynamic control over metabolic flux during fermentation. Furthermore, multiplex CRISPR systems enable simultaneous modification of several genes or pathways, accelerating the development of optimized strains. By significantly reducing the time and resources required for strain engineering, CRISPR-based tools have become necessary for constructing next-generation fungal cell factories with improved stress tolerance, product titres, and metabolic efficiency (Ma et al. 2025).

15.3.3 Pathway Reconstruction and Carbon Flux Redirection

Biosynthetic pathway reconstruction is a hallmark of modern metabolic engineering, enabling fungi to produce non-native compounds or greatly increase production of metabolites. Advances in synthetic biology allow the assembly of synthetic operons and codon-optimized gene clusters sourced from bacteria, plants, or other fungi. These reconstructed pathways can introduce capabilities such as the production of advanced biofuels (e.g., butanol and fatty alcohols), high-value terpenoids, or pharmaceutical precursors.

Carbon flux redirection is achieved by balancing metabolic load, optimizing enzyme content, modulating cofactor usage (NADH/NADPH), and removing metabolic blockages. Computational tools such as genome-scale metabolic models (GEMs) and flux balance analysis (FBA) support the identification of key metabolic pathways and predict the effect of engineering interventions. This integrative approach ensures rational design of metabolic pathways that maximize carbon conversion efficiency and minimize by-product formation.

Pathway reconstruction, when combined with targeted control of flux distribution, enables the creation of highly efficient fungal production systems capable of converting diverse feedstocks into renewable fuels and chemicals with minimal metabolic burden (Lee et al. 2018).

15.3.4 Regulatory Network Optimization

Beyond metabolic pathway modification, the optimization of regulatory networks that govern gene expression, enzyme activity, and cellular responses is equally critical. Fungi possess complex regulatory mechanisms that respond to environmental cues such as nutrient availability, oxygen levels, temperature, pH, and stress signals. Modifying these networks enables precise control over metabolic pathways and ensures stable performance during industrial fermentation.

Synthetic promoters, including constitutive, inducible, or condition-responsive types, facilitate tailored regulation of gene expression to achieve desired expression levels. Engineered transcription factors and synthetic regulatory elements enable coordinated activation of multi-gene pathways and help avoid metabolic imbalances that lead to growth defects or reduced productivity. Post-translational regulation through protein engineering and signaling pathway modulation can further enhance metabolic control.

15.4 Types of Biofuels Produced by Engineered Fungi

Engineered fungi have emerged as highly versatile platforms for the production of diverse biofuels, owing to their metabolic flexibility, robust growth on heterogeneous feedstocks, and strong enzyme secretion systems that efficiently break down complex biomass. Advances in synthetic biology, metabolic engineering, and systems biology have enabled the development of fungal strains capable of producing a broad selection of energy carriers ranging from traditional alcohols to advanced biofuels. Different types of biofuels have been illustrated in Fig. 15.1. Fungi such as *Saccharomyces cerevisiae*, *Yarrowia lipolytica*, and various filamentous species are now widely modified to enhance product yields, improve tolerance to inhibitors,

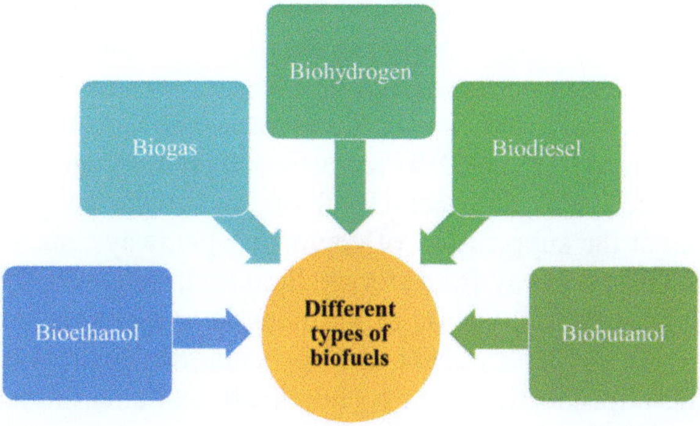

Fig. 15.1 Different types of biofuels derived from fungi

and expand substrate utilization. These engineered fungal systems are critical in transitioning from first-generation biofuels based on food crops to second and third-generation fuels derived from lignocellulosic biomass, waste streams, and non-food carbon sources (Dhiman et al. 2024).

15.4.1 Bioethanol Production

Bioethanol remains the leading biofuel produced using engineered fungi, primarily due to the exceptional fermentative capacity of yeasts like *S. cerevisiae*. Traditional yeast strains naturally ferment hexose sugars; however, metabolic engineering efforts have expanded their capabilities to utilize pentose sugars (xylose and arabinose) present in lignocellulosic biomass. This improvement significantly enhances the efficiency of second-generation bioethanol production. Strategies such as introducing xylose isomerase pathways, optimizing sugar transporters, and engineering redox cofactor balance have dramatically increased ethanol yields from non-conventional substrates. Moreover, improved inhibitor tolerance toward acetic acid and phenolic compounds has allowed engineered fungi to more efficiently ferment hydrolysates derived from harsh pretreatments. Enhanced thermotolerance and osmotolerance also allow longer fermentation cycles, decrease cooling costs, and improve yield at industrial scales. As a result, engineered yeasts serve as indispensable biocatalysts for sustainable and cost-effective bioethanol production (Zhang et al. 2023).

15.4.2 Biodiesel Production

Oleaginous fungi have emerged as powerful cell factories for biodiesel production due to their ability to accumulate massive amounts of lipids, often exceeding 60%–70% of their dry biomass under nitrogen-limited conditions. These lipids or single-cell oils (SCOs) have fatty acid profiles similar to plant-derived oils, making them an excellent renewable alternative for biodiesel synthesis. Species like *Y. lipolytica*, *Mortierella alpina*, and *Rhodosporidium toruloides* have been extensively engineered to enhance lipid synthesis pathways by increasing acetyl-CoA availability, improving NADPH regeneration, and upregulating key enzymes such as ATP-citrate lyase, fatty acid synthase, and malic enzyme. Genetic modifications also target the suppression of competing pathways, such as β-oxidation, which diverts fatty acids away from storage lipid accumulation. Additionally, engineering efforts support the utilization of inexpensive waste-derived carbon sources, such as glycerol, volatile fatty acids, and lignocellulosic hydrolysates, making fungal-based biodiesel production both economically and environmentally sustainable (Caporusso et al. 2021).

15.4.3 Biobutanol and Other Higher Alcohols

Higher-chain alcohols such as biobutanol, isobutanol, and long-chain alcohols possess superior energy density and combustion characteristics compared to ethanol. Fungi typically do not produce these alcohols in significant quantities; therefore, extensive metabolic pathway reconstruction is required. Synthetic biology efforts have introduced the *Clostridial* acetone-butanol-ethanol (ABE) pathway into fungal hosts, enabling them to carry out multi-step enzymatic reactions for butanol production. Engineering strategies include improving butanol tolerance, which is crucial because higher alcohols exert greater toxicity on cellular membranes compared to ethanol. Isobutanol production achieved through modification of the Ehrlich pathway or the keto-acid pathway has seen significant optimization through enhanced mitochondrial targeting of enzymes, cofactor engineering, and dynamic pathway regulation to prevent metabolic burden. These advances are bringing engineered fungi closer to becoming competitive platforms for next-generation alcohol biofuels (Serrano-Echeverry et al. 2025).

15.4.4 Biogas, Hydrogen, and Other Energy-Rich Compounds

Fungi contribute to biogas and hydrogen production largely through co-culture systems and enzyme-mediated biomass degradation. In anaerobic digesters, fungi break down lignocellulosic components into simpler substrates, thereby enhancing methane yields from methanogenic archaea. Some fungal species also produce hydrogen under dark fermentation conditions by utilizing pyruvate-ferredoxin oxidoreductase pathways. Additionally, engineered fungi can synthesize value-added energy compounds such as terpenes, organic acids (e.g., succinate or acetate), and polyketides, which can be upgraded into drop-in fuels. These alternative biofuels broaden the energy portfolio and reduce reliance on fossil fuels while enabling innovative routes for renewable energy generation (Harirchi et al. 2022).

15.5 Waste Materials as Feedstocks for Fungal Bioprocesses

Waste valorization is a central pillar of sustainable bioprocessing, and fungi are among the most effective microorganisms for converting heterogeneous waste materials into high-value biofuels and biochemicals (Fig. 15.2). Fungi possess exceptional enzymatic capabilities that allow them to degrade complex polymers such as cellulose, hemicellulose, lignin, starch, pectin, and lipids. Using waste feedstocks not only reduces production costs but also supports environmental goals by diverting waste from landfills and lowering greenhouse gas emissions. The ability of fungi to grow on diverse substrates such as agricultural residues, food waste, lignocellulosic biomass, industrial effluents, and municipal waste offers unmatched flexibility for large-scale biorefinery systems (Atta et al. 2025).

15.5.1 Agricultural Waste

Agricultural residues, including rice husk, wheat straw, corn stover, sugarcane bagasse, banana waste, coconut coir, and oilseed cakes, represent some of the most abundant and cost-effective feedstocks. Filamentous fungi such as *T. reesei* and *A. niger* naturally secrete large amounts of cellulases and hemicellulases, making

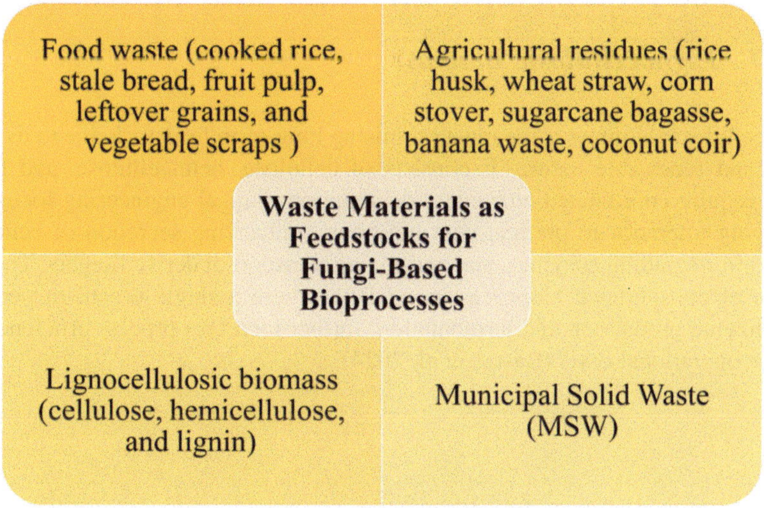

Fig. 15.2 Waste materials used as feedstocks for fungal bioprocesses

them ideal for hydrolyzing plant biomass. Engineered strains with improved enzyme cocktails or enhanced lignin-degrading capabilities accelerate saccharification and fermentable sugar release. These sugars are subsequently converted into ethanol, butanol, lipids, or organic acids. Moreover, agricultural wastes rich in lipids and proteins, such as oilseed cakes and fruit peels, serve as excellent precursors for single-cell oil accumulation in oleaginous fungi (Da Silva et al. 2025a).

15.5.2 Food and Industrial Waste Streams

Food waste is rich in carbohydrates, proteins, and lipids, making it a nutrient-dense substrate for fungal fermentation. Engineered fungi can convert cooked rice, stale bread, fruit pulp, leftover grains, and vegetable scraps into ethanol, lipids, and organic acids. Fungi like *Rhizopus oryzae* thrive on starch-rich waste streams, producing both ethanol and lactic acid (Da Silva et al. 2025b).

Industries such as breweries, distilleries, paper mills, biodiesel facilities, and dairy plants generate nutrient-rich waste streams. Fungi can metabolize spent grain, whey, molasses, lignin liquor, and glycerol into biofuels. Engineered strains show improved tolerance to the chemical complexity of industrial waste and can convert mixed substrates into desired fuels with high efficiency. Such valorization not only enhances sustainability but also reduces waste disposal costs for industries. Moreover, lipid-rich waste, such as used cooking oil and dairy waste, is suitable for microbial oil production. Utilizing food waste not only contributes to sustainable fuel production but also addresses the global challenge of food waste management (Borkertas et al. 2025).

15.5.3 Lignocellulosic Biomass

Lignocellulosic biomass is the most promising long-term feedstock due to its abundance and renewable nature. It consists of cellulose, hemicellulose, and lignin, which require coordinated enzymatic degradation. Fungal engineering focuses on improving tolerance to pretreatment inhibitors, enhancing secretion of cellulases and lignin-degrading enzymes, and optimizing transport of derived sugars. The integration of consolidated bioprocessing (CBP), where a single organism performs both enzyme production and fermentation, further increases process efficiency and reduces operational costs (Basera et al. 2024).

15.5.4 Municipal Solid Waste (MSW)

Municipal waste presents an enormous untapped resource for fungal bioprocessing. Organic fractions consisting of food scraps, yard waste, and paper can be pretreated and fermented by engineered fungi. The use of MSW supports zero-waste strategies and circular bioeconomy models, particularly in urban environments. Fungal systems also help detoxify harmful organic compounds, making MSW a practical substrate for integrated biorefineries.

15.6 Bioprocess Design and Optimization

Effective biofuel production depends not only on engineering fungal strains but also on optimizing fermentation processes, bioreactor designs, operational parameters, and downstream recovery strategies. Bioprocess optimization ensures high yields, reduced costs, and reliable scale-up from laboratory to industrial levels.

15.6.1 Fermentation Systems: Batch, Fed-Batch, and Continuous

Batch fermentation is widely used for small-scale production due to its simplicity, but it has limitations in productivity. Fed-batch systems allow controlled nutrient addition, reducing substrate inhibition and improving product yield. Continuous fermentation, which maintains a steady-state culture, offers the highest productivity and is suited for large-scale industrial operations. Engineered fungi with stable performance enable consistent output in continuous systems. Integration of sensors, automated monitoring, and feedback control further enhances fermentation efficiency (Dhiman et al. 2024).

15.6.2 Co-culture and Consolidated Bioprocessing (CBP)

Co-culture systems exploit the complementary metabolic functions of different microbial species. For example, fungi may degrade lignocellulose while yeast ferments the resulting sugars into ethanol. Consolidated bioprocessing (CBP) integrates enzyme production, biomass saccharification, and fermentation into a single step, reducing operational costs and simplifying bioprocess design. Engineered fungi with multi-functional enzymatic capabilities are ideal candidates for CBP-based biofuel production (Vieira et al. 2025).

15.6.3 Bioreactor Design Considerations

Fungal morphology, either yeast-like or filamentous, plays a crucial role in bioreactor performance. Filamentous fungi require gentle mixing to avoid shear damage, whereas yeasts tolerate higher agitation. Airlift bioreactors, stirred-tank reactors, and packed-bed systems are commonly used, depending on oxygen demand and viscosity. Optimal aeration, pH control, and temperature regulation are essential for achieving maximum biomass and product yields (Ogawa et al. 2024).

15.6.4 Downstream Processing and Product Recovery

Downstream processing involves the separation, purification, and concentration of biofuels. Ethanol is recovered through distillation, while lipids are extracted using solvent extraction or mechanical methods. Advanced techniques such as membrane filtration, adsorption, and chromatographic purification improve recovery efficiency and reduce energy consumption. Innovative in situ product removal technologies minimize product inhibition and increase overall yield (Janković et al. 2023).

15.7 Integration of Systems Biology and Computational Tools

Systems biology offers powerful approaches to understand, model, and optimize complex fungal metabolic networks. Integrating genomic, transcriptomic, proteomic, and metabolomic data enables rational strain engineering and predictive bioprocess design.

15.7.1 Genome-Scale Metabolic Models (GEMs)

GEMs reconstruct entire metabolic networks based on genomic information. These models identify metabolic bottlenecks, essential genes, and optimal engineering targets. GEM-guided strain engineering has significantly improved the production of ethanol, lipids, and organic acids in fungi. Advanced GEMs incorporate regulatory and kinetic information for more accurate predictions (Passi et al. 2021).

15.7.2 Flux Balance Analysis (FBA)

FBA is a constraint-based approach used to simulate metabolic flux distribution. It enables researchers to predict how genetic modifications will affect product yield. FBA has been instrumental in designing fungal strains with improved acetyl-CoA flux, enhanced lipid accumulation, and optimized redox balance (Ravi and Gunawan 2021).

15.7.3 Multi-Omics Approaches

Integrating genomics, transcriptomics, proteomics, and metabolomics provides a holistic understanding of fungal metabolism. Omics-guided engineering allows the identification of differentially expressed genes, metabolic pathway shifts, and stress-response mechanisms. These insights are essential for developing robust fungal strains suitable for industrial conditions (Xie et al. 2025).

15.7.4 AI-Based Optimization and Predictive Metabolic Engineering

AI and machine learning models accelerate strain design by predicting genetic modifications, optimizing fermentation conditions, and identifying unknown regulatory interactions. AI-driven metabolic engineering allows rapid prototyping of fungal cell factories, significantly reducing research timelines and improving success rates.

15.8 Challenges and Limitations in Fungal Metabolic Engineering

Despite substantial progress, several challenges still hinder the widespread commercial adoption of engineered fungi for biofuel production. Genetically engineered traits may be lost over successive generations, especially under industrial fermentation conditions. Genome instability, metabolic burden, and evolutionary drift can affect product consistency. Strategies such as chromosome integration, dynamic regulation, and stress-adapted evolution help maintain long-term strain robustness (Zhang et al. 2025).

Lignocellulosic and waste-derived feedstocks contain inhibitory compounds like phenolics, furans, and organic acids that reduce fungal growth and productivity. Engineering detoxification pathways, improving membrane transporters, and using adaptive laboratory evolution (ALE) can enhance resistance to inhibitors. Scaling up from laboratory to industrial reactors introduces challenges such as uneven mixing, oxygen transfer limitations, and shear stress. Economic feasibility depends on reducing energy inputs, improving yields, and utilizing inexpensive feedstocks. Continuous fermentation and automation help reduce operational costs (Adebule et al. 2025).

The use of genetically modified (GM) fungi often raises regulatory issues related to biosafety, environmental containment, and public acceptance. Strict guidelines govern large-scale use of GM microbes, particularly in open systems. Improved containment strategies and transparent regulatory frameworks are essential for commercialization.

15.9 Future Prospects and Emerging Trends

The future of fungal metabolic engineering is driven by rapid advances in synthetic biology, computational design, and sustainable biomanufacturing. Emerging technologies promise to unlock the full potential of fungi as next-generation biofuel producers.

Synthetic biology tools, including modular genetic parts, synthetic promoters, CRISPR-based regulation, and biosensors, enable precise control of fungal metabolism. These tools allow the construction of programmable fungal chassis with customizable metabolic pathways and enhanced stress tolerance. Future bioprocesses will increasingly rely on engineered microbial consortia where different fungi or bacteria share metabolic functions. Division of labor improves substrate utilization, reduces metabolic burden, and increases overall productivity. Modular pathway engineering enables assembly of complex biosynthetic routes with minimal cellular stress (Lv et al. 2022).

Fungal systems will play a critical role in circular bioeconomy models, where waste materials are fully valorized into fuels, chemicals, and biomaterials. Zero-waste biorefineries integrate multiple product streams such as biofuels, enzymes, organic acids, single-cell protein, etc., maximizing resource utilization and minimizing environmental impact. Next-generation fungal cell factories will combine AI-guided metabolic engineering, adaptive evolution, and synthetic organelle engineering to produce high-value biofuels with exceptional efficiency. These advanced strains will feature enhanced carbon utilization pathways, superior stress tolerance, and programmable metabolic circuits, positioning fungi at the forefront of sustainable energy biotechnology.

15.10 Conclusion

The escalating global energy demand and the environmental consequences of fossil fuel dependence have intensified the need for sustainable, renewable, and circular energy solutions. Metabolically engineered fungi have emerged as highly promising platforms for converting waste into valuable biofuels. Fungi have the inherent metabolic flexibility, enzyme secretion mechanism, and ability to degrade complex lignocellulosic substrates, therefore acting as indispensable cell factories for next-generation biorefineries.

This chapter highlights the advances in metabolic engineering, ranging from gene overexpression and pathway knockouts to CRISPR–Cas-based precision editing, regulatory network optimization, and carbon flux redistribution that have significantly enhanced fungal biofuel production. Engineered fungi produce a wide spectrum of biofuels, including bioethanol, biodiesel, biogas, and other energy-rich compounds, often using low-cost agricultural, food, industrial, and municipal wastes as feedstocks. The integration of systems biology, multi-omics analyses, computational modeling, and AI-driven optimization has further enabled strain improvement and pilot-scale development, accelerating the transition from laboratory-scale innovation to industrial application.

Despite these advancements, challenges related to strain development, scalability, economic feasibility, and regulatory acceptance remain key barriers to commercialization. Addressing these limitations and challenges through strain improvement, dynamic metabolic control, improved pilot-scale design, and robust regulatory frameworks will be critical for the widespread deployment of fungal biofuel technologies. The convergence of synthetic biology, artificial intelligence, and bioeconomy principles is expected to unlock the full potential of fungi as sustainable bioenergy producers.

Overall, metabolically engineered fungi serve as a powerful and versatile biological tool for waste valorization and renewable energy generation. Continued bioprospecting and technological development will play a pivotal role in enhancing resource efficiency, reducing environmental pollution, and supporting the transition toward green energy and low-carbon emissions.

References

Adebule AP, Eregie SB, Sanusi IA, Kana GE (2025) Progress in the development of detoxification strategies and the current application of nanoparticle-based detoxification of pretreated lignocellulosic hydrolysate for bioprocessing in a biorefinery scenario. Biomass Bioenergy 206:108628. https://doi.org/10.1016/j.biombioe.2025.108628

Adegboye MF, Ojuederie OB, Talia PM, Babalola OO (2021) Bioprospecting of microbial strains for biofuel production: metabolic engineering, applications, and challenges. Biotechnol Biofuels 14(1):5. https://doi.org/10.1186/s13068-020-01853-2

Atta OM, Chen L, Mohamed H, Ullah S, Abo-Kadoum M, Abdel-Aziz AE, Taleb K, Mosa MA, Hassane AM (2025) Biotechnological approaches concerning valorisation of fungal enzymes in agro-industrial waste management. Biomass Bioenergy 206:108585. https://doi.org/10.1016/j.biombioe.2025.108585

Basera P, Chakraborty S, Sharma N (2024) Lignocellulosic biomass: insights into enzymatic hydrolysis, influential factors, and economic viability. Discov Sustain 5(1):311. https://doi.org/10.1007/s43621-024-00543-5

Borkertas S, Viskelis J, Viskelis P, Streimikyte P, Gasiunaite U, Urbonaviciene D (2025) Fungal biomass fermentation: valorizing the food industry's waste. Fermentation 11(6):351. https://doi.org/10.3390/fermentation11060351

Caporusso A, Capece A, De Bari I (2021) Oleaginous yeasts as cell factories for the sustainable production of microbial lipids by the valorization of agri-food wastes. Fermentation 7(2):50. https://doi.org/10.3390/fermentation7020050

Contato AG, Conte-Junior CA (2025) Fungal lytic polysaccharide monooxygenases (LPMOs): functional adaptation and biotechnological perspectives. Eng 6(8):177. https://doi.org/10.3390/eng6080177

Da Silva BP, Gonçalves JO, Brandão PM, De Almeida Pinto LA, Christ-Ribeiro A (2025a) Plant sources and fermentation in cellular agriculture and alternative proteins: a review. Trends Food Sci Technol 167:105455. https://doi.org/10.1016/j.tifs.2025.105455

Da Silva SB, Silva RK, De Abreu ÍBS, De Sousa MH, Dutra ED, Albuquerque AA, De Morais MA Jr, De Souza RB (2025b) Integrated bioprocesses for urban food waste: insights into biological pathways, process integration, and circular economy perspectives. Recycling 10(5):188. https://doi.org/10.3390/recycling10050188

Dagariya S, Bhatankar J, Dakal TC, Gadi BR, Giudici P (2025) Metabolic and evolutionary engineering of food yeasts. Processes 13(6):1852. https://doi.org/10.3390/pr13061852

Deshpande P, Dhakephalkar PK (2025) From gut to green energy: the essential role of anaerobic fungi in sustainable biofuel production. Sustain Energy Fuels 9(24):6642–6664. https://doi.org/10.1039/d5se00240k

Dhiman S, Kaur P, Narang J, Mukherjee G, Thakur B, Kaur S, Tripathi M (2024) Fungal bioprocessing for circular bioeconomy: exploring lignocellulosic waste valorization. Mycology 15(4):1–26. https://doi.org/10.1080/21501203.2024.2316824

Garg S, Kim M, Romero-Suarez D (2024) Current advancements in fungal engineering technologies for sustainable development goals. Trends Microbiol 33(3):285–301. https://doi.org/10.1016/j.tim.2024.11.001

Harirchi S, Wainaina S, Sar T, Nojoumi SA, Parchami M, Parchami M, Varjani S, Khanal SK, Wong J, Awasthi MK, Taherzadeh MJ (2022) Microbiological insights into anaerobic digestion for biogas, hydrogen or volatile fatty acids (VFAs): a review. Bioengineered 13(3):6521–6557. https://doi.org/10.1080/21655979.2022.2035986

Holechek JL, Geli HME, Sawalhah MN, Valdez R (2022) A global assessment: can renewable energy replace fossil fuels by 2050? Sustainability 14(8):4792. https://doi.org/10.3390/su14084792

Janković T, Straathof AJ, Kiss AA (2023) Advanced downstream processing of bioethanol from syngas fermentation. Sep Purif Technol 322:124320. https://doi.org/10.1016/j.seppur.2023.124320

Lee H, Vo PNL, Na D (2018) Advancement of metabolic engineering assisted by synthetic biology. Catalysts 8(12):619. https://doi.org/10.3390/catal8120619

Leong YK, Varjani S, Lee D, Chang J (2022) Valorization of spent mushroom substrate for low-carbon biofuel production: recent advances and developments. Bioresour Technol 363:128012. https://doi.org/10.1016/j.biortech.2022.128012

Lv X, Hueso-Gil A, Bi X, Wu Y, Liu Y, Liu L, Ledesma-Amaro R (2022) New synthetic biology tools for metabolic control. Curr Opin Biotechnol 76:102724. https://doi.org/10.1016/j.copbio.2022.102724

Ma B, Li Y, Wang T, Li D, Jia S (2025) Advances in CRISPR/CAS9-based gene editing in filamentous fungi. J Fungi 11(5):350. https://doi.org/10.3390/jof11050350

Ogawa M, Moreno-García J, Barzee TJ (2024) Filamentous fungal pellets as versatile platforms for cell immobilization: developments to date and future perspectives. Microb Cell Factories 23(1):280. https://doi.org/10.1186/s12934-024-02554-3

Pabbathi NPP, Velidandi A, Pogula S, Gandam PK, Baadhe RR (2025) Advancements in metabolic engineering: enhancing biofuel production through *Escherichia coli* and *Saccharomyces cerevisiae* models. Processes 13(7):2115. https://doi.org/10.3390/pr13072115

Panahi HKS, Dehhaghi M, Guillemin GJ, Gupta VK, Lam SS, Aghbashlo M, Tabatabaei M (2022) A comprehensive review on anaerobic fungi applications in biofuels production. Sci Total Environ 829:154521. https://doi.org/10.1016/j.scitotenv.2022.154521

Passi A, Tibocha-Bonilla JD, Kumar M, Tec-Campos D, Zengler K, Zuniga C (2021) Genome-scale metabolic modeling enables in-depth understanding of big data. Metabolites 12(1):14. https://doi.org/10.3390/metabo12010014

Ramachandraiah A, Kaushik J, Gowda L, Chigadannavar PS, Krishnappa R, Venkatesh S, Sivadas S, Judith S, Ramesh UM, Quadri Z, Dey P, Chinthala M, Gola D, Chauhan N, Bharti RK (2020) Prospect of biofuel production by fungus. In: Fungi bio-prospects in sustainable agriculture, environment and nano-technology. Elsevier, pp 285–313. https://doi.org/10.1016/b978-0-12-821394-0.00011-1

Ravi S, Gunawan R (2021) ΔFBA-predicting metabolic flux alterations using genome-scale metabolic models and differential transcriptomic data. PLoS Comput Biol 17(11):e1009589. https://doi.org/10.1371/journal.pcbi.1009589

Ren F, Wu F, Wu X, Bao T, Jie Y, Gao L (2024) Fungal systems for lignocellulose deconstruction: from enzymatic mechanisms to hydrolysis optimization. GCB Bioenergy 16(5):e13130. https://doi.org/10.1111/gcbb.13130

Salazar-Cerezo S, De Vries RP, Garrigues S (2023) Strategies for the development of industrial fungal producing strains. J Fungi 9(8):834. https://doi.org/10.3390/jof9080834

Saye LMG, Navaratna TA, Chong JPJ, O'Malley MA, Theodorou MK, Reilly M (2021) The anaerobic fungi: challenges and opportunities for industrial lignocellulosic biofuel production. Microorganisms 9(4):694. https://doi.org/10.3390/microorganisms9040694

Serrano-Echeverry VA, Guerrero-Fajardo CA, Castro-Tibabisco KT (2025) A review of biobutanol: eco-friendly fuel of the future—history, current advances, and trends. Fuels 6(3):55. https://doi.org/10.3390/fuels6030055

Singhania RR, Patel AK, Kumar P, Perumal PK, Chen C, Dong C (2025) Bioprospecting of cellulases from marine fungi for macro-algal biomass degradation for biofuel application. Int J Biol Macromol 307(Pt 1):141935. https://doi.org/10.1016/j.ijbiomac.2025.141935

Tamano K, Yoshimi A (2021) Metabolic engineering techniques to increase the productivity of primary and secondary metabolites within filamentous fungi. Front Fungal Biol 2:743070. https://doi.org/10.3389/ffunb.2021.743070

Vieira RIM, Da Silva Peixoto A, Monclaro AV, Ricart C a O, Filho EXF, Miller RNG, Gomes TG (2025) Fungal coculture: unlocking the potential for efficient bioconversion of lignocellulosic biomass. J Fungi 11(6):458. https://doi.org/10.3390/jof11060458

Xie M, Wang J, Wang F, Wang J, Yan Y, Feng K, Chen B (2025) A review of genomic, transcriptomic, and proteomic applications in edible fungi biology: current status and future directions. J Fungi 11(6):422. https://doi.org/10.3390/jof11060422

Zhang H, Zhang P, Wu T, Ruan H (2023) Bioethanol production based on *Saccharomyces cerevisiae*: opportunities and challenges. Fermentation 9(8):709. https://doi.org/10.3390/fermentation9080709

Zhang C, Singh RP, Yadav P, Kumar I, Kaushik A, Roychowdhury R, Mubeen M, Singh SK, Kumar A, Wang J (2025) Recent advances in biotechnology and bioengineering for efficient microalgal biofuel production. Fuel Process Technol 270:108199. https://doi.org/10.1016/j.fuproc.2025.108199

Chapter 16
Fungal-Based Nanotechnological Applications in Developing Biomaterial and Bioenergy: Recent Progress and Future Outlooks

Fareha Rayeen, Abhilasha Mishra, Shifa Suhail, Anmol Gupta, Manikant Tripathi, Pankaj Singh, Pradeep Kumar Singh, and Neelam Pathak

Abstract The development of safe, eco-friendly, and cost-efficient methods for nanoparticle production is a crucial objective of sustainable nanotechnology. Mycosynthesis, the biological synthesis of nanoparticles utilizing fungi, in particular, is uniquely suited for this task because of their capacity to secrete numerous enzymes and simplify downstream processing. As natural bioengineers, fungi facilitate the creation of nanoparticles with high biocompatibility and minimal environmental impact. This chapter explores how mycosynthesized nanoparticles address the efficiency gaps inherent in traditional energy conversion. Mycosynthesis nanotechnology addresses the efficiency gaps in the traditional conversion. With world energy demands outperforming fossil fuel reserves, fungal-mediated nanoparticles serve as a high-efficiency catalyst to enhance bioenergy production by increasing active reaction sites and improving the breakdown of lignocellulosic biomass. Potential applications in the development of new biomaterials derived from fungal mycelium have captured the attention of both the scientific community and society.

Fareha Rayeen, Abhilasha Mishra and Shifa Suhail contributed equally with all other contributors.

F. Rayeen · A. Mishra · P. K. Singh · N. Pathak (✉)
Department of Biochemistry, Dr. Rammanohar Lohia Avadh University, Ayodhya, Uttar Pradesh, India

S. Suhail
Department of Biosciences, Faculty of Sciences, Integral University, Lucknow, Uttar Pradesh, India

A. Gupta
MPA Research, Gautam Budh Nagar, Gaur City Centre, Greater Noida, Uttar Pradesh, India

M. Tripathi · P. Singh
Biotechnology Program, Dr. Rammanohar Lohia Avadh University, Ayodhya, Uttar Pradesh, India

M. Tripathi et al. (eds.), *Recent Advancements in Fungal Biotechnology Volume 1*, Fungal Biology, https://doi.org/10.1007/978-981-92-0339-0_16

The notable ability of mycelium networks to self-construct and aggregate can be used to produce diverse biomaterials. This chapter comprehensively explores mycosynthesis-based nanoparticle-mediated natural polymer-based materials, focusing on their characteristics, applications, and innovations across different sectors, including medicine, the environment, energy, textiles, and construction. With increasing concern about resource depletion and pollution, biomaterials offer a sustainable alternative to fossil-derived products. While this chapter enables us to propose a novel green pathway for industrial innovation, challenges remain regarding the processing, structure, and standardization of this emerging transdisciplinary field of knowledge. Future trends and research should focus on the optimization of fungal strains and the implementation of a standardized system to bridge the gap between laboratory success and commercial viability.

Keywords Mycosynthesis · Bioengineers · Fossil fuels · Transdisciplinary field · Commercial viability

16.1 Introduction

The modern industrial landscape is heavily reliant on conventional fossil fuels, including coal, oil, and gas (Senjyu and Howlader 2016). This dependence has led to multifaceted problems such as environmental degradation, depletion of resources, and economic instability (Lijó et al. 2019). Fossil fuel extraction and processing account for roughly around 55% of global greenhouse gas (GHG) emissions and global climate change, as well as intense energy demand (Strzalka et al. 2017). With the material utilization project reaching 160 billion tons annually, around 2.3% at the current growth rates, the scarcity of non-renewable resources is becoming critical (United Nations Environment Programme, International Resource Panel, Global Resources Outlook 2024), and heavy reliance on petroleum leads to significant market cost fluctuation and energy insecurity. In response to these issues, there is a global shift away from the linear economy toward a circular economy. This prioritized renewable biological resources to create sustainable products and services (Tan and Lamers 2021). Bioenergy currently represents about 80% of global renewable energy production, and it is one of the solutions to overcome fossil fuel exploitation, which is defined as produced from biological origin. Unlike intermittent sources, bioenergy can be stored and deployed as solid, liquid, and gaseous fuel for heating and electricity. Bio-based materials are defined as materials that are fully or partially of biological origins (Le et al. 2023). They offer a lower carbon footprint, natural abundance, and most importantly, biodegradability, which helps to mitigate the global crisis of plastic waste (Nwabike Amitaye et al. 2024; Fusteş-Dămoc et al. 2023; Pandit et al. 2018).

As we seek eco-friendly solutions parallel to the rise of the bioeconomy industrial solution, nanotechnology has emerged as a transformative tool with applications of bioenergy and biomaterials ranging from medicine to environmental remediation. However, traditional chemical synthesis of nanoparticles (NPs) often involves toxic reagents. Green synthesis addresses this by using natural biological processes to achieve precise control over size, shape, and functionality (Malik et al. 2023; Salem and Fouda 2021; Huston et al. 2021). Biosynthesized NPs often demonstrated enhanced stability, biocompatibility, reduced toxicity, and lower risk to the environment. The capacity to generate NPs has been discovered in all categories of organisms. Numerous investigations have indicated that NPs can be successfully synthesized from plants, animals, bacteria, fungi, actinomycetes, algae, lichens, and viruses (Loshchinina et al. 2022). The importance of fungus-mediated synthesis of NPs cannot be overstated.

Myconanotechnology, as it is commonly called, is thus defined as the interface between nanotechnology and mycology (Hanafy 2018; Sousa et al. 2020). The field of myconanotechnology is of great potential, partly due to the wide diversity of fungi (Khandel and Shahi 2018). Mycology is the study of fungi, while nanotechnology is the study involving the design, synthesis, and application of technologically important particles of nanoscale (Sousa et al. 2020). In this light, the design and synthesis of metal NP using fungi is termed "mycofabrication." In recent times, the fungal system has emerged as an important bionanofactory for the synthesis of NP of gold, silver, CdS, platinum, and other metals. Fungal bionanofactories have produced important NPs of good dimensions and monodispersity (Guilger-Casagrande and de Lima 2019). Additionally, fungi demonstrate remarkable versatility in their ability to synthesize NPs of various shapes, sizes, and compositions. This versatility opens numerous possibilities for tailoring NPs to meet specific industrial demands, ranging from drug delivery systems to catalysis and beyond (Adebayo et al. 2021). Furthermore, fungi are highly adaptable and can thrive in various environmental conditions, providing researchers with the flexibility needed for scalable NP production. By utilizing these distinctive properties, scientists can unlock new possibilities in NP synthesis, paving the way for groundbreaking applications across different industries (Prasher et al. 2024).

16.2 Fungal-Based Nanoparticles

In nanotechnology, fungal-derived biomolecules have emerged as a sustainable alternative to conventional chemical methods for NP synthesis (Meera et al. 2025; Rai et al. 2025). These biomolecules secrete abundant proteins, achieve high synthesis efficiency, simplify downstream processing, and reduce toxic by-product generation (Meera et al. 2025). Microorganism-mediated NP synthesis is an eco-friendly approach that utilizes bacteria, fungi, viruses, and their metabolites for nanomaterial fabrication (Kulkarni et al. 2023). These microorganisms act as biological scaffolds that facilitate the synthesis and assembly of structurally defined NPs. Fungi have emerged as effective biological manufacturing systems for NP production because of their controlled growth characteristics and metabolic diversity, including enzymes, proteins, peptides, and polysaccharides (Wadhwa et al.

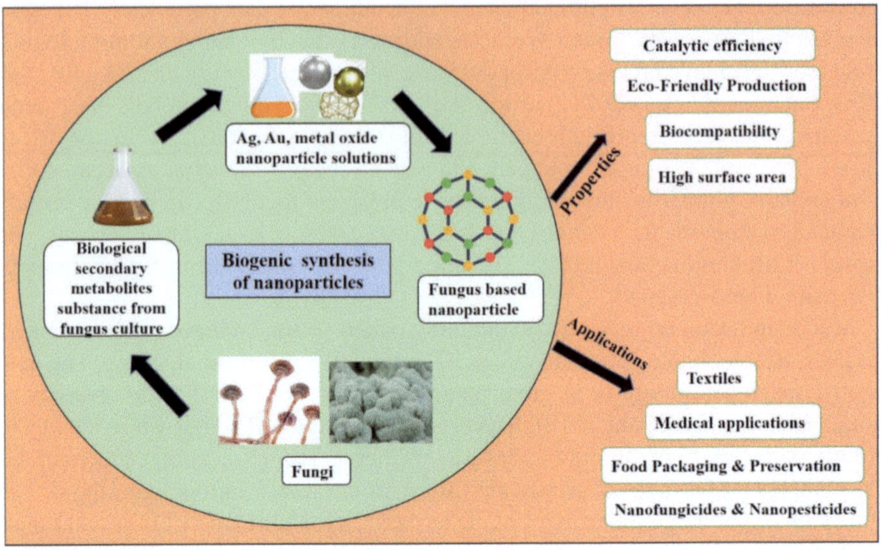

Fig. 16.1 Schematic representation of fungus-based nanoparticle synthesis and its multifunctional applications

2024). The ability to utilize different fungal species and adjust environmental parameters enables the synthesis of NPs with modified physicochemical properties (Saravanan et al. 2021). As a result, fungal nanotechnology offers promising pathways for the development of novel applications in agriculture, medicine, and industry, as shown in Fig. 16.1.

16.2.1 Diverse Properties of Fungai-Based Nanoparticles

Fungi constitute a unique biological platform for NP synthesis, distinguished by attributes not found in other biological systems (Wadhwa et al. 2024). They serve as reservoirs of diverse bioactive metabolites with significant industrial applications. Microscopic filamentous fungi, including ascomycetes, deuteromycetes, and other fungal taxa, have been documented to produce approximately 6400 bioactive compounds (Abo Nouh et al. 2022). These organisms demonstrate inherent tolerance to heavy metals and can internalize and bioaccumulate these elements (Ibrar et al. 2020; Abo Nouh et al. 2022). Fungi offer specific advantages over other microbial systems due to their high production of enzymes and proteins that promote NP synthesis (Adeleke et al. 2024). Consequently, they are effective in both the reduction and stabilization phases of nanomaterial fabrication. Fungi can be cultivated on a large scale, which enables the consistent production of NPs with uniform morphology and dimensions. Fungal-mediated NPs biogenic origin confers superior biocompatibility, making them suitable for various biomedical applications and reducing the cytotoxic effects commonly observed with chemically synthesized NPs (Kulkarni et al. 2023). In the agricultural sector, fungus-derived NPs function as nanopesticides and nanofertilizers, thereby decreasing reliance on hazardous agrochemicals (Anjum et al. 2023). Mycogenic NPs exhibit fungicidal, bactericidal, larvicidal, and nematocidal activities, providing effective strategies for controlling pests and phytopathogens, including resistant strains (Ali et al. 2024). They also contribute significantly to mycoremediation by enabling fungi to convert metal and metalloid compounds into less toxic forms and sequester them as NPs within the mycelium (Cherian et al. 2025). This process allows for the elimination and recycling of toxic elements from soil and water. Immobilizing fungal biomass with NPs supports the development of hybrid biosorbents for hazardous element removal (Yadav et al. 2024). Mycogenic NPs are effective in degrading industrial and agricultural pollutants. In addition, fungus-derived NPs have proven successful in pesticide remediation (Wadhwa et al. 2024) and in treating wastewater contaminated with microbial pathogens (Sridhar et al. 2024).

16.2.2 Mycosynthesis of Metal Nanoparticles (Ag, Au)

Metal NPs made by standard physical and chemical methods are expensive and often have toxic residues on their surfaces. As a result, green and simple methods are now used for making metal NPs, since these are less toxic and better for the environment. The use of fungi has been very effective in creating many types of metal NPs, especially silver (Ag) and gold (Au). These NPs have unique features, such as sizes from 10 to 100 nm, high surface area-to-volume ratios, certain pore sizes, surface charges and densities, crystalline and non-crystalline forms, round or tube shapes, and sensitivity to their environment.

16.2.2.1 Silver Nanoparticle Synthesis

The unique physicochemical characteristics of silver nanoparticles (AgNPs)—including their shape, size, and morphology—enable effective interactions with a wide range of microorganisms, animals, and plants (Vanlalveni et al. 2024). Fungi serve as efficient agents for biogenic AgNP synthesis owing to their high metal tolerance and ease of cultivation (Adeleke et al. 2024). In comparison to bacterial systems, fungal cultures provide superior biomass production and eliminate the need for additional filtrate extraction steps (Dey et al. 2022). Furthermore, fungus-mediated AgNPs are distinguished by their high monodispersity, well-defined composition, and narrow size distribution (Anjum et al. 2023) (Fig. 16.1). For example, Durán et al. (2023) synthesized AgNPs using *Fusarium oxysporum* and proposed that silver ion reduction occurred through the action of the nitrate reductase enzyme and anthraquinones. Similarly, Srivastava and Bhargava (2021) synthesized NPs using the fungus *Aspergillus fumigatus*, achieving a particle size of 322.8 nm at

Table 16.1 Shows some studies in which gold and silver nanoparticles synthesized from different fungal species were employed for various applications in different fields

Fungi	Application field	Application	References
Trichoderma spp.	Medicine	Antibacterial against pathogenic bacteria	Singh et al. (2016)
Fusarium oxysporum	Medicine	Antibacterial against *Staphylococcus aureus*	Naimi-Shamel et al. (2019)
Glomus aureum	Environment	Catalyzes the degradation of methyl orange dyes under light	Ali et al. (2020)
Aspergillus fischeri	Catalyst	Degrade methylene blue	Banerjee and Ravishankar Rai (2016)
Rhizopus oryzae	Agriculture	Treat chlorinated pesticides	Das et al. (2009)
Trichoderma atroviride	Agriculture	Inhibit the tea plant pathogen *Phomopsis theae*	Ponmurugan (2017)
Penicillium	Delivery agent	Effectively combine with DNA and RNA as carriers in gene delivery	Mishra et al. (2012)
Nigrospora oryzae	Medicine	Alter the physiological functions of parasites by causing female paralysis and death	Kar et al. (2014)
Macrophomina phaseolina	Health	Antibacterial (multidrug-resistant strains of *Escherichia coli*)	Chowdhury et al. (2014)
Pleurotus cornucopiae var. *citrinopileatus*	Health	Antifungal against *Candida* species	Owaid et al. (2015)
Pleurotus djamor var. *roseus*	Medicine	Antiproliferative in cancer cells	Raman et al. (2015)
Alternaria alternata	Agriculture	Inhibit the proliferation of microorganisms on the cotton plant	Ibrahim and Hassan (2016)
Guignardia mangiferae	Agriculture	Control the phytopathogenic fungi *Colletotrichum* sp., *Rhizoctonia solani*, and *Curvularia lunata*.	Balakumaran et al. (2015)

25 °C, with increasing particle dimensions as temperature was elevated, reaching 1073.45 nm at 55 °C. Nanoparticle synthesis using fungal species is presented in Table 16.1.

16.2.2.2 Gold Nanoparticle Synthesis

Biological techniques have gained popularity since microorganisms can not only metabolize Au (III) but also stabilize gold nanoparticles (AuNPs) due to their unique structure (Chowdhury et al. 2022). Microorganisms have been widely studied and applied as a widely available, inexpensive, and environmentally adaptable reducing agent (Jorge et al. 2023). While the biosynthesis of AuNPs in bacteria is well understood, fewer than 30 fungal species have been examined for AuNP biosynthesis.

Compared to bacteria, fungi have several advantages in AuNPs biosynthesis, including (1) fungi have higher metal ion binding capacity and higher tolerance concentration; (2) fungi can exude vast volumes of extracellular organic materials, which can be employed as reducing media and capping agents; (3) fungal mycelium is more stable under agitation and high flow pressure; and (4) smaller fungal-mediated synthesis of AuNPs can be achieved, resulting in greater quantum yields and biocompatibility (Robinson et al. 2021; Xu et al. 2024). The presence of AuNPs biosynthetic capability in fungi suggests that the reduction of Au^{3+} to form protein-metal nanoconjugates represents a common response to toxic stress, where the enzymatic machinery required is readily available in environmental microorganisms (Chowdhury et al. 2022). Several fungal species, including *Phoma glomerata*, *Aspergillus fumigatus*, *Coriolus versicolor*, *Fusarium semitectum*, *Fusarium oxysporum*, *Trichoderma viride*, *Trichothecium* sp., *Phanerochaete chrysosporium*, *Trichoderma asperellum*, and *Colletotrichum* sp., have been identified as suitable organisms for this purpose (Srivastava and Bhargava 2021), shown in Fig. 16.1.

16.2.3 Mechanism of Fungal-Mediated Nanoparticle Formation

Biogenic synthesis of NPs, especially through fungi, has significantly advanced nanotechnology. Fungi efficiently convert metals and metalloids into functional NPs through complex biochemical processes, which occur via intracellular and extracellular pathways (Elazab et al. 2023; Panda et al. 2025).

16.2.3.1 Recognition of Precursors

Fungi, especially those in the Ascomycota and Basidiomycota groups, can recognize and interact with various metal and metalloid precursors, including salts and organic compounds (Gajewska et al. 2022). For silver and gold NP synthesis, fungi typically process compounds such as silver nitrate or gold chloride. These ions are absorbed into the fungal biomass, triggering a series of biochemical reactions (Gajewska et al. 2022).

16.2.3.2 Intracellular and Extracellular Pathways

Fungi use both intracellular and extracellular mechanisms to reduce and stabilize metal or metalloid ions, producing NPs with distinct morphologies and properties (Durán et al. 2023). Intracellularly, specific enzymes such as reductases convert metal ions into elemental forms (Durán et al. 2023). Extracellular synthesis occurs in the surrounding environment, often mediated by secreted metabolites, and results in NP accumulation on cell surfaces or in the culture medium (Murthy et al. 2024). NADPH-dependent reductases are key enzymes that transfer electrons from NADPH to metal ions, reducing them to zero-valent NPs (Murthy et al. 2024). Nanoparticles are extracted by disrupting the biomass through chemical treatment, centrifugation, and filtration (Murthy et al. 2024). Fungal proteins and peptides act as both reducing and stabilizing agents, preventing aggregation through electrostatic stabilization (Patel et al. 2025). Factors such as pH, temperature, and growth medium composition significantly influence extracellular synthesis and affect NPs' size and morphology (Patel et al. 2025).

16.2.4 Fungal-Based Nanotechnological Applications in Biomaterials

Several key principles guide the use of fungal-based NPs in biomaterials. Biocompatibility ensures these NPs integrate with biological systems without adverse effects (Varshan and Namasivayam 2024). Surface modification enables functionalization with specific biomolecules for targeted applications. Controlled release supports the creation of smart materials that respond to environmental changes (Sroka et al. 2025). Hierarchical assembly leverages the self-organizing properties of fungal NPs to form complex structures, which is especially valuable in tissue engineering where scaffolds must replicate natural tissue architecture (Mulay et al. 2024).

The improvement of mechanical properties is among the most desirable objectives when optimizing fungal-based biomaterials (Mulay et al. 2024; Liu et al. 2025). Most basidiomycete fungi used to generate mycomaterials are white-rot decayers with di- or trimitic hyphal systems (Luangharn et al. 2020; Singh et al. 2025a, b). Building on this, genomic data mining enables researchers to identify and characterize protein-coding genes involved in the morphogenesis of skeletal and binding hyphae, providing insights that could support novel approaches to reinforced mycomaterials (Xie et al. 2025). Additionally, understanding gene families encoding proteins that impact fungal colonization and adherence, such as hydrophobins and those involved in cell wall composition, like ergosterol, is crucial (Barcan et al. 2023). These genetic features should be considered for their impacts on the mechanical properties of bio-based materials and their roles in substrate degradation regulation. For example, understanding peroxidases (Barcan et al. 2023) and the influence of genetic mutations on improving fungal biomass production, like elevated chitin production (Xie et al. 2025), is important.

Beyond genetics, fungal-mediated NPs are effective across various scientific disciplines, including medicine, pharmaceuticals, agriculture, and electronics (Adeleke et al. 2024). Accordingly, several studies evaluate mycogenic NPs in plant disease control, post-harvest pathogen management, mycotoxin mitigation, plant pest management, and combating certain animal pathogens (Salem and Mekky 2024;

Dadayya et al. 2025). Fungal nanomaterials also show significant potential for diagnostics, biosensors, precision agriculture, and targeted smart delivery systems (Chaudhari et al. 2024). For instance, soil mycobiota can influence zinc mobilization from ZnO NPs in soils, affecting zinc mobility and bioavailability (Jing et al. 2025). In order to examine these interactions, *Aspergillus niger*, a common soil fungus, was chosen as a test organism to study its relationship with ZnO NPs (Kumawat et al. 2025). In parallel, researchers are developing antifungal nanohybrid agents composed of organic or inorganic compounds, biological components, and biopolymers to create more cost-effective, reliable, and efficient products against fungal infections of plants and animals (Ghareeb et al. 2024).

Fungal-based nanotechnology offers numerous advantages over conventional synthetic approaches (Panda et al. 2025). Environmental sustainability represents a primary advantage, as fungal synthesis eliminates the need for harsh chemicals and high-energy processes typically required in chemical synthesis (Wadhwa et al. 2024). The process operates under ambient conditions with minimal energy requirements, contributing to a reduced carbon footprint (Wadhwa et al. 2024).

Cost-effectiveness is another significant advantage, as fungi can be cultivated using agricultural waste and inexpensive substrates, making large-scale production economically viable. The biocompatibility of fungal NPs eliminates concerns about toxicity and adverse biological reactions, expanding their applicability in biomedical fields.

16.2.5 Applications

The applications of fungal-based NPs span multiple industries, with particular emphasis on sustainable technologies and biomedical innovations (Adeleke et al. 2024; Srivastava et al. 2025). The antimicrobial properties of fungal NPs make them valuable in healthcare, food safety, and environmental remediation (Srivastava et al. 2025). Eco-friendly synthesized NPs demonstrate excellent antimicrobial activity against various pathogenic bacteria and fungi. They offer a safer alternative to chemically synthesized antimicrobial agents. Additionally, their environmental sustainability addresses concerns about chemical contamination in agricultural and industrial processes.

16.2.5.1 Biomedical

Nanotechnology is increasingly accepted in biomedicine, driving significant advances in fungal nanobiotechnology. Research in nanomedicine has progressed steadily, particularly in diagnosis, treatment, and drug delivery (Wang et al. 2024). Fungal nanobiotechnology is now applied in fluorescent labeling, gene and drug delivery, pathogen detection, tissue engineering, hyperthermia-based tumor treatment, magnetic resonance imaging (MRI) contrast enhancement, and phagokinetic studies (Tarannum et al. 2024). Nanoparticles have also been studied for their ability to boost the antimicrobial activity of antibiotics against both Gram-positive and Gram-negative bacteria (Tavares et al. 2020). Silver nanoparticles, in particular, have enhanced the antibacterial effects of ampicillin, kanamycin, erythromycin, and chloramphenicol, with the greatest improvement seen in ampicillin (Ali et al. 2021). These results suggest that combining antibiotics with AgNPs can improve antimicrobial efficacy (Ali et al. 2021).

16.2.5.2 Textile Industry

Fungal NPs enhance textile functionality and sustainability (Fernandes et al. 2022). Chitosan nanofibers can be processed into high-surface-area structures for smart textiles, adding multifunctional properties (Nirwan et al. 2025). Integrating fungal AgNPs into fibers delivers durable antimicrobial protection without compromising comfort or appearance (Gulati et al. 2022). This treatment remains effective after multiple washes, making it well-suited for healthcare and sportswear textiles (Gulati et al. 2022).

Fungal NPs support the creation of responsive textiles that adapt to environmental changes. These materials offer UV protection, moisture control, and temperature regulation through precise nanoparticle distribution (Kiran et al. 2025). Using fungal chitin as a sustainable alternative to synthetic fibers addresses environmental concerns. Fungal fibers require minimal chemical processing and can be produced from agricultural waste (Kiran et al. 2025).

Fungal nanoparticle coatings offer water repellency, stain resistance, and wrinkle reduction while preserving breathability (Thapliyal et al. 2025). The natural origin of these coatings aligns with consumer preferences for environmentally friendly textile products (Thapliyal et al. 2025). These innovations demonstrate the significant potential of fungal nanotechnology to create sustainable, high-performance materials with reduced environmental impact.

16.2.5.3 Sustainable Packaging

Fungal functionalized nanostructures can be utilized in food at various stages, including processing, packaging, and safety (Thapliyal et al. 2025). However, few reports exist on using fungal biomass for the synthesis of NPs in food packaging (Thapliyal et al. 2025). Nanoparticles are effective at controlling the growth of foodborne pathogens and could be used in food packaging (Anvar et al. 2021). When the AgNPs-enhanced starch/agar film was tested, it showed strong antibacterial properties against *E. coli* and *Listeria monocytogenes* (Trotta et al. 2023). As a result, starch/agar films exhibit improved physical (mechanical strength and water vapor barrier) and functional (antibacterial activity) properties, making them suitable for active food packaging applications (Trotta et al. 2023).

16.2.5.4 Fungal Chitin Nanofibers or Microfilaments

Chitin microfibrils and nanofibrils represent structural components extracted from chitin that are covalently associated with glucan, galactomannans, and proteins, including mannoproteins and hydrophobins within fungal cell walls, thereby constituting native biocomposite materials (Mautner 2023). The primary biological role of chitin involves providing structural integrity, with this function conserved across diverse organisms, including crustaceans, insect exoskeletons, nematode eggs, and protozoan cysts (Zhang et al. 2021).

The conventional extraction and defibrillation processes for obtaining chitin from crustacean sources necessitate shell demineralization through acidic treatment, which, combined with specialized equipment requirements, significantly diminishes the commercial viability of this approach (Ali et al. 2024). These limitations stem from the generation of environmentally hazardous chemical waste and increased energy consumption (Ali et al. 2024). Given these constraints, fungal-derived chitin has emerged as a sustainable alternative that offers renewable sourcing, facile isolation procedures, and abundant availability through established production and fermentation systems that yield mycelium as the primary raw material (Zannat et al. 2025).

16.3 Fungal-Based Nanotechnological Applications in Biomaterials

In the evolving landscape of material sciences and nanotechnology, there is a significant shift toward integrating living organisms into manufacturing processes. This nanobiofabrication approach allows for the development of natural resources in a highly regulated environment, bridging the gap between nanotechnology and microorganisms (Singh et al. 2024). By manipulating material at the molecular level, these biological systems enable the design of innovative materials with desired, precise, and customizable attributes (Taylor et al. 2024; Singh et al. 2024). As the global population rises, so does the demand for sustainable alternatives in construction, textiles, packaging, medicine, and others. Traditional synthetic products are often too difficult to repair or dispose of without causing environmental degradation.

Mycelium-based materials offer circular solutions, transforming organic waste into high-performance products. At the heart of this revolution is mycelium, which is produced by filamentous fungi. This biomass grows as an intricate, interconnected web of microscopic cells. In its natural habitat, this heterotrophic organism acts as the planet's primary recycler, utilizing powerful secretory mechanisms to decompose and recycle organic material (Vandelook et al. 2021).

16.3.1 Advantages

Mycelium materials from filamentous fungi are increasingly viewed as a viable replacement for toxic and petroleum-based plastics. These biomaterials are inherently sustainable, biocompatible, adaptable, compatible, abundant, cost-effective, and biodegradable, offering a unique suite of physical and functional properties. Their physical resilience properties include non-flammability (natural fire resistance), elasticity, porosity, and hydrophobicity. These biomaterials exhibit biological intelligence, self-repairing, self-organizing, and self-healing. Functional properties are natural antimicrobial, antioxidant, and skin whitening characteristics, which make them ideal for medical and cosmetic uses. These biomaterials also have production efficiency because they are fast-growing, cost-effective, and highly tunable, based on the feedstock providing system (Vandelook et al. 2021). This transition toward the development of these nano-based biomaterials is driving a significant shift across the food industry, medicine, agriculture, biotechnology, and other sectors. By leveraging the fungal adaptability and abundance, we could move from synthetic toxic material toward a bio-integrated, harmonious future.

16.3.2 Applications

16.3.2.1 Sustainable Packaging

Due to the recent globalization of the economy, the packaging market has expanded. The internationalization of business has raised the demand for wrapping materials recently. Extruded polystyrene and expanded polyethylene foams, which are the two most common industrial packaging materials today, are mostly derived from petroleum fuels, which have several drawbacks, including high power consumption during production, complex degradation, and consequent environmental damage (Cerimi et al. 2019). It is neither recyclable nor biodegradable; therefore, it poses a severe threat to the environment. Furthermore, due to the emission of GHGs, the process of producing polystyrene requires a lot of energy and is not ecologically friendly (Abhijith et al. 2018). Due to their lightweight and non-toxic properties, mycelium-based materials are capable of being utilized for a variety of packaging applications, including electronics, fragile objects, and the food sector. In a variety of industries, the green-biocomposites made from only bio-based components can replace standard petroleum-based packaging materials (Verma et al. 2023).

Innovative strategies in Polyhydroxyalkanoates (PHA) production by utilizing GHGs for sustainable bioplastics. PHAs are biopolymers that serve as sustainable alternatives to fossil fuel-based plastics, as these have comparable plastic properties like high elasticity, tensile strength, and air resistance, with the added advantages of biocompatibility and biodegradability. PHAs serve as microorganisms' energy and carbon storage and are produced via fermentation of sugars and fatty acids from waste feedstocks. Despite their advantages, PHA production is greatly hindered by the high production cost, with carbon sources accounting for 35–40% of the total cost (Ma et al. 2024). Utilizing GHGs like methane (CH_4) and carbon dioxide (CO_2) as carbon sources offers several benefits of carbon sequestration, promotes a bio-based circular economy, and reduces costs. This is an underexplored field, with fewer than 300 publications indexed in the Web of Science Core Collection (keywords used: polyhydroxyalkanoates/PHAs/PHB, carbon dioxide/CO_2, methane/CH_4, fermentation/bioconversion for analyzing PHA production from CO_2, and fermentation/bioconversion for analyzing PHA production from CH_4). However, it is gaining traction, with over 80% of these studies published within the last decade (Ochoa and Iszauk 2025). Challenges identified include low productivity, limited microbial species, and inefficient gas-to-liquid transfer rates, and may be addressed by stress-induced PHA production, the use of thermophilic microbial communities, and advancements in co-culturing and genetic engineering. Techniques such as using water-in-oil emulsions have improved PHA synthesis, significantly increasing microbial growth and PHA accumulation. Mixed carbon sources, like biogas, also enhance polymer properties, particularly poly(3-hydroxybutyrate) (P(3HB)), a widely studied PHA. This technology is exemplified by the company Mango Materials, which partners with and capitalizes on methane-producing facilities such as landfills and wastewater treatment plants. The company uses a consortium of methanotrophs to primarily produce P(3HB) pellets that have been optimized for versatility, finding applications in injection molding, fiber extrusion, additive manufacturing, and other plastic-substitute uses. Within approximately 6 weeks, the material disintegrates due to microbial activity in the ocean, without leaving harmful residues. This makes the material environmentally sustainable and suitable for reducing plastic pollution in aquatic ecosystems (Mango Materials 2025).

16.3.2.2 Environmental Waste Management

Agricultural waste burning is a significant cause of pollution in underdeveloped nations. Agricultural waste/residues, being a rich source of lignocellulosic content, may indeed serve as great substrates for biomaterial manufacturing. By putting microbes onto waste substrates, researchers created agro-based biomaterials. The lignocellulosic component of agro-waste functions as a feed for microorganisms and offers enough nutrients for growth. To create sustainable biomaterials, researchers studied the growth morphology and adhesion capabilities of several fungi on diverse agricultural wastes (Joshi et al. 2020; Vu et al. 2022). The mycelium growth rate is strongly dependent on the kind of agricultural waste. The produced mycelium fibers were homogenous, thermally stable, and had outstanding hydrophobic, adhesive, and mechanical characteristics. The manufactured bio-blocks outperform the standard packaging material, polystyrene, in terms of heat resistance and compressive strength by five to sixfold. Aside from the intrinsic physicochemical qualities of these bio-blocks, the green synthesis technique, simplicity of manufacturing, non-toxicity, and biodegradability are the additional and most important advantages. These mycelium-based bio-blocks can replace non-biodegradable materials such as plastic, which have a negative environmental effect (Joshi et al. 2020).

Eco-friendly cellulose aerogel composites for efficient dye and oil removal from wastewater innovation, as applied in the environmental sector, also explore the synthesis and usage of cellulose aerogel composites derived from pineapple leaf fibers and cotton waste to address the pressing issue of wastewater pollution caused by dyes and oils. The technology utilizes the sol–gel method in an alkali–urea solution, where cellulose fibers are dissolved in the alkali–urea solution (typically NaOH-urea) to break down the structure, allowing the formation of a homogenous solution and subsequent formation of a gel under suitable conditions, followed by freeze-drying to create aerogels with remarkable properties, including low density (0.053–0.069 g/cm^3) and high porosity (nearly 95%). The mechanical performance exhibited had between a 5 and 9 times higher Young's modulus compared to previous aerogel composites, underscoring the feasibility of using agricultural waste to produce effective water treatment materials and the potential of these cellulose aerogel composites to contribute to sustainable environmental solutions (Vu et al. 2022).

16.3.2.3 Biomedical

Multiple fungi are also used as a starting point for the creation of nanomaterials for a variety of purposes. The fungi-mediated NPs are created by attaching metal precursors to the fermenting culture, which are then absorbed by the mycelium mass and converted into the desired NPs (Guilger-Casagrande and de Lima 2019). The integration of collagen, fibrin, and glycoproteins into the development of organs-on-chips (OoCs) for drug testing, screening, disease modeling, and personalized medicine. Drug discovery is one of the most important aspects of the pharmaceutical industry. However, there are some shortcomings in the drug evaluation stage, including animal welfare protections and organ-on-a-chip technology, with the ability to simulate human organ systems, enabling the drug evaluation process that includes testing for its toxicity, activity, safety, metabolism, and pharmacological efficacy; drug pharmacokinetic/pharmacodynamic (PK/PD) modeling; disease modeling; and personalized medicine. These OoCs can be single or multiple organs on a chip, and this may include lung, heart, liver, kidney, gut, bone marrow, skin, and muscles (Deng et al. 2023). The oxidized sodium alginate/gelatin/halloysite nanotubes (OSA/Gel/HNTs) hydrogel was used as an injectable, adhesive, and antibacterial dressing for hemostasis.

Wound care faces problems like hemorrhages and infections that may cause shock and disability, and sometimes lead to death. Healing requires four main stages: (1) the hemostasis stage, being the most effective step to reduce the death rate; (2) the inflammation stage; (3) the proliferation stage; and (4) the wound structure remodeling stage. Thus, wound dressing with hemostatic and antibacterial properties is of great importance, but this is limited in conventional dressing due to insufficient mechanical strength, antibacterial efficacy, and biocompatibility. Hydrogels are polymers with 3D structures that have been applied in wound dressing due to their strong network formed by physically or chemically crosslinked chains. Lin et al. (2024) researched an oxidized sodium alginate/gelatin/halloysite hydrogel (OSA/Gel/HNTs) that has been applied for wound healing. This combines OSA and gelatin as the primary matrix, combined with halloysite nanotubes. This hydrogel forms a self-crossed network through Schiff base reactions, ensuring excellent mechanical properties, injectability, and skin adhesion. HNTs facilitate crosslinking, reducing the gelation time and providing antibacterial activity against some pathogens like *Escherichia coli* and *Staphylococcus aureus*. This hydrogel has been proven to also reduce clotting time, minimize bleeding, and demonstrate effective hemostatic and antibacterial properties for advanced wound dressing. Cancer and dementia are successfully treated using triterpenoid chemicals (TCs) extracted from Morchella mycelium (Wang et al. 2020). The TCs have potent anti-inflammatory and anticancer properties. A different investigation found that *Pleurotus djamor* acetylated mycelia polysaccharides (AMPS) have strong antioxidant and anti-aging properties. Additionally, the kidney, brain, and liver are protected by the mycelia of many fungal species (Zhao et al. 2023).

16.3.2.4 Textile Industry

The textile industry is one of the most environmentally harmful industries, accounting for 20% of water pollution and 10% of GHG production globally. Additionally, it reportedly utilizes an average of 400 m^2 of land, 9 m^3 of water, and 391 kg of raw materials while generating 270 kg of CO_2 emissions per year per capita. During use, microfibers are released during washing, and less than 1% are recycled, with the remainder often ending up in landfills, thereby exacerbating waste and pollution concerns. Despite this, the demand for textiles continues to rise, with projections indicating a global production of 145 million tons by 2030, driving the demand for alternative, animal and petrochemical-free biotextiles (European Parliament 2024). Companies such as TômTex, Modern Synthesis, and MycoWorks™ are driving innovation in sustainable biotextiles. TômTex creates 100% certified bio-based textiles that have a much lower environmental impact by requiring only 0.1% of the water used in conventional leather production and 80–90% less GHG emissions. They offer biodegradable, petrochemical-free, and customizable products made of sustainably sourced chitosan from mushrooms for their TômTex Series MTM and Series WSTM, respectively (Hahn 2020). MycoWorks™ creates leather-like materials with specific strength, performance, and aesthetic requirements by leveraging its patented Fine Mycelium™ technology, which involves the precise engineering and monitoring of mycelium growth. The biomaterial still undergoes tanning and finishing processes, but the company employs a proprietary chrome-free tanning and dyeing technology designed to minimize environmental impact (Hahn 2020).

In the context of shifting social standards and increasing emphasis on environmental sustainability, fungal-derived leather is now emerging as a promising alternative to replace not only bovine leather but also synthetic ones, which require the use of hazardous chemicals derived from fossil fuels. This results in a lack of biodegradability, presenting the same limited end-of-life options as most plastics (Jones et al. 2020). Mycelium leathers, also known as mycelium-based leather, mainly originated from *Polyporales* species (Basidiomycota) that use lignocellulosic substrates as a source. The biomass generated by the development of hyphae might have light or rigid consistency, depending on their chemical structure, which is formed by flexible structures composed of chitin, glucans, and glycoproteins (Raman et al. 2022). Agricultural and forest wastes are converted into substrates for most species, which allows the bioconversion of these resources without causing any imbalance or degradation of the natural ecosystem (Mengqi et al. 2023). Mostly, fungal-derived leather can be developed in two ways: growing hyphal filaments in a liquid medium (liquid fermentation), where hyphae are pressed into pure fiber pulp, and a solid-state fermentation process, with physical and chemical treatments, where the mycelium layers are formed on a surface composed of solid lignocellulosic nutrients (Ross et al. 2020; Jones et al. 2020). Different fermentation processes are commonly mentioned in literature, and the liquid state is based on standard media or low-cost agricultural waste, such as blackstrap molasses, as a raw material for fungal growth. Moreover, they can be separated into fibers and are processed by traditional papermaking techniques such as fiber suspension, filtration, pressing, and drying (Nawawi et al. 2020; Jones et al. 2020). The industrial and construction materials sectors have shown interest in mycelium-based technologies (Raman et al. 2022). However, experimental raw data is hardly accessible, whether from individual knowledge or intellectual property, avoiding the development of innovative technologies in different sectors, varying from mycology to engineering. Commercial startups and established companies include Ecovative, MycoWorks, Bolt Threads, Mycotech Lab, and Mycel (Ballen Sierra et al. 2023).

16.4 Fungal-Based Nanotechnological Applications in Bioenergy

Fungi play an important role in developing ecological and environmental benefits to humans, along with offering health and nutritional advantages. They are also essential to various industries. Fermented foods and beverages, which are products derived from fungi, are available in the market. However, fungi hold the greatest potential financial value in blue carbon trading due to their capacity to store significant amounts of carbon in the soil (Tiquia-Arashiro and Grube 2019). They can also play an important role in enhancing resource efficiency by serving as renewable alternatives to fossil resources. In addition, they serve as important sources of nourishment, including nutraceuticals, cosmeceuticals, proteins, amino acids, carbohydrates, vitamins, fats, and minerals. Nanotechnology is a powerful scientific field. Nanoparticles' surface area is greater, and they possess distinctive properties such as solubility, toxicity, strength, magnetism, diffusivity, optics, color, and thermodynamics. This technology has been applied across various fields, including agriculture, health diagnostics, the sports industry, clothing, cosmetics, space elevators, the food industry, and for enhancing processing speeds in mobile phones and computers (Khan et al. 2022).

16.4.1 Advantages

Fungal-based technology is cheaper than traditional chemical synthesis and can develop a wide range of NPs with different shapes and sizes. It has the potential to develop in large-scale production using bioreactors, reduced energy consumption in NP synthesis, and biocompatibility of fungal-derived nanomaterials (Bansal et al. 2011).

16.4.2 Applications

Nanoparticles can immobilize fungal enzymes, which enhance the stability and reusability in the biofuel production process. Nanoparticles, which are derived from fungus, can improve the efficiency of biomass pretreatment and facilitate better access to cellulose and hemicellulose for biofuel production. Nanoscale fungal structures serve as efficient biocatalysts for many bioenergy-related reactions, like biodiesel synthesis. Also, fungal-based nanosensors can be developed for monitoring and optimizing bioenergy production processes. Fungal NPs act as carriers for enzymes or microorganisms used in biofuel production, enhancing their performance and stability (Chettri et al. 2025). Fungal nanostructures enhance hydrogen production by improving substrate utilization and fermentation efficiency, enhancing electrode potential, and increasing power output. It also enhances anaerobic digestion processes, leading to increased biogas productivity, is used to remove contaminants from biomass feedstocks for the quality of biofuels produced, and lastly, develops advanced materials for bioenergy applications (Arif et al. 2018).

16.4.2.1 Lignocellulose Biomass to Biofuel Production

Lignocellulosic biomass mainly comprises lignin, cellulose, and hemicellulose. These residues represent a substantial and readily present biomass resource, serving as important feedstock for biofuel fermentation. The conversion of lignocellulosic biomass into products for sustainable biofuel production, meant for transportation fuel, is primarily achieved through gasification and enzymatic hydrolysis (Fatma et al. 2018). Many sources of lignocellulosic biomass exist, including agricultural waste, waste energy crops, aquatic plants, forest leftovers, and organic components of municipal waste. Biofuels are produced from energy crops and agricultural residues, while corn residue is one of the major sources for the production of ethanol. The pre-treatment step plays an important role in disrupting the recalcitrant structures and enhancing the accessibility of cellulose to enzymes. Pre-treatment techniques involving physical, chemical, biological, and combined approaches have been implemented on a pilot scale. These techniques develop significant opportunities to reduce the total cost and make large-scale production of cellulosic ethanol possible. Many challenges in lignocellulosic biofuel production include biomass variability, high pre-treatment and enzyme costs, inhibitor formation during processing, and the need for efficient microbial strains.

Emerging technologies like consolidated bioprocessing, where a single microorganism performs hydrolysis and fermentation, hold promise for reducing costs and simplifying production. Overall, lignocellulosic biomass conversion to biofuels offers a sustainable alternative to fossil fuels, contributing to energy security and GHG reduction. Advanced technologies are important to overcome current limitations and enable large-scale, economically viable biofuel production from this abundant renewable resource (Mignogna et al. 2024).

16.4.2.2 Renewable Energy Systems

Renewable energy systems integrating fungal-based nanotechnology show a transformative approach toward sustainable energy production. This technology makes it more durable, efficient, and eco-friendly by using the special qualities of NPs generated from fungi (Velamakanni et al. 2022). These systems encompass a wide range of applications, including energy storage, biofuel production, and conversion devices. Fungal NP enhances the breakdown of complex biomass into fermented sugars and increases bioethanol and biodiesel production. This catalytic enhancement makes biofuel systems more economically and environmentally viable by consuming less energy and processing time. In addition, this fungal nanotechnology contributes to electron transfer in microbial fuel cells, which converts organic waste into electrical energy, and integrates waste management with the production of renewable energy. Devices like supercapacitors and batteries also benefit from fungal nanotechnology. The higher surface area and conductivity allow the fabrication of electrodes with superior charge capacity and cycling stability (Ojha and Pradhan 2025). Advancement supports the production of renewable energy storage solutions, which are both efficient and eco-friendly, addressing the intermittency issues with solar and wind power. In addition, using minimal hazardous reagents and reducing the consumption of energy, the green synthesis of NP using fungi is consistent with sustainable manufacturing standards. This environmentally friendly production method supports the wider goals of renewable energy systems to reduce carbon footprints and enhance the circular economy principles. Continual innovations of this technology are essential for the development of clean energy solutions and protecting the environment (Abady et al. 2025).

16.4.2.3 Biogas Enhancement

Biogas enhancement is one of the primary benefits of fungal NPs, which promotes the degradation of complex organic matter such as lignocellulosic biomass. These NPs act as enzyme carriers involved in catalytic reactions, which promote the hydrolysis and acidogenesis stages of anaerobic digestion (Tamilselvan and Selwynraj 2024). Consequently, the process to optimize methane production by increasing the bioavailability of simpler compounds required by methanogenic archaea. This technology also improves the stability and efficiency of microbial consortia involved in biogas generation. Additionally, the fungal nanomaterials help to reduce inhibitory substances that are present in feedstock, such as heavy metals or ammonia, by adsorption and thus protecting microbial communities (Ghosh et al. 2023).

16.4.2.4 Biosensing

Integration of biosensors with fungal-based nanomaterials shows a cutting-edge approach to monitoring and optimizing the production of bioenergy. These biosensors utilize the high surface area and reactivity of fungal NPs to identify specific environmental factors and biochemical markers. For example, fungal-based nanoparticles functionalized with an enzyme or antibodies can specifically sense substrates or inhibitors in biofuel reactors, allowing for perfect control over fermentation conditions. Fungal-based biosensors are being developed to improve process monitoring, early detection of contaminants, and enhanced operational stability in bioenergy systems (Table 16.2). Integration of this technology with biosensing platforms thus holds great potential for advancing bioenergy technologies by ensuring efficient, safe, and cost-effective production. Biosensors have a wide range of applications in healthcare, environmental monitoring, food safety, and biotechnology (Mehrotra 2016).

16.5 Ecological and Economic Benefits Associated with Myconanotechnology in Bioenergy and Biofuel Production

Myconanotechnology, which is introduced with fungal biotechnology and nanotechnology, offers significant ecological and economic benefits in developing bioenergy and biofuel production. This multidisciplinary approach leverages the unique biological properties of fungi along with nanoscale materials to improve the efficiency, sustainability, and cost-effectiveness of transforming biomass into renewable energy sources (Patil et al. 2021). Myconanotechnology contributes to bioenergy by enhancing environmentally friendly processes. Fungi have a natural enzymatic process that

Table 16.2 Fungal-based nanotechnological approaches with their action mechanisms, properties, and applications

Nanotechnology approach	Mechanism of action	Properties	Application	References
Metal NPs like Ag, Au, and Fe_3O_4 (fungal-synthesized)	Fungal enzymes and metabolites minimize the metal ion to NP	Eco-friendly, low-cost, stable, and scalable nanoparticle	Catalyst for biofuel production, improves anaerobic digestion, and biogas upgrading	Rai et al. (2021)
Fungal nano-biocatalysts	Nanostructured fungal enzymes promote reaction surface area	Promotes reaction rate, minimizes loss of enzymes	Biodiesel transesterification, bioethanol production	Sheldon and Woodley (2018)
Nanocellulose (fungal-derived)	Degradation of fungi transforms biomass into nano-sized cellulosic fibers	Renewable, high-strength, and biodegradable	Biofuel carrier material bioenergy storage matrices	Klemm et al. (2011)
Fungal enzyme-immobilized NPs	Enzymes like cellulase, laccase, and xylanase immobilized on NPs promote catalytic efficiency	Reusability, high enzyme stability, enhanced hydrolysis	Saccharification of lignocellulosic biomass for bioethanol and biobutanol	Singh et al. (2021)
Fungal bio-nanocomposites	A mixture of fungal biomass with NPs enhances functional properties	Thermal stability, high adsorption	Storage of bioenergy, pellets, biofuel	Singh et al. (2025a, b)
Nanocarbon materials (fungal-based)	Pyrolysis of fungal biomass produces nanocarbon structures	Renewable source, high conductivity	Electrodes in microbial fuel cells	Ingle et al. (2023)
Fungal-encapsulated nano-enzymes	Nano-encapsulation protects the enzymes from harsh conditions	Thermal resistance, Long shelf-life	Industrial bioenergy processes	Ramya et al. (2023)
Fungal nano-enhanced anaerobic digestion	Nanoparticles synthesized by fungi enhance microbial metabolism	Higher methane yield, reduced lag phase	Biogas production	Keerio et al. (2025)

is capable of breaking down complex forms of lignocellulosic biomass into fermentable sugars, an important step in biofuel production. The application of NPs can improve these enzymatic activities by increasing substrate accessibility and stability, thereby accelerating biomass degradation. This lowers reliance on intense chemical pre-treatments that produce hazardous byproducts and consume a lot of energy, lowering energy input and pollution in the environment. Moreover, many chemically synthesized nanomaterials are toxic to the environment, but fungal-based nanomaterials are non-toxic and easily biodegradable. Their utilization in biocatalysts and biofilms for the production of bioenergy promotes green chemistry principles, lowers toxic waste generation, and environmental contamination.

Furthermore, fungi may eliminate contaminants and heavy metals from biomass feedstocks, providing cleaner raw materials for the production of biofuel and reducing the negative environmental effects. Myconanoparticles can improve the surface area and catalytic efficiency of fungal enzymes, which leads to higher biofuel production from the same biomass quantity (Rajak et al. 2021). This synergistic effect of fungi and NPs shortens the reaction times and lowers the energy consumption during pretreatment and fermentation stages. These nanomaterials are cost-effective and use sustainable substrates like agricultural waste, making the process economically viable and scalable. Myconanotechnology integration facilitates the creation of multipurpose bioreactors and biosensors that monitor the production of biofuel in real time, lowering downtime and increasing quality control, resulting in increased productivity and economic benefits. The ability of fungi to adapt to diverse environmental conditions, combined with the precision of nanotechnology, allows for the utilization of a wide range of biomass types, including waste or low-value materials, enhancing resource availability and lowering competition with food crops. By developing value-added goods from local biomass and generating employment opportunities in bioprocessing and nanomaterial production industries, this diversification improves rural economies.

In conclusion, by enhancing biomass conversion efficiency, reducing environmental impact, and reducing production costs, myconanotechnology in bioenergy and biofuel production serves as a sustainable pathway. Its ecological benefits arise from biodegradability, pollutant remediation, and lower chemical use, while economic benefits arise from increased yields, process optimization, and resource diversification (Panda et al. 2025).

16.6 Challenges and Future Outlooks

While the fungal-mediated synthesis of NPs is a burgeoning field, several hurdles remain for industrial integration. Optimization at large scale for species-specific production is inherently difficult (Rezghi Rami et al. 2024). Variations in substrate composition, fungal metabolic activity, environmental growth factors like temperature, pH, humidity, and light can lead to inconsistent NP shape, size, and yield. Furthermore, as production scales up, the risk of microbial contamination increases, which may lower the quality and cost (Liu et al. 2023). The next challenge is purification of fungal biomass to extract NPs without damaging their functional properties, which is complex and expensive. Standard laboratory techniques like centrifugation and filtration are often difficult or costly to scale to an industrial level. Consequently, significant investment is required to build the infrastructure necessary for commercial-scale cultivation and processing to compete with established chemical and green synthesis methods. Regulatory and research gaps present another challenge: robust testing or clear regulatory frameworks are essential to ensure the safety of fungal-derived NPs in agricultural, medical, and environmental sectors (Rezghi Rami et al. 2024). There is also a significant gap in our understanding of metabolic pathways involved in NP synthesis, which may hinder process optimization. Additionally, concerns regarding long-term environmental and biological impact may slow public and market acceptance.

The integration of fungal-mediated nanotechnological applications in the production of biofuels and biomaterials aligns with global sustainability goals and the circular economy principle. Using inexpensive agricultural waste as a substrate for fungal growth reduces reliance on raw materials (Sidhu et al. 2025). This synergy supports the development of eco-friendly agrochemicals like soil enhancers, fertilizers, and pesticides as an alternative to synthetic chemicals. Biomedical applications include diagnostics, antimicrobial therapy, and targeted drug delivery using biocompatible NPs. NPs are utilized for water purification and contaminant removal (An et al. 2022). To achieve commercial viability, comprehensive techno-economic assessments must account for the costs of NP synthesis within the bioenergy and biomaterials production chain (Soni and Soni 2025). Pilot-scale research is vital to test the integration of NPs in large-scale energy systems. In addition, safety assessments, particularly in vivo studies, are necessary to assess the toxicity and capability of these NPs on human health, the environment, and the microbial system. Future research should be expanded into the use of NPs as fuel additives. Currently, there is a lack of practical data regarding fluid dynamics, heat transfer, and efficiency. Hence, fungal systems offer a revolutionary pathway toward sustainable bioenergy and biomaterials development. Challenges in scalability, economic feasibility, and environmental risk could persist in the synthesis of bioenergy and biomaterial production. By optimizing processing techniques and fostering collaboration between industry and policymakers, fungal-mediated nanotechnology can become a primary driver of environmental and economic progress, mitigating climate change and reducing fossil fuel dependency.

16.7 Conclusion

Myconanotechnology is fundamentally reshaping the landscape of sustainable agriculture and renewable energy. By leveraging the natural biosynthetic capabilities of fungi, researchers are developing high-performance, biocompatible NPs that serve as eco-friendly alternatives to traditional industrial materials. Recent progress demonstrates that biosynthesized NPs derived from fungi act as nanofungicides, providing targeted protection against pathogens; nanofertilizers, which enhance nutrient delivery and crop growth; and nano-agrochemicals, which reduce the global reliance on synthetic herbicides and insecticides, thereby minimizing environmental runoff. This shift toward myconanotechnology aligns with the urgent global demand for agricultural solutions that increase crop yields without compromising soil health or biodiversity. As fossil fuel reserves deplete, the transition to biomass-based bioenergy and biomaterials has become essential. Fungi-derived nanomaterials address critical technical bottlenecks in bioenergy and biomaterial production, improving both the quality and quantity of outputs. Hence, the convergence of technological innovation and fungal biology holds the key to a sustainable future. By bridging current technical gaps and expanding the library of fungal species used in synthesis, myconanotechnology will likely become a cornerstone of the global circular economy.

References

Abady MM, Mohammed DM, Soliman TN, Shalaby RA, Sakr FA (2025) Sustainable synthesis of nanomaterials using different renewable sources. Bull Natl Res Cent 49(1):24

Abhijith R, Ashok A, Rejeesh CR (2018) Sustainable packaging applications from mycelium to substitute polystyrene: a review. Mater Today Proc 5(1):2139–2145. https://doi.org/10.1016/j.matpr.2017.09.211

Abo Nouh FA, Gezaf SA, Abo Nahas HH, Abo Nahas YH, Vargas-De-La-Cruz C, Solorzano Acosta RA et al (2022) Bioprospecting for biomolecules from different fungal communities: an introduction. In: Industrially important fungi for sustainable development: volume 2: bioprospecting for biomolecules. Springer International Publishing, Cham, pp 1–71

Adebayo EA, Azeez MA, Alao MB, Oke AM, Aina DA (2021) Fungi as veritable tool in current advances in nanobiotechnology. Heliyon 7(11):e08480. https://doi.org/10.1016/j.heliyon.2021.e08480

Adeleke BS, Olowe OM, Ayilara MS, Fasusi OA, Omotayo OP, Fadiji AE et al (2024) Biosynthesis of nanoparticles using microorganisms: a focus on endophytic fungi. Heliyon 10(21):e39636

Ali S, Ali H, Siddique M, Gulab H, Haleem MA, Ali J (2020) Exploring the biosynthesized gold nanoparticles for their antibacterial potential and photocatalytic degradation of the toxic water wastes under solar light illumination. J Mol Struct 1215:128259

Ali HR, Emam AN, Hefny EG, Koraney NF, Mansour AS, Salama AM et al (2021) Silver nanoparticles enhance the effectiveness of traditional antibiotics against S. aureus causing bovine mastitis within the safety limit. J Nanopart Res 23(11):243

Ali J, Alam D, Noori R, Faridi S, Sardar M (2024) Metal nanoparticles: management and control of phytopathogenic fungi. In: Advances in antifungal drug development: natural products with antifungal potential. Springer Nature Singapore, Singapore, pp 411–437

An C, Sun C, Li N, Huang B, Jiang J, Shen Y, Wang C, Zhao X, Cui B, Wang C, Li X, Zhan S, Gao F, Zeng Z, Cui H, Wang Y (2022) Nanomaterials and nanotechnology for the delivery of agrochemicals: strategies towards sustainable agriculture. J Nanobiotechnol 20(1):11. https://doi.org/10.1186/s12951-021-01214-7

Anjum S, Vyas A, Sofi T (2023) Fungi-mediated synthesis of nanoparticles: characterization process and agricultural applications. J Sci Food Agric 103(10):4727–4741

Anvar AA, Ahari H, Ataee M (2021) Antimicrobial properties of food nanopackaging: a new focus on foodborne pathogens. Front Microbiol 12:690706

Arif S, Liaquat R, Adil M (2018) Applications of materials as additives in anaerobic digestion technology. Renew Sust Energ Rev 97:354–366

Balakumaran MD, Ramachandran R, Kalaichelvan PT (2015) Exploitation of endophytic fungus, Guignardia mangiferae for extracellular synthesis of silver nanoparticles and their in vitro biological activities. Microbiol Res 178:9–17

Ballen Sierra LA, Mendes-Pereira T, García GJY, Werkhaizer CQ, de Rezende JB, Rodrigues TAB, Badotti F, de Carvalho Cardoso ES, da Costa AM, Uetanabaro AP, Aguilar MT, Góes-Neto A (2023) Current situation and future perspectives for the use of fungi in the biomaterial

industry and proposal for a new classification of fungal-derived materials. PeerJ Mater Sci 5:e31. https://doi.org/10.7717/peerj-matsci.31
Banerjee K, Ravishankar Rai V (2016) Study on green synthesis of gold nanoparticles and their potential applications as catalysts. J Clust Sci 27(4):1307–1315
Bansal V, Ramanathan R, Bhargava SK (2011) Fungus-mediated biological approaches towards 'green' synthesis of oxide nanomaterials. Aust J Chem 64(3):279–293
Barcan AS, Dinu LD, Vălu V, Vamanu E (2023) Influences of edible mushrooms on dynamic and diversity of gut microbiota. In: Mushrooms with therapeutic potentials: recent advances in research and development. Springer Nature Singapore, Singapore, pp 113–146
Cerimi K, Akkaya KC, Pohl C, Schmidt B, Neubauer P (2019) Fungi as source for new bio-based materials: a patent review. Fungal Biol Biotechnol 6(1):17. https://doi.org/10.1186/s40694-019-0080-y
Chaudhari V, Vairagade V, Thakkar A, Shende H, Vora A (2024) Nanotechnology-based fungal detection and treatment: current status and future perspective. Naunyn Schmiedeberg's Arch Pharmacol 397(1):77–97
Cherian B, Cherian T, Merlin T, Jose S (2025) Investigation of antibiosis, anti-diabetic, antioxidant, anti-inflammatory, molecular docking and dye degradation potential of green synthesized copper ferrite (CuFe2O4) nanoparticles using mushroom Pleurotus florida. Discov Nano 20(1):1–28
Chettri D, Boro M, Malik AA, Verma AK (2025) Nanotechnology for a greener future: sustainable energy and biofuel advances. Biofuels 16(6):622–637
Chowdhury S, Basu A, Kundu S (2014) Green synthesis of protein capped silver nanoparticles from phytopathogenic fungus Macrophomina phaseolina (Tassi) Goid with antimicrobial properties against multidrug-resistant bacteria. Nanoscale Res Lett 9(1):365
Chowdhury NK, Choudhury R, Gogoi B, Chang CM, Pandey RP (2022) Microbial synthesis of gold nanoparticles and their application. Curr Drug Targets 23(7):752–760
Dadayya M, Veeranna SH, Basaiah T (2025) Detection of fungal phytopathogens by the nanotechnology method. In: Molecular approaches for the detection of fungal phytopathogens. CRC Press, Boca Raton, pp 169–190
Das SK, Das AR, Guha AK (2009) Gold nanoparticles: microbial synthesis and application in water hygiene management. Langmuir 25(14):8192–8199
Deng S, Li C, Cao J, Cui Z, Du J, Fu Z, Yang H, Chen P (2023) Organ-on-a-chip meets artificial intelligence in drug evaluation. Theranostics 13(13):4526–4558. https://doi.org/10.7150/thno.87266
Dey P, Rangarajan V, Singh J, Nayak J, Dilip KJ (2022) Current perspective on improved fermentative production and purification of fungal cellulases for successful biorefinery applications: a brief review. Biomass Convers Biorefinery 12(3):967–995
Durán N, De Jesus MB, Tasic L, Fávaro WJ, Nakazato G (2023) Mechanism of synthesis of metal nanoparticles by fungi. In: Mycosynthesis of nanomaterials. CRC Press, Boca Raton, pp 254–280
Elazab NT, Younis SA, Abdelgalil SA (2023) Biogenic synthesis of nanoparticles mediated by fungi. In: Plant mycobiome: diversity, interactions and uses. Springer, Cham, pp 241–265
European Parliament (2024) The impact of textile production and waste on the environment (infographics). European Parliament. Available online: https://www.europarl.europa.eu/topics/en/article/20201208STO93327/the-impact-of-textile-productionand-waste-on-the-environment-infographics. Accessed 16 Feb 2025
Fatma S, Hameed A, Noman M, Ahmed T, Shahid M, Tariq M et al (2018) Lignocellulosic biomass: a sustainable bioenergy source for the future. Protein Pept Lett 25(2):148–163
Fernandes M, Padrão J, Ribeiro AI, Fernandes RD, Melro L, Nicolau T et al (2022) Polysaccharides and metal nanoparticles for functional textiles: a review. Nanomaterials 12(6):1006
Fusteș-Dămoc I, Dinu R, Măluțan T, Mija A (2023) Valorisation of chitosan natural building block as a primary strategy for the development of sustainable fully bio-based epoxy resins. Polymers 15(24):4627. https://doi.org/10.3390/polym15244627

Gajewska J, Floryszak-Wieczorek J, Sobieszczuk-Nowicka E, Mattoo A, Arasimowicz-Jelonek M (2022) Fungal and oomycete pathogens and heavy metals: an inglorious couple in the environment. IMA Fungus 13(1):6

Ghareeb A, Fouda A, Kishk RM, El Kazzaz WM (2024) Unlocking the potential of titanium dioxide nanoparticles: an insight into green synthesis, optimizations, characterizations, and multifunctional applications. Microb Cell Factories 23(1):341

Ghosh S, Rusyn I, Dmytruk OV, Dmytruk KV, Onyeaka H, Gryzenhout M, Gafforov Y (2023) Filamentous fungi for sustainable remediation of pharmaceutical compounds, heavy metal and oil hydrocarbons. Front Bioeng Biotechnol 11:1106973

Guilger-Casagrande M, de Lima R (2019) Synthesis of silver nanoparticles mediated by fungi: a review. Front Bioeng Biotechnol 7. https://doi.org/10.3389/fbioe.2019.00287

Gulati R, Sharma S, Sharma RK (2022) Antimicrobial textile: recent developments and functional perspective. Polym Bull 79(8):5747–5771

Hahn J (2020) TômTex™: leather alternative biomaterial from seafood shells and coffee grounds. Dezeen. Available online: https://www.dezeen.com/2020/08/22/tomtex-leather-alternative-biomaterial-seafood-shells-coffee/. Accessed 11 Oct 2024

Hanafy MH (2018) Myconanotechnology in the veterinary sector: status quo and future perspectives. Int J Vet Sci Med 6(2):270–273. https://doi.org/10.1016/j.ijvsm.2018.11.003

Huston M, DeBella M, DiBella M, Gupta A (2021) Green synthesis of nanomaterials. Nanomaterials 11(8):2130. https://doi.org/10.3390/nano11082130

Ibrahim HM, Hassan MS (2016) Characterization and antimicrobial properties of cotton fabric loaded with green synthesized silver nanoparticles. Carbohydr Polym 151:841–850

Ibrar M, Ullah MW, Manan S, Farooq U, Rafiq M, Hasan F (2020) Fungi from the extremes of life: an untapped treasure for bioactive compounds. Appl Microbiol Biotechnol 104(7):2777–2801

Ingle PU, Banode K, Mishra S, Gade AK, Rai M (2023) Fungi-mediated synthesis of carbon-based nanomaterials. In: Mycosynthesis of nanomaterials. CRC Press, Boca Raton, pp 193–214

Jing YANG, Yuanzhe MA, Yunmei WU, Yifan LIU, Huini DU, Fuyong WU (2025) Influence of arbuscular mycorrhizal fungi and zinc oxide nanoparticles on cadmium uptake and accumulation in winter wheat growing in naturally Cd-contaminated soil. Pedosphere. https://doi.org/10.1016/j.pedsph.2025.01.005

Jones M, Gandia A, John S, Bismarck A (2020) Leather-like material biofabrication using fungi. Nat Sustain 4(1):9–16. https://doi.org/10.1038/s41893-020-00606-1

Jorge AM, Athira KK, Alves MB, Gardas RL, Pereira JF (2023) Textile dyes effluents: a current scenario and the use of aqueous biphasic systems for the recovery of dyes. J Water Process Eng 55:104125

Joshi K, Meher MK, Poluri KM (2020) Fabrication and characterization of bioblocks from agricultural waste using fungal mycelium for renewable and sustainable applications. ACS Appl Bio Mater 3(4):1884–1892. https://doi.org/10.1021/acsabm.9b01047

Kar PK, Murmu S, Saha S, Tandon V, Acharya K (2014) Anthelmintic efficacy of gold nanoparticles derived from a phytopathogenic fungus, Nigrospora oryzae. PLoS One 9(1):e84693

Keerio HA, Alnadish AM, Panhwar S, Ali Z, Kumar S (2025) Nanotechnologies transforming the future of biogas production: a current trend. Biofuels 17(3):1–23

Khan Y, Sadia H, Ali Shah SZ, Khan MN, Shah AA, Ullah N et al (2022) Classification, synthetic, and characterization approaches to nanoparticles, and their applications in various fields of nanotechnology: a review. Catalysts 12(11):1386

Khandel P, Shahi SK (2018) Mycogenic nanoparticles and their bio-prospective applications: current status and future challenges. J Nanostructure Chem 8(4):369–391. https://doi.org/10.1007/s40097-018-0285-2

Kiran S, Ijaz MU, Amjad A, Rasool M, Hussain M, Mateen A (2025) Application of nanoparticles in textiles for moisture management. In: Nanoparticles integrated functional textiles. Springer Nature Switzerland, Cham, pp 271–296

Klemm D, Kramer F, Moritz S, Lindström T, Ankerfors M, Gray D, Dorris A (2011) Nanocelluloses: a new family of nature-based materials. Angew Chem Int Ed 50(24):5438–5466

Kulkarni D, Sherkar R, Shirsathe C, Sonwane R, Varpe N, Shelke S et al (2023) Biofabrication of nanoparticles: sources, synthesis, and biomedical applications. Front Bioeng Biotechnol 11:1159193

Kumawat G, Rajpurohit D, Vyas D, Bhojiya AA, Upadhyay SK, Jain D (2025) Characterization of green-synthesized zinc oxide nanoparticles and its influence on post-harvest shelf-life of garlic against black mold disease caused by *Aspergillus niger*. Front Microbiol 16:1532593

Le DL, Salomone R, Nguyen QT (2023) Circular bio-based building materials: a literature review of case studies and sustainability assessment methods. Build Environ 244:110774. https://doi.org/10.1016/j.buildenv.2023.110774

Lijó L, González-García S, Lovarelli D, Moreira MT, Feijoo G, Bacenetti J (2019) Life cycle assessment of renewable energy production from biomass, pp 81–98. https://doi.org/10.1007/978-3-319-93740-3_6

Lin X, Zhang D, Chen X, Liu M (2024) Self-cross-linked oxidized sodium alginate/gelatin/halloysite hydrogel as injectable, adhesive, antibacterial dressing for hemostasis. ACS Sustain Chem Eng 12(31):11739–11753. https://doi.org/10.1021/acssuschemeng.4c03668

Liu S, Xiong C, Lin L, Keyhani NO, Zhu M, Zhao Z, Zhang W, Yang C, Su H, Liu P, Guan X, Qiu J (2023) Assessing the structure and diversity of fungal community in plant soil under different climatic and vegetation conditions. Front Microbiol 14. https://doi.org/10.3389/fmicb.2023.1288066

Liu Z, Yu C, Li Z, Wang X, Shang D, Dong W (2025) Matrix metalloproteinase-triggered self-assembling peptides for biomedical applications. J Mater Chem B 13(28). https://doi.org/10.1039/D5TB00625B

Loshchinina EA, Vetchinkina EP, Kupryashina MA (2022) Diversity of biogenic nanoparticles obtained by the fungi-mediated synthesis: a review. Biomimetics 8(1):1. https://doi.org/10.3390/biomimetics8010001

Luangharn T, Karunarathna SC, Mortimer PE, Hyde KD, Xu J (2020) Morphology, phylogeny and culture characteristics of *Ganoderma gibbosum* collected from Kunming, Yunnan Province, China. Phyton 89(3):743

Ma R, Li J, Tyagi R, Zhang X (2024) Carbon dioxide and methane as carbon source for the production of polyhydroxyalkanoates and concomitant carbon fixation. Bioresour Technol 391:129977. https://doi.org/10.1016/j.biortech.2023.129977

Malik S, Muhammad K, Waheed Y (2023) Nanotechnology: a revolution in modern industry. Molecules 28(2):661. https://doi.org/10.3390/molecules28020661

Mango Materials (2025) The future is biodegradable. Mango Materials. Available online: https://www.mangomaterials.com/. Accessed 18 Jan 2025

Mautner A (2023) Fungi-derived composite nanopapers. In: Bioprospects of macrofungi. CRC Press, Boca Raton, pp 233–257

Meera CR, Zaynitdinova L, Great U (2025) Sustainable production of nanoparticles from fungi and their agricultural applications. In: Nano-microbiology for sustainable development. Springer Nature Switzerland, Cham, pp 71–103

Mehrotra P (2016) Biosensors and their applications–a review. J Oral Biol Craniofac Res 6(2):153–159

Mengqi Z, Shi A, Ajmal M, Ye L, Awais M (2023) Comprehensive review on agricultural waste utilization and high-temperature fermentation and composting. Biomass Convers Biorefinery 13(7):5445–5468. https://doi.org/10.1007/s13399-021-01438-5

Mignogna D, Szabó M, Ceci P, Avino P (2024) Biomass energy and biofuels: perspective, potentials, and challenges in the energy transition. Sustainability 16(16):7036

Mishra A, Tripathy SK, Yun SI (2012) Fungus mediated synthesis of gold nanoparticles and their conjugation with genomic DNA isolated from Escherichia coli and Staphylococcus aureus. Process Biochem 47(5):701–711

Mulay AR, Hwang J, Kim DH (2024) Microphysiological blood-brain barrier systems for disease modeling and drug development. Adv Healthc Mater 13(21):2303180

Murthy HN, Joseph KS, Paek KY, Park SY (2024) Nanomaterials as novel elicitors of pharmacologically active plant specialized metabolites in cell and organ cultures: current status and future outlooks. Plant Growth Regul 104(1):5–30

Naimi-Shamel N, Pourali P, Dolatabadi S (2019) Green synthesis of gold nanoparticles using Fusarium oxysporum and antibacterial activity of its tetracycline conjugant. J Mycol Med 29(1):7–13

Nawawi WMFW, Jones MP, Kontturi E, Mautner A, Bismarck A (2020) Plastic to elastic: fungi-derived composite nanopapers with tunable tensile properties. Compos Sci Technol 198:108327. https://doi.org/10.1016/j.compscitech.2020.108327

Nirwan VP, Amarjargal A, Hengsbach R, Fahmi A (2025) Electrospun smart hybrid nanofibers for multifaceted applications. Macromol Rapid Commun 46(13):2400617

Nwabike Amitaye A, Elemike EE, Akpeji HB, Amitaye E, Hossain I, Mbonu JI, Aziza AE (2024) Chitosan: a sustainable biobased material for diverse applications. J Environ Chem Eng 12(4):113208. https://doi.org/10.1016/j.jece.2024.113208

Ochoa C, Iszauk E (2025) Bringing life to materials to replace plastics and leather. Extantia Capital Blog. Available online: https://medium.com/extantia-capital/bringing-life-to-materials-to-replace-plastics-and-leather-why-we-investedin-modern-synthesis-cb4a2c022fe9. Accessed 11 Oct 2024

Ojha R, Pradhan D (2025) The potential of microbial fuel cell for converting waste to energy: an overview. Sustain Chem Environ 9:100196

Owaid MN, Raman J, Lakshmanan H, Al-Saeedi SSS, Sabaratnam V, Abed IA (2015) Mycosynthesis of silver nanoparticles by Pleurotus cornucopiae var. citrinopileatus and its inhibitory effects against Candida sp. Mater Lett 153:186–190

Panda J, Nongbet A, Avula SK, Nayak D, Rustagi S, Panda BP, Mohanta YK (2025) Macro-fungi mediated nanoparticles for sustainable agriculture: recent advancement and future strategies. Discov Sustain 6(1):375

Pandit P, Nadathur GT, Maiti S, Regubalan B (2018) Functionality and properties of bio-based materials. In: Bio-based materials for food packaging. Springer Singapore, pp 81–103. https://doi.org/10.1007/978-981-13-1909-9_4

Patel M, Parikh D, Parihar A, Prajapati B, Patel MB, Salave S et al (2025) Protein and peptide therapeutics: stability challenges, regulatory demands, and innovative formulation solutions for enhanced clinical effectiveness. Protein Pept Lett 32(7):490–507

Patil AG, Kounaina K, Aishwarya S, Harshitha N, Satapathy P, Hudeda SP et al (2021) Myconanotechnology for sustainable agriculture: challenges and opportunities. In: Recent trends in mycological research: volume 1: agricultural and medical perspective, pp 457–479. https://doi.org/10.1007/978-3-030-60659-6_20

Ponmurugan P (2017) Biosynthesis of silver and gold nanoparticles using Trichoderma atroviride for the biological control of Phomopsis canker disease in tea plants. IET Nanobiotechnol 11(3):261–267

Prasher P, Sharma M, Kumar T (2024) Bacteriogenic metal nanoparticles: novel green fungicides. In: Nanofungicides. Elsevier, pp 85–102. https://doi.org/10.1016/B978-0-323-95305-4.00005-4

Rai M, Bonde S, Golinska P, Trzcińska-Wencel J, Gade A, Abd-Elsalam KA et al (2021) Fusarium as a novel fungus for the synthesis of nanoparticles: mechanism and applications. J Fungi 7(2):139

Rai M, Gade A, Zimowska B, Shende S, Elias De Matos JM, Ingle P et al (2025) Harnessing pycnidia-forming fungi for eco-friendly nanoparticle production, applications, and limitations. Front Microbiol 16:1603728

Rajak RC, Saha P, Singhvi M, Kwak D, Kim D, Lee H et al (2021) An eco-friendly biomass pretreatment strategy utilizing reusable enzyme mimicking nanoparticles for lignin depolymerization and biofuel production. Green Chem 23(15):5584–5599

Raman J, Reddy GR, Lakshmanan H, Selvaraj V, Gajendran B, Nanjian R et al (2015) Mycosynthesis and characterization of silver nanoparticles from Pleurotus djamor var. roseus and their in vitro cytotoxicity effect on PC3 cells. Process Biochem 50(1):140–147
Raman J, Kim D-S, Kim H-S, Oh D-S, Shin H-J (2022) Mycofabrication of mycelium-based leather from brown-rot fungi. J Fungi 8(3):317. https://doi.org/10.3390/jof8030317
Ramya RK, Theraka K, Ramprasadh SV, Bharathi SV, Srinivasan S, Jacob S, Kuila A (2023) Pragmatic treatment strategies for polyaromatic hydrocarbon remediation and anti-biofouling from surfaces using nano-enzymes: a review. Appl Biochem Biotechnol 195(9):5479–5496
Rezghi Rami M, Meskini M, Ebadi Sharafabad B (2024) Fungal-mediated nanoparticles for industrial applications: synthesis and mechanism of action. J Infect Public Health 17(10):102536. https://doi.org/10.1016/j.jiph.2024.102536
Robinson JR, Isikhuemhen OS, Anike FN (2021) Fungal–metal interactions: a review of toxicity and homeostasis. J Fungi 7(3):225
Ross P, Scullin M, Wenner N, Chase J, Miller Q, Saltidos R, McGaughy P (2020) Mycelium growth bed with perforation layer and related method for creating a uniform sheet of mycelium from a solid-state medium. US20200196541A1. Available at https://patents.google.com/patent/US20200196541A1/en. Accessed 21 Dec 2022
Salem SS, Fouda A (2021) Green synthesis of metallic nanoparticles and their prospective biotechnological applications: an overview. Biol Trace Elem Res 199(1):344–370. https://doi.org/10.1007/s12011-020-02138-3
Salem SS, Mekky AE (2024) Biogenic nanomaterials: synthesis, characterization, and applications. In: Biogenic nanomaterials for environmental sustainability: principles, practices, and opportunities. Springer International Publishing, Cham, pp 13–43
Saravanan A, Kumar PS, Govarthanan M, George CS, Vaishnavi S, Moulishwaran B et al (2021) Adsorption characteristics of magnetic nanoparticles coated mixed fungal biomass for toxic Cr (VI) ions in aquatic environment. Chemosphere 267:129226
Senjyu T, Howlader AM (2016) Operational aspects of distribution systems with massive DER penetrations. In: Integration of distributed energy resources in power systems. Elsevier, pp 51–76. https://doi.org/10.1016/B978-0-12-803212-1.00003-9
Sheldon RA, Woodley JM (2018) Role of biocatalysis in sustainable chemistry. Chem Rev 118(2):801–838
Sidhu AK, Agrawal SB, Verma N, Kaushal P, Sharma M (2025) Fungal-mediated synthesis of multimetallic nanoparticles: mechanisms, unique properties, and potential applications. Front Nanotechnol 7. https://doi.org/10.3389/fnano.2025.1549713
Singh P, Kim YJ, Zhang D, Yang DC (2016) Biological synthesis of nanoparticles from plants and microorganisms. Trends Biotechnol 34(7):588–599
Singh R, Liu H, Shanklin J, Singh V (2021) Hydrothermal pretreatment for valorization of genetically engineered bioenergy crop for lipid and cellulosic sugar recovery. Bioresour Technol 341:125817
Singh NB, Kumar B, Usman UL, Susan MABH (2024) Nano revolution: exploring the frontiers of nanomaterials in science, technology, and society. Nano-Struct Nano-Objects 39:101299. https://doi.org/10.1016/j.nanoso.2024.101299
Singh S, Singh NP, Agrawal S, Kumar A (2025a) Unleashing the potential of white-rot fungi mycelium for functional biomaterials development. Discov Mater 5(1):96
Singh R, Maurya R, Mishra AK, Kumar B, Shafi H, Kumar V, Jagtap CY (2025b) 11 Biocomposite nanomaterials. In: Nanomaterials in environmental remediation: developments and challenges, pp 191–212. https://doi.org/10.1201/9781003484189-11
Soni SK, Soni R (2025) Techno-economic and environmental assessment, pp 291–314. https://doi.org/10.1007/978-981-96-9013-8_9
Sousa F, Ferreira D, Reis S, Costa P (2020) Current insights on antifungal therapy: novel nanotechnology approaches for drug delivery systems and new drugs from natural sources. Pharmaceuticals 13(9):248. https://doi.org/10.3390/ph13090248

Sridhar A, Sunil CK, Sarkar R, Sarojini S (2024) Greener assembly of nano catalysts and sustainable applications of magnetically retrievable and plasmonic nano catalysts. Top Catal 67(1):265–279

Srivastava S, Bhargava A (2021) Biological synthesis of nanoparticles: fungi. In: Green nanoparticles: the future of nanobiotechnology. Springer Singapore, Singapore, pp 101–137

Srivastava S, Mathur P, Prakash P, Falletta E, Katha U, Pagani A et al (2025) Mushroom-derived innovations: sustainable biomaterials for biomedical engineering. Biomed Mater Devices 3(1):381–395

Sroka M, Zaborniak I, Chmielarz P (2025) Development of smart surfaces for medicine and biotechnology: advances in glass functionalization through RDRP techniques. ACS Biomater Sci Eng 11(8):4694–4713

Strzalka R, Schneider D, Eicker U (2017) Current status of bioenergy technologies in Germany. Renew Sust Energ Rev 72:801–820. https://doi.org/10.1016/j.rser.2017.01.091

Tamilselvan R, Selwynraj AI (2024) Enhancing biogas generation from lignocellulosic biomass through biological pretreatment: exploring the role of ruminant microbes and anaerobic fungi. Anaerobe 85:102815

Tan ECD, Lamers P (2021) Circular bioeconomy concepts—a perspective. Front Sustainability 2. https://doi.org/10.3389/frsus.2021.701509

Tarannum N, Pooja K, Jakhar S, Mavi A (2024) Nanoparticles assisted intra and transdermic delivery of antifungal ointment: an updated review. Discov Nano 19(1):11

Tavares TD, Antunes JC, Padrão J, Ribeiro AI, Zille A, Amorim MTP et al (2020) Activity of specialized biomolecules against gram-positive and gram-negative bacteria. Antibiotics 9(6):314

Taylor RA, Jalali M, Marti J, Booth TJ, Baker MAB (2024) Nanotechnology: creating, manipulating, and observing nanostructured systems in biology and medicine. In: Pharmacognosy. Elsevier, pp 727–747. https://doi.org/10.1016/B978-0-443-18657-8.00007-4

Thapliyal D, Verros GD, Arya RK (2025) Nanoparticle-doped antibacterial and antifungal coatings. Polymers 17(2):247

Tiquia-Arashiro SM, Grube M (2019) Fungi in extreme environments: ecological role and biotechnological significance. Springer, Cham

Trotta F, Da Silva S, Massironi A, Mirpoor SF, Lignou S, Ghawi SK, Charalampopoulos D (2023) Silver bionanocomposites as active food packaging: recent advances & future trends tackling the food waste crisis. Polymers 15(21):4243

United Nations Environment Programme, International Resource Panel (2024) Global resources outlook 2024—bend the trend: pathways to a liveable planet as resource use spikes. United Nations Environment Programme, Nairobi. Available online: https://wedocs.unep.org/handle/20.500.11822/44901. Accessed 1 Aug 2024

Vandelook S, Elsacker E, van Wylick A, de Laet L, Peeters E (2021) Current state and future prospects of pure mycelium materials. Fungal Biol Biotechnol 8(1):20. https://doi.org/10.1186/s40694-021-00128-1

Vanlalveni C, Ralte V, Zohmingliana H, Das S, Anal JMH, Lallianrawna S, Rokhum SL (2024) A review of microbes mediated biosynthesis of silver nanoparticles and their enhanced antimicrobial activities. Heliyon 10(11):e32333

Varshan GA, Namasivayam SKR (2024) A green chemistry principle for the biotransformation of fungal biomass derived chitosan into versatile nano scale materials with high biocompatibility and potential biological activities—a review. BioNanoScience 14(4):4145–4166

Velamakanni RP, Gothalwal R, Velamakanni RS, Ayinampudi SR, Vuppugalla P, Merugu R (2022) Fungi-mediated green synthesis of nanoparticles and their renewable energy applications. In: Green nano solution for bioenergy production enhancement. Springer Nature Singapore, Singapore, pp 201–224

Verma N, Jujjavarapu SE, Mahapatra C (2023) Green sustainable biocomposites: substitute to plastics with innovative fungal mycelium based biomaterial. J Environ Chem Eng 11(5):110396. https://doi.org/10.1016/j.jece.2023.110396

Vu PV, Doan TD, Tu GC, Do NHN, Le KA, Le P (2022) A novel application of cellulose aerogel composites from pineapple leaf fibers and cotton waste: removal of dyes and oil in wastewater. J Porous Mater 29(4):1137–1147. https://doi.org/10.1007/s10934-022-01238-w

Wadhwa K, Kapoor N, Kaur H, Abu-Seer EA, Tariq M, Siddiqui S et al (2024) A comprehensive review of the diversity of fungal secondary metabolites and their emerging applications in healthcare and environment. Mycobiology 52(6):335–387

Wang Z, Wang H, Kang Z, Wu Y, Xing Y, Yang Y (2020) Antioxidant and anti-tumour activity of triterpenoid compounds isolated from Morchella mycelium. Arch Microbiol 202(7):1677–1685. https://doi.org/10.1007/s00203-020-01876-1

Wang J, Zhao W, Zhang Z, Liu X, Xie T, Wang L et al (2024) A journey of challenges and victories: a bibliometric worldview of nanomedicine since the 21st century. Adv Mater 36(15):2308915

Xie M, Wang J, Wang F, Wang J, Yan Y, Feng K, Chen B (2025) A review of genomic, transcriptomic, and proteomic applications in edible fungi biology: current status and future directions. J Fungi 11(6):422

Xu F, Li Y, Zhao X, Liu G, Pang B, Liao N et al (2024) Diversity of fungus-mediated synthesis of gold nanoparticles: properties, mechanisms, challenges, and solving methods. Crit Rev Biotechnol 44(5):924–940

Yadav N, Saini O, Debnath N, Singh S, Thakur TK, Rajendra K et al (2024) Optimizing enhanced heavy metal detoxification by novel hybrid fungal hyphae-nano-biocomposite functionalized with graphene oxide: unravelling process parameters & adsorption modelling. Process Saf Environ Prot 188:917–928

Zannat A, Eason I, Wylie B, Rogers RD, Berton P, Shamshina JL (2025) Comparative analysis of chitin isolation techniques from mushrooms: toward sustainable production of high-purity biopolymer. Green Chem 27(12):3217–3233

Zhang X, Yuan J, Li F, Xiang J (2021) Chitin synthesis and degradation in crustaceans: a genomic view and application. Mar Drugs 19(3):153

Zhao P, Guan M, Tang W, Walayat N, Ding Y, Liu J (2023) Structural diversity, fermentation production, bioactivities and applications of triterpenoids from several common medicinal fungi: recent advances and future perspectives. Fitoterapia 166:105470. https://doi.org/10.1016/j.fitote.2023.105470